# THERMAL ENGINEERING

**HARRY L. SOLBERG**

*Associate Dean of Engineering*
*Purdue University*

**ORVILLE C. CROMER**

*Professor of Mechanical Engineering*
*Purdue University*

**ALBERT R. SPALDING**

*Head, Department of Freshman Engineering*
*Purdue University*

New York · London · John Wiley & Sons, Inc.

# THERMAL ENGINEERING

5 6 7 8 9 10

COPYRIGHT © 1960 BY JOHN WILEY & SONS, INC.

Library of Congress Catalog Card Number: 60–11730

PRINTED IN THE UNITED STATES OF AMERICA

ISBN 0 471 81147 5

# Preface

This book, *Thermal Engineering,* is an outgrowth of a textbook by the same authors, *Elementary Heat Power,* the first edition of which was published in 1946.

In recent years the objectives of engineering education have undergone critical evaluation as a result of which curricula and course content have been revised to place increased emphasis on the engineering sciences and to de-emphasize instruction in the art of engineering. We believe that this change in emphasis has been correct but feel that in some instances it has been carried too far. It has been stated frequently that the distinguishing characteristic of the engineer is *ability to design.* It is doubtful that an engineering student can study intelligently the principles involved in the design of mechanical engineering systems and components if such common items of equipment as heat exchangers, prime movers, pumps, and compressors are looked upon as mysterious black boxes or as circles or rectangles in a flow diagram. We believe that the student must acquire an understanding of the major principles of construction, operation, and performance of some of the types of common mechanical equipment and that, from the standpoint of motivation and conservation of time, such knowledge should be acquired as early as possible in his scholastic program.

*Thermal Engineering* has been written with specific attention to the following considerations:

1. The book should serve as an introduction to a rigorous sequence of courses in thermodynamics, fluid mechanics, and heat transfer.

2. The student should acquire sufficient understanding of the construction, principles of operation, and performance of the major components of a thermal power-generating system so that no time need be spent in a discussion of such subject matter in subsequent courses.

3. Major emphasis should be placed on the energy and material balances and their application to the thermal power system and the major components of the system.

4. Instruction should be given in the use of consistent systems of units.

5. The student should develop familiarity with the thermodynamic properties of matter and facility in using the perfect-gas relation and the tables of thermodynamic properties of vapors in the solution of engineering problems.

6. A balanced treatment should be presented of fossil and nuclear fuels as sources of energy, the reactions by which energy is released in fossil and nuclear fuels, and the manner in which part of this energy is converted into work.

7. The student should be introduced to the basic principles of heat transfer with emphasis on the manner in which these principles affect the economical design of heat-transfer apparatus.

8. The book should include material suitable for a terminal course on the first law of thermodynamics and its application to the generation of power and refrigeration.

9. The book should serve as a supplement to a textbook on thermo-dynamics in which inadequate reference is made to the equipment used for the generation of power and for refrigeration.

H. L. Solberg
O. C. Cromer
A. R. Spalding

*Lafayette, Indiana*
*May, 1960*

# Acknowledgments

We gratefully acknowledge the helpful suggestions that have been received from the members of the staff of the School of Mechanical Engineering. We are particularly indebted to Professors V. E. Bergdolt, C. L. Brown, J. B. Jones, R. J. Grosh, and C. F. Warner for their criticism of parts of the manuscript.

We wish to thank the following companies for their courtesy and cooperation in making illustrations available for inclusion in this book:

Allis-Chalmers Mfg. Co.: Figs. 9.8, 9.9, 10.24, 10.32, 10.33, 11.25.
American Arch Co.: Fig. 6.1.
American Blower Corp.: Figs. 12.8, 12.9, 12.10, 12.14, 12.15.
American Bosch Arma Corp.: Figs. 5.37, 5.38.
Babcock & Wilcox Co.: Figs. 6.2, 6.5, 6.9, 6.14, 6.15, 6.16, 6.17, 6.18, 6.20, 6.21, 6.22, 6.23, 6.24, 8.3, 8.5, 8.9, 8.13, 8.20, 8.23, 8.25.
Bacharach Industrial Instrument Corp.: Fig. 5.9.
Bendix Aviation Corp.: Figs. 5.33, 5.35, 5.36.
Buffalo Forge Co.: Figs. 12.4, 12.5, 12.7, 12.12, 12.13.
Carrier Corp.: Fig. 14.7.
Chevrolet Division, G.M.C.: Fig. 5.26.
Chicago Pneumatic Tool Co.: Figs. 12.26, 12.27, 12.28, 12.29, 12.30.
Combustion Engineering, Inc.: Figs. 6.10, 6.19, 8.7, 8.8, 8.12, 8.14, 8.21, 8.22.
Cooper-Bessemer Corp.: Figs. 5.39, 5.46.
Cummins Engine Co., Inc.: Fig. 5.21.
Cyclotherm Corp.: Fig. 8.2.
Dean Brothers Pumps, Inc.: Figs. 11.15, 11.16.
De Laval Steam Turbine Co.: Figs. 10.2, 11.27, 12.20, 12.21, 12.22, 12.23.
Delco-Remy Division, G.M.C.: Fig. 5.42.
Detroit Stoker Co.: Figs. 6.8, 6.12, 6.13.
Dynamatic Corp.: Fig. 5.12.
Ex-Cell-O Corp.: Fig. 5.40.
Fairbanks-Morse & Co.: Fig. 5.24.
Foster Wheeler Corp.: Figs. 6.3, 8.6.
General Electric Co.: Fig. 10.5.
Hays Corp.: Fig. 4.9.

Heat Exchanger Inst.: Fig. 9.13.

Ingersoll-Rand Co.: Figs. 9.11, 9.12, 12.18, 12.19.

Le Roi Co.: Fig. 5.43.

Lummus Co.: Fig. 9.3.

National Advisory Committee for Aeronautics: Fig. 10.34.

Parr Instrument Co.: Fig. 4.8.

Roots-Connersville Blower Corp.: Figs. 12.16, 12.17.

Studebaker-Packard Corp.: Fig. 5.19.

C. J. Tagliabue Mfg. Co.: Fig. 5.16.

Tecumseh Products Co.: Fig. 14.6.

Vilter Mfg. Co.: Fig. 14.4.

Westinghouse Electric Corp.: Figs. 9.6, 9.7, 9.10, 10.16, 10.21, 10.25, 10.26, 10.29, 10.30, 10.31, 10.35, 11.24, 11.26.

Willys-Overland Motors, Inc.: Fig. 5.20.

Worthington Pump & Machinery Corp.: Figs. 11.28, 11.31, 11.32, 11.33.

H. L. S.
O. C. C.
A. R. S.

# Contents

# Abbreviations

| | |
|---|---|
| brake horsepower | bhp |
| brake horsepower-hour | bhp-hr |
| brake mean effective pressure | bmep |
| brake specific fuel consumption | bsfc |
| British thermal unit | Btu |
| cubic feet per minute | cfm |
| cubic foot | cu ft |
| degrees centigrade | C |
| degrees Fahrenheit | F |
| degrees Kelvin | K |
| degrees Rankine | R |
| electromotive force | emf |
| feet per minute | fpm |
| feet per second | fps |
| foot | ft |
| foot-pound | ft-lb |
| gallon | gal |
| gallons per minute | gpm |
| horsepower | hp |
| horsepower-hour | hp-hr |
| hour | hr |
| inch | in. |
| indicated horsepower | ihp |
| indicated horsepower-hour | ihp-hr |
| kilowatt | kw |
| kilowatt-hour | kw-hr |
| mean effective pressure | mep |
| minute | min |
| outside diameter | OD |
| pound | lb |
| pounds of force | $lb_f$ |
| pounds of mass | $lb_m$ |
| pounds per square foot | psf |
| pounds per square foot absolute | psfa |

| | |
|---|---|
| pounds per square inch | psi |
| pounds per square inch absolute | psia |
| pounds per square inch gage | psig |
| revolutions per minute | rpm |
| second | sec |
| square foot | sq ft |
| square inch | sq in. |

# Symbols

| | |
|---:|:---|
| area | $A$ |
| acceleration | $a$ |
| specific heat | $c$ |
| specific heat at constant pressure | $c_p$ |
| specific heat at constant volume | $c_v$ |
| constant | $C$ |
| diameter | $d$ |
| draft | $D$ |
| piston displacement | $D_p$ |
| force | $F$ |
| local acceleration of gravity | $g$ |
| dimensional constant | $g_c$ |
| head | $h$ |
| enthalpy, per unit mass; $h = u + pv/J$ | $h$ |
| enthalpy of compressed liquid | $h_c$ |
| enthalpy of saturated liquid | $h_f$ |
| enthalpy of evaporation | $h_{fg}$ |
| enthalpy of dry saturated vapor | $h_g$ |
| enthalpy of wet vapor | $h_x$ |
| enthalpy of superheated vapor | $h_s$ |
| height of chimney | $H$ |
| mechanical equivalent of heat | $J$ |
| specific heat ratio; $k = c_p/c_v$ | $k$ |
| length | $L$ |
| mass | $m$ |
| molecular weight | $M$ |
| exponent of polytropic expansion | $n$ |
| number of revolutions per minute | $n$ |
| number of power impulses per minute | $N$ |
| pressure, absolute or gage | $p$ |
| brake mean effective pressure | $P_b$ |
| indicated mean effective pressure | $P_i$ |
| rate of heat transfer | $q$ |
| quantity of heat | $Q$ |

| | |
|---|---|
| higher heating value | $Q_H$ |
| lower heating value | $Q_L$ |
| heating value of refuse | $Q_r$ |
| compression ratio | $\tau$ |
| radius | $r$ |
| gas constant in equation $pv = RT$ | $R$ |
| entropy per unit mass | $s$ |
| temperature in degrees Fahrenheit or centigrade | $t$ |
| torque | $t$ |
| absolute temperature in degrees Rankine or Kelvin | $T$ |
| internal energy per unit mass | $u$ |
| overall coefficient of heat transfer | $U$ |
| specific volume; volume per unit mass | $v$ |
| total volume | $V$ |
| velocity | $V$ |
| weight; gravitational force acting on a mass | $w$ |
| work | $W$ |
| quality of vapor | $x$ |
| distance measured from a datum level | $z$ |
| compressibility factor | $Z$ |
| specific weight; weight per unit volume | $\gamma$ |
| density; mass per unit volume | $\rho$ |
| efficiency | $\eta$ |
| time | $\tau$ |
| temperature difference | $\theta$ |
| angular velocity | $\omega$ |
| absorptivity | $\alpha$ |
| reflectivity | $\rho$ |
| transmissivity | $\tau$ |

# CHAPTER 1

# Matter and Energy

## 1.1 Introduction

At the beginning of the nineteenth century, the stagecoach and the saddle horse were still the principal means of travel on land. Freight was transported on land in wagons drawn by horses or oxen. Barge canals were built where practical, and the barges were drawn by horses. At sea the sailing ship had not reached the zenith of its development, and man was to be dependent upon the wind for another half-century. Although the "overshot" water wheel had been in use for several centuries to drive flour mills and small manufacturing establishments, man was still largely dependent for his livelihood upon his own physical exertions, the efforts of his family including his children, and the work of his domesticated animals. However, during the latter half of the eighteenth century, the development of an efficient steam engine, the application of this engine to the operation of factories, the introduction of wrought iron, and the development of machine tools and machinery accelerated the industrial revolution and ultimately resulted in our modern industrial civilization which is founded upon the low-cost mass production of goods that can be sold cheaply throughout the world.

The Newcomen steam engine was invented in 1705 to pump water from the English coal mines. It was fairly well developed by 1720 and remained in extensive use for the next 50 years. As illustrated in Fig. 1.1, a piston was hung in a steam cylinder from one end of an overhead oscillating or "walking" beam by a chain and piston rod. The piston of the water cylinder was attached to the other end of the walking beam, also by a chain and piston rod. When the steam valve $A$ was opened with the piston near the bottom of the cylinder, steam from the boiler filled the steam cylinder at atmospheric pressure as the counterweight $W$ caused the steam piston to rise. When the steam piston reached the top position in its stroke

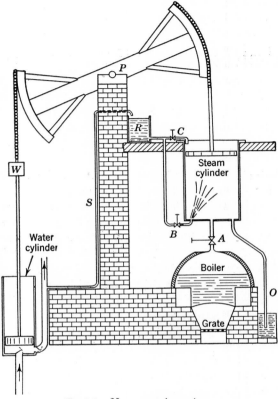

**Fig. 1.1.** Newcomen's engine.

as shown in Fig. 1.1, the steam valve $A$ was closed, and cold water
was sprayed into the cylinder by opening the injection-water valve
$B$. The vacuum created by the condensation of the steam caused
the steam piston to descend, thus raising the water piston and filling
the water cylinder with water. The water was delivered by the pump
on the downward stroke of the water piston which coincided with
the upward stroke of the steam piston. Part of the water delivered
by the pump was discharged to the reservoir $R$ through the supply
line $S$. Water from the reservoir was sprayed into the steam cylinder
at the proper time by manipulation of valve $B$. Valve $C$ was regulated to produce a layer of water on the top side of the piston and
thus prevent air leakage into the steam cylinder between the cylinder
walls and the poorly fitted piston.

The original engine had hand-operated valves. The story is told that a boy named Humphrey Potter became tired of manipulating the valves on one of these engines and arranged a mechanism for operating them from the walking beam. Mechanically operated valves were soon applied to all Newcomen engines. The boiler, which was little more than a large kettle with a fire under it on a grate, was provided with a safety valve as a protection against excessive pressure. The Newcomen engine was used extensively to pump water from mines, to supply water for cities and for the operation of canals, and to fill reservoirs from which the water could be used in water wheels to drive machinery.

James Watt, an instrument maker at the University of Glasgow, while repairing a model of a Newcomen engine in 1763, noticed the large waste of energy due to alternately heating the steam cylinder with steam and cooling it with injection water. He realized that this loss could be reduced by keeping the cylinder as hot as possible with insulation and by using a separate condenser or water-cooled chamber which could be connected to the steam cylinder at the proper time by a valve. He patented the idea of the separate condenser in 1769. Subsequently, he closed the top of the steam cylinder with a cover or cylinder head, introduced steam alternately on both sides of the piston, and thus made the engine double acting. He invented a governor to regulate the speed of the engine; a slide valve to control the admission, expansion, and exhaust of the steam; a pump to remove the air and condensate from the condenser, and, in fact, brought the steam engine to a fairly high state of development.

In 1774 Wilkinson patented a boring machine with which he was able to machine quite accurately the steam cylinders required by Watt. Without this invention it is doubtful if Watt's engine would have been successful. About 1780 methods were developed for producing "puddled" or wrought iron, a material that soon became available in reasonable quantities for the fabrication of machinery. Improvements in materials and manufacturing processes made it possible to replace wooden shafts, gears, wheels, and other machine parts with parts constructed of cast or wrought iron. These developments not only made the steam engine possible but also created a market for it. As stated by Wood, "Coal could not have been won without the steam engine; the steam engine could not have been worked without coal. Automatic machinery could not have been developed without steam power; the steam engine could not have been constructed without machine tools. Neither engine nor machinery could have been developed without iron, and the iron could not have been wrought without mechanism and power. It was one great cycle of interacting and reacting forces, no one of which could have come to perfection without the aid of the rest." *

By 1790 the use of Watt's engine had spread throughout England and made available an efficient machine for producing power wherever needed in sufficient amounts to operate large factories. Machinery had already been invented for the spinning of thread and the weaving of cloth, and the great textile industry of England developed rapidly under the impetus given to it by the steam engine.

It was natural that efforts should be made to apply the steam engine to the field of transportation. In 1804 Trevithick built a self-propelled steam carriage to run on a horse tramway in Wales. In 1807 Robert Fulton, using engines built to Watt's designs, successfully operated his steamboat, the *Clermont,* on the Hudson River. In 1829 George Stephenson built the famous *Rocket,* a locomotive that traveled at a speed of 30 miles per hr in competitive trials. The application of the steam engine to the ship and to the locomotive ultimately made it possible to move raw materials cheaply and quickly to manufacturing plants and to distribute the finished products throughout the world.

In 1882 Thomas Edison started the Pearl Street Station in New York for the purpose of supplying electricity to the users of the new incandescent lamp, thus laying the foundation for the great central-station industry which now supplies the general public with electric light and power. Parsons patented a reaction turbine in 1884, and in 1889 de Laval was granted patents on an impulse turbine. By 1910 the steam turbine had replaced the reciprocating steam engine in the central-station industry.

In 1860 Lenoir patented a gas engine which operated without compression, and it is reported that 400 Lenoir engines were operating in Paris in 1865. In 1876 Otto developed a successful four-stroke-cycle gas engine. Daimler in 1887 applied a light-weight high-speed Otto cycle engine to a motor car, and the development of the automobile followed. In 1892 Dr. Rudolph Diesel patented the Diesel cycle engine for burning nonvolatile oil in an internal-combustion engine at high efficiency. In 1903 the Wright brothers made the first successful flight of a heavier-than-air plane powered by a gasoline engine; thus began the development of the aircraft industry.

* Wood, *Industrial Engineering in the Eighteenth Century,* pp. 17–18.

During the last decade, the gas turbine in the form of the turbo-jet and turboprop engines has replaced the reciprocating internal-combustion engine in the military combat airplane and in the faster and larger commercial aircraft. The gas turbine is also being used in such applications as electric power generation, natural gas transmission line pumping, and locomotives.

The recent development of the rocket threatens to revolutionize warfare with guided missiles and earth satellites. Since the rocket carries its own supply of oxygen for the burning of its fuel, it is capable of operating at altitudes where the earth's atmosphere is highly rarefied.

Today airplanes, automobiles, buses, railroads, and steam- and Diesel-powered ships provide quick and cheap transportation. Electric power is available at low rates and in ample quantities in all sections of the United States, and its consumption is doubling every 10 or 11 years. The productive capacity of American industry is due directly to a high degree of mechanization and automation which in turn is dependent upon an adequate supply of cheap power. The rapidly increasing demands for energy for transportation, manufacturing, and use in our homes, offices, and stores are being met by consumption of our fossil fuels—coal, oil, and gas—at a rate which threatens their extinction within 400 years. Fortunately, man has recently learned to utilize the energy released by controlled nuclear fission and has thereby greatly extended the known reserves of energy. The commercial exploitation of nuclear energy for power generation is just beginning and promises to expand rapidly. If man can learn to control and utilize the energy released by nuclear fusion, he will then unlock an almost unlimited source of energy with which to provide power, light, and heat.

This book, *Thermal Engineering,* is concerned with the principles of operation, details of construction, and actual performance of the major types of equipment that are used for the generation of power from fossil fuels and nuclear reactions. The subject matter includes a study of the properties of fluids used for power generation and refrigeration, fuels and their combustion, nuclear reactions, internal-combustion engines, equipment for burning fuels for steam generation and industrial uses, nuclear reactors, heat transfer, steam generation, heat exchangers, turbines, pumps, fans and compressors. power-plant cycles, and refrigeration. The student who wishes to pursue the subject further should follow a study of this book by a thorough course in *thermodynamics,* a subject that is concerned with the laws governing the transformation of energy from one form to another. Engineering thermodynamics deals with the properties of the fluids used in power-producing machinery, the energy relationships and the behavior of these fluids when subjected to changes under specified conditions, the theoretical performance of such machinery, and the factors that affect and limit this performance. Further study in the field of thermal engineering is possible in such specialized areas as internal-combustion engines, steam power plants, steam and gas turbines, compressors and pumps, combustion of fuels, steam generation, refrigeration and air conditioning, and nuclear energy.

### 1.2   The Steam Power Plant

The function of a steam power plant is to convert the energy in nuclear reactions or in coal, oil, or gas into mechanical or electric energy through the expansion of steam from a high pressure to a low pressure in a suitable *prime mover* such as a turbine or engine. Where the output of the plant is electric energy distributed for general sale to all customers who wish to purchase it, the plant is called a *central station*. If the plant is operated by a manufacturing company which takes the output of the plant for its own use, it is called an *industrial plant*. A *noncondensing* plant discharges the steam from the prime mover at an exhaust pressure equal to or greater than atmospheric pressure. A *condensing* plant exhausts from the prime mover into a *condenser* at a pressure *less than* atmospheric pressure. In general, central-station plants are condensing plants since their sole output is electric energy and a reduction in the exhaust pressure at the prime mover decreases the amount of steam required to produce a given quantity of electric energy. Industrial plants are frequently noncondensing plants because large quantities of low-pressure steam are required for manufacturing operations. The power required for operation of a manufacturing plant may often be obtained as a by-product by generating steam at high pressure and expanding this steam in a prime mover to the back pressure at which the steam is needed for manufacturing processes.

Figure 1.2 illustrates diagrammatically the major pieces of equipment that are installed in a condensing steam power plant fired by pulverized coal. The *steam-generating unit* consists of a *furnace* in which the fuel is burned, a *boiler, superheater,* and *economizer,* in which high-pressure steam is generated, and an *air heater* in which the loss of the energy due to combustion of the fuel is reduced to a minimum. The *boiler* is composed of a drum, in which a water level is maintained at about the mid-point so as to permit separation of the steam from the water, and a bank of inclined tubes, connected to the drum in such a manner as to permit water to circulate from the drum through the tubes and back to the drum. The hot products of combustion from the furnace flow across the boiler tubes and evaporate part of the water in the tubes. The furnace walls are composed of tubes which are also connected to the boiler drum to form very effective steam-generating surfaces. The steam which is separated from the water in the boiler drum then flows through a *superheater* which is in effect a coil of tubing surrounded by the hot products of combustion. The temperature of the steam is increased in the superheater to perhaps 800 to 1100 F, at which temperature the high-pressure superheated steam flows through suitable piping to the turbine.

Since the gaseous products of combustion leaving the boiler tube bank are at a relatively high temperature and their discharge to the chimney would result in a large loss in energy, an *economizer* may be used to recover part of the energy in these gases. The economizer is a bank of tubes through which the boiler feedwater is pumped on its way to the boiler drum. Inasmuch as this feedwater will be at a temperature considerably below that of the water in the boiler tubes, the temperature of the products of combustion may be reduced in the economizer to less than the boiler exit-gas temperature. A further reduction in gas temperature may be made by passing the products of combustion through an *air heater* which is a heat exchanger cooled by the air required for combustion. This air is supplied to the air heater at normal room temperature and may leave the air heater at 400 to 600 F, thus returning to the furnace energy that would otherwise be wasted up the chimney. The products of combustion are usually cooled in an air heater to an exit temperature of 275 to 400 F, after which they may be passed through a *dust collector* which will remove objectionable dust and thence through an *induced-draft fan* to the chimney. The function of the induced-draft fan is to pull the gases through the heat-transfer surfaces of the boiler, superheater, economizer, and air heater and to maintain a pressure in the furnace that is slightly less than atmospheric pressure. A *forced-draft fan* forces the combustion air to flow through the air heater, duct work, and *burner* into the furnace.

Coal is delivered to the plant in railroad cars or barges which are unloaded by machinery. The coal may be placed in storage or may be crushed and elevated to the overhead raw-coal bunker in the boiler room. As shown in Fig. 1.2, the coal flows by gravity from the overhead bunker to the *pulverizer* or *mill* through a *feeder* which automatically maintains the correct amount of coal in the mill. In the mill the coal is ground to a fine dust. Some of the hot

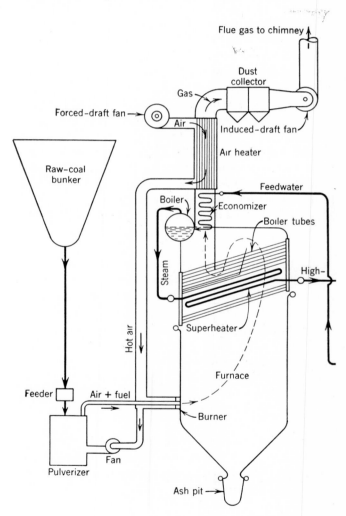

**Fig. 1.2.** Diagrammatic arrangement

air from the air heater is forced through the mill to dry the coal and to pick up the finely pulverized particles and carry them in suspension to the burner where they are mixed with the air required for their combustion and discharged into the furnace at high velocity to promote good combustion.

The high-pressure, high-temperature steam is expanded in a *steam turbine* which is generally connected to an electric generator. From 3 to 5 per cent of the output of the generator is needed to light the plant and to operate the many motors required for fans, pumps, etc., in the plant. The rest of the generator output is available for distribution outside the plant.

In a central-station plant, the exhaust steam from the turbine is delivered to a *condenser,* the function of which is to convert this steam into condensate (water) at the lowest possible pressure. The condenser is a large gas-tight chamber filled with tubes through which cold water is pumped. Under average conditions, about 800 tons of cooling water are required in the condenser for each ton of coal burned. Consequently, large power plants must be located on lakes, rivers, or the seacoast where plenty of cool water is available for use in the condenser. The exhaust steam is condensed for two reasons: (1) it was distilled in the boiler and is therefore free of scale-forming

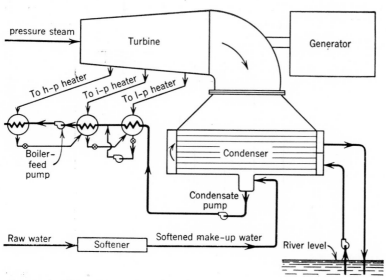

of a condensing steam power plant.

material and should be retained in a closed system, and (2) the efficiency of the plant is increased substantially by reducing the exhaust pressure at the turbine to as low a pressure as possible. For instance, it is estimated that a reduction in the exhaust pressure from 1 psia (lb per sq in. abs) to ½ psia will reduce the coal consumption of the average plant by 4 or 5 per cent.

The condensed steam, which is normally at a temperature of 70 to 100 F, is pumped out of the condenser by means of a *hot-well pump* and is discharged through several *feed-water heaters* to a *boiler feed pump* that delivers the water to the economizer. Figure 1.2 shows a high-pressure heater, an intermediate-pressure heater, and a low-pressure heater, all supplied with steam which is extracted from the turbine at appropriate pressures after having done some work by expansion to the extraction pressure in the turbine.

Figure 1.2 also shows one method by which raw water may be passed through a *softener* to remove the scale-forming impurities, after which it is admitted to the condenser at such a rate as to keep the system full of scale-free water.

Most steam power plants of large size are now being built for operation at steam pressures of 1500 to 2400 psi, and in some plants pressures up to 5000 psi are being used. Steam temperatures of 1000 to 1100 F are in general use. Turbine-generator capacities of 250,000 kw (1 kilowatt = 1.34 horsepower) are common, and units of 500,000

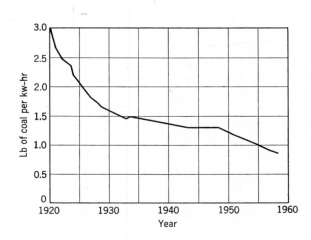

**Fig. 1.3.** Average coal consumption per kilowatt-hour of electric energy generated in the United States.

kw are in operation. Steam-generating units capable of delivering 3,000,000 lb of steam per hr are now in operation. Overall efficiency of the plant from raw coal supplied to electric energy delivered to the transmission line depends upon size, steam pressure, temperature, and other factors, and 40 per cent is now being realized on the basis of a full year of operation. Figure 1.3 shows the decrease in coal required to turn out a kilowatt-hour (kw-hr), the unit of electric energy, over the past 35 years. Since the best of the modern power plants will generate 1 kw-hr on 0.70 lb of coal, the downward trend in fuel consumption may be expected to continue as the older and less efficient power plants are replaced by new plants.

### 1.3   The Internal-Combustion-Engine Power Plant

The internal-combustion-engine power plant including essential auxiliaries is shown diagrammatically in Fig. 1.4. The fuel is burned directly in the cylinder of the *engine* or *prime mover,* and the high pressure thus generated drives the *piston* downward and rotates a *crankshaft.*

Air is supplied to the engine through a *silencer* and *cleaner,* the function of which is to reduce noise and remove dust which would accelerate cylinder and piston wear if allowed to enter the cylinder. Figure 1.4 shows a *supercharger* installed in the air-intake system. The function of the supercharger is to increase the amount of air supplied to the cylinder by acting as an air pump. This in turn permits burning more fuel and obtaining more power from a given size of cylinder. An *intake manifold* is used to distribute the air equally from the supercharger to the various cylinders of a multicylinder engine.

The exhaust system consists of an *exhaust manifold* for collecting the discharge gases from each of the cylinders into a common exhaust line, an exhaust silencer or *muffler* for reducing noise, and the *exhaust stack* for disposing of the exhaust gases to the atmosphere without creating a public nuisance.

The cooling system includes a *pump* for circulating water through the cylinder jackets and heads of each cylinder and a *heat exchanger* to remove the energy absorbed in the engine by the cooling water. The heat exchanger may be air-cooled as in the automobile radiator, or it may be water-cooled as shown in Fig. 1.4. Seldom is raw water fit to circulate directly through the jackets of an internal-combustion engine.

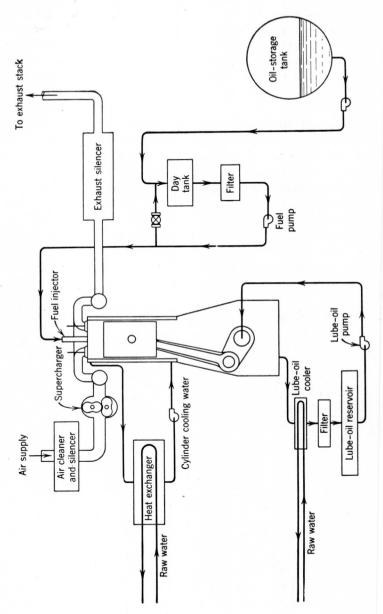

**Fig. 1.4.** The internal-combustion-engine power plant.

The lubricating oil may be passed through a *cooler, filter,* and *reservoir* and is supplied to the engine under pressure by means of an *oil pump,* usually to a hollow crankshaft. The oil serves as a lubricant for the rubbing surfaces of the engine and also as a coolant.

The fuel system consists of a storage tank from which the fuel may be supplied to a small day tank or reservoir. The oil is filtered and pumped as needed to the fuel-injection system which is an integral part of the engine.

Since the fuel is burned directly in the cylinder of the prime mover, the internal-combustion-engine power plant is simpler and more compact than the steam power plant. It is seldom built in engine sizes of more than 4000 hp, whereas a 300,000-hp steam turbine is common. It is more efficient than a steam power plant of comparable size but not so efficient as large steam central-station plants, which moreover can burn a cheaper grade of fuel. Consequently, the internal-combustion engine is used primarily in the transportation field for driving automobiles, buses, trucks, tractors, locomotives, ships, and airplanes where a compact, light-weight, efficient power plant of relatively small size is necessary.

### 1.4   The Gas-Turbine Power Plant

The essential components of the gas-turbine power plant are illustrated diagrammatically in Fig. 1.5. Air is compressed in an *axial-flow compressor* from atmospheric pressure to a pressure which is usually between the limits of 75 and 120 psi. The compressed air may then flow through a *regenerator* or *heat exchanger* in which the hot exhaust gas from the *turbine* is utilized to increase the temperature of the air, thereby recovering energy that would otherwise be lost to the atmosphere. Fuel is sprayed into the *combustor* in which it combines chemically with the oxygen in the air to produce a hot gas leaving the combustor at some temperature between 1200 and 1700 F. The pressure of the air decreases slightly between the compressor discharge and turbine inlet because of friction, but the increase in temperature in the regenerator and combustor results in more than doubling the volume. The hot gas then expands in the *turbine* in which it does enough work to drive the compressor as well as an electric generator or some other suitable machine. The exhaust gases leaving the turbine are cooled in the regenerator before being discharged to the atmosphere.

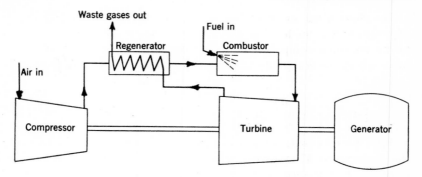

**Fig. 1.5.**  The gas-turbine power plant.

Where space and weight limitations are critical or fuel is cheap, the regenerator may be omitted with a substantial decrease in efficiency.  The *turboprop engine* as applied to the airplane operates without a regenerator and with a geared propeller as the load.  In the *turbojet engine* as applied to the airplane, the turbine develops only enough power to drive the compressor and exhausts into a nozzle at a back pressure considerably in excess of atmospheric pressure. The rearward expansion of the exhaust gases from the nozzle at high velocity creates the thrust which propels the airplane.

### 1.5   The Nuclear Power Plant

In the nuclear power plant, energy is released in a *reactor* by nuclear fission, a process that is discussed in Chapter 4.  A coolant is pumped through the reactor to absorb and remove this energy and thereby prevent an excessive temperature in the reactor.  In the more common types of nuclear power plants, the high-temperature coolant that leaves the reactor flows through a *heat exchanger* in which steam is generated.  The nuclear reactor, heat exchanger, and pump are shown diagrammatically in Fig. 1.6.  Extensive provisions are made to protect the operating personnel and the general public from the hazards of radioactivity by the installation of radiation and containment shields which enclose all radioactive components of the system.

The heat exchanger serves as a steam boiler.  The steam flows through a turbine and associated equipment that are identical in design and arrangement with similar equipment in a conventional steam power plant as illustrated on the right page of Fig. 1.2. In other words, the nuclear reactor, heat exchanger, and pump replace the fuel-burning equipment and the steam generator of the conventional steam power plant that are illustrated on the left page of Fig. 1.2.

### 1.6  Mechanical Refrigeration

In the power plant, a high-pressure gas or vapor is expanded in a prime mover to a low pressure with the result that a part of the energy in the high-pressure fluid is converted into work. The reverse process is used in mechanical refrigeration as is illustrated diagrammatically in Fig. 1.7. Some liquid, called a *refrigerant,* which has the property of boiling and absorbing its latent heat at a suitable low temperature, is allowed to do so in an *evaporator* which may be located in an insulated compartment called a *refrigerator.* For example, the refrigerant might evaporate at some temperature between 0 and 20 F. As a result of the absorption of its latent heat of evaporation, the insulated space in the refrigerator is cooled. The vapor

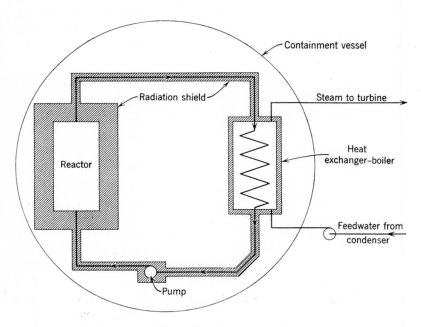

**Fig. 1.6.** Steam-generating system of a nuclear power plant.

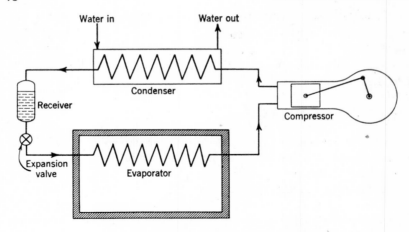

**Fig. 1.7.** The compression system of refrigeration.

from the evaporator is compressed in a *compressor* to a pressure sufficiently high so that the vapor can be condensed or liquefied in a *condenser*. The condenser may be cooled by water or, in small units such as the domestic refrigerator, by air. The high-pressure liquid drains from the condenser into a *receiver* from which it flows through an *expansion valve* in which the pressure is reduced to the evaporator pressure.

For air-conditioning applications, the air may be cooled by blowing it directly over the evaporator coils or the evaporator may be used as a heat exchanger to cool water which may be circulated to suitable conveniently located air coolers.

### 1.7   System Components

The steam-turbine power plant, the internal-combustion-engine power plant, the gas-turbine power plant, the nuclear power plant, and the refrigeration system have much in common. All of them have heat exchangers such as evaporators, boilers, superheaters, regenerators, coolers, or condensers in which heat is transferred to or from the fluid that is flowing through them. Work is done by the expansion of this fluid through a turbine or engine. Energy is supplied to the system through the burning of fuel or nuclear reactions, or in the refrigeration system, as electric energy required by a motor-driven compressor. These heat exchangers and prime movers or compressors are connected by suitable piping. Fans, pumps, and blowers circulate the fluid through the system of heat exchangers, pipes, etc.

Heat exchangers, pumps, fans, and compressors are used also in many applications not associated with power generation. Since the characteristic function of an engineer is ability to design, considerable attention is given in this book to the important features of construction of these essential components of mechanical engineering systems, the reasons why they are so constructed, their overall performance, and the methods of determining performance.

## 1.8   Dimensions and Units

A physical quantity is something that can be measured. A dimension is a property or quality of the quantity or entity that characterizes or describes that quantity. Thus, distance has the characteristic property or dimension of length. Area is a function of length and breadth or, dimensionally, $[A] = [L^2]$. Also, volume is a function of length $\times$ breadth $\times$ height or, dimensionally, $[V] = [L^3]$.

The dimension of length may be expressed in various *units*, such as inches, feet, yards, miles, centimeters, meters, and kilometers. Any physical quantity is exactly specified by a dimension and by a multiple or fraction of a defined unit, such as a length of 14 ft.

A dimensional system that will completely describe an event can be constructed from a relatively small number of fundamental dimensions. One of these fundamental dimensions usually is *time*, $[\tau]$. The unit of time is the *second*, which is defined as 1/86,400 part of a mean solar day. Length, $[L]$, is usually considered to be another of the fundamental dimensions. The standard *meter*, which is the basic unit of length, is the distance between two lines, measured at 32 F, on a particular bar of platinum-iridium that is kept at the International Bureau of Weights and Measures at Sèvres, France. The *centimeter* (cm) is $\frac{1}{100}$ of the meter. In the systems of units used in most engineering calculations in the United States, the unit of length is the *foot*, which is $\frac{1200}{3937}$ of the standard meter.

Matter is that which occupies space and has inertia or resistance to change of motion. Mass may be defined as the quantity of matter. If a mass, $m$, is acted upon by an unbalanced force, $F$ (a force being a push or pull), then, according to Newton's second law of motion,

$$F \propto ma$$

or
$$F = \frac{1}{g_c} ma \qquad (1.1)$$

where $a$ represents the acceleration imparted to the mass and $1/g_c$ is a constant of proportionality. Acceleration may be expressed dimensionally as follows:

$$\text{Velocity, } [V] = \frac{[L]}{[\tau]}$$

$$\text{Acceleration} = \frac{[V]}{[\tau]} = \frac{[L]}{[\tau^2]} = [L\tau^{-2}]$$

Therefore Equation 1.1 may be written dimensionally as follows:

$$[F] = [M][L\tau^{-2}] = [ML\tau^{-2}] \qquad (1.2)$$

If three of the dimensions in Equation 1.2 are defined, the fourth dimension may be expressed in terms of the known three. The dimensions of length and time have already been defined. The system can be fully specified by defining *either* mass or force as the third dimension.

Five different systems of units, based on Equations 1.1 and 1.2, are in common use, together with many hybrids or mixtures of these systems. These five systems of units are discussed in the following paragraphs.

**The English Dynamical System.** In this system, as in the English gravitational and English engineering systems, the unit of time is the second, the unit of length is the foot, and acceleration is therefore expressed in ft per sec². Mass is *defined* as the third dimension, and the unit of force is derived. The international standard of *mass* is the *kilogram*, which is defined as the mass of a particular piece of platinum-iridium located at Sèvres, France. The standard *pound* of mass (lb$_m$) is *defined* as 0.4535924+ of the standard kilogram. The unit of force is selected in such a manner that the factor of proportionality, $g_c$, in Equation 1.1 will be equal numerically to unity. Then an unbalanced force of one unit will impart to a mass of one standard pound an acceleration of 1 ft per sec². Such a force is called a *poundal*, or

$$1 \text{ poundal} = \frac{1}{g_c} \times 1 \text{ lb}_m \times 1 \text{ ft per sec}^2 \qquad (1.3)$$

where the symbol $lb_m$ means mass in standard pounds and $g_c$ is the dimensional proportionality factor of Equation 1.1 and has a numerical value of 1.0, or

$$g_c = 1.0 \frac{lb_m \text{ ft}}{\text{poundal sec}^2}$$

**Example 1.** What force in poundals is required to accelerate 100 $lb_m$ at the rate of 10 ft per sec²?
*Solution:*

$$F = \frac{1}{g_c} ma = \frac{1}{1.0 \dfrac{lb_m \text{ ft}}{\text{poundal sec}^2}} \times 100 \; lb_m \times \frac{10 \text{ ft}}{\text{sec}^2} = 1000 \text{ poundals}$$

Note that by proper use of the units for $g_c$, $m$, and $a$, the answer is given in correct units.

**The cgs System.** In this system of units, the *centimeter* (cm) is used as the unit of distance, the *second* as the unit of time, and the *gram* (1/1000 of the standard kilogram) as the unit of mass. The unit of force is *selected* in such a manner that the factor of proportionality, $g_c$, in Equation 1.1 will again be equal numerically to unity. Then an unbalanced force of one unit will impart to a mass of one gram an acceleration of 1 cm per sec². Such a force is called a *dyne*, or

$$1 \text{ dyne} = \frac{1}{g_c} \times 1 \text{ gm} \times 1 \text{ cm per sec}^2 \tag{1.4}$$

and

$$g_c = 1.0 \frac{\text{gm cm}}{\text{dyne sec}^2}$$

**Example 2.** What force in dynes will be required to accelerate a mass of 100 grams at the rate of 20 cm per sec²?
*Solution:*

$$F = \frac{1}{g_c} ma = \frac{1}{1.0 \dfrac{\text{gm cm}}{\text{dyne sec}^2}} \times 100 \text{ gm} \times \frac{20 \text{ cm}}{\text{sec}^2} = 2000 \text{ dynes}$$

**The MKS System.** In this system of units, which is used extensively in Continental Europe, the unit of time is the *second,* the unit of mass is the *kilogram,* and the unit of distance is the *meter.* The unit of force is again *selected* in such a manner that the factor of proportionality, $g_c$, in Equation 1.1 will be equal numerically to unity. Then an unbalanced force of one unit will impart to the standard kilogram of mass an acceleration of 1 meter per sec². Such a force is called a *newton,* or

$$1 \text{ newton} = \frac{1}{g_c} \times 1 \text{ kg} \times 1 \text{ meter per sec}^2 \tag{1.5}$$

and
$$g_c = 1.0 \frac{\text{kg meter}}{\text{newton sec}^2}$$

**Example 3.** What force in newtons will impart to a mass of 100 kg an acceleration of 10 meters per sec²?

*Solution:*

$$F = \frac{1}{g_c} ma = \frac{1}{1.0 \dfrac{\text{kg meter}}{\text{newton sec}^2}} \times 100 \text{ kg} \times \frac{10 \text{ meter}}{\text{sec}^2}$$

$$= 100 \text{ newtons} \qquad \textit{WHY NOT 1000 NEWTONS?}$$

**The English Gravitational System.** In the three preceding systems, mass has been *defined* in terms of the standard kilogram at Sèvres, France, and the unit of force has been *selected* in terms of mass, distance, and time. In the English gravitational system, the unit of force which is called the pound of force and given the symbol $\text{lb}_f$ has been *defined* as the force of gravity acting on one standard pound of mass at sea level and 45 degrees of latitude. The unit of mass is then *selected* in such a manner that the factor of proportionality, $g_c$, in Equation 1.1 will be numerically equal to unity. Then an unbalanced force of one pound ($\text{lb}_f$) will accelerate a unit of mass at the rate of 1 ft per sec². Such a unit of mass is called the *slug,* or

$$1 \text{ lb}_f = \frac{1}{g_c} \times 1 \text{ slug} \times 1 \text{ ft per sec}^2 \tag{1.6}$$

and
$$g_c = 1.0 \frac{\text{slug ft}}{\text{lb}_f \text{ sec}^2}$$

**Example 4.** What force will accelerate a mass of 100 slugs at the rate of 10 ft per sec²?

*Solution:*

$$F = \frac{1}{g_c} ma = \frac{1}{1.0 \dfrac{\text{slug ft}}{\text{lb}_f \text{ sec}^2}} \times 100 \text{ slugs} \times \frac{10 \text{ ft}}{\text{sec}^2} = 1000 \text{ lb}_f$$

**The English Engineering System.** In this system of units, which is used in most engineering calculations in the United States and will be used in this book, the standard pound of mass ($lb_m$) from the English dynamical system and the standard pound of force ($lb_f$) from the English gravitational system are used with the foot and second as the units.

Any body falling freely in a vacuum at sea level and 45 degrees of latitude will be accelerated by the force of gravity at the rate of 32.174 ft per sec². This constant is known as the standard acceleration of gravity. Since one pound of force has been defined as the gravitational force acting on one pound of mass at sea level and 45 degrees of latitude, therefore one pound of force will accelerate one pound of mass at the rate of 32.174 ft per sec².

Or, applying Equation 1.1,

$$1 \text{ lb}_f = \frac{1}{g_c} \times 1 \text{ lb}_m \times 32.174 \text{ ft per sec}^2 \qquad (1.7)$$

and

$$g_c = 32.174 \frac{\text{lb}_m \text{ ft}}{\text{lb}_f \text{ sec}^2}$$

**Example 5.** What force in $lb_f$ will be required to accelerate 100 $lb_m$ at the rate of 20 ft per sec²?

*Solution:*

$$F = \frac{1}{g_c} ma = \frac{1}{32.174 \dfrac{\text{lb}_m \text{ ft}}{\text{lb}_f \text{ sec}^2}} \times 100 \text{ lb}_m \times \frac{20 \text{ ft}}{\text{sec}^2} = 62.2 \text{ lb}_f$$

The units used in each of these five common systems of units and the corresponding numerical value and units for the proportionality factor, $g_c$, in the equation of Newton's second law are shown in Table 1.1.

Since

$$F = \frac{1}{g_c} ma \quad \text{or} \quad g_c = \frac{ma}{F}$$

it should be noted that $g_c$ always has the dimensions of

$$\frac{ML\tau^{-2}}{F} = \frac{ML}{F\tau^2}$$

and that $g_c$ may or may not be equal numerically to unity, depending upon the system being used.

**Example 6.** Calculate $g_c$ for a system in which the unit of mass is the kilogram, the unit of force is the pound of force, the unit of length is the yard, and the unit of time is the minute.

## TABLE 1.1

### Common Systems of Units

| System | Mass | Force | Distance | Time | $g_c$ |
|---|---|---|---|---|---|
| English dynamical | $lb_m$ | poundal | ft | sec | $1.0 \dfrac{lb_m \text{ ft}}{\text{poundal sec}^2}$ |
| English gravitational | slug | $lb_f$ | ft | sec | $1.0 \dfrac{\text{slug ft}}{lb_f \text{ sec}^2}$ |
| English engineering | $lb_m$ | $lb_f$ | ft | sec | $32.174 \dfrac{lb_m \text{ ft}}{lb_f \text{ sec}^2}$ |
| cgs | gram | dyne | cm | sec | $1.0 \dfrac{\text{gm cm}}{\text{dyne sec}^2}$ |
| MKS | kg | newton | meter | sec | $1.0 \dfrac{\text{kg meter}}{\text{newton sec}^2}$ |

*Solution:* Any of the five systems of units outlined above could be used in the solution. Let us use the English engineering system. Then

$$g_c = 32.174 \frac{lb_m \text{ ft}}{lb_f \text{ sec}^2}$$

Since 1 $lb_m$ = 0.4535924 kg, 1 ft = ⅓ yd, and 1 sec = ¹⁄₆₀ min, then

$$g_c = 32.174 \frac{lb_m \text{ ft}}{lb_f \text{ sec}^2} \times 0.45359 \frac{kg}{lb_m} \times \frac{1 \text{ yd}}{3 \text{ ft}} \times (60)^2 \frac{\sec^2}{\min^2}$$

$$= 17{,}500 \frac{\text{kg yd}}{lb_f \min^2}$$

A comparison of $g_c$ for the English gravitational and English engineering systems of units shows that

$$1 \text{ slug} = 32.174 \ lb_m \tag{1.8}$$

Weight ($w$) is defined as the gravitational force acting on a given mass in *any* gravitational field. In accordance with the universal law of gravitation, the weight of a given mass on or near the surface of the earth varies inversely as the square of the distance between the centers of the earth and the body.

The local acceleration of gravity, $g$, is that acceleration which will be imparted to any body which is allowed to fall freely in a vacuum

### TABLE 1.2

#### Effect of Location on Weight and Local Acceleration of Gravity

| Location | Weight of 100 $lb_m$ | $g$ |
| --- | --- | --- |
| At sea level at the Equator | 99.733 | 32.088 |
| At sea level at 45° of latitude | 100.000 | 32.174 |
| At sea level at the North Pole | 100.261 | 32.258 |
| 40,000 ft above sea level at 45° of latitude | 99.615 | 32.050 |
| 500 miles above sea level | 78.849 | 25.369 |
| 2000 miles above sea level | 44.169 | 14.211 |
| At surface of the moon | 17.001 | 5.47 |

at any location. The accelerating force is the weight of the body. Table 1.2 shows the variation in the local acceleration of gravity and the weight of a mass of 100 $lb_m$ at various locations on or near the surface of the earth.

The variation in the local acceleration of gravity and the variation in the weight of a given mass with location on the surface of the earth are generally less than the inaccuracies introduced into engineering calculations through the use of the 10-inch slide rule, uncertainties in the values of physical properties of substances, and errors in engineering instrumentation. Consequently, an approximate value of 32.2 ft per $sec^2$ is often used for the local acceleration of gravity. Also, it is frequently assumed that the weight of a body in $lb_f$ is numerically equal to its mass in $lb_m$. However, in this day of intercontinental ballistic missiles, earth satellites, etc., the engineer may be called upon to solve problems in which the variation in weight and local acceleration of gravity with location must be considered, and it is important that he understand the distinction between mass and weight.

Since, in accordance with Newton's second law of motion,

$$F = \frac{1}{g_c} ma \tag{1.1}$$

then, if the only force acting on a body is the force of gravity (its weight, $w$), the resulting acceleration will be the local acceleration of gravity, $g$, or

$$w = \frac{1}{g_c} mg$$

or

$$\frac{w}{g} = \frac{m}{g_c} \qquad (1.9)$$

Combining Equations 1.1 and 1.9 gives the following important relation:

$$F = \frac{w}{g} a \qquad (1.10)$$

Because of the importance of Equation 1.9, the ratio $g/g_c$ will appear in many equations in this book. It will be noted from Table 1.2 that on the surface of the earth the average value of the local acceleration of gravity is about 32.2 ft per sec$^2$ with a maximum deviation from this value of less than $\pm 0.5$ per cent. Therefore, for slide-rule calculations involving the English engineering system of units for problems on the surface of the earth,

$$\frac{g}{g_c} = \frac{32.2 \dfrac{\text{ft}}{\text{sec}^2}}{32.174 \dfrac{\text{lb}_m \text{ ft}}{\text{lb}_f \text{ sec}^2}} = 1.00 \frac{\text{lb}_f}{\text{lb}_m} \text{ (approximately)} \qquad (1.10a)$$

In the English gravitational system, weight is expressed in $\text{lb}_f$ and local acceleration of gravity is expressed in ft per sec$^2$. Mass is expressed in slugs and

$$g_c = 1.0 \frac{\text{slug ft}}{\text{lb}_f \text{ sec}^2}$$

In the English engineering system, weight is also expressed in $\text{lb}_f$ and local acceleration of gravity is expressed in ft per sec$^2$. Mass is expressed in $\text{lb}_m$ and

$$g_c = \frac{32.174 \text{ lb}_m \text{ ft}}{\text{lb}_f \text{ sec}^2}$$

The relation between weight in $\text{lb}_f$ and mass in either slugs (English gravitational system) or $\text{lb}_m$ (English engineering system) will be illustrated by several examples.

**Example 7.** A body weighs 100 $\text{lb}_f$ at sea level and 45 degrees of latitude. What is its mass (a) in slugs and (b) in $\text{lb}_m$?

*Solution:* At sea level and 45 degrees of latitude, the local acceleration of gravity, $g = 32.174$ ft per sec$^2$.

Since
$$\frac{w}{g} = \frac{m}{g_c} \qquad\qquad (1.9)$$

then
$$m = \frac{w}{g} g_c$$

(a) In the English gravitational system,

$$\text{Mass in slugs} = \frac{w}{g} g_c = \frac{100 \text{ lb}_f}{32.174 \dfrac{\text{ft}}{\text{sec}^2}} \times \frac{1.0 \text{ slug ft}}{\text{lb}_f \text{ sec}^2} = 3.11 \text{ slugs}$$

(b) In the English engineering system,

$$\text{Mass in lb}_m = \frac{w}{g} g_c = \frac{100 \text{ lb}_f}{32.174 \dfrac{\text{ft}}{\text{sec}^2}} \times \frac{32.174 \text{ lb}_m \text{ ft}}{\text{lb}_f \text{ sec}^2} = 100 \text{ lb}_m$$

**Example 8.** A body weighs 100 lb$_f$ at a position where the local acceleration of gravity is 20 ft per sec$^2$. What is its mass (a) in slugs and (b) in lb$_m$?
*Solution:*

$$m = \frac{w}{g} g_c$$

(a) In the English gravitational system,

$$\text{Mass in slugs} = \frac{w}{g} g_c = \frac{100 \text{ lb}_f}{20 \dfrac{\text{ft}}{\text{sec}^2}} \times \frac{1.0 \text{ slug ft}}{\text{lb}_f \text{ sec}^2} = 5.0 \text{ slugs}$$

(b) In the English engineering system,

$$\text{Mass in lb}_m = \frac{w}{g} g_c = \frac{100 \text{ lb}_f}{20 \dfrac{\text{ft}}{\text{sec}^2}} \times \frac{32.174 \text{ lb}_m \text{ ft}}{\text{lb}_f \text{ sec}^2} = 160.9 \text{ lb}_m$$

**Example 9.** What unbalanced force expressed in lb$_f$ is required to accelerate a body at the rate of 30 ft per sec$^2$ if the body weighs 100 lb$_f$ at a place where the local acceleration of gravity is 20 ft per sec$^2$?
*Solution:* From Example 8, this body which weighs 100 lb$_f$ was found to have a mass of 160.9 lb$_m$ or 5.0 slugs. The problem may be solved in three ways as follows:

(a) From Equation 1.10,
$$F = \frac{w}{g} a$$

$$F = \frac{100 \text{ lb}_f}{20 \dfrac{\text{ft}}{\text{sec}^2}} \times 30 \frac{\text{ft}}{\text{sec}^2} = 150 \text{ lb}_f$$

(b)  Using the English gravitational system of units and Equation 1.1,

$$F = \frac{1}{g_c} ma$$

$$F = \frac{1}{\dfrac{1.0 \text{ slug ft}}{\text{lb}_f \text{ sec}^2}} \times 5 \text{ slug} \times 30 \frac{\text{ft}}{\text{sec}^2} = 150 \text{ lb}_f$$

(c)  Using the English engineering system of units and Equation 1.1,

$$F = \frac{1}{g_c} ma$$

$$F = \frac{1}{32.174 \dfrac{\text{lb}_m \text{ ft}}{\text{lb}_f \text{ sec}^2}} \times 160.9 \text{ lb}_m \times 30 \frac{\text{ft}}{\text{sec}^2} = 150 \text{ lb}_f$$

The normal methods that are used for "weighing" a body do not determine weight; instead they measure mass. In the chemical balance a beam is supported at its mid-point on a knife-edge about which it is free to oscillate through a small angle. An unknown mass is suspended in a pan from a knife-edge at the left end of the beam, and known masses are placed in a similar pan suspended on a knife-edge at an equal distance to the right from the center point. Thus the unknown mass is balanced with known masses. Since the same gravitational field acts on both the known and unknown masses, the strength of this field does not affect the result and masses are compared directly.

Figure 1.8 shows diagrammatically a system of levers and knife-edges such as might be used in a conventional platform scale. An unknown mass is placed on the platform. The gravitational field produces a downward force (weight of the body) which is counter-balanced by the weight of known masses that are placed on the scale beam and by moving a constant mass closer to or further away from the knife-edge support of the scale beam. Thus the weight of the known masses on the scale beam produces a force which is equalized through the multiplying linkage of the lever system with the weight of the unknown mass. Since the same gravitational field acts on both the known and the unknown masses, the result is independent of the strength of the gravitational field and masses are being compared.

Weight or gravitational pull could be measured by hanging an unknown mass from a vertical steel helical spring that is supported at its upper end and has a pointer attached to its lower end so as to

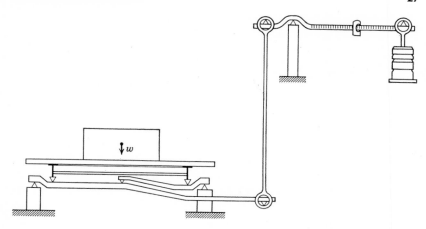

**Fig. 1.8.** Diagrammatic arrangement of a platform scale.

indicate the deflection of the spring. The deflection is directly proportional to the applied force or the weight of the body that is supported by the spring. However, for accurate results, the scale of the spring has to be calibrated or checked from time to time by hanging known masses on the spring. Thus, even the spring scale, through the process of calibration, becomes a device for comparing masses rather than for measuring weight or gravitational pull.

It may be concluded that ordinary methods of "weighing" do not measure weight or gravitational pull. Instead, they determine masses by comparison with known masses. After the mass has been measured, then the weight can be calculated from the local acceleration of gravity by use of Equation 1.9, $w/g = m/g_c$.

**Example 10.** A piece of metal is "weighed" at sea level and 45 degrees of latitude where $g = 32.174$ ft per sec$^2$ by means of a platform scale and also by means of a calibrated spring scale and is found by each method to "weigh" 50 lb. The piece of metal together with the spring scale and the platform scale and associated standard masses are transported to the moon where the acceleration of gravity is 5.47 per ft sec$^2$. What will be the "weight" of the block of matter at the moon as determined by (a) the platform scale, and (b) the spring scale?

*Solution:* (a) Platform scale. The block of metal will be placed on the platform and the standard masses which belong to the scale and were used for performing the weighing operation on the earth will be placed on the scale beam. Since the same local acceleration of gravity acts on the standard scale masses and on the mass to be weighed, masses are being compared, and the platform scale will give the same reading of 50 lb as was obtained at the surface of the earth.

(b) Spring scale. In accordance with Equation 1.10, $F/a = w_e/g_e = w_m/g_m$, where the subscripts $e$ and $m$ refer to the earth and moon, respectively. Then,

$$w_m = w_e \times \frac{g_m}{g_e} = 50 \text{ lb}_f \times \frac{5.47 \dfrac{\text{ft}}{\text{sec}^2}}{32.174 \dfrac{\text{ft}}{\text{sec}^2}} = 8.5 \text{ lb}_f$$

However, if the standard masses which were used to calibrate the spring scale on the surface of the earth are used on the moon to obtain a new calibration for the spring, it will be found that the piece of metal will cause the spring to deflect the same amount as the deflection caused by hanging 50 lb of mass on the spring.

It is unfortunate that the term *weight* is commonly used to mean mass as well as the gravitational force acting on a mass. In this book, with only one exception, the word *weight* will be used to mean gravitational force acting on a body. Weight will usually be expressed in pounds of force, lb$_f$. The term *mass* will be used to express quantity of matter. Mass will usually be expressed in pounds of mass, lb$_m$, or simply pounds, where no confusion should result. The only exception to this practice will be in the use of the well-established term *molecular weight* in connection with the chemistry of combustion.

A unit of mass that is used extensively, particularly in chemical calculations, is the *mole*. Depending upon the system of units in use, this might be the gram-mole or the pound-mole. In this book, the pound-mole will be used and, in accordance with usual practice, will be called a mole. The mole is that quantity of matter that has a mass in pounds numerically equal to the so-called "molecular weight" of the element or compound. Since the molecular weights of oxygen, hydrogen, and carbon dioxide are 32, 2, and 44, respectively, the mass of 1 mole of each of these substances is 32, 2, and 44 lb$_m$, respectively.

Also, 1 pound-mole of any substance consists of $2.7319 \times 10^{26}$ molecules.

### 1.9 Matter

The molecule is the smallest division of matter that has all the chemical and physical properties of a large quantity of matter. Molecules in turn are composed of atoms, of which there are 92 different kinds, called elements, found in nature, in addition to several elements that have been made by man. Each atom consists of a nucleus in which most of the mass is concentrated and around which electrons move in orbits in much the same manner as the planets revolve about our sun.

When chemical reactions occur, the atoms are regrouped to form molecules of different substances without basic changes in the atoms themselves. In the combustion process, this rearrangement of atoms into different molecules is accompanied by the release of energy, a part of which may be converted into work in a suitable engine. When this energy is released by chemical reactions, there is a very slight decrease in mass but the change is too small to be measured by any known instruments. Accordingly, chemical reactions may be said to take place in accordance with the *law of conservation of matter,* which states that matter is indestructible, that the mass of material is the same before and after the chemical reactions occur, and that the mass of each of the chemical elements remains unchanged during the chemical reactions. Thus, if air and fuel are supplied to the cylinder of an automobile engine, the mass of the exhaust gases leaving the cylinder in a given period of time equals the mass of air and fuel supplied; also, the mass of carbon, hydrogen, oxygen, nitrogen, and other elements entering the cylinder is equal to the mass of these same elements leaving the cylinder, although an entirely new set of chemical compounds may result from the combustion of the fuel in the cylinder.

When nuclear reactions occur, a very large and complex atom may be split into two smaller atoms whose combined mass is significantly less than the mass of the original atom. A small amount of mass is converted into a tremendous amount of energy. Thus, for nuclear reactions, the law of conservation of matter does not apply.

Also, as the velocity of a body approaches the velocity of light, its mass decreases. However, such velocities are not encountered in engineering problems. Consequently, it may be concluded that the law of conservation of matter is valid in all *engineering* problems except those involving nuclear reactions.

Matter ordinarily exists as a solid, liquid, or gas. In the solid phase or condition, the molecules are held in fixed positions by powerful forces. However, the atoms may vibrate about mean positions within the molecule. Solids, therefore, have definite shape and volume, and relatively great external forces are required to deform them. In the liquid phase, the molecules have a motion of translation with frequent collisions from which they rebound as perfectly elastic bodies. They occupy a definite volume at a given temperature but conform to the shape of the confining vessel. In the gaseous phase, the molecules are far apart in comparison to their size, move in straight lines between collisions, and will escape into space unless confined within a closed vessel.

Many substances, such as water, may exist in the solid, liquid, or gaseous phase, depending upon the pressure and temperature. Thus, ice may melt to form water, and water may evaporate to form steam. When a substance is in the gaseous phase but at a temperature and pressure not far from the temperature and pressure at which it can be liquefied, it is often referred to as a *vapor*. A *perfect gas* is at a temperature and pressure far removed from the temperature and pressure at which it can be liquefied.

### 1.10 Properties of Matter

Work is obtained from an engine by causing a fluid such as steam or gas to undergo heating, cooling, expansion, and compression processes or changes in a suitable mechanism. It is important that the *state* or molecular condition of the fluid be known while it is being subjected to these changing conditions. Among the easily measured characteristics or properties that define the state or condition of the fluid at any instant are volume per unit of mass, temperature, and pressure. Other properties such as internal energy, enthalpy, and entropy will be considered later. In general, if a proper combination of two of these properties is known, the state or molecular condition of the fluid is defined, and the values of the other properties are fixed. Also, if a given mass of fluid initially occupying some known volume at a known pressure and temperature undergoes some change or process after which the final volume, pressure, and temperature can be determined, then the change in volume, pressure, and temperature may be computed by subtracting the final values of these properties from the initial values without regard to the kind of change that may have occurred between the initial and final states.

Thus, referring to Fig. 1.9, let the coordinates of the point $p_1$, $V_1$ represent to scale the pressure and volume of a given mass of gas behind a gas-tight frictionless piston in a cylinder. If the piston moves to the right to a final position represented by the dotted lines, the final pressure and volume of the gas are indicated by the coordinates of the point $p_2$, $V_2$. During the expansion process, the rela-

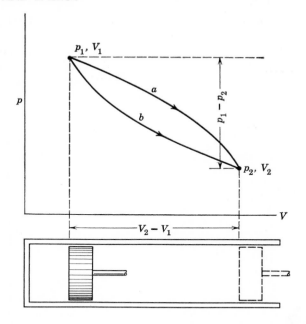

**Fig. 1.9.** Change in pressure and volume during expansion of a gas.

tionship between the pressure and volume of the gas at each successive position of the piston might be represented by some curve such as $a$ or $b$ which is called the path of the process. The increase in volume, $V_2 - V_1$, during the process depends only on the initial and final positions of the piston and is independent of how the pressure varied with volume, that is, the path. Also, the change in pressure for the process is dependent only on the difference between the initial and final pressures and is independent of the path. Such properties are known as *point functions* since a change in their values during any process depends upon the initial and final values of the properties and is independent of the path or kind of changes that take place between the initial and final states.

### 1.11   Specific Volume

The specific volume, $v$, is defined as the volume occupied by a unit quantity of matter. In the English engineering system of units, it is usually expressed as cubic feet per pound of mass (cu ft per $lb_m$).

Density, $\rho$, is the reciprocal of specific volume and may be expressed as pounds of mass per cubic foot or slugs per cubic foot.

Specific weight, $\gamma$, is defined as the weight of 1 cu ft of material (lb$_f$ per cu ft).

In the English engineering system of units, where

$v$ = specific volume, cu ft per lb$_m$
$\rho$ = density, lb$_m$ per cu ft
$\gamma$ = specific weight, lb$_f$ per cu ft
$m$ = mass, lb$_m$
$V$ = total volume, cu ft
$w$ = weight, lb$_f$

then $v = \dfrac{V}{m}$

$$\rho = \frac{m}{V} = \frac{1}{v}$$

$$\gamma = \frac{w}{V} = \frac{m \dfrac{g}{g_c}}{V} = \rho \frac{g}{g_c}$$

$$(1.10\,b)$$

$$(1.10\,b)$$

$$(1.10)$$

**Example 11.** Determine the specific volume, density, and specific weight of a gas when 5 lb$_m$ of the gas occupy 75 cu ft at a location where the local acceleration of gravity, $g$ (a) is 32.2 ft per sec$^2$, (b) is 30 ft per sec$^2$.

*Solution:* Specific volume and density are functions of mass and are therefore independent of the local acceleration of gravity.

$$\text{Specific volume, } v = \frac{V}{m} = \frac{75 \text{ ft}^3}{5 \text{ lb}_m} = 15 \frac{\text{ft}^3}{\text{lb}_m}$$

$$\text{Density, } \rho = \frac{m}{V} = \frac{5 \text{ lb}_m}{75 \text{ ft}^3} = 0.0667 \frac{\text{lb}_m}{\text{ft}^3} = \frac{1}{v}$$

(a) For $g = 32.2$ ft per sec$^2$,

$$\gamma = \frac{g}{g_c}\rho = \frac{32.2 \dfrac{\text{ft}}{\text{sec}^2}}{32.174 \dfrac{\text{lb}_m \text{ ft}}{\text{lb}_f \text{ sec}^2}}\left(0.0667 \frac{\text{lb}_m}{\text{ft}^3}\right) = 0.0667 \frac{\text{lb}_f}{\text{ft}^3}$$

(b) For $g = 30$ ft per sec$^2$,

$$\gamma = \frac{g}{g_c}\rho = \frac{30 \dfrac{\text{ft}}{\text{sec}^2}}{32.174 \dfrac{\text{lb}_m \text{ ft}}{\text{lb}_f \text{ sec}^2}}\left(0.0667 \frac{\text{lb}_m}{\text{ft}^3}\right) = 0.063 \frac{\text{lb}_f}{\text{ft}^3}$$

### 1.12 Temperature

It is a common experience that bodies feel hot or cold to the sense of touch, depending upon their temperatures. In comparing the temperatures of two bodies, $A$ and $B$, a third body called a thermometer which has some observable property that changes with temperature is brought into intimate contact with body $A$ until the thermometer and that body reach thermal equilibrium, that is, until there is no further change in the reading of the thermometer. Then the thermometer may be brought into intimate contact with body $B$ until thermal equilibrium is attained as evidenced by no further change in the reading of the thermometer. The difference in the two readings of the thermometer is then taken as an indication of the difference in temperature of bodies $A$ and $B$.

In order to set up a scale of temperatures to which numbers can be assigned, it is customary to use for one reference point the equilibrium temperature of a mixture of cracked ice and air-saturated water. For a second reference temperature, the equilibrium temperature of pure water and its vapor at 14.696 psia is used. On the Fahrenheit (F) scale, the numbers 32 and 212 are assigned to these reference temperatures. The numbers 0 and 100 are assigned to these temperatures on the centigrade (C) scale. Then, as seen by reference to Fig. 1.10, 180 F is equivalent to 100 C, 1 F is equal to ⅝ C, or

$$t_C = \tfrac{5}{9}(t_F - 32) \tag{1.11}$$

where $t_C$ = temperature, C
$t_F$ = temperature, F

The most common thermometer is the mercury-in-glass thermometer in which the relative expansion of mercury and glass with change in temperature is used as the variable property. The change in electrical resistance with temperature of a given length of wire or the electromotive force produced when the junctions of two dissimilar metals are maintained at different temperatures may also be used as the basis for temperature measurement. However, it will be found that, if these various kinds of thermometers are calibrated at the ice point and at the steam point, they will not give an intermediate reading exactly half-way between the readings obtained at these end points when in contact with a body at a temperature exactly half-way between these reference temperatures. In other words, their

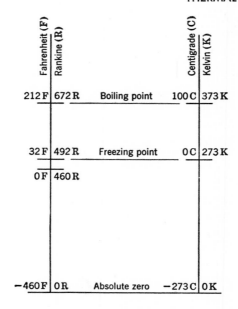

**Fig. 1.10.**  Temperature scales.

F = Fahrenheit scale
R = absolute temperature, degrees Rankine
C = centigrade scale
K = absolute temperature, degrees Kelvin

readings are not directly proportional to or linear with respect to temperature.

The constant-volume gas thermometer contains a constant volume of a gas such as hydrogen or helium under a relatively low pressure which can be measured accurately. The pressure of the gas will be found to vary in a linear manner with temperature, and this type of thermometer is therefore used as the basis for temperature measurement against which other types of thermometers can be compared. If the constant-volume gas thermometer is used to measure successively lower temperatures, it will be found that an extrapolation of a curve of pressure versus temperature will be a straight line and will give a reading of −459.6 F or −273 C at zero pressure. This temperature is called the zero of absolute temperature. Consequently, on the Fahrenheit scale, absolute zero is −459.6 F (approximately −460 F) as indicated in Fig. 1.10. Absolute temperatures measured in Fahrenheit degrees are called degrees Rankine (R). On the centigrade scale, absolute zero is $-273$ C, and absolute temperatures measured in degrees centigrade are called degrees Kelvin (K). Then,

$$T_R = t_F + 460 \text{ (approximately)} \tag{1.12}$$

$$T_K = t_C + 273 \tag{1.13}$$

where $T_R$ = absolute temperature, R
$T_K$ = absolute temperature, K

### 1.13 Pressure

Pressure is the push or force exerted per unit of area on the confining or supporting surface. For gases, it may be considered as the effect of the impacts of moving molecules upon the walls of the confining vessel. If more air is put into a steel tank, the pressure is

**Fig. 1.11.** Bourdon-type pressure gage.

Fig. 1.12. Barometer.

increased because of the great number of molecular impacts per unit area. If the temperature of the air in the tank is increased, the mean molecular velocity is increased also, the molecules bombard the walls of the tank more frequently and at higher mean velocities, and the pressure is therefore higher.

The Bourdon type of pressure gage (Fig. 1.11) is the commonest type of gage that is used for the measurement of pressures of considerable magnitude. The tube is elliptical in cross section as shown in section A–A, bent into the arc of a circle, rigidly attached to the case of the gage at one end, and free to move at the other end. When a pressure is applied to the inside of the Bourdon tube, the elliptical cross section tends to become circular, and the free end of the tube moves outward. By means of a suitable link, pivoted gear sector, and pinion, a needle is moved around a graduated scale on the face of the gage to indicate the pressure.

If the pressure within the Bourdon tube is less than the atmospheric pressure on the outside of the tube, the elliptical cross section tends to be flattened, the free end of the tube moves nearer the center, and the pointer is moved in the opposite direction. Thus, the gage is operated by the *difference* between atmospheric pressure on the outside of the Bourdon tube and the pressure within the tube.

Pressure readings obtained with the gage are called *gage pressures* and may be expressed in pounds per square inch gage, usually abbreviated to read psig.

Atmospheric pressure is measured by means of a barometer as illustrated in Fig. 1.12. If a glass tube, over 30 in. long and closed at its lower end, is filled with mercury, closed with a stopper, inverted, and mounted with its lower end in a dish of mercury, and the stopper is then removed, the mercury will fall in the tube until the weight of the column of mercury is balanced by the pressure of the atmosphere. The height of the column of mercury is then a measure of the atmospheric or barometric pressure. Standard atmospheric pressure may be expressed in any of the following units:

$$760 \text{ mm of Hg} = 29.92 \text{ in. of Hg} = 14.696 \text{ psia} \qquad (1.14)$$

The following relationship between pressure in pounds per square inch and inches of mercury may be derived from Equation 1.14 and should be memorized:

$$1 \text{ in. of Hg} = 0.491 \text{ psi} \qquad (1.15)$$

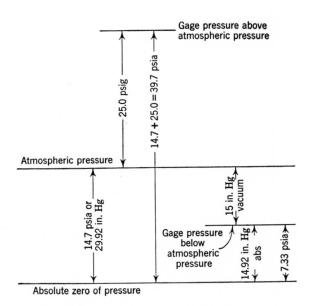

**Fig. 1.13.** Gage and absolute pressure.

The scale of the Bourdon gage is usually graduated to read pressures above atmospheric pressure in pounds per square inch and pressures below atmospheric pressure in inches of mercury. Since atmospheric pressure varies with altitude above sea level and with constantly changing atmospheric conditions, it is necessary to use *absolute* pressures when dealing with the properties of substances. As indicated in Fig. 1.13, the absolute pressure in pounds per square inch (psia) may be determined for gage pressures above atmospheric pressure by adding to the gage reading in pounds per square inch the barometric pressure in pounds per square inch absolute. For pressures below atmospheric pressure, the absolute pressure in inches of mercury may be obtained by subtracting the gage reading in inches of mercury from the barometric pressure in inches of mercury. The result, which will be in inches of mercury absolute, may be converted to pounds per square inch absolute by using the conversion factors in Equation 1.15.

The use of manometers for the measurement of small differences in pressure will be discussed in Chapter 11.

**Example 12.** What is the pressure in psia if the gage reads 25.0 psig and the barometer reads 29.5 in. of Hg?

*Solution:* Since the pressure is greater than atmospheric pressure, the pressure in psia is equal to the gage reading in psi plus the barometric pressure in psi. See Fig. 1.13.

$$\text{Barometric pressure in psi} = 29.5 \text{ in. of Hg} \times \frac{0.491 \text{ psi}}{\text{in. of Hg}} = 14.5 \text{ psi}$$

Then

$$\text{Absolute pressure} = 25.0 \text{ psi} + 14.5 \text{ psi} = 39.5 \text{ psia}$$

**Example 13.** What is the absolute pressure in in. of Hg and in psia if a vacuum gage reads 15 in. of Hg and the barometer reads 29.8 in. of Hg?

*Solution:* Since the vacuum gage reads a negative pressure with respect to atmospheric pressure (see Fig. 1.13), the absolute pressure is the difference between the barometric pressure and the gage reading.

$$\text{Absolute pressure} = 29.8 \text{ in. of Hg} - 15 \text{ in. of Hg} = 14.8 \text{ in. of Hg abs.}$$

Also,

$$\text{Absolute pressure in psia} = 14.8 \text{ in. of Hg abs.} \times \frac{0.491 \text{ psi}}{\text{in. of Hg}} = 7.27 \text{ psia}$$

## 1.14 Energy

A body is said to possess energy when it is capable of doing work. In more general terms, energy is "capacity for producing an effect." Energy may be classified as *stored energy* or *energy in transition*.

Chemical energy is *stored* in high explosives and fuels such as coal or gasoline and may be released by the process of combustion. Energy is *stored* in the water behind a dam and may be converted into work as the water flows through a hydraulic turbine to a lower level. Sufficient energy must be *stored* in an earth satellite to keep it in an orbit as it travels around our earth. The energy that is *stored* in the automobile battery is used to operate the starter.

When the switch in a suitable electric circuit is closed, electric energy may *flow* from a distant power plant to a motor, an electric toaster, or a radio, in which it is converted into work, heat, or sound, respectively. In a ship, energy *flows* through the rotating drive shaft from the engine to the propeller. Energy is being received by our earth from the sun, and life would soon cease if this *flow* of energy were to be interrupted.

The power engineer is concerned with the generation of shaft work from the stored chemical energy in fuels or the nucleus of certain atoms or the stored mechanical potential energy in a body of water behind a dam. The electrical engineer is interested in the conversion of shaft work into electric energy in a generator and the transmission of the electric energy over transmission lines to convenient locations where it can be transformed into work, heat, light, or sound. The refrigerating engineer deals with the removal of energy from an insulated low-temperature space and the discharge of this energy to the atmosphere or to cooling water. The metallurgical engineer requires large quantities of energy at high temperature for the smelting and refining of metals. The chemical engineer requires large amounts of energy in the form of heat and work for manufacturing processes.

The complete or partial transformation of energy from one of the many forms in which it may exist into other forms of energy takes place in accordance with the *law of conservation of energy.* This law states that *energy can be neither created nor destroyed;* therefore, when energy is transformed either completely or partially from one form to another, the total amount of energy remains unchanged.

There is one important exception to the law of conservation of energy and this exception occurs in nuclear reactions. In nuclear reactions, a significant quantity of mass disappears and is converted into energy.

In all types of engineering problems except those involving nuclear reactions, the laws of conservation of energy and conservation of matter apply. They are among the most useful laws in the physical sciences and form the foundation upon which most of the subject matter of this book rests.

### 1.15  Mechanical Potential Energy

The stored energy associated with the position of tangible bodies is called mechanical potential energy.

Any tangible body on or near the surface of the earth is attracted to the earth by the gravitational force which is called the weight of the body. If a building stone is lifted from the ground to the top of a building, the energy required to lift it is stored in the stone as potential energy or energy of position. This stored energy remains unchanged as long as the stone retains its position in the building. The potential energy that has been stored in the stone as a result of the lifting operation is the product of the force required to lift it, that is, its weight, expressed in pounds of force, $lb_f$, and the vertical distance it was lifted, expressed in feet and designated by the symbol $z$. Then

$$\text{Potential energy} = wz \text{ ft-}lb_f \tag{1.16}$$

It is probable that the distance above sea level of the ground from which the stone was lifted was different from that of the quarry from which the stone was obtained. Thus the stored potential energy in the stone after being placed in the building has a different value with respect to the ground level from that with respect to its original position in the quarry. Consequently, in evaluating potential energy, the vertical distance must be measured above some selected datum plane. Engineering problems that involve potential energy normally are concerned with the *change* in potential energy during a process, and this change can always be evaluated by measuring the vertical distances in the initial and final positions from the same datum plane.

It should be noted that potential energy is a function of gravitational force or weight. If the mass of a body is known rather than its weight (as pointed out in Article 1.8, it is mass and not weight that is measured by a platform scale), then the weight can be calculated from Equation 1.9, $w/g = m/g_c$, and the potential energy for a mass of $m$ lb is then given in ft-$lb_f$ by the expression

$$PE = \frac{g}{g_c} mz \tag{1.17}$$

**Example 14.** Determine the potential energy in a mass of 100 lb located 50 ft above a datum plane where the local acceleration of gravity is 30 ft per sec².

*Solution:* The problem may be solved directly by use of Equation 1.17.

$$PE = \frac{g}{g_c}\, mz = \frac{30\ \dfrac{\text{ft}}{\text{sec}^2}}{32.174\ \dfrac{\text{lb}_m\ \text{ft}}{\text{lb}_f\ \text{sec}^2}} \times 100\ \text{lb}_m \times 50\ \text{ft} = 4670\ \text{ft-lb}_f$$

### 1.16 Mechanical Kinetic Energy

Reference to any textbook on physics will show that, when a body is caused to move as a result of an unbalanced force, the work done, $W$, is equal to the product of the displacement of the body and the component of the force, $F_x$, in the direction of the displacement, or

$$W = F_x\, dx$$

Let Fig. 1.14 represent a body of mass $m$ which is initially at rest at position $x_1$ on a frictionless horizontal plane. Let this body be acted upon by a horizontal force in the $x$ direction. Since

$$F = \frac{1}{g_c}\, ma = \frac{1}{g_c}\, m\, \frac{dV}{d\tau}$$

where $V$ is velocity, then

$$W = \int_{x_1}^{x_2} \frac{1}{g_c}\, m\, \frac{dV}{d\tau}\, dx = \int_{V_1}^{V_2} \frac{1}{g_c}\, m\, \frac{dx}{d\tau}\, dV = \int_{V_1}^{V_2} \frac{1}{g_c}\, mV\, dV$$

For a body initially at rest,

$$W = \frac{1}{g_c}\, m\, \frac{V^2}{2}$$

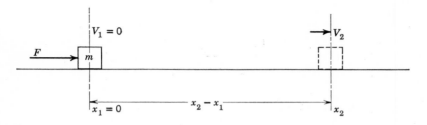

**Fig. 1.14.** Effect of a force acting on a mass.

It should be noted that neither the force nor the distance through which it acts appears in the final equation for work except as it affects the velocity. Since it was assumed that the motion occurred on a horizontal frictionless plane, all the work done has been stored in the body and has resulted in a change in velocity. The energy so stored is called *kinetic energy*.

Therefore
$$KE = \frac{1}{2g_c} mV^2 = \frac{mV^2}{2g_c} \tag{1.18}$$

It should be noted also that kinetic energy is a function of mass and velocity squared and, unlike potential energy, is independent of the strength of the gravitational field.

The kinetic energy of a body having a weight $w$ in $lb_f$ may be determined by modifying Equation 1.18 as follows:

$$\frac{w}{g} = \frac{m}{g_c} \tag{1.9}$$

Then
$$KE = \frac{m}{g_c} \frac{V^2}{2} = \frac{w}{g} \frac{V^2}{2} = \frac{wV^2}{2g} \tag{1.19}$$

**Example 15.** Determine the kinetic energy stored in a mass of 100 $lb_m$ which is moving at a velocity of 100 ft per sec at a place where the local acceleration of gravity is 30 ft per sec$^2$.

*Solution:* Since $KE = mV^2/2g_c$, the local acceleration of gravity, $g = 30$ ft per sec$^2$, does not enter into the problem.

$$KE = \frac{mV^2}{2g_c} = \frac{100 \ lb_m \times 100 \ \dfrac{ft}{sec} \times 100 \ \dfrac{ft}{sec}}{2 \times 32.174 \ \dfrac{lb_m}{lb_f} \dfrac{ft}{sec^2}} = 15{,}500 \ \text{ft-lb}_f$$

*Alternate Solution:*

$$w = \frac{g}{g_c} m = \left( \frac{30 \ \dfrac{ft}{sec^2}}{32.174 \ \dfrac{lb_m \ ft}{lb_f \ sec^2}} \right) 100 \ lb_m = 93.3 \ lb_f$$

$$KE = \frac{wV^2}{2g} = \frac{93.3 \ lb_f \times 100 \ \dfrac{ft}{sec} \times 100 \ \dfrac{ft}{sec}}{2 \times 30 \ \dfrac{ft}{sec^2}} = 15{,}500 \ \text{ft-lb}_f$$

## 1.17  Internal Energy

Molecules, like tangible bodies, have mass. In the solid state, they are held in fixed positions by powerful forces. However, the atoms of which they are composed vibrate about mean positions. In the liquid and gaseous states, the molecules have motion of translation and rotation or spin. Consequently, because of their mass and motion, the molecules have kinetic energy stored within them. Since the mean molecular velocity of translation is some function of temperature, any change in temperature is accompanied by a change in the kinetic energy stored in the molecules. The entire mass of the substance may be at rest, as, for instance, a stationary tank of water or gas, and the mechanical kinetic energy is therefore zero, but the molecules have motion with respect to each other within the substance and therefore possess *molecular kinetic energy.*

Also, molecules are attracted to each other by forces that are very large in the solid phase but less in the liquid phase and tend to vanish in the perfect-gas phase where the molecules are far apart in comparison to their size. In the melting of a solid or the vaporization of a liquid, it is necessary that these powerful molecular attractive forces be overcome. The energy required to bring about this change of phase is stored in the molecules as *molecular potential energy* and will be released when the substance returns to its initial state.

Thus, molecules have stored in them potential energy of position and kinetic energy of motion in much the same manner that tangible bodies possess mechanical potential and kinetic energy. This stored molecular energy is generally referred to as *internal energy.* The internal energy per unit of mass is designated by the symbol $u$. The symbol $U$ designates the internal energy of $m$ lb of a substance.

At any given temperature, pressure, and specific volume, the internal or molecular energy has a definite value. When the fluid undergoes some change such as an expansion and then returns to its initial pressure, temperature, and specific volume, the internal energy will again have the same value. Internal energy, like pressure, temperature, and specific volume, is therefore a property of the substance.

### 1.18    Enthalpy

Temperature, pressure, specific volume, and internal energy are properties which define the state of a substance. In general, if a proper combination of two of these properties is known for a given state, the others are fixed and can be computed. As will be seen in Chapter 2, the internal-energy term and the product of the pressure and specific volume appear in some of the basic and most frequently used equations in thermodynamics. Consequently, it is convenient to group them into a new term that is called enthalpy, $h$, and is defined as follows:

$$h = u + pv \qquad (1.20)$$

If $p$ is expressed in $\dfrac{lb_f}{ft^2}$ and $v$ is expressed in $\dfrac{ft^3}{lb_m}$, then the product $pv$ has the units of $\dfrac{lb_f}{ft^2} \times \dfrac{ft^3}{lb_m} = \dfrac{ft\text{-}lb_f}{lb_m}$. To be consistent, the internal energy and the enthalpy must be expressed in the same units.

## 1.19  Summary

This book is concerned with fossil fuels and nuclear reactions as sources of energy, the properties of fluids that are used in systems for power generation and refrigeration, the principles governing the transfer of energy in heat exchangers, the principles governing the conversion of energy in prime movers and compressors, and the construction and performance of the important types of equipment that are used in power and refrigerating plants.

A clear understanding of the meaning of basic concepts and terms is essential as a foundation for further study. Some of these concepts are summarized in the following paragraphs:

*Matter:* that which occupies space and has inertia; the substance of which physical objects are composed.

*Mass:* a quantity of matter. The standard unit of mass is the international kilogram which is the mass of a particular piece of platinum-iridium that is located at Sèvres, France. In the English engineering system of units, the standard pound of mass ($lb_m$) is 0.4535924+ of the mass of the international kilogram.

*Force:* a push or pull. In the English engineering system of units, the standard pound of force ($lb_f$) is that unbalanced force which will cause a standard pound of mass to be accelerated at the rate of 32.174 ft per sec². It is also the gravitational force acting on one standard pound of mass at sea level and 45 degrees of latitude.

*Slug:* a unit of mass which is accelerated at the rate of 1 ft per sec² by an unbalanced force of one standard pound. One slug equals 32.174 $lb_m$.

*Mole or pound-mole:* that quantity of matter which has a mass in pounds numerically equal to the so-called "molecular weight" of the element or compound. One pound-mole of any substance consists of $2.7319 \times 10^{26}$ molecules.

*Weight:* the force of gravity acting on a body in any gravitational field.

*Newton's second law:* $F = \dfrac{1}{g_c} ma$, where $g_c$ is a dimensional constant, the units and numerical value of which depend upon the system of units used. In the English engineering system of units,

$$g_c = 32.174 \frac{lb_m \text{ ft}}{lb_f \text{ sec}^2}$$

*Property:* an observable characteristic that helps to define the state of a substance. Common properties are pressure, temperature, specific volume, internal energy, enthalpy, and entropy. In general, the state of a substance is determined when two properly selected properties are known.

*Specific volume:* the volume of a unit mass of material. In the English engineering system of units, specific volume is expressed in cubic feet per pound of mass (cu ft per $lb_m$).

*Density:* mass per unit of volume; the reciprocal of specific volume. In the English engineering system of units, density is expressed in pounds of mass per cubic foot ($lb_m$ per $ft^3$). In the English gravitational system, it is expressed as slugs per cubic foot.

*Specific weight:* the weight of a unit volume of material. In the English engineering system of units, specific weight is expressed in pounds of force per cubic foot ($lb_f$ per $ft^3$).

*Process:* a physical or chemical occurrence during which an effect is produced by the exchange, transformation, or redistribution of energy or mass.

*Energy:* the capacity to produce a change or effect.

*Mechanical potential energy:* energy that is stored in a body of matter because of its position in a gravitational field.

$$PE = wz = m \frac{g}{g_c} z$$

*Mechanical kinetic energy:* energy that is stored in a body of matter because of its motion.

$$KE = \frac{mV^2}{2g_c} = \frac{wV^2}{2g}$$

*Internal energy:* energy that is stored within a mass of molecules by reason of their random motion and configuration.

*Law of conservation of matter:* except for nuclear reactions, matter is indestructible; during chemical reactions, the chemical composition may change, but the mass of each element involved in the reaction remains constant.

*Law of conservation of energy:* except for nuclear reactions, during any engineering process involving energy transformation or transfer, the total amount of energy remains constant.

*Enthalpy:* a property of matter which is the sum of the internal energy stored within a unit mass of molecules and the product of the pressure and specific volume of the substance, expressed in consistent units.

## PROBLEMS

**1.** What force in dynes will accelerate a mass of 2 kilograms at the rate of 30 cm per sec$^2$?

**2.** A body has a mass of 300 lb$_m$. What is the mass in slugs?

**3.** What force in lb$_f$ will accelerate a mass of 40 slugs at the rate of 5 ft per sec$^2$?

**4.** What force in lb$_f$ will accelerate a mass of 500 lb$_m$ at the rate of 7 ft per sec$^2$?

**5.** A body weighs 300 lb$_f$ at a location where the local acceleration of gravity is 25 ft per sec$^2$. What is the mass of the body (*a*) in lb$_m$, (*b*) in slugs? (*c*) What force in lb$_f$ will be required to impart to this body an acceleration of 6 ft per sec$^2$?

**6.** Determine the value of $g_c$ in a system of units in which the force is expressed in lb$_f$, the mass is expressed in grams, the unit of length is the inch, and the unit of time is the minute.

**7.** A satellite has a mass of 1000 lb. What is its weight in lb$_f$ at an elevation of 500 miles above sea level?

**8.** A body has a mass of 20 slugs. What is its weight in lb$_f$ at the surface of the moon?

**9.** What force in lb$_f$ will be required to accelerate a body having a weight of 100 lb$_f$ on the surface of the moon at the rate of 5 ft per sec$^2$?

**10.** On the surface of the moon, 5 lb$_m$ of gas occupy a volume of 40 cu ft. Determine (*a*) the specific volume of the gas, (*b*) the density of the gas, and (*c*) the specific weight of the gas.

**11.** A storage tank having a volume of 10,000 cu ft contains 1500 lb$_m$ of helium. Determine (*a*) the specific volume and (*b*) the density of the helium.

**12.** The specific volume of steam at 1000 psia and 900 F is 0.7604 cu ft per lb$_m$. At a location where the local acceleration of gravity is 30 ft per sec$^2$, (*a*) what is the specific weight of the steam, and (*b*) what is the density of the steam in (1) lb$_m$ per cu ft and (2) slugs per cu ft?

**13.** The specific volume of water at 70 F is 0.01606 lb$_m$ per cu ft. At a location where the local acceleration of gravity is 25 ft per sec$^2$, determine (a) the specific weight of the water, (b) the density in $lb_m$ per cu ft, and (c) the density in slugs per cu ft.

**14.** During a test of a domestic heating plant, 30 lb of coal were fired, 4 lb of dry refuse were collected from the ashpit, and 417 lb of gaseous products of combustion were discharged up the chimney. How much air was supplied?

**15.** A total of 80 lb of gasoline and 1200 lb of dry air was supplied to an internal-combustion engine, and 1092 lb of dry gaseous products of combustion were discharged from the exhaust pipe. How much water vapor was discharged in the exhaust gases?

**16.** Convert the following temperatures from Fahrenheit to Rankine: 400 F, −10 F.

**17.** Convert the following temperatures from centigrade to Kelvin: 300 C, −100 C.

**18.** Convert the following Fahrenheit temperatures to centigrade: 200 F, −30 F.

**19.** Convert the following centigrade temperatures to Fahrenheit: −40 C, 200 C.

**20.** Convert the following pressures in in. of Hg to psi and psf: 30, 10.

**21.** Convert the following pressures from psi to in. of Hg: 2, 14.7, 20.

**22.** A pressure gage reads 200 psi, and the barometric pressure is 29.6 in. of Hg. What is the absolute pressure in psi and psf?

**23.** A pressure gage reads 20 in. of Hg vac. If the barometric pressure is 29.7 in. of Hg, determine the absolute pressure in in. of Hg, psi, and psf.

**24.** A body which weighs 1000 $lb_f$ is located 10 ft above a datum plane. What potential energy in ft-$lb_f$ is stored in this body with respect to the datum plane?

**25.** A body having a mass of 1000 $lb_m$ is located 20 ft above a datum plane in a location where the local acceleration of gravity is 25 ft per sec². What potential energy in ft-$lb_f$ is stored in the body with respect to the datum plane?

**26.** A body having a mass of 50 slugs is located 40 ft above a datum plane in a location where the acceleration of gravity is 20 ft per sec². What potential energy in ft-$lb_f$ is stored in the body with respect to the datum plane?

**27.** A body which has a mass of 100 $lb_m$ is moving in a straight line with a uniform velocity of 100 ft per sec. If it is located at sea level and 45 degrees of latitude, how much kinetic energy in ft-$lb_f$ is stored in the body?

**28.** A body which weighs 1000 $lb_f$ at a location where the acceleration of gravity is 25 ft per sec² is moving in a straight line at a uniform velocity of 120 ft per sec. How much kinetic energy in ft-$lb_f$ is stored in the body?

**29.** A body which has a mass of 80 slugs is located close to the surface of the moon and is moving in a straight line with a uniform velocity of 200 ft per sec. How much kinetic energy in ft-$lb_f$ is stored in the body?

# CHAPTER 2

# The Energy Balance

## 2.1 Introduction

Several types of power plants were described briefly in Chapter 1. Their components consist of heat exchangers, prime movers, pumps, compressors, etc., through which a fluid flows and in which heat is transferred or work is done. The transfer and conversion of energy in such equipment take place in accordance with the law of conservation of energy. This chapter deals with the law of conservation of energy as applied to nonflow and flow processes.

## 2.2 The System

In analyzing the flow of mass and energy in some apparatus, it is convenient to define the *system* as a particular region that is surrounded by real or imaginary surfaces that can be specified. The real or imaginary surfaces that enclose this space constitute the *boundaries* of the system. The region outside the boundaries that may be affected by the transfer of energy or mass is called the *surroundings*.

Systems may be divided into two types: (*a*) the *closed* or fixed-mass system and (*b*) the *open* system.

Perhaps the simplest closed system is illustrated in Fig. 2.1 by a steel tank which contains some fluid at some pressure and temperature. The space within the tank is the system, and the inner surface of the tank is considered to be the boundary. The tank itself and anything outside the tank that would be affected by energy transfer constitute the surroundings. Although it is customary to consider the space within the tank as the system, a constant mass is contained within the system boundaries and this fixed mass is sometimes referred to as the system.

Another very important type of closed system is the volume within

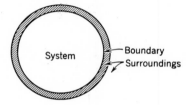

**Fig. 2.1.** Constant-volume closed system.

a cylinder and frictionless, gas-tight movable piston as illustrated in Fig. 2.2. Obviously, pistons in real machinery are neither frictionless nor gas-tight. However, in order to analyze many problems, it is necessary to assume idealized conditions and then to modify the final results if necessary to consider imperfections. In the system illustrated in Fig. 2.2, the inside surfaces of the cylinder, cylinder head, and piston are the boundaries of the system. This system has a variable volume depending upon the position of the piston, whereas the system illustrated in Fig. 2.1 has a constant volume. Both systems contain a constant or fixed mass.

In the two closed systems that have been described, the boundaries consist of the inside surfaces of metal walls. In the general case, a closed system may be considered as some region surrounded by stationary or moving real or imaginary boundaries across which no mass is transferred.

The open system is illustrated in Fig. 2.3. It consists of a region surrounded by specified boundaries so arranged that a fluid may enter and leave the system. The system consists of the space contained within the inner surfaces of the apparatus including the volume within the inlet and outlet pipes out to some specified planes or sections in those pipes such as section 1–1 and 2–2 in Fig. 2.3. As an illustration of such a system, consider one cylinder of an automobile engine with the inlet and outlet planes located in the inlet and exhaust manifolds. Then a fluid consisting of a mixture of air and fuel passes the inlet section intermittently as the engine rotates. The products of combustion pass the outlet section intermittently. The volume of the

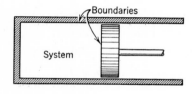

**Fig. 2.2.** Variable-volume closed system.

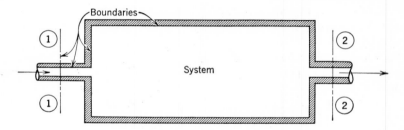

**Fig. 2.3.** Open or flow system.

system changes as the piston moves in the cylinder. Energy is re-leased by combustion within the system, heat is transferred through the cylinder walls, and work is done on the moving piston. In the general case of the open system, the mass within the system and the conditions at inlet and outlet may vary with respect to time.

In heat exchangers, turbines, pumps, etc., operating under steady conditions, the flow of fluid in the inlet and outlet pipes occurs at constant velocity and the mass within the system is constant. Where such factors as velocity, pressure, temperature, and flow rate are constant with respect to time, the system is said to be a *steady-flow system*. The steady-flow system is therefore a special case of the open system. Much of this book will be concerned with the transfer and conversion of energy that occur in steady-flow systems.

### 2.3 Work

When a force causes a displacement of a body, the work that is done is the product of the displacement and the component of the force acting in the direction of the displacement. If $F$ is the component of the force acting in the direction of the displacement, $dx$, then, since in the general case the force may be a variable,

$$W = \int_{x_1}^{x_2} F \, dx \tag{2.1}$$

where $W$ is the work done.

Let Fig. 2.4 illustrate a mass, $m$, that is suspended by a cord which is wrapped around a pulley on a horizontal shaft having at its other end a paddle wheel or fan in a gas-tight insulated box. The system under consideration is the space within the box which is filled with some fluid. The gravitational field acting on the suspended mass produces a downward force that is called the weight of the mass. If the mass is allowed to fall under the influence of this force, work is done which is equal to the product of the weight of the body and the distance it falls. The shaft rotates, the fluid in the system is stirred, and the resulting fluid friction causes an increase in the temperature of the system.

The following points should be noted:

1. Work is done only when the weight moves. Since $W = \int F\,dx$, work ceases when movement ceases.

2. Considering the boundaries of the system as the inside surfaces of the box, work has flowed or been transferred from the surroundings (the suspended mass) across the boundaries of the system to the fluid within the system. Work transferred in this manner is often called shaft-work since a turning effort or torque rotates the shaft.

3. The work that has entered the system is used to stir the fluid in the system with the result that the temperature rises. This means that the average velocity of the molecules has been increased and the internal energy of the fluid has therefore increased. In other words, the energy that entered the system as work, since it cannot be destroyed, is stored within the system as an increase in internal energy.

4. The energy that has been transferred to the system came from the surroundings and resulted in a decrease in the mechanical potential energy of the surroundings due to the change in the position of the weight.

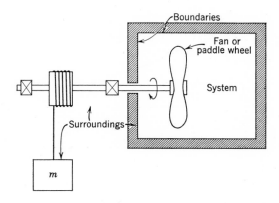

**Fig. 2.4.** Transfer of work across boundaries of a closed system.

Since we are concerned with the transfer, storage, and conversion of energy in a system, work will be defined as follows:

*Work is energy that is being transferred across the boundaries of a system because of a force acting through a distance.*

Work flows from the surroundings to the system or vice versa only when a force produces some movement or displacement. The driving potential that produces the flow of energy is a *force.*

### 2.4  Work Transferred across the Moving Boundary of a Closed System in a Frictionless Process

Referring to Fig. 2.5, let it be assumed that a cylinder is closed with a frictionless gas-tight movable piston. The system consists of the space bounded by the inside surfaces of the cylinder walls, cylinder head, and piston. It contains a gas at some initial pressure $p_1$. On the $pV$ diagram of Fig. 2.5, the point 1 is so located as to represent to some scale the initial pressure of the gas and the initial volume of the system. If the piston is allowed to move to some final position represented by point 2 under such conditions that the pressure remains constant at its initial value of $p_1$ while the volume of the system increases, then the line 1–2 on the $pV$ diagram is a locus of points representing the relationship between pressure and volume of the gas in the system during the process.

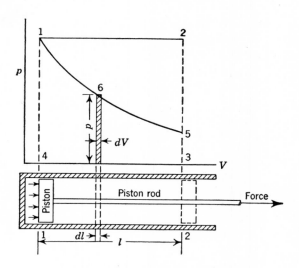

**Fig. 2.5.**  Pressure–volume diagram.

The constant force acting on the piston during this expansion is

$$F = p_1 A$$

where $F$ = force acting on the piston, $lb_f$

$p_1$ = pressure of the gas on the piston, psfa

$A$ = area of the piston, sq ft

Since $W = \int F \, dx$, then for the constant-pressure process 1–2 of Fig. 2.5,

$$W = F \times l = (p_1 \times A) \times l = p_1(Al) = p_1(V_2 - V_1) \quad (2.2)$$

where $l$ is the stroke of the piston, ft

$V_1$ and $V_2$ are the initial and final volumes of the system, cu ft

Work is expressed in foot-pound units. Since the foot is the unit of length, volume must be expressed in cubic feet, area in square feet, and pressure in pounds per square foot absolute.

It should be noted on Fig. 2.5 that the product $p_1 \, (V_2 - V_1)$ of Equation 2.2 is equal to the area 1–2–3–4 which therefore represents graphically the work done by the gas during the constant-pressure expansion from 1 to 2.

In the general case, the pressure of the gas varies as the volume changes. A variable-pressure expansion is represented on Fig. 2.5 by the curve 1–5. The equation of the curve 1–5 can often be expressed in the form $p V^n = C$, where $C$ is a constant and $n$ is some exponent between 0 and $\infty$. When the pressure is changing and the equation showing the relation between $p$ and $V$ is known, it is possible to evaluate the work by considering a small movement of the piston during which the change in pressure is negligible. Assume that the piston is at point 6 on Fig. 2.5. If the piston then moves a small distance $dl$, the work done may be written as follows:

$$\delta W = p \times A \times dl = p \times (A \, dl) = p \, dV \quad (2.3)$$

where $dV = A \, dl$ = increase in the volume of the system in cubic feet. The elementary cross-hatched area on Fig. 2.5 is $p \, dV$ and represents the work done during the small movement of the piston. By the use of the integral calculus the area under the curve 1–5 can be determined as follows:

$$W = \int_{V_1}^{V_2} p \, dV \quad (2.4)$$

Equation 2.4 is the general equation for calculating work done by or on a fluid *during a frictionless process in a closed system* when the algebraic relation between the pressure and volume is known. Equation 2.2 can be derived from Equation 2.4 by making $p_2 = p_1$.

**Example 1.** A cylinder contains 1 cu ft of gas at a pressure of 100 psia. If the gas expands at constant pressure until the volume is doubled, how much work is done?

*Solution:* For the conditions stated in the problem, $V_2 = 2V_1 = 2$ cu ft. Also, $p_2 = p_1 = 100$ psia or $100 \times 144$ psfa.

Furthermore, the process is illustrated by the constant pressure line 1–2 on Fig. 2.5.

$$W = \int_{V_1}^{V_2} p\, dV = p\int_{V_1}^{V_2} dV = p(V_2 - V_1)$$

and
$$W = \left(100\ \frac{\text{lb}_f}{\text{in.}^2} \times 144\ \frac{\text{in.}^2}{\text{ft}^2}\right)(2\ \text{ft}^3 - 1\ \text{ft}^3) = 14{,}400\ \text{ft-lb}_f$$

**Example 2.** A cylinder contains 1 cu ft of gas at an initial pressure of 100 psia. If this gas expands in such a manner that the curve 1–5 of Fig. 2.5 is a straight line, the final pressure is 50 psia, and the final volume is 3 cu ft, how much work is done?

*Solution:* Since the expansion line is a straight line between the initial coordinates of $p_1 = 100$ psia and $V_1 = 1$ cu ft and the final coordinates of $p_5 = 50$ psia and $V_5 = 3$ cu ft, then the average pressure acting on the piston is (100 psia + 50 psia)/2 = 75 psia as illustrated in the sketch. The cross-hatched area shows the work done. Therefore

$$W = \int_{V_1}^{V_5} p\, dV = (p_{\text{average}})(V_5 - V_1) = \left(75\ \frac{\text{lb}_f}{\text{in.}^2} \times 144\ \frac{\text{in.}^2}{\text{ft}^2}\right)$$

$$(3\ \text{ft}^3 - 1\ \text{ft}^3) = 21{,}600\ \text{ft-lb}_f$$

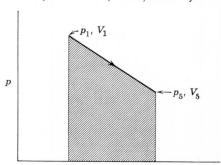

For the conditions discussed in Article 2.3 and illustrated in Fig. 2.4, work is supplied to a *constant-volume* system but the use of Equation 2.4, $W = \int p\, dV$, would give zero for the work done. In the process described in Article 2.3, the work that is transferred to the system is stored in the system as increased internal energy through the mechanism of fluid friction. Equations 2.2 and 2.4 are valid only for a *frictionless* process as indicated by the heading to this article.

Attention is again called to the definition of work as *energy that is being transferred across the boundaries of a system because of a force acting through a distance.* Referring to Fig. 2.5, as the piston moves to the right, the force could rotate a shaft through a suitable mechanism and drive an electric generator or lift a weight. Energy is leaving the system as work. This flow of energy stops the instant the piston ceases to move.

When work is done by the system as it expands and forces the piston to move to the right, $V_2$ is greater than $V_1$ and Equation 2.4 gives a positive result. During compression, an external force pushes the piston to the left, work is supplied from the surroundings which perform work on the system, $V_2$ is less than $V_1$, and Equation 2.4 gives a negative result. Consequently, it is customary to consider the work term as positive $(+W)$ where the system is expanding and is transferring work from the system to the surroundings. Conversely, the work term is negative $(-W)$ when the system is undergoing compression and work is being transferred from the surroundings to the system.

In Article 1.10 is was pointed out that properties of a substance such a temperature, specific volume, pressure, internal energy, and enthalpy are point functions; that is, the differences between their initial and final values in any process or change of state are independent of the path followed during the process. The work that is done during a process is dependent upon the path, that is, how the properties varied during the process. For a frictionless process, the area under the curve obtained by plotting $p$ versus $V$ for the process represents the work done. It will be noted by referring to Fig. 2.5 that the area under the curve 1–6–5 is different from the area under the lines representing a constant-pressure expansion from 1 to 2 followed by a constant-volume process from 2 to 5. The end states 1 and 5 are the same but the work is different. Similarly, in Fig. 1.9, the work done during expansion along path $a$ is different from the work done during expansion along path $b$. Such expressions as $p_2 - p_1$, $V_2 - V_1$, or $T_2 - T_1$ for a given process represent definite and specific changes in the value of the properties of the system, depend only on initial and final states, and are independent of the path. However, the work done during a given process cannot be evaluated until the path is known. It cannot be expressed as $W_2 - W_1$ since it occurs during the process only and is dependent on the nature of the process. Consequently, the symbol $_1W_2$ will be used to represent the work done when the state of the system changes from an initial state 1 to a final state 2.

In mathematical terms, the differentials of the properties, $dp$, $dT$, $dV$, etc., are exact differentials. However, work being a path function, the differential of work is inexact and is written as $\delta W$. As will be shown later, heat is also a path function; the differential is therefore inexact and is written as $\delta Q$.

In the actual engine, the curve representing the relation between $p$ and $V$ must be obtained experimentally by means of an instrument called an *indicator*, as explained in Chapter 5. Let Fig. 2.6 represent a simple engine in which the piston is connected by means of a connecting rod and crank to a rotating shaft on which is mounted a flywheel. As the piston moves to the right on the expansion or power stroke, a plot of the pressure of the gas in the cylinder against the volume of the gas gives the curve 1–2–3–4. The work done *by the gas on the piston* as it moves to the right is represented by the area under this curve on the $pV$ diagram, that is, the cross-hatched area 1–2–3–4–a–b. Part of this work is used to turn the shaft and drive a generator or some other machine, and part of the work is stored in

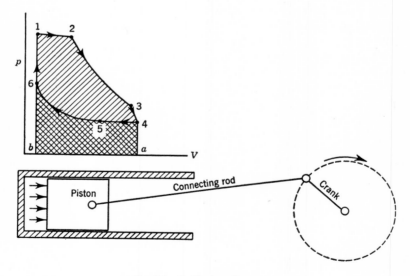

**Fig. 2.6.** Indicator diagram.

the flywheel by reason of acceleration of its angular velocity. On the return or compression stroke of the piston, while it is traveling toward the left, the energy stored in the flywheel is used to return the piston to the starting point against the pressure of the gas in the cylinder. The curve 4–5–6–1 represents the relation between $p$ and $V$ during the compression stroke, and the cross-hatched area under this curve, or 4–5–6–$b$–$a$, represents the work *done on the gas* in returning the piston to the starting point. The net work done by the gas during one revolution of the engine is the difference between the two areas, or the area 1–2–3–4–5–6. It is apparent that work is done because a higher average pressure is maintained in the cylinder during the expansion or power stroke than during the compression or return stroke of the piston. The methods used to measure the net area within the $pV$ diagram and to compute the work done are discussed in Chapter 5.

*Power* is the *rate* of doing work. The horsepower (hp) and the kilowatt (kw) are the units of power used by the engineer and may be defined as follows:

$$1 \text{ hp} = 33,000 \text{ ft-lb of work per min}$$

$$1 \text{ hp} = 0.746 \text{ kw}$$

One horsepower-hour (hp-hr) is defined as the *quantity* of work done in 1 hr if work is performed continuously at an average rate of 33,000 ft-lb per min for a period of 1 hr. Therefore

$$1 \text{ hp-hr} = 33,000 \frac{\text{ft-lb}_f}{\text{min}} \times 60 \text{ min} = 1,980,000 \text{ ft-lb}_f$$

**Example 3.** During the expansion stroke, the average pressure of the gas on the piston of an engine is 100 psia, whereas, on the return or compression stroke, the average pressure is 25 psia. Compute the work done per revolution of the engine if the volume displaced by the piston in traveling from one end of its stroke to the other end is 2 cu ft.

*Solution:*

Net work = work of expansion minus work of compression

$$= \left(100 \frac{\text{lb}_f}{\text{in.}^2} \times 144 \frac{\text{in.}^2}{\text{ft}^2} \times 2 \text{ ft}^3\right) - \left(25 \frac{\text{lb}_f}{\text{in.}^2} \times 144 \frac{\text{in.}^2}{\text{ft}^2} \times 2 \text{ ft}^3\right)$$

$$= 21,600 \text{ ft-lb}_f$$

**Example 4.** If the engine of Example 3 makes 100 rpm, what hp is developed?

*Solution:*

$$\text{Hp} = \frac{21{,}600 \dfrac{\text{ft-lb}_f}{\text{stroke}} \times 100 \dfrac{\text{strokes}}{\text{min}}}{33{,}000 \dfrac{\text{ft-lb}_f}{\text{min} \times \text{hp}}} = 65.8 \text{ hp}$$

## 2.5   Heat

Let Fig. 2.7 represent a closed system containing a gas and having its surroundings at a temperature higher than that of the system. It is common experience that in such a situation the temperature of the system will increase. Also, in general, the temperature of the surroundings will decrease. If conditions are allowed to proceed to equilibrium, the system and the surroundings will attain a uniform temperature. Energy has been transferred across the boundaries of the system. *Heat is defined as energy that is being transferred across the boundaries of a system because of a temperature difference.* The energy that is transferred was stored initially in the molecules of the surroundings as internal energy. After the transfer has been completed, it is stored in the system as internal energy. The temperature of gas in the system has increased, which means that the average velocity of translation of the molecules in the system or, in other words, the internal energy of the gas, has increased.

The similarity in the definitions of heat and work should be noted. Both are forms of energy in the process of *being transferred* across the system boundaries. In the case of heat, the driving potential is a *temperature difference.* In the case of work, the driving potential is a *force.* Since both heat and work are energy in *transition,* the quantity of energy transferred as heat or as work may be expressed in the same units.

Since heat, like work, is energy that is being transferred across the

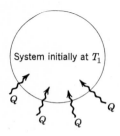

**Fig. 2.7.** Heat transfer from surroundings to the system across system boundaries.

boundaries of a system, the quantity of heat transferred depends on the path of the process. Therefore, heat is a path function; the differential of heat is inexact and is written as $\delta Q$. The symbol $_1Q_2$ is often used to designate the quantity of heat transferred when the state of the system changes from an initial state 1 to a final state 2.

The unit of heat is the British thermal unit (Btu) which is defined as $\frac{1}{180}$ of the quantity of heat required to change the temperature of 1 $lb_m$ of water from 32 to 212 F, a temperature increase of 180 F.

Inasmuch as energy may be transferred across the boundaries of a system as either heat or work, the units in which heat and work are expressed must be consistent in any equation in which both terms appear. Heat is normally expressed in British thermal units (Btu). Work is normally expressed in ft-lb$_f$ units. Throughout this book, this practice will be followed consistently.

The relationship between the Btu and the ft-lb$_f$ is *defined* as follows:

$$1 \text{ Btu} = 778.26 \text{ ft-lb}_f \qquad (2.5)$$

The conversion factor for changing ft-lb$_f$ to Btu is given the symbol $J$ in honor of Dr. Joule, an English physicist who first measured the relationship between the ft-lb$_f$ and the Btu. $J$ is defined as follows:

$$J = \frac{778.26 \text{ ft-lb}_f}{\text{Btu}} \qquad (2.6)$$

For slide-rule calculations, a value of 778 may be used.

**Example 5.** Express 100,000 ft-lb$_f$ of work in British thermal units.
*Solution:*

$$\frac{W}{J} = \frac{100,000 \text{ ft-lb}_f}{778 \dfrac{\text{ft-lb}_f}{\text{Btu}}} = 128.5 \text{ Btu}$$

For the particular cases illustrated by Figs. 2.4 and 2.7, if the quantities of energy transferred as work and heat, respectively, are the same, the effect on the properties of the system, such as temperature, would be identical.

## 2.6 Specific Heat

In engineering literature, the specific heat ($c$) is defined as the quantity of heat that must be transferred across the boundaries of a system to cause a temperature rise of 1 degree per unit of mass in the system *in the absence of friction.* Frictional effects are excluded because it will be noted by reference to Fig. 2.4 that work can produce a temperature change in a system through the mechanism of fluid friction. In the English system of units, the specific heat is the number of Btu required to change the temperature of 1 $lb_m$ 1 F. Thus, if it is found experimentally that the temperature of 2 $lb_m$ of a substance is increased 10 F by the addition of 5 Btu, the specific heat,

$$c = \frac{5 \text{ Btu}}{2 \text{ lb}_m \times 10 \text{ F}} = 0.25 \text{ Btu per lb}_m \text{ per F}$$

In general, the specific heat is a variable. Therefore

$$c = \frac{\delta Q}{m \, dt} \quad \text{or} \quad \delta Q = mc \, dt$$

Then
$$Q = m \int_{t_1}^{t_2} c \, dt \tag{2.7}$$

where $Q$ = heat transferred, Btu
   $m$ = mass being heated, $lb_m$
   $c$ = specific heat, Btu per $lb_m$ per F
   $t_1$ = initial temperature, F
   $t_2$ = final temperature, F

For most substances, $c$ increases with temperature and can be expressed as some function of temperature such as $c = a + bt + et^2 \cdots$. For most of the problems encountered in this book, a mean value of the specific heat will be used which permits treating the specific heat as a constant, in which case Equation 2.5 reduces to the form

$$Q = mc(t_2 - t_1) \tag{2.8}$$

**Example 6.** Forty pounds of oil which has a mean specific heat of 0.6 Btu per lb per F are heated from 70 to 200 F. How much heat is transferred to the oil?

*Solution:* Since a mean value is given for the specific heat, it may be treated as a constant and $Q = mc(t_2 - t_1)$

$$Q = 40 \text{ lb}_m \times \frac{0.6 \text{ Btu}}{\text{lb}_m \text{ F}} (200 \text{ F} - 70 \text{ F}) = 3120 \text{ Btu}$$

**Example 7.** Eighty pounds of a substance which has a specific heat given by the equation $c = 0.28 + 0.00005t$ are heated from 100 to 800 F. How much heat is transferred to the substance?

*Solution:* Since the specific heat is a function of temperature, then

$$Q = m \int_{t_1}^{t_2} c\, dt = m \int_{t_1}^{t_2} (0.28 + 0.00005t)\, dt$$

$$= 80 \left[ 0.28t + \frac{0.00005}{2}\, t^2 \right]_{100}^{800}$$

$$= 80 \left[ 0.28(800 - 100) + \frac{0.00005}{2}\, (640{,}000 - 10{,}000) \right]$$

$$= 80[196 + 16] = 16{,}960 \text{ Btu}$$

If heat is added to a compressible fluid such as a gas, the volume is either constant or changing during the process. If the volume is constant, all the heat that is supplied is used to increase the temperature of the fluid. The heat required to change the temperature of 1 lb of a compressible fluid 1 degree at constant volume is known as the constant-volume specific heat and is given the symbol $c_v$.

A fluid may expand at a constant pressure during the heating process as would be the case if a gas were heated in a vertical cylinder closed by a gas-tight frictionless piston as illustrated in Fig. 2.8. Then work would be done by the expansion of the gas and

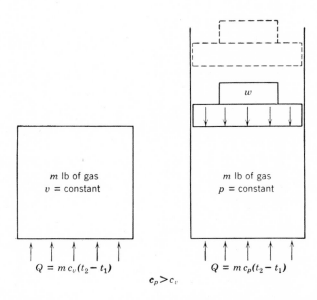

**Fig. 2.8.** Constant-volume and constant-pressure heating of a compressible fluid.

the work so done would be stored as an increase in the mechanical potential energy of the piston and weight in a final position shown by the dotted line in Fig. 2.8. Sufficient heat must be added to increase the temperature of the gas and also to perform the work of expansion. The heat required to raise the temperature of 1 lb of fluid 1 degree at constant pressure is known as the constant-pressure specific heat and is given the symbol $c_p$. It is obvious that $c_p$ is greater than $c_v$, because no work is done during constant-volume heating, but work is done during constant-pressure heating. $c_p$ and $c_v$ are determined experimentally, and their mean values are given in Table 3.1 for a number of common gases.

### 2.7 Energy Balance for the Nonflow Process

Consider a closed variable-volume system bounded by a cylinder, cylinder head, and frictionless gas-tight piston as illustrated in Fig. 2.9. The following assumptions are made:

1. The system contains some compressible fluid such as air or steam.

2. The motion of the piston is slow enough so that, at every instant, properties such as temperature, pressure, and specific volume are uniform throughout the system.

3. The cylinder is at rest so that changes in mechanical potential and kinetic energies may be neglected.

4. Energy crosses the boundaries of the system only as heat and work.

Let the piston move to the right a small distance $dx$ so that a small amount of work $\delta W$ is done on the surroundings by the expansion of the system. At the same time, let a small amount of heat $\delta Q$ be transferred to the system. Since in accordance with the law

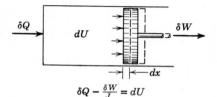

**Fig. 2.9.** Energy balance for a closed system.

of conservation of energy, energy can be neither created nor destroyed but remains constant in amount, an energy balance for the process may be written as follows:

Energy transferred to the system as heat minus energy leaving the system as work equals the change in energy stored in the system, or, mathematically,

$$\delta Q - \frac{\delta W}{J} = dU$$

Then
$$\delta Q = dU + \frac{\delta W}{J}$$

and
$$_1Q_2 = U_2 - U_1 + \frac{_1W_2}{J} \tag{2.9}$$

where $_1Q_2$ = energy transferred to or from the system as heat during a process, expressed in Btu. A positive sign indicates heat supplied to the system. A negative sign indicates heat removed from the system.

$_1W_2$ = energy transferred from or to the system as work during a process, expressed in ft-lb$_f$. A positive sign indicates work done by the system (expansion). A negative sign indicates work done on the system (compression).

$U_2 - U_1$ = the change in the internal energy stored in the molecules within the system, expressed in Btu.

$J$ = 778 ft-lb$_f$/Btu, a conversion factor for changing ft-lb$_f$ to Btu to make all units in the equation consistent.

The principles involved in this energy balance will be illustrated by several examples.

**Example 8.** A system as illustrated in Fig. 2.9 performs 100,000 ft-lb$_f$ of work on the surroundings by expansion of the gas behind the piston. At the same time, 100 Btu are transferred to the system as heat. Is the internal energy increased or decreased and by how much?

*Solution:* The solution of problems involving energy balances will be simplified by making a sketch of the system and indicating thereon the quantities of energy being transferred and the direction of flow.

$+_1Q_2 = 100$ Btu $\qquad\qquad +_1W_2 = 100,000$ ft-lb$_f$

From Equation 2.9

$$_1Q_2 = U_2 - U_1 + \frac{_1W_2}{J}$$

$$U_2 - U_1 = {_1Q_2} - \frac{_1W_2}{J}$$

$$U_2 - U_1 = 100 \text{ Btu} - \frac{100{,}000 \text{ ft-lb}_f}{\dfrac{778 \text{ ft-lb}_f}{\text{Btu}}}$$

$$U_2 - U_1 = 100 \text{ Btu} - 128.5 \text{ Btu} = -28.5 \text{ Btu}$$

Therefore, during the process, the internal energy has decreased by 28.5 Btu. This is evident when it is considered that 128.5 Btu left the system as work at the same time that 100 Btu entered the system as heat.

**Example 9.** A system as illustrated in Fig. 2.9 is compressed and 150,000 ft-lb$_f$ of work are done on the system during the process. At the same time, 100 Btu are transferred from the system as heat. Is the internal energy of the system increased or decreased and by how much?

*Solution:* The following sketch illustrates the process:

$$-{_1Q_2} = 100 \text{ Btu} \qquad\qquad\qquad -{_1W_2} = 150{,}000 \text{ ft-lb}_f$$

Since

$$_1Q_2 = U_2 - U_1 + \frac{_1W_2}{J}$$

$$U_2 - U_1 = {_1Q_2} - \frac{_1W_2}{J}$$

$$U_2 - U_1 = -100 \text{ Btu} - \left(-\frac{150{,}000 \text{ ft-lb}_f}{\dfrac{778 \text{ ft-lb}_f}{\text{Btu}}}\right)$$

$$U_2 - U_1 = -100 \text{ Btu} + 193 \text{ Btu} = +93 \text{ Btu}$$

The internal energy has increased 93 Btu during the process. The energy supplied to the system as work exceeds the energy leaving the system as heat.

**Example 10.** A gas at a pressure of 500 psia and a temperature of 2000 F expands in the cylinder of a Diesel engine. During the expansion, 200,000 ft-lb$_f$ of work are done. At the same time, 20 Btu are transferred to the jacket cooling water. Is the internal energy increased or decreased and by how much?

*Solution:* The following sketch illustrates the process:

$$-{_1Q_2} = 20 \text{ Btu} \qquad\qquad\qquad +{_1W_2} = 200{,}000 \text{ ft-lb}_f$$

Since
$$_1Q_2 = U_2 - U_1 + \frac{_1W_2}{J}$$

$$U_2 - U_1 = {_1Q_2} - \frac{_1W_2}{J}$$

$$U_2 - U_1 = -20 \text{ Btu} - \frac{200,000 \text{ ft-lb}_f}{\dfrac{778 \text{ ft-lb}_f}{\text{Btu}}}$$

$$U_2 - U_1 = -20 \text{ Btu} - 257 \text{ Btu} = -277 \text{ Btu}$$

The internal energy of the system decreased 277 Btu because energy left the system both as heat and as work.

## 2.8   The Steady-Flow System

The open system as illustrated in Fig. 2.3 is representative of the component equipment of the common types of power plants such as boilers, superheaters, condensers, heaters, pumps, compressors, and turbines. A fluid flows through some apparatus, the inside surfaces of which constitute the system boundaries, and energy transfer and conversion take place in accordance with the law of conservation of energy. It will be assumed that steady-flow conditions exist, that is, the following conditions are fulfilled:

1. The mass of fluid in the system remains constant. When 1 lb of fluid enters the system, at the same time 1 lb leaves the system.

2. Pressure, temperature, specific volume, and velocity of flow are constant with respect to time at the entrance section and also at the exit section.

3. Energy is being transferred across the system boundaries as heat and work at a constant rate.

A process which fulfills these conditions is known as a steady-flow process. Because of its importance in the design and testing of power-plant equipment, most of the remainder of this chapter will deal with the energy balance for the steady-flow process.

## 2.9   Flow Energy

In the steady-flow process, there is a transfer of mass as well as energy across the system boundaries. Let Fig. 2.10 represent a tank into which a fluid is flowing from an inlet pipe. The head of the tank is attached to a cylinder which contains a frictionless gas-tight piston

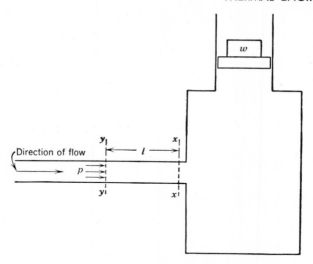

**Fig. 2.10.** Steady flow of a fluid past section $x$–$x$.

loaded by a mass having a weight $w$. The piston is free to move vertically as the volume of the fluid in the tank changes so as to maintain a constant pressure in the tank. Let it be assumed that the tank is well insulated so that no heat transfer occurs between the fluid in the tank and the surroundings.

The fluid in the inlet pipe is at a pressure of $p_1$ psfa and has a specific volume of $v_1$ cu ft per lb. The fluid is flowing past the entrance section $x$ because it is being pushed from behind by the fluid further upstream in the pipe; otherwise the fluid would not be flowing. Consider an imaginary plane at section $y$ so located with respect to section $x$ that 1 lb of fluid is contained within the cylinder bounded by these planes. If $A$ represents the internal cross-sectional area of the pipe in square feet, then the volume of this cylinder is $Al = v_1$ cu ft. Assume that the pipe contains an imaginary piston the face of which is located at the plane $y$. The force exerted by the imaginary piston on the fluid at section $y$ is $p_1 \times A$ lb. The work done in pushing 1 lb of fluid past section $x$ may be computed as follows:

$$W = \text{force} \times \text{distance} = (p_1 A) \times l$$

$$W = p_1(Al) = p_1 v_1 \text{ ft-lb}_f \text{ per lb}_m \text{ of fluid} \qquad (2.10)$$

Thus, when a fluid is flowing past a given plane or section, it is flowing because of work being done on it by the fluid behind it to push it past the section, and the amount of this work is $p_1v_1$ ft-lb$_f$ per lb$_m$ of fluid. Since this form of energy is always associated with the pushing of a fluid past a given section, it is called *flow energy* or flow work.

It should be noted that the fluid which is pushed into the tank as the result of $p_1v_1$ ft-lb of work being done on it has in turn caused the weighted piston to move upward to make room for the gas and has therefore done an equal amount of work on the piston in increasing the potential mechanical energy stored in the piston and mass.

### 2.10  The Energy Balance for the Steady-Flow Process

In accordance with the law of conservation of energy, the energy entering a system under steady-flow conditions must equal the energy leaving the system in the same interval of time. Also, since the mass in the system remains constant, the following equations may be written:

$$\text{Matter in} = \text{matter out}$$

$$\text{Energy in} = \text{energy out}$$

Because of these relationships, it is convenient to write the energy balance for a flow of one pound of mass. At the entrance section 1–1 of Fig. 2.11, which is located $z_1$ ft above some datum plane such as the

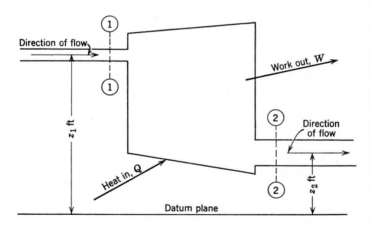

**Fig. 2.11.** Steady-flow apparatus.

floor, the fluid has a velocity of $V_1$ fps, a temperature $T_1$, a specific volume $v_1$, a pressure $p_1$, and an internal energy $u_1$. The various kinds of energy associated with 1 lb$_m$ as it flows past section 1–1 may be written as follows:

(1) Mechanical potential energy $\quad \dfrac{g}{g_c} z_1$ ft-lb$_f \qquad$ (1.17)

(2) Mechanical kinetic energy $\quad \dfrac{V_1{}^2}{2g_c}$ ft-lb$_f \qquad$ (1.18)

(3) Internal energy $\qquad\qquad u_1$ Btu

(4) Flow energy $\qquad\qquad\quad p_1 v_1$ ft-lb$_f \qquad$ (2.10)

Similarly, at section 2–2, the energy associated with 1 lb$_m$ of fluid as it leaves the system is:

(1) Mechanical potential energy $\quad \dfrac{g}{g_c} z_2$ ft-lb$_f$

(2) Mechanical kinetic energy $\quad \dfrac{V_2{}^2}{2g_c}$ ft-lb$_f$

(3) Internal energy $\qquad\qquad u_2$ Btu

(4) Flow energy $\qquad\qquad\quad p_2 v_2$ ft-lb$_f$

Between sections 1–1 and 2–2, energy may be transferred to or from the system as heat and work.

Since energy in = energy out, then for a flow of 1 lb$_m$ the energy balance may be written in British thermal units as follows:

$$\frac{gz_1}{g_c J} + \frac{V_1{}^2}{2g_c J} + \frac{p_1 v_1}{J} + u_1 + Q = \frac{gz_2}{g_c J} + \frac{V_2{}^2}{2g_c J} + u_2 + \frac{p_2 v_2}{J} + \frac{W}{J}$$

$$(2.11)$$

The same sign convention applies to the heat and work terms in Equations 2.11 and 2.9. The heat-transfer term is positive when heat is transferred to the system and negative when heat is transferred from the system. The work term is positive when work is done by the system (expansion) and negative when work is done on the system (compression).

Care must be exercised in the use of consistent units in Equation 2.11. Velocity is expressed in feet per second, pressure in pounds per square foot absolute, specific volume in cubic feet per pound. All energy terms must be expressed in British thermal units as in Equation 2.11 or in foot-pound units.

In the energy balance for the nonflow process, Equation 2.9, the symbols $_1Q_2$ and $_1W_2$ represent the energy transferred across the system boundaries as heat and work during the process; $U_1$ and $U_2$ represent the initial and final internal energies of the fluid in the system as determined by the states of the system at the beginning and end of the process. In the energy balance for the steady-flow process, Equation 2.11, the symbols $Q$ and $W$ represent the energy transferred across the system boundaries as heat and work per pound of mass flowing through the system. The subscripts 1 and 2 refer to the conditions at the entrance and exit sections.

In Article 1.18, enthalpy was defined as follows:

$$h = u + pv \qquad (1.20)$$

and it was pointed out that consistent units must be used. Wherever the product $pv$ occurs in this book, $p$ will be expressed in pounds of force per square foot and $v$ in cubic feet per pound of mass so that the product $pv$ is ft-lb$_f$/lb$_m$. Enthalpy and internal energy may be expressed in these units, but it is customary to express them in Btu per pound of mass. Therefore the conversion factor $J$ will be introduced into the equation and hereafter it will be written for 1 lb$_m$ as follows:

$$h = u + \frac{pv}{J} \qquad (2.12)$$

In the energy balance for the steady-flow process, Equation 2.11, the product $pv$ represents flow energy at entrance and exit as well as a product of two properties. The student should understand clearly that the product $pv$ represents flow energy only in a process involving the flow of mass into or out of a system.

By introducing the term enthalpy into Equation 2.11, the following simplified equation is obtained:

$$\frac{gz_1}{g_cJ} + \frac{V_1^2}{2g_cJ} + h_1 + Q = \frac{gz_2}{g_cJ} + \frac{V_2^2}{2g_cJ} + h_2 + \frac{W}{J} \qquad (2.13)$$

Tables and charts giving enthalpy values for steam and some refrigerants appear in the Appendix of this book. Their use will be explained in Chapter 3, where the equation for calculating enthalpy changes for a perfect gas will be developed also.

The energy balance will now be applied to the following power-system components, all of which will be discussed in subsequent chapters in the book:

1. Heat exchangers such as boilers, superheaters, economizers, air heaters, condensers, and feed-water heaters.

2. Prime movers such as steam and gas turbines and steam and internal-combustion engines.

3. Pumps.

4. Compressors and fans.

5. Nozzles.

6. The throttling process, governors, and calorimeters.

### 2.11 Application of the Energy Balance for the Steady-Flow Process to Heat Exchangers

Heat exchangers are a class of equipment in which steam or some other vapor is generated or condensed, or a liquid or gas such as oil or air is heated or cooled. A simple boiler in which water is being converted into steam is shown diagrammatically in Fig. 2.12.

For a *steady-flow* process in a *heat exchanger* the following assumptions may be made:

1. One pound of fluid leaves the apparatus for each pound of fluid entering the apparatus.

2. No work is done between the entrance and exit sections since no shaft is being rotated.

3. The velocities at the entrance and exit sections are low and of the same order of magnitude in order to keep the pressure drop in the piping system to a reasonably low value.

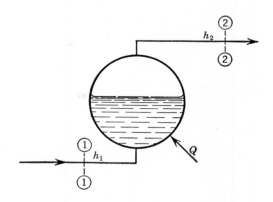

**Fig. 2.12.** Heat exchanger.

$$Q_{in} = h_2 - h_1$$

4. The change in mechanical potential energy of the fluid, $(z_1 - z_2)/J$, is in the order of a small fraction of 1 Btu and may be neglected. It may be noted that a difference in elevation of 77.8 ft between the entrance and exit sections will amount to only 0.1 Btu.

The energy balance may be written again for convenience with the negligible terms canceled as follows:

$$\frac{\cancel{gz_1}}{\cancel{g_cJ}} + \frac{\cancel{V_1^2}}{\cancel{2g_cJ}} + u_1 + \frac{p_1v_1}{J} + Q = \frac{\cancel{gz_2}}{\cancel{g_cJ}} + \frac{\cancel{V_2^2}}{\cancel{2g_cJ}} + u_2 + \frac{p_2v_2}{J} + \frac{\cancel{W}}{\cancel{J}}$$

(2.11)

Then,                        $h_1 + Q = h_2$

or                          $Q = h_2 - h_1$                      (2.14)

where $h_1$ and $h_2$ are the enthalpies per pound of fluid at entrance and exit, and $Q$ is the heat added or removed per pound of fluid.

## 2.12  Application of the Energy Balance for the Steady-Flow Process to Prime Movers

The function of a prime mover is to perform work. Let Fig. 2.13 represent diagrammatically a prime mover such as a steam turbine through which a fluid is flowing under steady-state conditions. The following assumptions may be made:

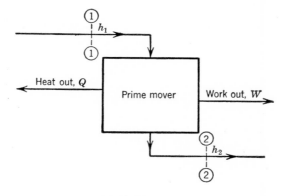

**Fig. 2.13.** Prime mover.

$$\frac{W}{J} = (h_1 - h_2) - Q_{\text{out}}$$

1. One pound of fluid leaves the apparatus for each pound of fluid entering the apparatus.

2. The velocities at the entrance and exit sections are low and of the same order of magnitude in order to keep the pressure drop in the piping system to a reasonably low value.

3. The change in mechanical potential energy of the fluid between entrance and exit, $(z_1 - z_2)/J$, may be neglected.

The energy balance for the steady-flow process may be written again for convenience with the negligible terms canceled as follows:

$$\frac{gz_1}{g_cJ} + \frac{V_1^2}{2g_cJ} + u_1 + \frac{p_1v_1}{J} = Q + \frac{gz_2}{g_cJ} + \frac{V_2^2}{2g_cJ} + u_2 + \frac{p_2v_2}{J} + \frac{W}{J}$$

(2.11)

Then,
$$h_1 = Q + h_2 + \frac{W}{J}$$

or
$$\frac{W}{J} = (h_1 - h_2) - Q \qquad (2.15)$$

where $h_1$ and $h_2$ are the enthalpies per pound of fluid at entrance and exit, $W$ is the work done per pound of fluid, and $Q$ is the heat removed per pound of fluid.

It is apparent by examination of Fig. 2.13 and Equation 2.15 that any energy removed from the system as heat cannot be converted into work, and thereby the efficiency of the machine is reduced. The ideal expansion in a prime mover should therefore be one in which there is *no heat transfer to the cooler surroundings*. The heat-transfer term $Q$ in Equation 2.15 would then be zero, and the equation would reduce to the form

$$\frac{W}{J} = h_1 - h_2 \qquad (2.16)$$

Such an expansion or compression which takes place under conditions of thermal isolation, that is, with no heat transfer between the working fluid and its surroundings, is called an *adiabatic process*.

As the working fluid flows through the prime mover shown in Fig. 2.13, the amount of work produced may be less than the work obtainable under ideal conditions because of mechanical friction in the mechanism or fluid friction due to high velocities in crooked passages in the machine. Such frictional effects reduce the work done and increase the enthalpy of the fluid at the point of exit from the apparatus in accordance with the law of conservation of energy. Consequently, the ideal expansion in a prime moved is a *frictionless adiabatic* expansion, that is, an expansion without friction and without heat transfer to the cooler surroundings. During a frictionless adiabatic expansion, a property of the substance called *entropy* is constant, and such a process is frequently called an *isentropic* (meaning constant-entropy) process. A discussion of the determination of entropy will be presented in Chapter 3. A full understanding of the significance of entropy requires a thorough study of the science of thermodynamics and is beyond the scope of this book. However, the student should understand that the *ideal* expansion in a prime mover is a *frictionless adiabatic* or *constant-entropy* (*isentropic*) process.

### 2.13  Application of the Energy Balance for the Steady-Flow Process to Pumps

A pump is a device for transferring a *liquid* from one place to another, usually against an external pressure, and is illustrated diagrammatically in Fig. 2.14. The following assumptions may be made concerning the flow of a liquid through the pump:

1. One pound of fluid leaves the apparatus for each pound of fluid entering the apparatus.

2. The liquid is at room temperature or the apparatus is well insulated so that there will be no significant transfer of heat between the entrance and exit sections.

3. The liquid is incompressible.

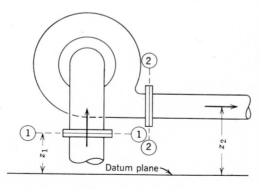

**Fig. 2.14.** Diagrammatic arrangement of a pump.

The energy balance may be written again for convenience with one of the terms canceled in accordance with the above assumptions:

$$\frac{gz_1}{g_c J} + \frac{V_1^2}{2g_c J} + u_1 + \frac{p_1 v_1}{J} + \cancel{Q} + \frac{W}{J} = \frac{gz_2}{g_c J} + \frac{V_2^2}{2g_c J} + u_2 + \frac{p_2 v_2}{J}$$

(2.11)

The work term normally appears on the right side of the equality sign. However, in a pump, work is supplied to the system, the sign is therefore negative, and the term has been transferred to the left side of the equality sign (the "energy-in" side) in the above equation.

This equation may be written in foot-pound units and the density, $\rho$, may be substituted for the specific volume, $v$, as follows:

$$\frac{g}{g_c} z_1 + \frac{V_1^2}{2g_c} + J u_1 + \frac{p_1}{\rho_1} + W = \frac{g}{g_c} z_2 + \frac{V_2^2}{2g_c} + J u_2 + \frac{p_2}{\rho_2}$$

Since a liquid is practically incompressible, it may be assumed that $\rho_1 = \rho_2$ and

$$W = \frac{g}{g_c} (z_2 - z_1) + \frac{V_2^2 - V_1^2}{2g_c} + \frac{p_2 - p_1}{\rho} + J(u_2 - u_1) \quad (2.17)$$

If water is being pumped, a change in internal energy of 1 Btu produced by friction results in a change in temperature of 1 F since the specific heat of water is approximately 1.0. Since 1 Btu $= 778$ ft-lb$_f$, a change in internal energy represented by a 1.0 F temperature rise would represent the same amount of energy as would be required to elevate 1 lb$_f$ of water through a vertical distance of 778 ft. Temperatures cannot be measured to $\pm 0.001$ F by even the best of engineering instrumentation; therefore Equation 2.17, although correct, is impractical to use.

If $W_0$ represents the *useful* work performed by the pump in creating flow and in changing the mechanical potential and kinetic energy of a pound of fluid being pumped, then Equation 2.17 may be written as follows:

$$W_0 = \frac{g}{g_c} (z_2 - z_1) + \frac{V_2^2 - V_1^2}{2g_c} + \frac{p_2 - p_1}{\rho}$$

(2.18)

The difference between Equations 2.17 and 2.18 represents the frictional effects in the pump between the inlet and outlet sections. The application of Equation 2.18 to the pumping of liquids will be discussed in Chapter 11.

### 2.14  Application of the Energy Balance for the Steady-Flow Process to Compressors and Fans

Gas compressors and fans are used to move or compress air and many industrial gases at pressures that vary from a fraction of 1 psi to more than 15,000 psi. The operating conditions vary over such wide limits that the analysis should start with the energy-balance Equation 2.11. Depending on the particular conditions of operation, some of the terms in this equation may be canceled to produce a simpler equation applicable to the particular equipment under consideration.

### 2.15  Application of the Energy Balance for the Steady-Flow Process to the Nozzle

A nozzle is a flow channel that is so proportioned as to permit a fluid to expand from its initial pressure to a lower pressure with a

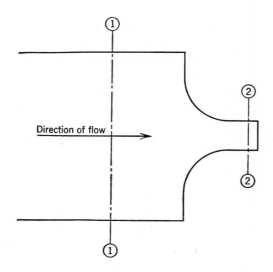

**Fig. 2.15.**  Nozzle.

$$\frac{V_2{}^2 - V_1{}^2}{2g_cJ} = h_1 - h_2$$

conversion of as much as possible of the energy stored in the fluid in other forms at the entrance section into the kinetic energy of a high-velocity jet. Let Fig. 2.15 represent a nozzle in which a fluid at the entrance section 1 expands to a lower pressure at the exit section 2. The following assumptions may be made:

1. One pound of fluid leaves the apparatus for each pound of fluid entering the apparatus.

2. The change in mechanical potential energy of the fluid between the entrance and exit sections may be neglected.

3. No work is done between the entrance and exit sections since no shaft is being rotated.

4. Because of the high final velocity and the short distance between entrance and exit sections, the heat transferred to the surroundings per pound of fluid may be neglected.

The energy balance may be written as follows with certain terms canceled because of the above-mentioned assumptions:

$$\frac{gz_1}{g_cJ} + \frac{V_1{}^2}{2g_cJ} + u_1 + \frac{p_1v_1}{J} + Q = \frac{gz_2}{g_cJ} + \frac{V_2{}^2}{2g_cJ} + u_2 + \frac{p_2v_2}{J} + \frac{W}{J} \quad (2.11)$$

Then,
$$\frac{V_1{}^2}{2g_cJ} + h_1 = \frac{V_2{}^2}{2g_cJ} + h_2$$

or
$$\frac{V_2{}^2 - V_1{}^2}{2g_cJ} = h_1 - h_2 \quad (2.19)$$

Equation 2.19 shows that the increase in kinetic energy as a fluid flows through a nozzle is equal to the decrease in enthalpy. The use of the equation in designing steam-turbine nozzles will be discussed in Chapter 10.

### 2.16   Application of the Energy Balance for the Steady-Flow Process to the Throttling Process

The high-velocity jet produced in the nozzle as discussed in the preceding article could be directed against suitably shaped blades mounted on the periphery of a rotating disk, and the kinetic energy of the jet could then be used to do work in rotating a shaft. This is the principle upon which the turbine operates. On the other hand, the high velocity may be dissipated in internal whirls and eddies with the resultant conversion of the kinetic energy into internal energy of the

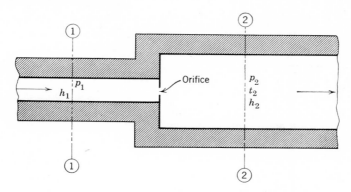

**Fig. 2.16.** The throttling process.
$$h_1 = h_2$$

fluid. Such a process is called a throttling process and may take place in the apparatus shown in Fig. 2.16.

The following assumptions are made in developing the energy equation for the throttling process:

1. One pound of fluid leaves the apparatus at section 2–2 for each pound of fluid that enters the apparatus at section 1–1.

2. The apparatus is well insulated, and the heat transfer to the cooler surroundings may be neglected.

3. No work is done between the entrance and exit sections; no shaft is being rotated.

4. The areas at the entrance and exit sections are such that the kinetic energy of the fluid is negligible or of the same order of magnitude so that the change in kinetic energy in the apparatus may be neglected.

5. The change in mechanical potential energy due to a difference in elevation between the entrance and exit sections may be neglected.

On the basis of the above assumptions, the energy balance may be written as follows with the negligible terms canceled:

$$\frac{gz_1}{g_cJ} + \frac{V_1^2}{2g_cJ} + u_1 + \frac{p_1v_1}{J} + Q = \frac{gz_2}{g_cJ} + \frac{V_2^2}{2g_cJ} + u_2 + \frac{p_2v_2}{J} + \frac{W}{J} \quad (2.11)$$

Then
$$h_1 = h_2 \quad (2.20)$$

The throttling calorimeter which is used for the determination of the quality of wet steam is discussed in Chapter 3. The steam-turbine governor uses the throttling process to control the amount of steam admitted to the turbine and is considered in Chapter 10.

### 2.17  The First Law of Thermodynamics

The energy balance for the nonflow process, Equation 2.9, and the energy balance for the steady-flow process, Equation 2.11, are based on the law of conservation of energy. The science of thermodynamics deals with the laws governing the conversion of energy from one form into another. The first law of thermodynamics is essentially the law of conservation of energy. Equations 2.9 and 2.11 may be taken as mathematical expressions of the first law, depending upon whether the nonflow or steady-flow process is being considered.

A closed cycle consists of a series of events that take place in a closed system in such a manner that the system returns to its initial state at the end of the cycle, after which the sequence of events is repeated over and over again. Let Fig. 2.17 represent on $pV$ coordinates the changes in pressure and volume that occur in a cycle for a closed system consisting of a cylinder and gas-tight, frictionless piston containing a

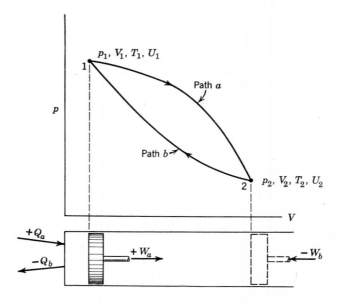

**Fig. 2.17.** Transferred energy in a closed cycle.

compressible fluid. At the start of the cycle, the fluid in the system has an initial state defined by $p_1$, $V_1$, $T_1$, and $U_1$. It expands along some path $a$ to a position where the state of the system is defined by $p_2$, $V_2$, $T_2$, and $U_2$. During this expansion, a quantity of energy is transferred from the system to the surroundings in the form of work which may be represented mathematically as $\int_1^2 \delta W_a$. At the same time the quantity of energy transferred to the system as heat is $\int_1^2 \delta Q_a$. Then let work be done on the system by the surroundings in compressing the fluid and forcing the piston to return to its initial position along some path $b$. The work done during this process is $\int_2^1 \delta W_b$ and the heat transferred is $\int_2^1 \delta Q_b$. The quantity of heat removed from the system is such that when the piston has returned to its initial position the system has returned to its initial conditions of $p_1$, $V_1$, $T_1$, and $U_1$.

The energy balance for the expansion and compression processes may be written as follows:

For path $a$: 
$$\int_1^2 \delta Q_a = (U_2 - U_1) + \frac{1}{J} \int_1^2 \delta W_a$$

For path $b$: 
$$\int_2^1 \delta Q_b = (U_1 - U_2) + \frac{1}{J} \int_2^1 \delta W_b$$

Adding these equations gives the following result:

$$\oint Q = 0 + \frac{1}{J} \oint W \tag{2.21}$$

where $\oint Q$ = net heat transferred during the cycle

$\oint W$ = net work done during the cycle

The net change in internal energy for the cycle was zero since the system was returned to its initial state at the end of the cycle.

Equation 2.21 is often taken as a statement of the first law of thermodynamics and may be expressed as follows: *the work done in a cyclical process is equal to the heat that disappears.*

Thermal efficiency ($\eta_t$) is defined as the ratio of the energy converted into work in a given interval of time to the energy supplied to the system in the same interval of time, both expressed in consistent units, or

$$\eta_t = \frac{\dfrac{W}{J}}{E_{\text{in}}} \tag{2.22}$$

where $\eta_t$ = thermal efficiency

$\quad W$ = work done by the system in a unit of time, in ft-lb$_f$

$\quad E_{\text{in}}$ = energy supplied to the system in the same unit of time, in Btu. In general, the methods of calculating $E_{\text{in}}$ are specified in the standard test codes pertaining to the particular type of equipment being tested.

Since $\quad$ 1 hp-hr $= \dfrac{33{,}000 \text{ ft-lb}_f}{\text{min}} \times 60 \text{ min} = 1{,}980{,}000 \text{ ft-lb}_f$

$$1 \text{ hp-hr} = \frac{1{,}980{,}000 \text{ ft-lb}_f}{778 \dfrac{\text{ft-lb}_f}{\text{Btu}}} = 2545 \text{ Btu} \tag{2.23}$$

and $\quad$ $1 \text{ kw-hr} = \dfrac{1 \text{ hp-hr}}{0.746} = \dfrac{2545 \text{ Btu}}{0.746} = 3413 \text{ Btu} \tag{2.24}$

then $\quad$ $\eta_t = \dfrac{\text{hp} \times 2545}{E_{\text{hr}}} = \dfrac{\text{kw} \times 3413}{E_{\text{hr}}} \tag{2.25}$

where $E_{\text{hr}}$ is the energy supplied to the system, in Btu per hour.

**Example 11.** An automobile engine is found by test to deliver 50 hp continuously for 1 hr during which time 30 lb of gasoline are burned. The energy released by the combustion of 1 lb of gasoline is 19,000 Btu. What is the thermal efficiency of the engine?

*Solution:* Since

$$\eta_t = \frac{\text{hp} \times 2545}{E_{\text{hr}}}$$

$$\eta_t = \frac{50 \text{ hp} \times 2545 \text{ Btu/hp-hr}}{30 \text{ lb of fuel/hr} \times 19{,}000 \text{ Btu/lb of fuel}} = 0.223 \text{ or } 22.3\%$$

## 2.18   The Second Law of Thermodynamics

The concepts involved in the second law of thermodynamics are broad and affect much of the fields of physics, chemistry, and engineering. Since many of the statements of the second law which appear in textbooks are written with a particular field of science in mind, they often appear to the student to be unrelated. In its general sense, the second law states that, when heat is transferred from the surroundings to a real system, it is impossible to reverse the process and return the system and all its surroundings to their initial conditions.

For the limited purposes of this book, the second law may be stated as follows: given a heat engine operating in a cycle between a source of energy at a constant absolute temperature $T_H$ and an environment at a lower absolute temperature $T_L$ to which the engine may reject that portion of the energy received from the source which it cannot convert into work, the maximum thermal efficiency is as follows:

$$\eta_{max} = 1 - \frac{T_L}{T_H} \tag{2.26}$$

Complete conversion into work of all the energy transferred to the system as heat from the source requires that the sink or receiver temperature $T_L$ be equal to absolute zero. Since the lowest temperature at which an engine can reject heat is normally the temperature of the atmosphere or a body of water such as a river or lake, it is apparent that no machine can ever be built with a thermal efficiency approaching 100 per cent. Inasmuch as man can do nothing to control $T_L$, he must concentrate his efforts on increasing $T_H$. Much of the effort to improve the efficiency of power-generating equipment has been devoted to the utilization of higher initial temperatures. The maximum temperature is limited by the strength, corrosion resistance, and stability of metals at high temperature. Great strides have been made in recent years in the utilization of higher temperatures and it may be assumed that the pressure for increased efficiency will cause this trend to continue.

### 2.19  Summary

The energy balance is a statement of the law of conservation of energy.

*System:* a definite region or space enclosed within specified boundaries or surfaces that may be real or imaginary, fixed or movable.

*Closed system:* a system that contains a fixed mass. The constant mass in a closed system is sometimes referred to as the system.

*Open system:* a region or space through which mass is flowing.

*Surroundings:* the space outside the boundaries of the system; normally the space in the immediate vicinity of the system which is affected by energy or mass transfer to or from the system.

*Work:* energy that is being transferred across the boundaries of a system because of a force acting through a distance. In a frictionless process for a nonflow system, $W = \int p\,dV$.

*Power:* the rate of doing work; work done per unit of time. In the English engineering system of units, one horsepower (hp) = 33,000 ft-lb$_f$ per min and one kilowatt (kw) = 1.34 hp. A horsepower-hour (hp-hr) is the amount of work done in 1 hr at an average rate of 33,000 ft-lb$_f$ per min = 33,000 × 60 = 1,980,000 ft-lb$_f$.

*Heat:* energy that is being transferred across the boundaries of a system because of a temperature difference.

*Specific heat:* the amount of heat that is transferred across the boundaries of a system in the absence of friction to cause a temperature rise of 1 degree per unit of mass in the system.

The *British thermal unit (Btu)* is $\frac{1}{180}$ of the heat required to raise the temperature of 1 lb of pure water from 32 to 212 F. 1 Btu = 778.26 ft-lb$_f$ = $J$ ft-lb$_f$.

*Flow energy:* energy transferred across the boundaries of an open system because of the work done upon the moving fluid in pushing it across the system boundaries.

The *energy balance for the nonflow process* may be written as follows:

$$_1Q_2 = U_2 - U_1 + \frac{_1W_2}{J} \qquad (2.9)$$

The *energy balance for the steady-flow process* is expressed mathematically in Equations 2.11 and 2.13. These equations may be simplified when applied to various types of equipment by cancelling the terms that involve negligible changes in value.

The *first law of thermodynamics* states that the work done in a cyclical process is equal to the heat that disappears. Essentially it is the law of conservation of energy, and Equations 2.9 and 2.11 may be considered as statements of the first law.

*Thermal efficiency:* the ratio of the energy converted into work in a given interval of time to the energy supplied to the system in the same interval of time.

The *second law of thermodynamics:* for the limited purposes of this book, the second law of thermodynamics may be stated as follows: No engine operating in a cycle can continuously convert into work all the heat supplied to it. Given a source of heat at a constant absolute temperature $T_H$ and an environment or sink at a constant lower temperature $T_L$ to which the system may reject that portion of the energy supplied to it which it cannot convert into work, the maximum thermal efficiency is given by the equation

$$\eta_{max} = 1 - \frac{T_L}{T_H} \tag{2.26}$$

Since $T_L$ is for all practical purposes the absolute temperature of the atmosphere or a large body of water, improved thermal efficiency results from increased initial temperature but can never approach 100 per cent.

## PROBLEMS

1. Ten cubic feet of air at 100 psia and 80 F expand at constant pressure until the volume is doubled. Compute the work done.

2. Five cubic feet of air at 80 psig and 100 F are compressed at constant pressure until the volume is 2 cu ft. Barometric pressure is 29.8 in. of Hg. Compute the work done.

3. If a gas expands from $p_1$ and $V_1$ to $p_2$ and $V_2$ according to the relation $pV = C$, prove that $W = p_1V_1 \ln (V_2/V_1)$.

4. If a gas expands from $p_1$ and $V_1$ to $p_2$ and $V_2$ according to the relation $pV^n = C$, where $n \neq 1$, prove that $W = (p_1V_1 - p_2V_2)/(n-1)$.

5. Ten cubic feet of $CO_2$ at 500 psia and 100 F are heated at constant volume until the pressure is doubled. Compute the work done.

6. The piston of a single-cylinder engine has a diameter of 10 in. and travels 15 in. in moving from one end of the cylinder to the other end. If the average gas pressure on the piston during the expansion or working stroke is 130 psia and the average pressure on the return or compression stroke is 25 psia, compute (a) the work done by the gas during the expansion stroke, (b) the work done on the gas during the compression stroke, (c) the net work done during one revolution of the engine, and (d) the horsepower developed if the engine runs at 300 rpm.

7. How many ft-lb of energy are equivalent to 1000 Btu?

8. If 25 kw-hr of electrical energy are supplied to a heater, how much energy is this in Btu? In ft-lb?

9. Compute the Btu equivalent of 40 hp-hr; of 50 kw-hr.

10. Eight pounds of air initially at 100 F and 20 psig are heated at constant volume until the temperature is 300 F. Barometric pressure is 29.9 in. of Hg. How much heat is transferred to the air?

11. Fifteen pounds of air initially at 80 F and 60 psig are heated at constant pressure until the temperature is 200 F. Barometric pressure is 29.8 in. of Hg. How much heat is transferred to the air?

12. The specific heat of a substance is 0.6 Btu per lb per F. Compute the quantity of heat in Btu that must be transferred to 20 lb of the substance to increase the temperature from 50 to 200 C.

13. Oil has a specific heat of 0.5 Btu per lb per F. Water has a specific heat of 1.0. Two hundred pounds of oil are to be cooled per minute from 180 to 140 F. Initial and final water temperatures are 60 F and 75 F. How many lb of water are required per min?

14. The specific heat of methane $(CH_4)$ is given by the equation $c = 0.264 + 0.514 \times 10^{-3}T$, where $c$ is in Btu per lb per F and $T$ is in degrees R. Compute the specific heat at 500 F.

15. Using the data from Problem 14, how much heat must be transferred to methane to raise the temperature of 20 lb from 40 to 540 F?

16. The specific heat of water in Btu per mole per F is given by the equation

$$c = 19.86 - \frac{597}{\sqrt{T}} + \frac{7500}{T}$$

where $T$ is in degrees R. Compute the heat that must be transferred to 10 lb of water to change its temperature from 200 to 400 F.

17. For hydrogen, $c_p = 3.35 + 0.000114T$, where $c_p$ is in Btu per lb per F and $T$ is in degrees R. How much heat is removed from 200 lb of hydrogen in cooling it from 640 to 140 F?

18. A total of 150,000 ft-lb of work was done in compressing a gas in a nonflow process during which the internal energy of the gas increased 90 Btu. Was heat transferred to or from the system, and how much in Btu?

19. In the closed system as illustrated in Fig. 2.4, 75,000 ft-lb of work are transferred to the system and, at the same time, 40 Btu are transferred from the system as heat. Is the internal energy increased or decreased, and by how much in Btu?

20. Air enters a compressor with a velocity of 20 fps and an enthalpy of 100 Btu per lb and leaves with a velocity of 40 fps and an enthalpy of 205 Btu per lb. Assuming adiabatic compression and negligible change in potential energy, how many ft-lb of work are required to compress 1 lb of air?

21. A total of 70,000 lb of steam per hr enters a steam turbine at a velocity of 90 fps and an enthalpy of 1320 Btu per lb. The steam leaves at a velocity of 400 fps and an enthalpy of 990 Btu per lb. The heat transferred to the surroundings is 3 Btu per lb of steam flowing through the turbine. Calculate the horsepower developed by the steam in flowing through the turbine.

22. An air compressor is supplied with air at 14.7 psia and a specific volume of 13.4 cu ft per lb. The air is discharged at 100 psia and a specific volume of 2.7 cu ft per lb. The initial and final internal energies of the air are 16 and 49 Btu per lb respectively. The jacket cooling water removes 2 Btu per lb of air compressed. The change in kinetic and potential energy may be neglected. Compute (a) the work required

to compress 1 lb of air, and (b) the hp required to compress air at the rate of 100 lb per min.

**23.** Feedwater is supplied to a steam-generating unit at 300 F, at which condition the enthalpy is 270 Btu per lb. The steam leaves the unit at 600 psia and 800 F, at which condition the enthalpy is 1408 Btu per lb. Assume negligible changes in kinetic and potential energy. (a) If 100,000 lb of steam are generated per hr, how much heat must be transferred to the system in Btu per min? (b) If the unit is fired with coal having an energy content of 13,000 Btu per lb and 80 per cent of the energy in the coal is transferred to the system, how much coal is burned per hr?

**24.** A gasoline engine burns 20 lb of gasoline in 15 min when doing work at the rate of 150 hp. If the energy content of the gasoline is 20,000 Btu per lb, what is the thermal efficiency of the engine?

**25.** A Diesel engine consumes 15 lb of fuel oil in 10 min when carrying a steady load of 205 hp. The energy content or heating value of the fuel oil is 19,000 Btu per lb. Calculate (a) the thermal efficiency of the engine, and (b) the total amount of energy lost in the hot exhaust gases, cooling water, oil, etc., during a 15-min test.

**26.** A power plant has an average monthly output of 1,730,000 kw-hr. (a) If the average plant thermal efficiency is 25 per cent when burning coal having a heating value of 12,500 Btu per lb, what is the average monthly coal consumption in tons? (b) If the coal costs $8.70 per ton, what is the fuel cost in cents per kw-hr?

**27.** Natural gas having a heating value of 1000 Btu per cu ft is used in a power plant which has an overall efficiency of 22 per cent. (a) What is the gas consumption in cu ft per kw-hr? (b) If the gas costs 20 cents per 1000 cu ft, what is the fuel cost in cents per kw-hr?

# CHAPTER 3

# Thermodynamic
# Properties of Fluids

### 3.1   Introduction

The law of conservation of energy (the first law of thermodynamics) is one of the most important laws that the engineer uses in his professional activities. In the preceding chapter, this law was expressed in Equation 2.9 for the closed system and in Equation 2.11 for the steady-flow system. Before these energy balances can be applied to the solution of engineering problems, the student must acquire an understanding of the thermodynamic properties of the fluids used in power-generating equipment and the methods that are employed to evaluate these properties. Therefore, this chapter is concerned with the thermodynamic properties of fluids.

### 3.2   Perfect Gases

A perfect gas, often called an "ideal gas," is defined as a substance for which the relationship among the properties of pressure, temperature, and specific volume may be expressed by the following characteristic equation:

$$pv = RT \tag{3.1}$$

where $p$ = absolute pressure (psfa, $lb_f$ per sq ft abs, in the English system of units)

$v$ = specific volume (cu ft per $lb_m$ in the English system of units)

$T$ = absolute temperature (degrees Rankine, $t_F + 460$, in the English system of units)

$R$ = a constant that has a specific value for each gas and system of units

Hydrogen, helium, oxygen, nitrogen, and many other gases are liquefied commercially at a combination of high pressure and low

## TABLE 3.1

### Gas Constants and Specific Heats of Some Common Gases

Average Values at Room Temperature and Atmospheric Pressure

| Gas | $R$ | $c_v$ | $c_p$ |
|---|---|---|---|
| Air | 53.3 | 0.171 | 0.237 |
| Oxygen ($O_2$) | 48.2 | 0.156 | 0.218 |
| Nitrogen ($N_2$) | 55.1 | 0.175 | 0.245 |
| Hydrogen ($H_2$) | 767 | 2.42 | 3.41 |
| Carbon dioxide ($CO_2$) | 34.9 | 0.154 | 0.203 |
| Carbon monoxide (CO) | 55.1 | 0.175 | 0.246 |

temperature. As gases approach the conditions of pressure and temperature required for liquefaction, they no longer obey the characteristic equation of the perfect gas and are called vapors. However, at temperatures equal to and above normal room temperature and at low to moderate pressures, many gases follow the perfect-gas equation of state with sufficient accuracy for most engineering calculations.

The value of the gas constant $R$ for any perfect gas may be computed from $R = pv/T$ by measuring the specific volume of the particular gas at a given pressure and temperature. The value of the gas constant $R$ and the mean value of the specific heats at constant volume and constant pressure for some common gases are given in Table 3.1.

The units for the gas constant $R$ in the English engineering system may be determined as follows:

$$R = \frac{pv}{T} = \frac{(lb_f/ft^2)(ft^3/lb_m)}{\deg R} = \frac{ft\text{-}lb_f}{(lb_m)(\deg R)}$$

If all gases followed the perfect-gas law precisely, then the gas constant $R$ could be computed for all gases as follows:

$$R = \frac{1544}{M} \tag{3.2}$$

where $M$ = molecular weight of the gas based on 32 as the molecular weight of oxygen.

Equation 3.1 may be multiplied by $m$, the quantity of the gas in $lb_m$, to obtain the following relation:

$$pmv = mRT$$

$$pV = mRT \tag{3.3}$$

where $V$ = the total volume of $m$ $lb_m$ of the gas or $mv$.

For the fixed-mass or closed system, the relationship between pressure, volume, and temperature at the beginning and the end of any process involving a perfect gas may be derived as follows:

For the initial conditions: $p_1V_1 = m_1RT_1$

For the final condition: $p_2V_2 = m_2RT_2$

But for the closed system, $m_1 = m_2$ and

$$\frac{p_1V_1}{T_1} = \frac{p_2V_2}{T_2} \tag{3.4}$$

### 3.3  Internal Energy of a Perfect Gas

Perfect gases are so highly rarefied that the mean distance between molecules is large in comparison to the diameter of the molecule, and the molecular attractive forces have for all practical purposes disappeared. As a result, changes in volume, although they affect the mean distance between molecules, do not change the amount of molecular potential energy stored in the system. On the other hand, any change in temperature does affect the mean velocity of translation of the molecules and therefore changes the molecular kinetic energy. Therefore, the change in internal energy of a perfect gas is independent of a change in volume but is affected by a change in temperature. These relationships may be stated mathematically as follows:

$$\left(\frac{\delta u}{\delta v}\right)_T = 0; \quad u = f(t) \tag{3.5}$$

It is not possible to determine the total amount of internal energy stored in a unit mass of any substance at a specified state without reducing the substance to a condition of zero internal energy, a condition which is physically impossible. Fortunately, the engineer is always concerned with changes in internal energy during a process. The equation for the change in internal energy during a process will now be developed. Let Fig. 3.1 represent a closed variable-volume system that contains 1 $lb_m$ of a perfect gas initially at some temperature $T_1$. Then, if this gas is expanded or compressed at a constant temperature $T_1$, it will obey the equation $pv = RT_1 = C_1$ and a plot of $p$ versus $v$ will give the lower curve marked $T_1$ on the $pv$ plane. Note that, for

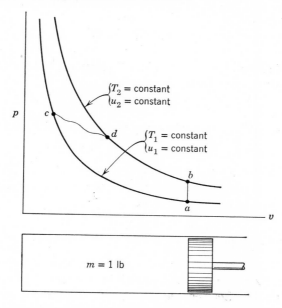

**Fig. 3.1.** Change in internal energy of a perfect gas.

every point on this curve, $T_1 =$ a constant and therefore $u_1 =$ a constant. Let the temperature of the perfect gas in the system be changed to some other temperature $T_2$ and let the gas then be expanded or compressed at this constant temperature $T_2$. A plot of $p$ versus $v$ for the process will give the upper curve. At all points on this curve, $T_2 =$ a constant, and $u_2 =$ a constant. The change in internal energy, $u_2 - u_1$, is the same for path $ab$ as for path $cd$ or for any other path between the two curves in Fig. 3.1. To evaluate the change in internal energy let us select the constant-volume path $ab$. Applying the energy balance for the frictionless nonflow process, Equation 2.9, to the constant-volume change gives

$$\delta Q = du + 0$$

since no work is done during the constant-volume process $ab$. But for a constant-volume process for 1 lb$_m$

$$\delta Q = c_v \, dT$$

then

$$du = c_v \, dT \qquad\qquad (3.6)$$

and
$$u_2 - u_1 = \int_{T_1}^{T_2} c_v \, dT \qquad\qquad (3.7)$$

If the specific heat at constant volume is assumed to be constant over the temperature range, then

$$u_2 - u_1 = c_v(T_2 - T_1) \qquad\qquad (3.8)$$

For a mass of $m$ lb, the equation may be written as follows:

$$U_2 - U_1 = mc_v(T_2 - T_1) \qquad\qquad (3.9)$$

Since the internal energy of a perfect gas is a function of temperature only, Equation 3.9 is a general equation that is applicable to *all* types of processes for perfect gases.

### 3.4   The Relation between $c_p$ and $c_v$ for a Perfect Gas

Let Fig. 3.2 represent a closed variable-volume system that contains 1 $lb_m$ of a perfect gas at an initial state represented by point 1 on the $pv$ plane. Let the system expand and perform work at the same time that the system receives heat at such a rate that the pressure remains *constant* until the final state 2 has been reached.

The energy balance for this process is

$$\delta Q = du + \frac{\delta W}{J} \qquad\qquad (2.9)$$

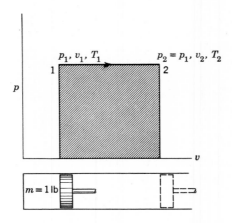

**Fig. 3.2.**   Constant-pressure expansion of a perfect gas.

but

$$\delta Q = c_p \, dT$$

$$du = c_v \, dT$$

$$\delta W = p \, dv$$

Then

$$c_p \, dT = c_v \, dT + \frac{p \, dv}{J}$$

Assuming constant specific heats, then

$$c_p(T_2 - T_1) = c_v(T_2 - T_1) + \frac{p(v_2 - v_1)}{J}$$

However,

$$pv_2 = RT_2 \quad \text{and} \quad pv_1 = RT_1$$

Then

$$c_p(T_2 - T_1) = c_v(T_2 - T_1) + \frac{R}{J}(T_2 - T_1)$$

and

$$c_p = c_v + \frac{R}{J} \tag{3.10}$$

It should be noted that, if sufficient heat is transferred to this closed system at constant pressure to change the temperature of 1 lb$_m$ of the system 1 degree, then $c_p$ is the quantity of heat transferred, $c_v$ is the amount of this energy that is stored in the system as an increase in internal energy, and $R$ is the ft-lb$_f$ of work done by expansion of the system.

### 3.5 Enthalpy of a Perfect Gas

Enthalpy, expressed in Btu per lb$_m$, has been defined as follows:

$$h = u + \frac{pv}{J} \tag{2.12}$$

Then, for a perfect gas, since $pv = RT$,

$$h = u + \frac{RT}{J}$$

Since $h$, $u$, and $T$ are properties and therefore point functions, this equation may be differentiated as follows:

$$dh = du + \frac{R}{J} \, dT$$

For a perfect gas, $du = c_v \, dT$;

therefore $\qquad dh = c_v \, dT + \dfrac{R}{J} dT = \left( c_v + \dfrac{R}{J} \right) dT$

But $\qquad\qquad\qquad\qquad c_v + \dfrac{R}{J} = c_p \qquad\qquad\qquad\qquad$ (3.10)

Therefore $\qquad\qquad\qquad\quad dh = c_p \, dT \qquad\qquad\qquad\qquad$ (3.11)

Assuming that the specific heat is constant, then

$$h_2 - h_1 = c_p(T_2 - T_1) \qquad\qquad (3.12)$$

Since enthalpy is a point function and the change in enthalpy is therefore independent of the path, Equation 3.12 is a general equation that is applicable to *all processes* involving *perfect gases.*

### 3.6   Application of the Energy Balance for the Frictionless Non-flow Progress to Perfect Gases

The basic relations that are involved may be summarized as follows:

The energy balance for the nonflow process: $\quad {}_1Q_2 = U_2 - U_1 + \dfrac{{}_1W_2}{J}$

$\qquad\qquad\qquad\qquad\qquad\qquad\qquad\qquad\qquad\qquad\qquad\qquad\qquad$ (2.9)

For a frictionless nonflow process: $\quad W = \displaystyle\int_{V_1}^{V_2} p \, dV \qquad$ (2.4)

For perfect gases: $\quad pV = mRT \qquad\qquad\qquad\qquad\qquad$ (3.3)

For a closed system for perfect gases: $\quad \dfrac{p_1 V_1}{T_1} = \dfrac{p_2 V_2}{T_2} \qquad$ (3.4)

For perfect gases: $\quad c_p = c_v + \dfrac{R}{J} \qquad\qquad\qquad\qquad$ (3.10)

For perfect gases: $\quad U_2 - U_1 = mc_v(T_2 - T_1) \qquad\qquad$ (3.9)

Since the internal energy of a perfect gas is a function of temperature only and is unaffected by volume changes, Equation 3.9 is a general equation that is applicable to *all* types of processes involving perfect gases.   Consequently, the internal-energy change in the energy balance, Equation 2.9, can always be evaluated by means of Equation 3.9 for any processes considered in this article.   Heat and work are path functions and depend on the nature of the process.

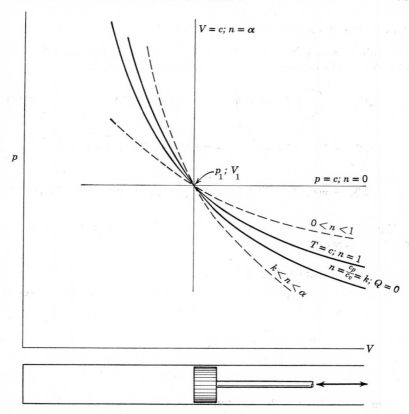

**Fig. 3.3.** Expansion or compression of a perfect gas.

After the internal-energy change has been evaluated, either the heat or work transferred must be calculated for the particular process; then the energy balance, Equation 2.9, may be used to evaluate the other energy term.

Let Fig. 3.3 represent a cylinder closed by a gas-tight frictionless piston behind which there is a constant mass of a perfect gas initially at $p_1$ and $V_1$ which may expand or be compressed without friction. Let it be assumed that the relationship between pressure and volume during any process may be expressed by the equation $pV^n = C$, where $n$ has some value between 0 and $\propto$. Several special cases will be considered.

### Case I. The Constant-Pressure Process; $n = 0$

It will be noted from Equation 3.4 that the volume varies directly with the absolute temperature when the pressure is constant.

$$_1W_2 = \int_{V_1}^{V_2} p \, dV = p \int_{V_1}^{V_2} dV = p(V_2 - V_1) \qquad (3.13)$$

$$_1Q_2 = U_2 - U_1 + \frac{_1W_2}{J} = mc_v(T_2 - T_1) + \frac{p(V_2 - V_1)}{J}$$

$$_1Q_2 = mc_v(T_2 - T_1) + \frac{mR}{J}(T_2 - T_1) = mc_p(T_2 - T_1) \quad (3.14)$$

Equation 3.14 may be written directly from the definition of the constant-pressure specific heat.

### Case II. The Constant-Volume Process; $n = \infty$

From Equation 3.4 it will be noted that the pressure varies directly with the absolute temperature for a constant-volume process.

$$_1W_2 = \int_{V_1}^{V_2} p \, dV = 0 \qquad \text{since } dV = 0$$

$$_1Q_2 = U_2 - U_1 + 0 = mc_v(T_2 - T_1) \qquad (3.15)$$

Equation 3.15 may be written directly from the definition of the constant-volume specific heat.

### Case III. The Constant-Temperature Process; $n = 1$

From Equation 3.4, $T_1 = T_2$ and $p_1 V_1 = p_2 V_2 = pV$.

$$_1W_2 = \int_{V_1}^{V_2} p \, dV = \int_{V_1}^{V_2} \frac{p_1 V_1}{V} \, dV = p_1 V_1 \int_{V_1}^{V_2} \frac{dV}{V} = p_1 V_1 \ln \frac{V_2}{V_1}$$

$$(3.16)$$

$$_1Q_2 = U_2 - U_1 + \frac{p_1 V_1 \ln \dfrac{V_2}{V_1}}{J} = 0 + \frac{p_1 V_1 \ln \dfrac{V_2}{V_1}}{J} = \frac{mRT \ln \dfrac{V_2}{V_1}}{J}$$

$$(3.17)$$

## Case IV. The Frictionless Adiabatic Process; $n = \dfrac{c_p}{c_v} = k$

An adiabatic process by definition is one in which there is no heat transfer. The energy balance may therefore be written in differential form as follows:

$$dU + \frac{\delta W}{J} = 0$$

Then

$$mc_v \, dT + \frac{p \, dV}{J} = 0 \tag{3.18}$$

Since, in the adiabatic process, $p$, $V$, and $T$ are all changing, the characteristic equation $pV = mRT$ may be differentiated to give

$$p \, dV + V \, dp = mR \, dT \tag{3.19}$$

Combining Equations 3.18 and 3.19 and multiplying by $R$ gives the following result:

$$mc_v R \, \frac{p \, dV + V \, dp}{mR} + \frac{Rp \, dV}{J} = 0$$

$$c_v(p \, dV + V \, dp) + \frac{R}{J} p \, dV = 0$$

$$\left(cv + \frac{R}{J}\right) p \, dV + c_v V \, dp = 0$$

$$c_p p \, dV + c_v V \, dp = 0$$

$$\frac{c_p}{c_v} p \, dV + V \, dp = 0$$

The ratio of the specific heats occurs in so many thermodynamic relations that it is convenient to introduce a new term, $k$, defined as follows:

$$k = \frac{c_p}{c_v}$$

Then

$$kp \, dV + V \, dp = 0$$

$$k \frac{dV}{V} + \frac{dp}{p} = 0$$

Integrating gives $\qquad k \ln V + \ln p = C_1$

or $\qquad\qquad\qquad\qquad pV^k = C \qquad\qquad\qquad (3.20)$

Equations 3.20 and 3.4 may be combined to produce the following relationship between the variables $p$, $V$, and $T$ for an adiabatic process:

$$\frac{T_2}{T_1} = \left(\frac{p_2}{p_1}\right)^{(k-1)/k} = \left(\frac{V_1}{V_2}\right)^{k-1} \qquad (3.21)$$

The work done during a frictionless adiabatic process may be evaluated by either of two methods as follows:

(a) From the energy balance, since $_1Q_2 = 0$:

$$\frac{_1W_2}{J} = -(U_2 - U_1) = mc_v(T_1 - T_2)$$

(b) From the work equation $W = \int_{V_1}^{V_2} p\, dV$:

The relation between the initial pressure and volume, $p_1$ and $V_1$, and the pressure and volume at any subsequent point during the process is as follows:

$$p_1 V_1{}^k = pV^k \quad \text{or} \quad p = p_1 V_1{}^k \frac{1}{V^k}$$

Then $\qquad W = \int_{V_1}^{V_2} p\, dV = \int_{V_1}^{V_2} p_1 V_1{}^k \frac{dV}{V^k} = p_1 V_1{}^k \int_{V_1}^{V_2} \frac{dV}{V^k}$

$$W = p_1 V_1{}^k \left[ \frac{V_2{}^{1-k} - V_1{}^{1-k}}{1 - k} \right]$$

Since $\qquad\qquad\qquad p_1 V_1{}^k = p_2 V_2{}^k$

then $\qquad\qquad W = \frac{p_2 V_2 - p_1 V_1}{1 - k} = \frac{p_1 V_1 - p_2 V_2}{k - 1} \qquad (3.22)$

### Case V. The Polytropic Process

Mathematically, the four processes that have been discussed in this article are special cases of the general case in which the expansion or compression may be expressed by the equation $pV^n = C$. In this general case $p$, $V$, and $T$ are all changing, heat and work are being transferred, and there is a change in internal energy. Such a process is called a polytropic process.

The relationship between $p$, $V$, and $T$ for a polytropic process may be obtained by substituting $n$ for $k$ in Equation 3.21:

$$\frac{T_2}{T_1} = \left(\frac{p_2}{p_1}\right)^{(n-1)/n} = \left(\frac{V_1}{V_2}\right)^{n-1} \tag{3.23}$$

The expression for work done during a frictionless polytropic expression may be derived in the same manner as Equation 3.22 was derived and may be stated as follows:

$$W = \frac{p_1 V_1 - p_2 V_2}{n - 1} \tag{3.24}$$

Then, for the polytropic process, the energy balance may be written as follows:

$$_1Q_2 = mc_v(T_2 - T_1) + \frac{p_1 V_1 - p_2 V_2}{J(n - 1)} \tag{3.25}$$

### 3.7  Real Gases

When a gas is reduced in temperature or compressed until the mean distance between molecules becomes small enough so that molecular attractive forces become appreciable and changes in volume alter internal energy, the perfect-gas law no longer applies. The usefulness of the characteristic equation may be extended by the introduction of a compressibility factor $Z$ as follows:

$$pv = ZRT \tag{3.26}$$

where $Z$ is a correction factor that varies with pressure and temperature and must be determined experimentally.

In 1873, van der Waals published his equation of state for real gases in the following form:

$$\left(p + \frac{a}{v^2}\right)(v - b) = RT \tag{3.27}$$

where $a$ and $b$ are constants. Several other widely used equations have been developed in an effort to modify the perfect-gas law to represent accurately the behavior of real gases over a wide range of pressure and temperature extending to the conditions under which the gases may be liquefied.

### 3.8   Liquids and Vapors

In the steam power plant and the refrigeration cycle the working fluid is a liquid in part of the system. The liquid is evaporated in an evaporator or boiler, and the vapor is condensed in a condenser and is expanded or compressed with the transfer of energy as work. The relations between pressure, temperature, specific volume, internal energy, and enthalpy during the energy transfers which result in evaporation, condensation, and the performance of work are too complex to be represented by usable equations. Instead, the thermodynamic properties are usually presented in tabular form. Tables A.1, A.2, and A.3 of the Appendix are tables of the thermodynamic properties of water and steam. Tables A.4, A.5, and A.6 are similar tables of the thermodynamic properties of ammonia ($NH_3$), one of the common refrigerants. These tables are typical of the tables of thermodynamic properties of a considerable number of fluids that may be used for refrigeration, power generation, or heat transfer. The student who understands the application of Tables A.1 to A.6 inclusive should have no difficulty in using the tables of properties of any other fluid for which similar data are available.

### 3.9   The Mechanism of Vapor Generation

Figure 3.4 illustrates a steel vessel partially filled with water and provided with a pressure gage, a thermometer well and thermometer, an adjustable discharge valve, and a source of heat. Let the discharge valve be wide open to the atmosphere which is assumed to be at standard atmospheric pressure or 14.696 psia, and let heat be transferred to the vessel. It will be observed that the thermometer reading will increase to 212 F, at which point it will remain constant while steam escapes to the atmosphere through the discharge valve. If the rate of heat transfer is increased or decreased, the rate of discharge of steam will be changed correspondingly but the temperature will remain constant at 212 F if the pressure in the vessel remains constant and equal to standard atmospheric pressure. If the walls of the vessel are of glass, it will be observed that the steam bubbles which are formed on the heated bottom of the vessel rise through the water, break at the liquid surface, and discharge their steam into the space above the surface. The steam in the space above the liquid surface is called *saturated steam*. If the source of heat is removed

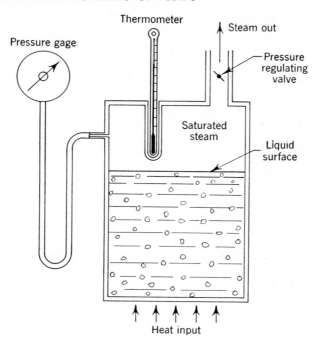

**Fig. 3.4.** Steam generation in a simple steam boiler.

momentarily so that all the steam bubbles rise to the surface of the water, the bubble-free water will be *saturated water,* and it will be found that the *saturated water* is at the same temperature as the *saturated steam.* This temperature is known as the *saturation temperature.*

If heat is transferred to the system and the discharge valve is partially closed so as to restrict the escape of saturated steam to the atmosphere, the pressure will rise, and the temperature of the saturated steam will increase also. It will be observed that at each pressure there is only one temperature of the saturated steam. This saturation temperature is plotted against steam pressure in Fig. 3.5. The relation between saturation temperature and pressure for a number of refrigerants is shown in Fig. 3.6. The principal difference in the curves for the steam and the refrigerants is in the lower saturation temperature of the refrigerants for a given pressure.

The thermal properties of saturated water and saturated steam

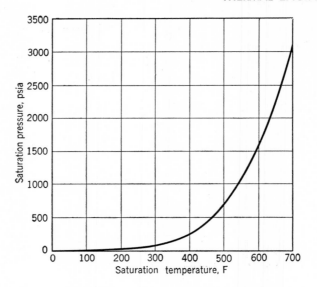

**Fig. 3.5.** Saturation temperature and pressure of steam.

have been measured accurately, and these properties are arranged
in brief tabular form in Table A.1 and Table A.2 of the Appendix
since they cannot be represented by simple equations. In Table A.1
the independent variable is the temperature of saturated steam; in
Table A.2 the independent variable is the absolute pressure of saturated steam. Identical results may be obtained by the use of Tables
A.1 and A.2, and the choice of the table will depend upon whether the
temperature or the pressure of saturated steam is specified.

The following data were obtained from these tables and should
be verified by the student for practice in using the tables:

| FROM TABLE A.1 | | FROM TABLE A.2 | |
|---|---|---|---|
| Temperature, F | Pressure, psia | Pressure, psia | Temperature, F |
| 170 | 5.992 | 6.0 | 170.06 |
| 212 | 14.696 | 14.696 | 212.0 |
| 250 | 29.825 | 30.0 | 250.3 |
| 280 | 49.203 | 50 | 281.01 |
| 320 | 89.66 | 90 | 320.27 |
| 400 | 247.31 | 250 | 400.95 |
| 705.4 | 3206.2 | 3206.2 | 705.4 |

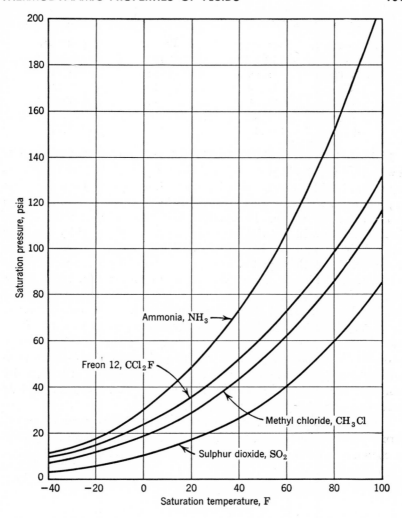

**Fig. 3.6.** Saturation pressure and temperature of some common refrigerants.

Tables A.4 and A.5 are tables of the properties of saturated liquid ammonia and saturated ammonia vapor. They are used in exactly the same way as the steam tables and differ from the steam tables only in the numerical values that appear in the tables. The following data were obtained from the ammonia tables and should be verified by the student:

| FROM TABLE A.4 | | FROM TABLE A.5 | |
| --- | --- | --- | --- |
| Temperature, F | Pressure, psia | Pressure, psia | Temperature, F |
| −60 | 5.55 | 5 | −63.11 |
| −20 | 18.30 | 20 | −16.64 |
| +10 | 38.51 | 40 | +11.66 |
| +80 | 153.0 | 140 | +74.79 |
| +120 | 286.4 | 260 | +113.42 |

If, in the apparatus shown in Fig. 3.4, the rate of heat transfer is low so that the liquid surface is only slightly disturbed by the bursting steam bubbles which are rising through the water, the steam that is produced will be free of entrained droplets of water and is known as *dry saturated steam*. However, at high rates of heat transfer, the liquid surface will be violently disturbed by the rapidly bursting steam bubbles, and droplets of water will be entrained with the steam and carried out through the discharge valve. Such steam is called *wet steam*. This wet steam will be at the saturation temperature corresponding to the steam pressure, because the saturated water and saturated steam are at the same temperature. The *quality* of wet steam may be defined as the mass of dry steam present in 1 $lb_m$ of the mixture of dry saturated steam and entrained water and is designated by the symbol $x$. Thus, if $x = 98$ per cent, the wet-steam mixture delivered from the boiler in Fig. 3.4 is composed of 2 parts by mass of saturated water, usually in the form of a fine mist, and 98 parts by mass of dry saturated steam.

Let Fig. 3.7 represent a simple steam boiler in the form of an externally heated horizontal cylindrical drum to which water is supplied at such a rate as to maintain a constant water level at about the mid-

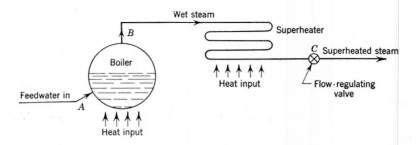

**Fig. 3.7.** Generation of superheated steam.

point of the drum. As heat is transferred to the system, the water is evaporated into steam. The steam so produced will normally be wet steam due to the entrainment of spray from the bursting steam bubbles. If the flow-regulating valve at point $C$ is adjusted to maintain a constant steam pressure, it will be found that the boiler is delivering wet steam at a constant temperature at the boiler outlet, point $B$. Variations in the rate of heat transfer to the boiler will change the rate of evaporation, but, no matter how rapidly the water is evaporated, the steam temperature will remain constant if the pressure is held at a uniform value by adjustment of the flow-regulating valve. If the wet steam is allowed to flow through an externally heated coil, called a *superheater* (Fig. 3.7), it will be found by suitable temperature-measuring devices that the temperature of the wet steam will remain constant (if pressure drop in the superheater is neglected) until all the entrained moisture has been evaporated and dry steam has been produced. Further transfer of heat in the superheater will cause the temperature of the steam to rise above the saturation temperature, and it is then said to be *superheated steam*. The temperature of the steam at the superheater outlet may be several hundred degrees above the saturation temperature of the steam if sufficient heat is supplied in the superheater. The difference between the temperature of the superheated steam and the saturation temperature at the pressure of the steam is called the *degrees of superheat*. It should be noted that superheating does *not* increase the pressure of the steam. In fact, in an actual case, it results in a decrease in pressure due to the frictional resistance to the flow of steam through the superheater coil. It should be noted further that the transfer of heat to saturated water or wet steam at constant pressure results in evaporation at the saturation temperature if any water is present, and that superheating will occur only when *heat is transferred to dry steam in the absence of water*. The thermal properties of superheated steam are given in brief form in Table A.3 of the Appendix. Similarly, the thermodynamic properties of superheated ammonia are given in Table A.6.

If heat is transferred from superheated vapor at constant pressure, the temperature will decrease to the saturation temperature at which point condensation will begin. No further reduction in temperature can occur at constant pressure until the vapor has been condensed to saturated liquid. Thus condensation takes place at the same temperature as evaporation occurs for any given pressure and is, in fact, the reverse process of evaporation.

### 3.10   The Specific Volume of Liquid and Vapor

The specific volume of a fluid is the volume occupied by a unit mass of the fluid and, in the English engineering system of units, is expressed in cubic feet per pound of mass and is given the symbol $v$. The following symbols will be used for the specific volume of liquid and vapor:

$v_f$ = specific volume of saturated liquid, cu ft per lb
$v_g$ = specific volume of dry saturated vapor, cu ft per lb
$v_{fg} = v_g - v_f$ = increase in specific volume when saturated liquid is converted into dry saturated vapor
$v_x$ = specific volume of wet vapor, cu ft per lb
$v_s$ = specific volume of superheated vapor, cu ft per lb

The thermodynamic properties are tabulated in Tables A.1 to A.6 inclusive for a mass of one pound.

In Table A.1 and Table A.2 of the Appendix, it may be noted that the specific volume of saturated water and dry saturated steam at 212 F are, respectively, as follows: $v_f = 0.01672$ cu ft per lb, and $v_g = 26.80$ cu ft per lb.  Similarly for any other temperature or pressure the specific volume of saturated water and dry saturated steam may be obtained from Table A.1 or Table A.2.

Likewise the specific volume of saturated liquid ammonia and dry saturated ammonia vapor may be obtained from Tables A.4 or A.5.

The specific volume of *wet vapor* may be computed by considering that 1.0 lb of wet vapor having a quality $x$ is composed of $x$ lb of dry saturated vapor and $(1 - x)$ lb of saturated liquid if $x$ is expressed as a decimal.

Then,
$$v_x = xv_g + (1 - x)v_f \tag{3.28}$$

**Example 1.**  Compute the specific volume of wet steam at 212 F if the quality is 98 per cent.

*Solution:* One pound of wet steam at 212 F and 98 per cent quality may be considered as a mixture of 0.98 lb of dry saturated steam and 0.02 lb of saturated water.   Then

$$v_x = 0.98(v_g) + 0.02(v_f)$$

or
$$v_x = 0.98(26.80) + 0.02(0.01672)$$

or
$$v_x = 26.284 + 0.000334$$

The volume of the saturated water in the wet-steam mixture is negligible in this problem.   For most engineering problems involving slide-rule computations and vapor of more than 95 per cent quality at ordinary pressures, the volume of the saturated liquid may be neglected, and the specific volume of wet vapor may be computed as follows:

$$v_x = xv_g \text{ (approximately)} \tag{3.29}$$

The specific volume of *superheated steam* may be determined from Table A.3 when the pressure and temperature are known. Thus, at a pressure of 100 psia, the specific volume of superheated steam, $v_s$, is found from Table A.3 to be 4.937 at 400 F, 5.589 at 500 F, 6.835 at 700 F, and 8.656 at 1000 F. Values for pressures and temperatures not given in the abridged Table A.3 may be found by interpolation or by reference to *Thermodynamic Properties of Steam* by Keenan and Keyes, published by John Wiley & Sons.

Likewise, the specific volume of superheated ammonia may be obtained from Table A.6 for a known pressure and temperature.

### 3.11  Enthalpy

In Articles 1.18 and 2.10, *enthalpy* was introduced as a property of the substance and was defined as follows:

$$h = u + \frac{pv}{J} \tag{3.30}$$

The following symbols will be used in designating the specific enthalpy or enthalpy per pound of fluid:

$h_f$ = enthalpy of saturated liquid, Btu per lb
$h_g$ = enthalpy of dry saturated vapor, Btu per lb
$h_{fg} = h_g - h_f$ = increase in enthalpy when saturated liquid is converted into dry saturated vapor = latent heat of evaporation, Btu per lb
$h_x$ = enthalpy of wet vapor at quality $x$, Btu per lb
$h_s$ = enthalpy of superheated vapor, Btu per lb
$h_c$ = enthalpy of compressed liquid, Btu per lb

Since the engineer is concerned with the change in enthalpy when a fluid flows through some apparatus, and since the total enthalpy of a substance could be evaluated only by reducing it to absolute zero, it is customary to select some convenient reference point as a zero of enthalpy and to compute enthalpies with respect to this reference point in much the same way as Fahrenheit temperatures may be measured above and below a convenient zero point. For water and steam, it is customary to select saturated water at 32 F as the zero or reference point and to compute the enthalpy of water and steam in Btu per pound from this reference state. For the common refrigerants, the zero of enthalpy is taken as saturated liquid at $-40$ F.

In Table A.1 for each of the temperatures listed and in Table A.2 for each of the pressures given, the enthalpy of saturated water, $h_f$; the enthalpy of dry saturated steam, $h_g$; and the latent heat of evaporation, $h_{fg}$, are given in Btu per pound of fluid. Thus, in Table A.1, at 212 F, $h_f = 180.07$ Btu per lb, $h_g = 1150.4$ Btu per lb, and $h_{fg} = 970.3$ Btu per lb. The same data will be found in Table A.2 at a pressure of 14.696 psia. Similarly, enthalpy values for saturated liquid ammonia and dry saturated ammonia vapor may be obtained from Tables A.4 and A.5.

The enthalpy of superheated steam, $h_s$, can be determined from Table A.3 when the pressure and temperature are known. Thus, at 100 psia, the enthalpy of superheated steam is 1279.1 Btu per lb at 500 F, 1428.9 Btu per lb at 800 F, and 1530.8 Btu per lb at 1000 F. Likewise, the enthalpy of superheated ammonia may be obtained from Table A.6 when the pressure and temperature are known.

### 3.12 Enthalpy and Quality of a Wet Vapor

A vapor having a quality, $x$, may be considered as a vapor in which each pound is composed of $x$ lb of dry saturated vapor and $(1 - x)$ lb of saturated liquid where $x$ is expressed as a decimal. Then

$$h_x = [xh_g] + [(1 - x)h_f] \tag{3.31}$$

or

$$h_x = xh_g + (1 - x)h_f \tag{3.32}$$

**Example 2.** Compute the enthalpy of 1 lb of wet steam at 100 psia and 97 per cent quality.

*Solution:* The following data are obtained from Table A.1 for 100 psia:

$h_f = 298.40$ Btu per $lb_m$; $\quad h_{fg} = 888.8$ Btu per $lb_m$; $\quad h_g = 1187.2$ Btu per $lb_m$

Considering that 1.0 lb of wet steam is a mechanical mixture of 0.97 lb of dry saturated steam and 0.03 lb of saturated water, then

$$h_x = xh_g + (1 - x)h_f = 0.97(1187.2) + 0.03(298.4) = 1160.5 \text{ Btu per lb}$$

Figure 3.8 is a plot on pressure-enthalpy coordinates of the values of $h_f$ and $h_g$ for saturated water and dry saturated steam. The curve $h_f$ may be considered to represent the enthalpy of steam having a quality

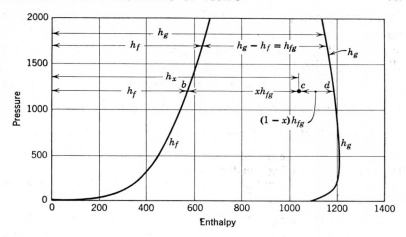

**Fig. 3.8.** Enthalpy of wet steam.

of zero per cent, while the curve $h_g$ represents the enthalpy of steam having a quality of 100 per cent. The horizontal distance between the curves of $h_f$ and $h_g$ at any pressure is the latent heat, $h_g - h_f$, or $h_{fg}$. Therefore, if steam has a quality of 50 per cent at a given pressure, the point representing the state of the steam is located on the appropriate constant-pressure line exactly half-way between the curves of $h_f$ and $h_g$. Let it be assumed that steam has a quality of 80 per cent at a particular pressure. Then, on Fig. 3.8 , the point $c$ would represent the state of the steam if it is located on the appropriate value of the ordinate and at a distance $bc$ which is 80 per cent of $bd$ from the curve of $h_f$ or a distance $cd$ which is 20 per cent of $bd$ from the curve of $h_g$. Since the enthalpy of saturated water (zero per cent quality) is $h_f$, then, from Fig. 3.8, it will be noted that

$$h_x = h_f + x(h_{fg}) \tag{3.33}$$

Equation 3.33 can be derived from Equation 3.32 as follows

$$h_x = xh_g + (1 - x)h_f = xh_g + h_f - xh_f$$

$$= h_f + x(h_g - h_f) = h_f + x(h_{fg})$$

**Example 3.** For the conditions specified in Example 2, calculate the enthalpy of wet steam, using Equation 3.33.

*Solution:* Using the steam-table data that have been tabulated for 100 psia in the solution to Example 2,

$$h_x = h_f + x(h_{fg}) = 298.4 + 0.97(888.8) = 1160.5 \text{ Btu per lb}$$

From Fig. 3.8, it will be noted that we may start with dry saturated steam and assume that part of the latent heat of evaporation is removed, resulting in a reduction in quality, and thus arrive at the following equation to describe the process:

$$h_x = h_g - (1 - x)h_{fg} \qquad (3.34)$$

Equation 3.34 can be derived from Equation 3.33 as follows:

$$h_x = h_f + x(h_{fg}) = (h_g - h_{fg}) + x(h_{fg}) = h_g - (1 - x)h_{fg}$$

**Example 4.** For the conditions specified in Example 2, calculate the enthalpy of wet steam, using Equation 3.34.

*Solution:* Using the steam-table data that have been tabulated in the solution to Example 2 for 100 psia,

$$h_x = h_g - (1 - x)h_{fg} = 1187.2 - (0.03)(888.8) = 1160.5 \text{ Btu per lb}$$

It should be noted that the same numerical result is obtained by Equations 3.32, 3.33, and 3.34. However, the quality of steam leaving a boiler is normally in excess of 98 per cent. Under these conditions, Equation 3.34 gives the highest order of accuracy in slide-rule computations and is to be preferred.

Where the quality of the steam is relatively high as in the case of steam generated in a modern boiler, a *throttling calorimeter* may be used to determine the enthalpy and quality. Before proceeding further, the reader should review thoroughly Article 2.16 dealing with the throttling process. Since no work is done in the process and it is assumed that the apparatus is so constructed and insulated that there is no significant change in kinetic energy of the fluid and no heat transfer, then

$$h_1 = h_2 \qquad (3.35)$$

The following data pertaining to the enthalpy of dry saturated steam at various pressures may be obtained from Table A.2:

| Steam Pressure, psia | Enthalpy, $h_g$ |
|---|---|
| 14.7 | 1150.4 |
| 100 | 1187.2 |
| 200 | 1198.4 |
| 400 | 1204.5 |

If dry saturated steam at ordinary boiler pressures is throttled to atmospheric pressure, the steam will become superheated since its enthalpy is greater than the enthalpy of dry saturated steam at atmospheric pressure. Also, if steam having only 2 or 3 per cent of

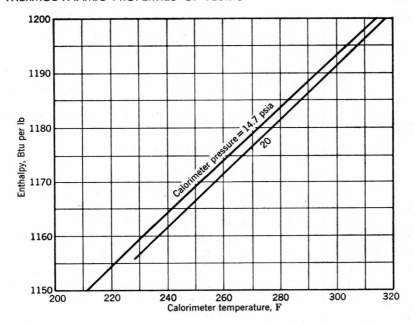

**Fig. 3.9.** Enthalpy of superheated steam in the normal range of the throttling calorimeter.

Plotted by permission of the authors from *Thermodynamic Properties of Steam* by Keenan and Keyes; John Wiley & Sons, publishers.

moisture, that is, a quality of 98 or 97 per cent, is throttled from ordinary boiler pressures to atmospheric pressure, its enthalpy is high enough to produce superheated steam at the lower pressure. If the steam after throttling is superheated, the enthalpy, $h_2$, can be found by measuring its pressure and temperature in the calorimeter and referring to the superheated-steam tables for the enthalpy at the measured pressure and temperature. Figure 3.9 is a plot of the enthalpy of superheated steam at temperatures and pressures normally found in throttling calorimeters and may be used in place of the abridged steam tables. Once the enthalpy of the steam at the lower pressure and temperature has been determined, the initial quality of the steam may be computed as follows:

Since
$$h_1 = h_2 \tag{3.35}$$

then
$$h_{f1} + x_1 h_{fg1} = h_2 \quad \text{and} \quad x_1 = \frac{h_2 - h_{f1}}{h_{fg1}} \tag{3.36}$$

where $h_2$ = enthalpy of superheated steam in the calorimeter

$h_{f1}$ = enthalpy of saturated water at the initial pressure $p_1$

$h_{fg1}$ = enthalpy of evaporation at the initial pressure $p_1$

$x_1$ = quality of the wet steam at the initial pressure $p_1$

**Example 5.** Determine the quality of steam at 200 psia if the steam after throttling to 20 psia is found to be at a temperature of 260 F.

*Solution:* From Fig. 3.9, $h_2$ at 20 psia and 260 F = 1172 = $h_1$. Then, from Table A.2 at 200 psia,

$$h_{f1} + x_1 h_{fg1} = 355.36 + x_1(843.0) = h_2 = 1172$$

$$x_1 = 96.9\%$$

Figure 3.10 illustrates a simple calorimeter which may be made from standard pipe fittings. It must be well insulated to reduce the transfer of heat to a minimum and must be suitably supported to prevent strain on the sampling tube. The sample of steam should be obtained from a vertical pipe at a point some distance from sharp turns or valves which would disturb the flow pattern in the pipe and cause an unequal distribution of the droplets of moisture across the pipe. A sampling tube consisting of a short length of pipe having holes drilled in its wall and plugged at the far end should be screwed into the pipe

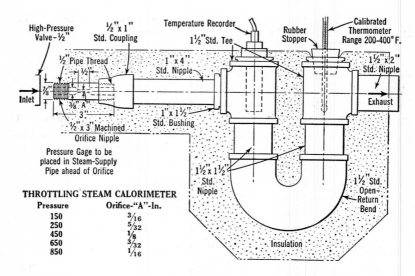

**Fig. 3.10.** Throttling calorimeter constructed from standard pipe fittings.

line and connected to the calorimeter through a shut-off valve.  A pressure gage must also be connected to the pipe line near the sampling tube for the purpose of measuring the steam pressure at the point where the quality and enthalpy are desired.  The steam is throttled through the orifice nipple to atmospheric pressure after which it flows through a U-shaped path to the thermometer and then to the atmosphere.  Since the steam is at atmospheric pressure at the thermometer, it is necessary only to read this thermometer and, from the upper curve of Fig. 3.9, to determine the enthalpy of the steam at the temperature indicated by the thermometer.  The temperature-recorder bulb shown in Fig. 3.10 may be omitted by plugging the opening in the tee with a standard pipe plug.

If the quality of the high-pressure steam is too low to produce superheated steam in the calorimeter, the throttling calorimeter cannot be used, and some other form of calorimeter must be employed. However, a properly operated steam boiler should produce steam of quality high enough to permit the use of the throttling calorimeter.

### 3.13  Enthalpy of a Compressed Liquid

Let it be assumed that water is at a temperature of 200 F and a pressure of 2000 psia.  The enthalpy of this water is required.  From Table A.1, it will be found that at 200 F saturated water has a saturation pressure of 11.526 psia and an enthalpy of 167.99 Btu per lb. However, we are concerned with water under a pressure of 2000 psia. From Table A.2 at 2000 psia, it will be found that the saturation temperature is 635.82 F and the enthalpy is 671.7 Btu per lb.  It is apparent that the enthalpy of water at 200 F and 2000 psia cannot be determined directly from either Table A.1 or Table A.2.

Water at 200 F and 2000 psia is called a *compressed liquid* since it is under a pressure in excess of the saturation pressure for the specified temperature.  It is sometimes called a *subcooled liquid* since the temperature is less than the saturation temperature corresponding to the specified pressure.

Considering the enthalpy equation, $h = u + pv/J$, the differences between saturated and compressed water at a given temperature are as follows:

1. The pressure term, $p$, is greater for compressed water than for saturated water.

2. The specific volume, $v$, is somewhat less for compressed water than for saturated water. The higher the temperature, the greater is the compressibility of water.

3. As water is compressed at a constant temperature, the resulting decrease in volume brings the molecules closer together and reduces their molecular potential energy or internal energy.

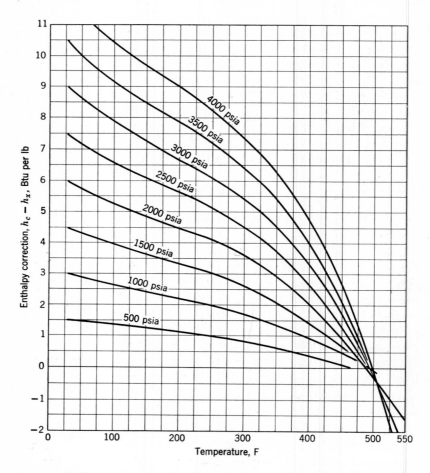

**Fig. 3.11.** Enthalpy correction for determination of the enthalpy of compressed water.

Plotted by permission of the authors from *Thermodynamic Properties of Steam* by Keenan and Keyes; John Wiley & Sons, publishers.

As saturated water is compressed at constant temperature to higher and higher pressures, the net effect is to increase the $pv$ product and decrease the internal energy. The net change in enthalpy from the enthalpy of saturated water during a constant-temperature compression is illustrated in Fig. 3.11.

In order to determine the enthalpy of compressed water, proceed as follows:

1. From Table A.1, determine the enthalpy of saturated water at the specified *temperature*, $h_f$.

2. From Fig. 3.11 (or Table 4 of *Thermodynamic Properties of Steam* by Keenan and Keyes), determine the increase in enthalpy above the enthalpy of the saturated water at the specified temperature, $h_c - h_f$.

3. Add the enthalpy correction from Fig. 3.11 to the enthalpy of saturated water at the specified temperature.

### 3.14   Internal Energy of a Vapor

Since, by definition, $h = u + pv/J$, the internal energy of a vapor may be computed by means of the following equation:

$$u = h - \frac{pv}{J} \tag{3.37}$$

For superheated vapor, the enthalpy and specific volume are obtained directly from the appropriate tables for the specified pressure and temperature. For a wet vapor, the enthalpy and specific volume must be calculated from the specified quality and the tabular values of enthalpy and specific volume of saturated liquid and saturated vapor at the specified pressure.

Tables A.2 and A.4 contain data on the internal energy of saturated liquid, $u_f$, and the internal energy of dry saturated vapor, $u_g$, from which the internal energy of wet vapor of quality $x$ may be calculated from the following equation, which is similar in form to Equation 3.32:

$$u_x = x u_g + (1 - x) u_f \tag{3.38}$$

### 3.15   Entropy

In Article 2.12 concerning the application of the energy balance for the steady-flow process to prime movers, it was pointed out that the ideal expansion in a prime mover occurs without heat transfer to or from the surroundings (an adiabatic process) and without friction. It was stated that for such a process a property known an entropy remains constant. A thorough discussion of entropy and its significance in the general field of thermodynamics is beyond the scope of this book. However, entropy will be used in connection with the design of turbine nozzles and the calculation of efficiencies for certain power-plant cycles. Therefore, for the limited use in this book, *entropy will be defined as a property that remains constant during a frictionless adiabatic process.*

In Tables A.1 and A.2 of the properties of saturated steam and in Tables A.4 and A.5 of the properties of saturated ammonia, values will be found for the entropy of saturated liquid, $s_f$; saturated vapor, $s_g$; and the change in entropy during evaporation, $s_{fg}$. Saturated water at 32 F and saturated ammonia at $-40$ F are assumed to have zero entropy as well as zero enthalpy. The entropy of wet vapor may be computed by means of equations having the same form as Equations 3.32, 3.33, or 3.34 by substituting entropy values for corresponding enthalpy values.

The entropies of superheated steam and superheated ammonia may be found in Tables A.3 and A.6, respectively, for the specified pressure and temperature.

The use of the steam tables to calculate the change in some of the thermodynamic properties of steam during an adiabatic expansion will be illustrated by Example 6.

**Example 6.** Steam initially at 1000 psia and 1000 F expands in a frictionless adiabatic (constant-entropy) process to a final pressure of 5.0 psia. Calculate the change in (*a*) enthalpy, (*b*) specific volume, and (*c*) internal energy.

*Solution:* From Table A.3 for 1000 psia and 1000 F, the initial conditions are found to be as follows: $h_1 = 1502.2$ Btu per $lb_m$; $v_1 = 0.7503$ cu ft per $lb_m$; $s_1 = 1.6525$. The initial internal energy may be calculated as follows:

$$u_1 = h_1 - \frac{p_1 v_1}{J} = 1502.2 - \frac{1000 \times 144 \times 0.7503}{778} = 1361 \text{ Btu per } lb_m$$

From Table A.2 for the final pressure of 5.0 psia, the entropy of dry saturated steam is found to be 1.8441.

For the conditions specified in the problem,

$$s_1 = s_2 = 1.6525$$

The final entropy is 1.6525, whereas for the final pressure $s_g = 1.8441$. Consequently, at the final conditions, the *steam must be wet.* The basic problem therefore is to determine the quality of the steam after expansion to 5 psia.

After this has been done, the final enthalpy, specific volume, and internal energy may be computed.

From Table A.2 at 5 psia, $h_f = 130.13$; $h_g = 1131.1$; $s_f = 0.2347$; $s_g = 1.8441$; $u_f = 130.12$; $u_g = 1063.1$; $v_f = 0.01640$; $v_g = 73.52$.

For the conditions of the problem, $s_1 = s_2 = 1.6525 = x_2 s_{g2} + (1 - x_2) s_{f1}$
$= x_2(1.8441) + (1 - x_2)(0.2347)$.

$$x_2 = 0.88 = 88\%$$

At 5 psia and $x_2 = 88\%$ quality,

$$h_2 = x_2 h_{g2} + (1 - x_2) h_{f2} = 0.88(1131.1) + 0.12(130.13)$$
$$= 1151 \text{ Btu per lb}_m$$

At 5 psia and $x_2 = 88\%$ quality,

$$v_2 = x_2 v_{g2} + (1 - x_2) v_{f2} = 0.88(73.52) + 0.12(0.0164)$$
$$= 64.7 \text{ cu ft per lb}_m$$

At 5 psia and $x_2 = 88\%$ quality,

$$u_2 = x_2 u_{g2} + (1 - x_2) u_{f2} = 0.88(1063.1) + 0.12(130.12)$$
$$= 1092 \text{ Btu per lb}_m$$

Then:
$$\text{The change in enthalpy} = h_2 - h_1 = 1151 - 1502$$
$$= -351 \text{ Btu per lb}_m \text{ (decrease)}$$

$$\text{The change in volume} = v_2 - v_1 = 64.7 - 0.75$$
$$= 64 \text{ cu ft per lb}_m \text{ (increase)}$$

$$\text{The change in internal energy} = u_2 - u_1 = 1092 - 1361$$
$$= -269 \text{ Btu per lb}_m \text{ (decrease)}$$

## 3.16 The Pressure-Volume Diagram for Steam

The relation between pressure and the specific volume of steam is illustrated in Fig. 3.12, which is drawn to scale. Other fluids such as ammonia have similar characteristics. The specific volumes of saturated water, $v_f$, and saturated steam, $v_g$, converge at what is known as the *critical point*.

For steam, the critical point has a pressure of 3206.2 psia and a temperature of 705.4 F. At the critical point, the specific volumes of saturated liquid and saturated vapor are equal. At any constant pressure below the critical pressure, say, 2000 psia, saturated water will evaporate at a constant temperature (the saturation temperature) during which process the specific volume increases from $v_f$ to $v_g$. If dry saturated steam at a constant pressure of, say, 2000 psia is superheated, the volume will increase with increase in temperature

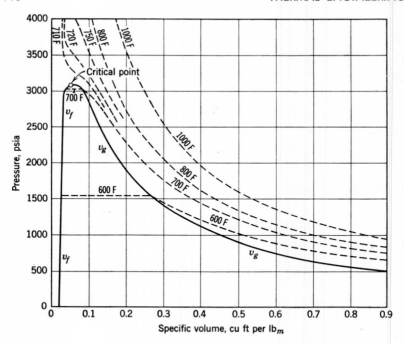

**Fig. 3.12.** Pressure-volume diagram for steam.

as indicated by the following appropriate constant-pressure line on Fig. 3.12. Lines of constant temperature are shown as dotted lines on Fig. 3.12.

At a pressure above the critical pressure, water will not boil. Assume that cold water is placed under a pressure of 3500 psia. Then it will be noted by reference to Fig. 3.12 that, as heat is transferred at this constant pressure, the volume and temperature increase but at no point does evaporation occur. Evaporation may be defined as an increase in volume at constant temperature (under a constant pressure) as a result of heat transfer. Several large steam power plants are now being constructed and operated at pressures above the critical point, thereby eliminating the boiler.

### 3.17  The Pressure-Enthalpy Diagram for Steam

The relation between pressure and enthalpy for steam is illustrated in Fig. 3.13, which is drawn to scale. The enthalpy of saturated water, $h_f$, and saturated steam, $h_g$, converge at the critical point where they are equal. Below the critical point, the horizontal distance between these curves represents the latent heat. At the critical point, the latent heat is zero. The diagram may be divided into three major areas as follows: the region of wet vapor, which is the area bounded by the curves of $h_f$ and $h_g$; the superheated-vapor region, which is the area to the right of the curve of $h_g$; and the compressed-liquid region, which is the area to the left of the curve of $h_f$. Lines of constant temperature are shown as dotted lines.

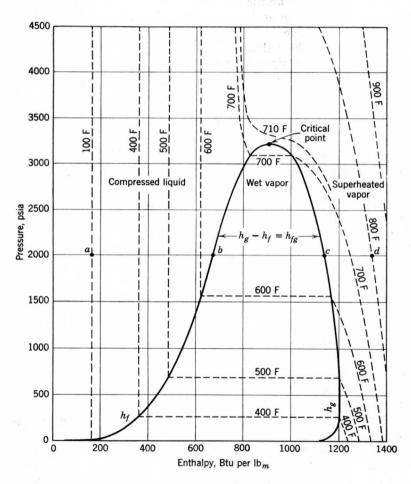

**Fig. 3.13.** Pressure-enthalpy diagram for water and steam.

Let it be assumed that water is initially at 2000 psia and 100 F. The state of the fluid is represented on Fig. 3.13 by point $a$. As heat is transferred at constant pressure, the temperature will increase until the saturation temperature has been reached at point $b$, where the enthalpy is 671.7 Btu and the temperature is 635.8 F. If the water is now evaporated to dry saturated steam at a constant pressure of 2000 psia, the state is represented by point $c$ on Fig. 3.13 and the distance $bc$ represents to scale the latent heat of evaporation, $h_{fg}$. Transfer of heat to the dry saturated vapor at constant pressure in a superheater will cause the temperature to rise. Point $d$ represents the state of the steam at a temperature of 800 F.

At a pressure above the critical pressure such as 4000 psia, reference to Fig. 3.13 will show that the temperature increases continuously with increase in enthalpy at constant pressure. Evaporation or boiling at constant temperature cannot take place above the critical pressure.

### 3.18   Summary

A perfect or ideal gas is defined as a fluid which obeys the law $pv = RT$, where the gas constant $R$ has a specific value for each gas and system of units. If all gases followed this law precisely, then $R$ would always be equal to 1544 divided by the molecular weight, or $R = 1544/M$ for the English engineering system of units. A perfect gas is so rarefied that attractive forces acting between molecules have practically vanished and changes in internal energy are independent of changes in volume. Therefore, the following equations apply:

$$du = c_v \, dT \qquad\qquad\qquad (3.6)$$

$$dh = c_p \, dT \qquad\qquad\qquad (3.11)$$

Real gases depart from the perfect-gas law as the pressure is increased and the temperature is decreased until the molecules are close enough so that changes in volume affect internal energy, that is, as they approach the conditions under which they can be liquefied. Modifications of the perfect-gas equation such as the van der Waals equation or the inclusion of a compressibility factor extend the useful range of the equation.

In steam power plants, most refrigeration systems, and many heat-transfer applications, the fluid in the system undergoes evaporation and condensation. At and near the conditions under which a change of phase occurs, the thermodynamic properties are too complex to be represented by usable equations. Instead, the data are presented in tabular form as in Tables A.1 to A.6 inclusive of the Appendix.

The *saturation temperature* is the temperature of the vapor that is produced during evaporation. For any fluid, at each pressure there is only one saturation temperature.

*Saturated liquid* is liquid at the saturation temperature.

*Saturated vapor* is moisture-free vapor at the saturation temperature.

The enthalpy, specific volume, and entropy of one pound of mass of saturated liquid and saturated vapor are tabulated in the saturation tables. Usually two sets of tables are available, one having temperature as the independent variable and the other having pressure as the independent variable.

A *wet vapor* may be considered as a mechanical mixture of $x$ parts of dry saturated vapor and $(1 - x)$ parts of entrained saturated liquid where $x$ is the quality of the wet vapor expressed in parts of dry saturated vapor per 100 parts of mixture.

A *superheated vapor* is a vapor at a temperature higher than the saturation temperature of the vapor at any given pressure. The properties of a superheated vapor are tabulated with pressure and temperature as independent variables.

A *compressed liquid* is a liquid that is under a pressure greater than the saturation pressure corresponding to its temperature.

The *critical point* is the condition of pressure and temperature above which a liquid cannot evaporate at constant temperature. At this point, the states of saturated liquid and saturated vapor coincide.

For the limited purposes of this book, *entropy* is defined as a property that remains constant during a frictionless adiabatic process.

### PROBLEMS

1. Compute the specific volume of air at 100 psia and 80 F.
2. Compute the specific volume of nitrogen at 60 psig and 100 F. Barometric pressure is 29.8 in. of Hg.
3. How many lb of air are contained in a room 22 ft wide by 40 ft long by 13 ft high if the room temperature is 72 F and the barometric pressure is 29.2 in. of Hg?
4. Oxygen is contained in a tank at a pressure of 10 in. of Hg gage and a temperature of 35 C. Barometric pressure is 29.0 in. of Hg. Compute the mass of oxygen in the tank.

**5.** A 100-cu-ft rigid steel tank contains air at 120 psig and 80 F. Some of the air escapes through a leak until the pressure has fallen to 30 psig, at which time the temperature is 70 F. Barometric pressure is 29.6 in. of Hg. (*a*) How much air escaped from the tank? (*b*) What volume does the air occupy after escaping from the tank if the room temperature is 60 F?

**6.** A balloon contains 30 lb of helium (molecular weight = 4). At a given elevation, the atmospheric pressure is 18 in. of Hg and the temperature is −10 F. Calculate the volume of helium in the balloon.

**7.** A 100-cu-ft steel tank contains methane ($CH_4$) at 40 psig and 80 F. Barometric pressure is 29.9 in. of Hg. Assuming that methane is a perfect gas, how many lb of methane are contained in the tank?

**8.** The specific volume of a gas is 15 cu ft at 32 F and 29.92 in. of Hg abs. (*a*) What is the gas constant, $R$? (*b*) What is the apparent molecular weight of the gas?

**9.** The specific volume of a gas is 20 cu ft per $lb_m$ at a location where the local acceleration of gravity is 25 ft per $sec^2$. (*a*) What is the density of the air? (*b*) What is its specific weight?

**10.** A closed system consisting of a cylinder and gas-tight piston contains 1 lb of air at 100 psia and 400 F. What is the volume of the system after expansion of the gas to 20 psia and 100 F?

**11.** For a gas, $c_p = 0.25$ and $c_v = 0.17$. What is the value of the gas constant, $R$?

**12.** A steel tank having a volume of 100 cu ft contains air at 50 psia and 80 F. Heat is transferred to the system until the pressure is 100 psia. (*a*) What is the final temperature? (*b*) What is the increase in internal energy? (*c*) How much work is done during the process? (*d*) What is the increase in enthalpy?

**13.** A closed system consisting of a cylinder and gas-tight piston contains 2 lb of nitrogen initially at 100 psia and 100 F. The system expands at constant pressure until the volume is doubled. Compute (*a*) the final temperature, (*b*) the change in internal energy, (*c*) the work done, (*d*) the heat transferred to the system, and (*e*) the change in enthalpy.

**14.** Ten cubic feet of air at 60 psia and 200 F expand at constant temperature in a frictionless nonflow process until the pressure is 20 psia. Comput (*a*) the work done, (*b*) the change in internal energy, and (*c*) the heat transferred.

**15.** Ten cubic feet of air at 100 psia and 300 F expand adiabatically in a frictionless nonflow process until the pressure is 20 psia. Compute (*a*) the final temperature, (*b*) the final volume, (*c*) the work done, (*d*) the change in internal energy, and (*e*) the heat transferred to the system.

**16.** Five cubic feet of air initially at 300 psia and 1000 F expand in a frictionless nonflow process according to the equation $pV^{1.3} = C$ until the pressure is 40 psia. Compute (*a*) the final temperature, (*b*) the final volume, (*c*) the work done, (*d*) the change in internal energy, and (*e*) the heat transferred.

17. Air at 14.7 psia and 80 F is compressed in a frictionless nonflow process according to the equation $pV^{1.35} = C$ until the pressure is 500 psia. (a) What is the final temperature? (b) Calculate the change in enthalpy per lb of air. (c) What is the change in internal energy per lb of air? (d) How much work is done to compress the air?

18. Three pounds of air initially at 20 psia and 80 F are compressed at constant temperature in a frictionless nonflow process until the pressure is 100 psia. Compute (a) the final volume, (b) the work done, (c) the change in internal energy, and (d) the heat transferred to or from the system.

19. What is the temperature of saturated steam at (a) 2 psia, (b) 100 psia, (c) 2500 psia?

20. What is the pressure of saturated steam at (a) 100 F, (b) 300 F, (c) 600 F?

21. What is the temperature of saturated ammonia at (a) 10 psia, (b) 30 psia, (c) 200 psia?

22. What is the pressure of saturated ammonia at (a) −30 F, (b) 80 F?

23. Determine the specific volume and enthalpy of dry saturated steam at (a) 100 psia, (b) 2000 psia.

24. Determine the specific volume, enthalpy, and entropy of dry saturated ammonia at (a) −10 F, (b) 100 F.

25. What are the enthalpy and specific volume of steam at (a) 400 psia and 700 F, (b) 1200 psia and 900 F, (c) 2500 psia and 1200 F?

26. What are the enthalpy and specific volume of ammonia at (a) 15 psia and 0 F, (b) 140 psia and 100 F?

27. Compute the enthalpy, specific volume, and internal energy of steam at 200 psia and 90 per cent quality.

28. Compute the enthalpy and specific volume of ammonia at 20 psia and 20 per cent quality.

29. Steam at 300 psia is throttled in a calorimeter to a pressure of 20 psia. The calorimeter temperature is 260 F. What are the quality, enthalpy, and specific volume of the steam at 300 psia?

30. Saturated liquid ammonia at 80 F is throttled in a valve to a pressure of 20 psia. What are the quality, specific volume, and enthalpy of the ammonia after throttling to 20 psia?

31. Steam at 200 psia is throttled in a calorimeter to 14.7 psia and a temperature of 240 F. Determine the enthalpy, quality, and specific volume of steam at 200 psia.

32. What is the enthalpy of water at (a) 3000 psia and 400 F, (b) 2000 psia and 300 F?

33. A pound of water initially at 1000 psia and 200 F is converted into steam at 1000 psia and 1000 F. Determine the change in enthalpy.

34. A pound of ammonia at 20 psia and 10 per cent quality is converted into ammonia at 20 psia and 10 F. Determine the change in (a) enthalpy, (b) specific volume.

**35.** Steam at 600 psia and 97 per cent quality is converted into steam at 600 psia and 800 F. Determine the change in (*a*) enthalpy, (*b*) specific volume, (*c*) internal energy, and (*d*) entropy.

**36.** Steam at 200 psia and 600 F expands in a frictionless adiabatic process to a pressure of 20 psia. Determine the change in (*a*) enthalpy, (*b*) specific volume, (*c*) internal energy, and (*d*) entropy.

**37.** Water at 4000 psi and 300 F is converted into steam at 4000 psi and 1000 F. Draw *pV* and *ph* diagrams of the process. Determine the increase in enthalpy. Did evaporation occur during the process? Explain.

# CHAPTER 4

# Fossil Fuels
# and Their Combustion;
# Nuclear Reactions

### 4.1  Introduction

Approximately 96 per cent of the energy used in the United States is obtained from fossil fuels, that is, coal, oil, and gas; water power accounts for most of the remaining 4 per cent. Several nuclear-energy power plants are in operation, and others are under construction. It is expected that during the next 25 years nuclear energy will be developed to supply a substantial portion of our total energy requirements.

This chapter is concerned with the classification of fossil fuels, the physical and chemical properties that determine the relative economic importance of various fossil fuels, the elementary principles of combustion of fossil fuels, the losses that occur when fossil fuels are burned, and the basic principles of nuclear reactions as sources of energy.

### 4.2  Classification of Fossil Fuels

Fossil fuels may be classified as shown in Table 4.1 into solid, liquid, and gaseous fuels. Each of these types of fuels may in turn be classified as natural fuels, that is, fuels found in nature, or manufactured fuels. Most of the manufactured fuels are made from natural fuels, sometimes as a by-product, but usually because they are better suited to a particular application than the natural fuel. Gasoline, which is refined from petroleum, is an example of a manufactured fuel that is produced specifically to meet the requirements of a fuel for automobile or aircraft engines.

TABLE 4.1

Classification of Fuels

| Type of Fuel | Natural Fuels | Manufactured Fuels |
|---|---|---|
| Solid | Coal<br>Lignite<br>Peat<br>Wood | Coke<br>Briquets<br>Charcoal<br>Waste fuels:<br>  Bagasse<br>  Sawdust<br>  Tanbark |
| Liquid | Petroleum | Gasoline<br>Kerosene<br>Fuel oil<br>Alcohol<br>Benzol<br>Shale oil |
| Gaseous | Natural gas | Producer gas<br>Blast furnace gas<br>Carbureted water gas<br>Coal gas<br>Acetylene |

Figure 4.1 shows the world production of the commercial sources of energy during the 40-year period ending in 1955. It will be noted that the amount of energy generated by water power is comparatively insignificant. Coal is the major source of energy but its production has increased only about 25 per cent, whereas the total amount of energy utilized has more than doubled. Most of the increase has been supplied by oil and gas.

In the United States, the coal reserves are large in comparison with the reserves of oil and gas and are widely distributed geographically. During the past 30 years, the proved supply of oil has been equal to only 12 to 20 times the annual consumption. Fortunately, new reserves have been discovered and improved methods of oil recovery have been developed so that there has been an ample supply of oil. However, most of the known oil reserves of the world are outside the boundaries of the United States, substantial quantities of oil are currently being imported, and the rate of consumption is increasing

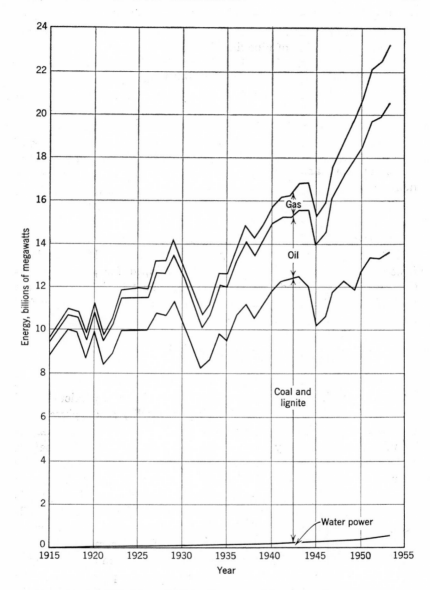

**Fig. 4.1.** World production of commercial sources of energy.

From *Proceedings of 1956 International Conference on Peaceful Use of Atomic Energy, United Nations,* Vol. 1.

rapidly. Natural-gas reserves probably exceed oil reserves. The consumption of natural gas has increased rapidly in recent years because of the extension of pipe lines into most of the densely populated areas of the country. Natural gas is the ideal fuel for domestic heating, is an important source of carbon and hydrogen for the chemical industry, and is burned extensively as an industrial fuel and a fuel for power generation.

The rate of utilization of energy is increasing rapidly. Many authorities believe that within 10 to 20 years the decreasing reserves of oil and natural gas in the United States will have to be supplemented by liquid and gaseous fuels made from coal and shale. Moreover, they estimate that at the increasing rate of energy consumption all our fossil fuels will be exhausted within a period of 250 to 400 years unless other sources of energy are found. Fortunately, the nuclear-fission process is being developed rapidly for the generation of power and this is opening up a very large source of energy. If the nuclear-fusion process can be controlled and developed for power generation, then an unlimited amount of energy can be released from sea water for utilization by man and all concern about exhaustion of fuel resources will disappear.

### 4.3   The Proximate Analysis of Coal

There is no single set of tests based upon physical or chemical properties which will accurately predict the value of a particular coal for a given use. The *proximate analysis* is readily made, gives much valuable informatioin, and is widely used. The proximate analysis includes the per cent by mass of the moisture, ash, volatile matter, and fixed carbon in the fuel. The heating value of the fuel in Btu per pound is normally reported with the proximate analysis. Although it is not part of the proximate analysis, the amount of sulphur is frequently reported with the proximate analysis because of the difficulties that are encountered in burning fuel that is high in sulphur.

*Moisture* * is defined as the loss in mass of a 1-gram sample when dried for 1 hr at 105 C in a constant-temperature oven through which dry preheated air is passed, and is expressed as a percentage of the mass of the sample tested. The analysis should report the moisture in

---

* Detailed methods of collecting coal samples and making analyses are given in the publications of the U. S. Bureau of Mines and the ASTM Standards on Coal and Coke.

the coal "as received," that is, the moisture in the coal at the point of sampling. The moisture content of coal will vary from 2 to 25 per cent or more, depending upon the rank or quality of the coal. Moisture is objectionable because it dilutes the combustible material in the coal and therefore reduces the heating value per pound of fuel. It must be transported from the mine to the point of use at considerable expense. When the coal is burned, the moisture must be evaporated before the coal can be ignited, thus causing a direct loss in energy and reducing the fuel-burning capacity of the equipment.

*Ash* is the incombustible residue that remains when the sample of the fuel is held at a temperature of 700 to 750 C in the presence of air until the combustible matter has been burned and a constant mass is attained. The ash constitutes from 4 to over 20 per cent of the coal.

The ash in coal comes from three sources: (1) the vegetable matter from which the coal was formed, (2) foreign material washed into the coal beds during their formation and often deposited in cracks or veins, and (3) material from the roof and floor of the mine. The ash due to items 2 and 3 can be partially removed by hand picking and mechanical cleaning of the coal at the mine, and most mines are equipped with machinery for reducing the ash content and grading the coal by size. Without mechanical cleaning, the ash content of the fine sizes of coal is normally higher than the ash content of lump coal.

Ash, like moisture, dilutes the combustible substance of the coal, thus reducing the heating value per pound, and not only must be shipped from the mine to the point of use at considerable expense but also will have to be removed after the coal is burned and disposed of, usually at added expense.

The principal constituents of ash are silica ($SiO_2$), alumina ($Al_2O_3$), iron oxide ($Fe_2O_3$), and calcium oxide ($CaO$). Depending upon the proportions of these compounds and others which are present in smaller amounts, the ash may melt at the temperatures that exist in the fuel bed and furnace. *Clinker* is ash that has fused or melted and then solidified. The temperature at which the ash fuses is one of the most important characteristics of coal and is determined by mixing finely ground ash with a suitable binder to form a cone that is mounted in a furnace and heated in a mildly reducing atmosphere. The *softening temperature* is the temperature at which the cone fuses to a spherical lump. In general, coals whose ash softens at temperatures above 2600 F give little trouble from clinkers, whereas coals

## TABLE 4.2

### Analyses of Representative Coals in the United States Compiled from Bulletins of the U. S. Bureau of Mines

Analysis of Mine Samples on the "As-Received" Basis

| Rank | Subgroup | State | County | Proximate Analysis | | | | Ultimate Analysis | | | | | Heating Value, btu per lb | Mineral-Matter-Free Basis | |
|---|---|---|---|---|---|---|---|---|---|---|---|---|---|---|---|
| | | | | Moisture | Volatile Matter | Fixed Carbon | Ash | Sulphur | Hydrogen | Carbon | Nitrogen | Oxygen | | Dry Fixed Carbon | Moist Btu |
| I. Anthracite | 1. Meta-anthracite | R. I. | Providence | 4.5 | 3.0 | 78.7 | 13.8 | 0.9 | 0.5 | 82.4 | 0.1 | 1.8 | 11,624 | 98.2 | |
| | 2. Normal anthracite | Pa. | Carbon | 4.1 | 3.5 | 81.7 | 10.7 | 0.5 | 2.2 | 81.6 | 0.6 | 4.4 | 12,590 | 97.1 | |
| | | Pa. | Luzerne | 3.6 | 4.7 | 76.1 | 15.6 | 1.1 | 2.7 | 74.3 | 0.8 | 5.5 | 12,050 | 96.1 | |
| | | Pa. | Schuylkill | 4.5 | 3.9 | 83.2 | 8.4 | 0.5 | 2.5 | 82.5 | 1.2 | 4.9 | 12,970 | 96.5 | |
| | 3. Semi-anthracite | Pa. | Northumberland | 1.5 | 10.1 | 76.5 | 11.9 | 0.8 | 3.7 | 78.7 | 1.4 | 3.5 | 13,390 | 89.7 | |
| | | Pa. | Sullivan | 2.0 | 9.6 | 77.8 | 10.6 | 0.6 | 3.7 | 79.5 | 0.9 | 4.7 | 13,520 | 90.1 | |
| | | Va. | Montgomery | 1.3 | 11.9 | 62.2 | 24.6 | 0.7 | 3.2 | 67.1 | 0.7 | 3.7 | 11,310 | 86.6 | |
| II. Bituminous | 1. Low volatile | W. Va. | McDowell | 1.7 | 18.1 | 76.0 | 4.2 | 0.6 | 4.7 | 85.4 | 1.2 | 3.9 | 14,720 | 81.2 | |
| | | Pa. | Cambria | 3.5 | 18.0 | 71.7 | 6.8 | 2.2 | 4.7 | 80.1 | 1.2 | 5.0 | 14,140 | 81.2 | |

| Class | County | State | | | | | | | | | | | | |
|---|---|---|---|---|---|---|---|---|---|---|---|---|---|---|
| 2. Medium volatile | Fayette | W. Va. | 1.7 | 26.7 | 69.0 | 2.6 | 0.8 | 5.0 | 85.1 | 1.6 | 4.9 | 14,930 | 72.5 | 14,020 |
| | Westmoreland | Pa. | 2.3 | 26.2 | 63.4 | 8.1 | 3.2 | 4.9 | 77.4 | 1.5 | 4.9 | 13,850 | 72.5 | 14,820 |
| | Buchanan | Va. | 3.1 | 21.8 | 67.9 | 7.2 | 1.0 | 5.0 | 80.1 | 1.5 | 5.2 | 14,030 | 76.4 | 14,780 |
| 3. High volatile A | Jackson | Ill. | 5.7 | 33.2 | 52.3 | 8.8 | 3.5 | 5.3 | 70.0 | 1.3 | 11.1 | 12,590 | | 13,140 |
| | Logan | W. Va. | 2.8 | 35.8 | 56.3 | 5.1 | 0.7 | 5.5 | 78.8 | 1.6 | 8.3 | 13,980 | | 13,920 |
| | Fayette | Pa. | 3.1 | 35.1 | 51.3 | 10.5 | 3.5 | 5.1 | 71.5 | 1.5 | 7.9 | 13,000 | | 13,200 |
| 4. High volatile B | Franklin | Ill. | 9.2 | 33.8 | 48.6 | 8.4 | 0.9 | 5.5 | 67.3 | 1.5 | 16.3 | 11,930 | | 13,140 |
| | Webster | Ky. | 5.4 | 34.9 | 50.4 | 9.3 | 1.1 | 5.1 | 70.4 | 1.6 | 12.5 | 12,501 | | 13,920 |
| | Williamson | Ill. | 9.8 | 27.3 | 55.4 | 8.1 | 0.9 | 5.1 | 68.5 | 1.1 | 16.3 | 12,015 | | 13,200 |
| 5. High volatile C | Logan | Ill. | 12.8 | 36.5 | 40.8 | 9.9 | 3.0 | 5.7 | 61.5 | 1.1 | 18.8 | 10,990 | | 12,370 |
| | Sullivan | Ind. | 13.5 | 32.5 | 48.4 | 5.6 | 1.1 | 5.9 | 66.0 | 1.5 | 19.9 | 11,788 | | 12,580 |
| | Polk | Iowa | 13.9 | 37.0 | 35.2 | 14.0 | 6.2 | 5.5 | 54.7 | 0.8 | 18.8 | 10,244 | | 12,200 |
| III. Subbituminous — 1. Subbituminous A | Carbon | Wyo. | 12.0 | 36.9 | 44.9 | 6.2 | 0.4 | 5.7 | 60.6 | 1.0 | 26.1 | 10,640 | | 11,440 |
| 2. Subbituminous B | Coss | Oreg. | 16.1 | 31.1 | 39.6 | 13.2 | 0.8 | 5.5 | 51.1 | 1.2 | 28.2 | 9,031 | | 10,550 |
| | Sheridan | Wyo. | 23.9 | 34.3 | 38.4 | 3.4 | 0.4 | 6.3 | 54.1 | 1.1 | 34.7 | 9,335 | | 9,700 |
| 3. Subbituminous C | Jackson | Colo. | 25.0 | 30.5 | 39.3 | 5.2 | 1.2 | 6.2 | 51.2 | 0.7 | 35.5 | 8,880 | | 9,450 |
| | Sweet Water | Wyo. | 19.8 | 35.7 | 37.8 | 6.7 | 0.7 | 5.1 | 48.1 | 1.3 | 38.1 | 7,830 | | 8,450 |
| IV. Lignite | Houston | Texas | 34.7 | 32.2 | 21.9 | 11.2 | 0.8 | 6.9 | 39.3 | 0.7 | 41.4 | 7,056 | | 8,040 |
| | Perkins | S. Dak. | 39.2 | 24.7 | 27.8 | 8.3 | 2.2 | 6.6 | 38.0 | 0.5 | 44.4 | 6,307 | | 6,890 |
| | Williams | N. Dak. | 42.1 | 25.0 | 24.4 | 8.5 | 1.3 | 7.1 | 35.2 | 0.5 | 47.5 | 5,994 | | 6,580 |

with ash-fusing temperatures below 2200 F will cause serious clinker troubles unless carefully handled.   Melted ash which flows readily may choke off the air supply in the fuel bed, is destructive to furnace brick work, and, if carried into boiler surfaces by the moving gas stream, may bridge across the tube bank and block the flow of gas, thus forcing a shutdown of the unit.   The ash-fusing temperature has more influence on the design of high-capacity furnaces than any other property or characteristic of coal.

*The volatile matter* in coal is determined by heating a 1-gram sample of fine coal for 7 min at 950 C in a platinum crucible that has a close-fitting cover to exclude air.   The loss in mass is volatile matter plus moisture.   The volatile matter is that part of the coal which can be gasified under the conditions of the test.   If the volatile matter could be collected and cooled to room temperature, it would be found to consist mainly of tars, oils, and gaseous hydrocarbon compounds.   The volatile matter in coal varies from almost nothing to about 50 per cent and is the smoke-producing constituent of coal. In general, coals high in volatile matter produce large quantities of combustible gases upon heating and require large combustion chambers or furnaces and careful firing for their complete and smokeless combustion.

*Fixed carbon* is determined by subtracting from 100 the sum of the moisture, ash, and volatile matter expressed as per cent.   From 25 to 90 per cent of the coal is fixed carbon, the content of fixed carbon increasing with the rank of the coal.

The *heating value* of the coal is normally reported with the proximate analysis and is the energy, expressed in Btu, which is released by the complete combustion of 1 lb of fuel.   The method of measuring the heating value is discussed in Article 4.11.

Table 4.2 gives the proximate analysis of representative samples of coal on an "as-received" basis.   Since moisture and ash are in a sense extraneous material whereas the fixed carbon and volatile matter are the useful coal substance, analyses that are used to compare coals are often reported on a "moisture-free" or "moisture-and-ash-free" basis. The method of converting the analysis from one basis to another can be illustrated by Example 1.

**Example 1.**   Convert the analysis of the Sullivan County, Ind., coal of Table 4.2 from an "as-received" basis to a "moisture-free" and to a "moisture-and-ash-free" basis.

*Solution:*

|  | A "As Received" | B "Moisture- Free" | C "Moisture- and-Ash-Free" |
|---|---|---|---|
| Volatile matter | 32.5 | 37.5 | 40.2 |
| Fixed carbon | 48.4 | 56.0 | 59.8 |
| Ash | 5.6 | 6.5 | ... |
| Moisture | 13.5 | ... | ... |
| Total | 100.0 | 100.0 | 100.0 |
| Btu per lb | 11,788 | 13,628 | 14,572 |

Column $A$ is obtained from Table 4.2. The items in column $B$ are computed by dividing the items of column $A$ by $(1.00 - 0.135) = 0.865$, since 1 lb of fuel contains 0.865 lb of volatile matter, fixed carbon, and ash. Similarly, column $C$ is obtained from column $A$ by dividing the volatile matter, fixed carbon, and heating value by $1.00 - (0.135 + 0.056) = 0.809$, since 1 lb of fuel contains 0.809 lb of volatile matter plus fixed carbon.

### 4.4 The Ultimate Analysis of Coal

The ultimate analysis of coal is an analysis on a percentage basis of the chemical elements that form the coal substance. Carbon, hydrogen, nitrogen, and sulphur are determined by the methods of organic chemical analysis while the ash is determined as in the proximate analysis. The oxygen is then calculated by difference. Table 4.2 gives the ultimate analyses of typical coals on the "as-received" basis. It should be noted that the column for "ash" in Table 4.2 applies to *both the proximate and ultimate analyses.*

In Table 4.2 the moisture in the coal is reported in the ultimate analysis as its chemical elements: that is, as hydrogen and oxygen. It is often desirable to report the ultimate analysis with the moisture listed as a separate item. Since hydrogen and oxygen combine to form water in the proportions of 1 lb of hydrogen + 8 lb of oxygen = 9 lb of water, then one ninth of the moisture is hydrogen and eight ninths of the moisture is oxygen. The moisture from the proximate analysis may be inserted into the ultimate analysis by deducting from the hydrogen and oxygen of the ultimate analysis the amount of hydrogen and oxygen in the moisture. The resulting analysis may be converted to a "moisture-free" or "moisture-and-ash-free" basis as in Example 1. This procedure may be illustrated by Example 2.

**Example 2.** Convert the ultimate analysis of the Sullivan, Ind., coal of Table 4.2 from the "as-received" basis to the "as-received" basis with the moisture as a separate item in the ultimate analysis. Then convert this analysis to a "moisture-free" and to a "moisture-and-ash-free" basis.

*Solution:*

| | | As Received | | | |
|---|---|---|---|---|---|
| Con-stituent | A<br>From<br>Table 4.2 | B<br>With Moisture as<br>Separate Item | | C<br>Moisture-<br>Free | D<br>Moisture-<br>and-Ash-Free |
| Carbon | 66.0 | | 66.0 | 76.3 | 81.5 |
| Nitrogen | 1.5 | | 1.5 | 1.7 | 1.9 |
| Sulphur | 1.1 | | 1.1 | 1.3 | 1.4 |
| Ash | 5.6 | | 5.6 | 6.5 | . . . |
| Hydrogen | 5.9 | $5.9 - \frac{1}{9} \times 13.5 =$ | 4.4 | 5.1 | 5.4 |
| Oxygen | 19.9 | $19.9 - \frac{8}{9} \times 13.5 =$ | 7.9 | 9.1 | 9.8 |
| Moisture | . . . | | 13.5 | . . . | . . . |
| Total | 100.0 | | 100.0 | 100.0 | 100.0 |

In column $A$, which is copied from Table 4.2, the moisture appears as hydrogen and oxygen. In column $B$ the moisture from the proximate analysis, amounting to 13.5 per cent, is inserted as a separate item. One ninth of this moisture or $\frac{1}{9} \times 13.5$ is hydrogen, which is subtracted from the hydrogen in column $A$ to get the hydrogen in column $B$. Similarly, eight ninths of the moisture or $\frac{8}{9} \times 13.5$ is oxygen, which is subtracted from the oxygen in column $A$ to get the oxygen in column $B$. Columns $C$ and $D$ are obtained from column $B$ as in Example 1 by dividing the items of column $B$ by $(1.00 - 0.135)$ and $1.00 - (0.135 + 0.056)$, respectively.

## 4.5  Origin of Coal

It is generally agreed that coal is of vegetable origin. In prehistoric times plant material accumulated in great swamps where the land was slowly sinking at a rate that covered the plant and tree growth with sufficient water to prevent normal decay without destroying plant life. Thus the remains of vegetable growth accumulated to considerable depth and were gradually converted to peat through biochemical decay. The peat was subsequently buried, in some cases to depths of several thousand feet, by sinking of the earth's crust and deposition of soil above the peat bed. Under the influence of time, moderate temperature, the weight of the overburden of soil, and thrust pressures created by movements of the earth's crust in regions where mountains were being formed, the peat was compacted

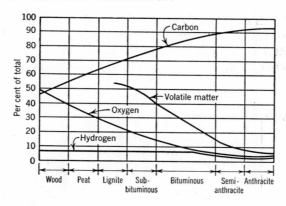

**Fig. 4.2.** Progressive transformation from wood into coal, based on moisture, ash, and sulphur-free coal.

and devolatilized. This brought about a transformation from woody material through the various *ranks* of coal such as lignite, subbituminous, and bituminous coal, to anthracite, as shown in Fig. 4.2. In general, the volatile matter and oxygen content were reduced, and the fundamental coal substance was concentrated into the carbon and ash of the anthracite coal. The "older" coals such as anthracite and low-volatile bituminous coals are found in the Appalachian region of the United States and in some localities in the Rocky Mountain region where earth-thrust pressures have been great. The deposits of lignite, subbituminous coal, and high-volatile bituminous coal are found in the Mississippi Valley regions where earth-thrust pressures due to mountain formation have been a minimum.

Figure 4.3 shows the changes in the proximate analysis and heating value of ash-free coal during progressive transformation from lignite to anthracite. It should be noted that the heating value of the "oldest" coals such as anthracite is lower than the heating value of the high-grade bituminous coals, because the devolatilization of the coal has progressed so far that much of the available hydrogen has been driven out of the coal, and hydrogen has a heating value over four times that of carbon.

## 4.6 Classification and Characteristics of Coals by Rank

For the purpose of classifying coals, mine samples are taken from freshly exposed faces of the coal seam in the mine. The proximate analyses of such samples of coal which contain the natural bed moisture are reported as being on the "moist" basis. Since moisture and mineral matter are to a certain extent extraneous material, the high-rank coals are classified on the basis of dry mineral-matter-free fixed

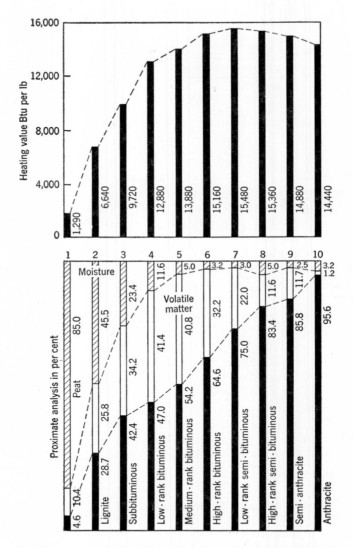

**Fig. 4.3.** Proximate analysis and heating value of ash-free coals.

carbon while the low-rank coals are classified on a moist mineral-matter-free Btu basis. Mineral matter in coal exceeds the ash content as reported in the proximate analysis, owing to changes that occur when the sample is heated during analysis. The following formulas, suggested by Parr, are used for coal classification:

Dry mineral-matter-free fixed carbon

$$= \frac{FC - 0.15S}{100 - (M + 1.08A + 0.55S)} \times 100 \quad (4.1)$$

Moist mineral-matter-free Btu

$$= \frac{\text{Btu} - 50S}{100 - (1.08A + 0.55S)} \times 100 \quad (4.2)$$

where $FC$, $M$, $A$, and $S$ refer to the percentages of fixed carbon, moisture, ash, and sulphur reported in the analysis of the coal containing its natural bed moisture.

Coals are classified as shown in Table 4.3 into lignite, subbituminous, and bituminous coals, and anthracite, with two or more groups under each major classification. They are classified by dry volatile matter and dry mineral-matter-free fixed carbon for values of 69 per cent or more fixed carbon, and by moist mineral-matter-free Btu and physical properties for coals having less than 69 per cent dry mineral-matter-free fixed carbon. Weathering coals are those that undergo a certain amount of degradation in size when dried under specified test conditions. Agglomerating coals are those in which the residue from the volatile-matter test is an agglomerate button of coke which will support a 500-gram mass or a button showing swelling or cell structure.

In Table 4.2 the analyses of the coals are grouped in accordance with the classification as given in Table 4.3.

*Lignite* has a distinct woody or clay-like structure with a moisture content of 30 to 45 per cent when mined. Upon drying, it disintegrates into small flakes. Because of the high moisture content, the heating value as mined is low, varying from 5500 to 8000 Btu per lb, and it is not economical to ship it far from the mine. Extensive lignite deposits exist in such states as Texas, North Dakota, and Montana, and this fuel will undoubtedly become important commercially as the supplies of high-grade coal are depleted.

*Subbituminous* coals are usually glossy black in color, do not have the woody appearance that characterizes much of the lignite, and have

## TABLE 4.3

### Classification of Coal by Rank *

| Class | Group | Limits of Dry Mineral-Matter-Free Fixed Carbon, Dry Volatile Matter, and Moist Mineral-Matter-Free Btu |
|---|---|---|
| I. Anthracite | 1. Meta-anthracite | Fixed carbon 98% or more; dry volatile matter 2% or less |
|  | 2. Anthracite | Fixed carbon 92–98%; dry volatile matter 2–8% |
|  | 3. Semi-anthracite | Fixed carbon 86–92%; dry volatile matter 8–14%; nonagglomerating |
| II. Bituminous | 1. Low-volatile bituminous coal | Fixed carbon 78–86%; dry volatile matter 14–22% |
|  | 2. Medium-volatile bituminous coal | Fixed carbon 69–78%; dry volatile matter 22–31% |
|  | 3. High-volatile A bituminous coal | Fixed carbon less than 69%; dry volatile matter over 31% Moist Btu 14,000 or more |
|  | 4. High-volatile B bituminous coal | Moist Btu 13,000 to 14,000 |
|  | 5. High-volatile C bituminous coal | Moist Btu 11,000–13,000 Either agglomerating or nonweathering |
| III. Subbituminous | 1. Subbituminous A coal | Moist Btu 11,000–13,000 Both agglomerating and weathering |
|  | 2. Subbituminous B coal | Moist Btu 9500–11,000 |
|  | 3. Subbituminous C coal | Moist Btu 8000–9500 |
| IV. Lignite | 1. Lignite | Moist Btu less than 8300.  Consolidated |
|  | 2. Brown coal | Moist Btu less than 8300.  Unconsolidated |

* ASTM Standards on Coal and Coke.

a moisture content when mined of 10 to 30 per cent. They "slack" or disintegrate upon exposure to the air and are therefore designated as nonweathering coals. Because of their high moisture content, low heating value, and nonweathering characteristics, they are not mined extensively although large deposits exist in the United States.

*Bituminous* or "soft" coals are widely distributed throughout the United States and are the principal fuels for steam generation, industrial purposes, and the manufacture of metallurgical coke and coal gas. They vary in composition as shown in Table 4.2 from the high-volatile coals of the Mississippi Valley to the low-volatile coals of the Appalachian region. The high-volatile coals, having a volatile-matter content as mined of 25 to 40 per cent, burn with a long yellow flame. They will produce objectionable quantities of smoke unless properly fired in a furnace of sufficient size to burn the volatile gases.

The low-volatile bituminous coals are relatively smokeless, are low in moisture and ash, have the highest heating value of any coals, and are the premium industrial coals of the country.

Bituminous coals are designated as *free-burning*, or *caking* coals. Caking coals soften and swell upon heating and solidify into masses of coke or carbon after the volatile matter has been driven off. Free-burning coals do not soften and swell on heating, do not adhere to adjacent pieces, and produce a noncoherent carbon residue after the volatile matter has been expelled.

*Semi-anthracite* coals occur in small quantities, are smokeless fuels having 8 to 14 per cent volatile matter, and are sold primarily for domestic use.

*Anthracite* or "hard" coal is a free-burning smokeless fuel having less than 8 per cent of volatile matter and consisting mainly of fixed carbon and ash. It is slow to ignite and burns with a short bluish flame. It commands a premium price as a domestic fuel, and only those sizes too small for domestic consumption are available for steam generation. Most anthracite is mined in three counties in eastern Pennsylvania, and the reserves are strictly limited.

Figure 4.4 shows the distribution of coal fields in the United States. These coal fields constitute over 50 per cent of the known coal reserves of the world and are widely distributed geographically, an im-

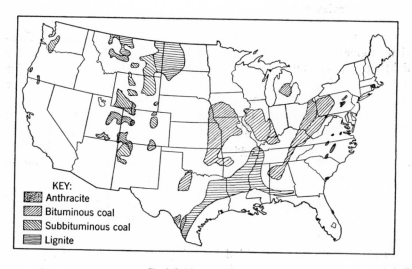

KEY:
Anthracite
Bituminous coal
Subbituminous coal
Lignite

**Fig. 4.4.** Coal fields of the United States.

portant factor in the industrial development of the United States. Nearly one half of these reserves are lignite and subbituminous coal, and the highest grades of coal such as anthracite will be exhausted in 100 years at the present rate of consumption.

### 4.7 Solid Fuels Other than Coal

*Coke* is the strong porous coherent mass, composed principally of carbon, which is produced when a coking coal is heated to about 2000 F in the absence of air. Coal gas and benzol are by-products of coke manufacture. Coke is the basic fuel of the metallurgical industries and finds a limited market as a domestic smokeless fuel. The fine sizes (coke breeze) which are too small for other uses are burned for steam generation.

*Briquets* are produced by mixing the fine sizes of high-grade coal with a suitable binder and pressing the mixture into lumps about the size of a cake of soap. The briquets have sufficient strength to resist breakage during shipment. A high-priced domestic fuel is thus produced from fine sizes of coal that normally have a low market value.

*Sawdust,* spent *tanbark,* and *bagasse* (sugar-cane refuse) are high-moisture waste fuels which, like wood, are of importance as industrial fuels in limited areas.

### 4.8 Petroleum

Crude petroleum is refined to produce gasoline for internal-combustion engines such as automobile and aircraft engines; kerosene for lighting; Diesel fuel for Diesel engines; lubricating oils; fuel oil for domestic heating, steam generation, and industrial furnaces; and oil for the manufacture or enrichment of gas. Petroleum is an intersolution of a mixture of hydrocarbon compounds of the following families: $C_nH_{2n+2}$, $C_nH_{2n}$, $C_nH_{2n-2}$, $C_nH_{2n-4}$, $C_nH_{2n-6}$, $\cdots$, $C_nH_{2n-14}$. Thus, in the paraffin series, $C_nH_{2n+2}$, compounds are found to vary from gaseous methane $(CH_4)$ to solid paraffin $(C_{30}H_{62})$. The ultimate analysis of crude petroleum falls within the following limits:

| | |
|---|---|
| Carbon | 80–87% |
| Hydrogen | 11–15% |
| Oxygen, nitrogen, and sulphur | 0.1–11% |

Crude petroleum is classified into the following three groups, depending upon the type of residue that remains upon distillation: (1)

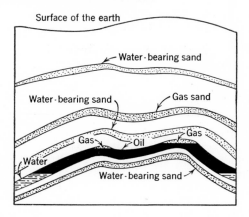

**Fig. 4.5.** Occurrence of petroleum and natural gas in a typical rock formation.

paraffin-base crudes, (2) asphalt-base crudes, and (3) mixed-base crudes, the last having a residue composed of a mixture of paraffin and asphalt.

Although there is no generally accepted theory that explains the origin of petroleum, it is believed to have resulted from the decomposition of marine vegetation or marine animals. Figure 4.5 shows a typical geological formation in which petroleum and natural gas are found. It consists of a porous stratum of rock overlaid with an impervious cover, all so inclined as to produce a pocket into which the oil and gas have collected from adjacent areas. The oil is often found over salt water.

The analyses of typical crude petroleums are given in Table 4.4.

Crude petroleum is refined by distillation into gasoline, kerosene, lubricating oils, distillates, and residual-fuel oils. Because of the demand for gasoline, much of the distillates or residues from the distillation process may be subjected to thermal cracking or partial decomposition and recombination to produce gasoline and other lighter fractions. Fuels for internal-combustion engines will be considered in more detail in Chapter 5.

Fuel oils are used for domestic and industrial heating and for steam generation. Average analyses and relative costs of the five standard grades of fuel oil are shown in Table 4.5. Numbers 1 and 2 fuel oils are distillates which are obtained by condensing the hydrocarbon vapors from the crude stills and are therefore free from ash. They are used primarily as domestic fuels. Numbers 4, 5, and 6 fuel oils are

### TABLE 4.4

#### Analyses of Typical Crude Oils

| Source of Oil | De-grees API | Carbon, per cent | Hydro-gen, per cent | Oxygen, per cent | Sulphur, per cent |
|---|---|---|---|---|---|
| California, Kern River | 15 | 86.80 | 11.57 | 0.74 | 0.89 |
| California, Sunset | 14 | 86.73 | 11.37 | 0.84 | 1.06 |
| Pennsylvania crude | 39 | 82.0 | 14.8 | 3.20 * | . . . |
| West Virginia crude | 36 | 84.3 | 14.1 | 1.60 * | . . . |
| Ohio crude | 28 | 84.2 | 13.1 | 2.7 * | . . . |
| Texas crude | 22 | 84.6 | 10.9 | 2.87 | 1.63 |
| Oklahoma crude | 25 | 87.93 | 11.47 | 0.19 | 0.41 |
| Mexican crude | 22 | 85.65 | 10.2 | . . . | 4.15 |
| Russia, Baku | 17 | 86.6 | 12.3 | 1.1 * | . . . |

* Oxygen plus nitrogen.

### TABLE 4.5

#### Typical Analyses and Relative Costs of Fuel Oils *

| Grade | No. 1 | No. 2 | No. 4 | No. 5 | No. 6 |
|---|---|---|---|---|---|
| Analysis, per cent: | | | | | |
|   Sulphur | 0.1 | 0.3 | 0.8 | 1.0 | 2.3 |
|   Hydrogen | 13.8 | 12.5 | . . . | . . . | 9.7 |
|   Carbon | 86.1 | 87.2 | . . . | . . . | 85.6 |
|   Ash | . . . | . . . | 0.03 | 0.03 | 0.12 |
| Gravity, deg API | 42 | 32 | 20 | 19 | . . . |
| Pour point, F | −35 | −5 | +20 | +30 | . . . |
| Viscosity, centistokes at 100 F | 1.8 | 2.4 | 27.5 | 130 | . . . |
| Water and sediment, volume % | . . . | . . . | 0.2 | 0.3 | 0.74 |
| Higher heating value, Btu/lb | 19,810 | 19,430 | 18,860 | 18,760 | 18,300 |
| Relative cost per Btu | 115 | 100 | 77 | 57 | 44 |

* From *Steam, Its Generation and Use*, The Babcock & Wilcox Company, 1955.

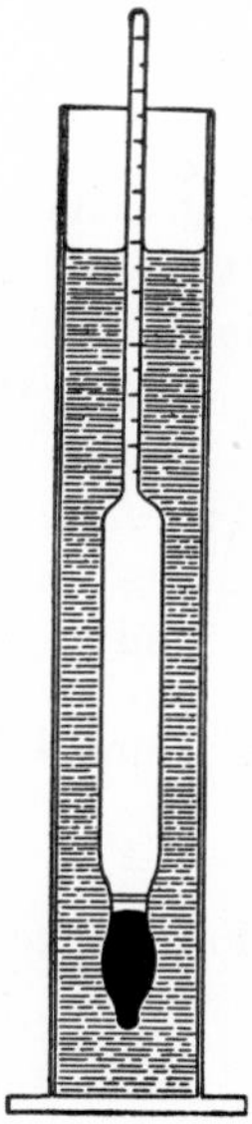

**Fig. 4.6.** Hydrometer.

residual oils which remain after the lighter portions of the crude have been removed by distillation. They contain the ash that was present in the crude. Numbers 5 and 6 oils are intended for use in equipment provided with heaters which will increase the fluidity of the oil and are employed extensively for steam generation in stationary power plants and naval and merchant marine vessels.

### 4.9  Properties of Fuel Oil

Among the important properties of fuel oil are (1) specific gravity, (2) heating value, (3) flash and burning points, (4) congealing point (pour test), and (5) viscosity.

The specific gravity of fuel oil is determined by placing a hydrometer in a sample of the oil at 60 F. The hydrometer (Fig. 4.6) is a standard weighted glass bulb with a graduated rod indicating the depth to which it sinks in the fluid under test. The hydrometer is graduated in degrees API (American Petroleum Institute), and the gravity of the oil is determined by reading the scale of the hydrometer at the level of the liquid surface. The specific gravity or the ratio of the density of fuel oil at 60 F to the density of water at 60 F may be computed as follows:

$$\text{Specific gravity} = \frac{141.5}{131.5 + \text{degrees API}} \qquad (4.3)$$

A gravity of 10 degrees API corresponds to a specific gravity of 1.00. It should be noted that, the higher the API gravity reading, the lower is the specific gravity of the fuel oil.

The heating value of fuel oil, expressed in Btu per pound, is determined by means of a bomb calorimeter as explained in Article 4.11. The American Petroleum Institute has accepted the following formulas for approximating the heating value of fuel oil in Btu per pound:

For uncracked oil,

$$Q_H = 17,660 + (69 \times \text{API gravity}) \qquad (4.4)$$

For cracked oil,

$$Q_H = 17,780 + (54 \times \text{API gravity}) \qquad (4.5)$$

The heating value per pound increases with the API gravity. However, fuel oil is purchased by the gallon or barrel (42 gal per barrel), and the number of pounds per gallon decreases with increased API gravity. Consequently, the number of Btu per gallon decreases as the API gravity increases, as shown in Fig. 4.7.

The *flash point* of an oil is the temperature at which the oil produces vapors in sufficient quantity to form a momentary flash under certain standardized conditions when a flame is brought near the surface of the oil. Since the flash point is the temperature at which combustible vapors are produced, it is an index of the relative safety with which the fuel oil may be handled and stored. A minimum flash point of 150 F is frequently specified. The *burning point* is the temperature at which the oil will produce combustible vapors rapidly enough to burn continuously and is usually about 20 F above the flash point.

The *pour point* is the lowest temperature at which oil will flow and is important in the purchase of oil that must be handled in cold weather.

*Viscosity* is an index of resistance to flow or a measure of internal friction. The viscosity is usually expressed as the time in seconds required for a specified quantity of oil to flow through an orifice of specified size. Since there are several different orifices in use, it is necessary to specify not only the time but also the temperature and the

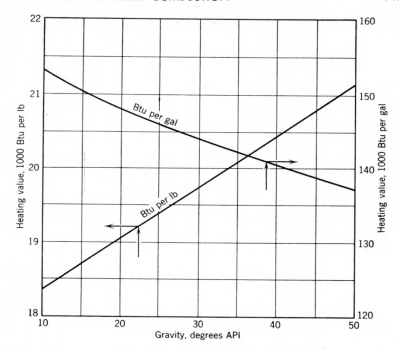

**Fig. 4.7.** Approximate heating values of uncracked fuel oil.

$$Q_H = 17,660 + (69 \times \text{API gravity})$$

name of the standard test apparatus. Oils of high viscosity require heating if they are to be pumped or burned readily.

## 4.10 Natural and Manufactured Gaseous Fuels

Gaseous fuels are ideal fuels from the standpoint of ease of handling, ease of control, and cleanliness, and are used extensively in many industrial applications where the product must not be contaminated, the furnace atmosphere must be regulated closely, or uniform heating is required.

*Natural gas* is found in the petroleum fields and may also be found in locations where petroleum is absent. Natural gas is transmitted long distances through high-pressure pipe lines which are provided with compressor stations at intervals to restore the gas to the high pressure which it gradually loses as a result of pipe-line friction.

## TABLE 4.6

### Analyses of Typical Gaseous Fuels

| Gas | Composition by Volume, per cent | | | | | | | | Higher Heating Value, btu per cu ft at 60 F and 30 in. Hg |
|---|---|---|---|---|---|---|---|---|---|
| | $H_2$ | $CH_4$ | $C_2H_4$ | $C_2H_6$ | CO | $CO_2$ | $O_2$ | $N_2$ | |
| Natural gas: | | | | | | | | | |
| California | ... | 77.5 | ... | 16.0 | ... | 6.5 | ... | ... | 1123 |
| Ohio | ... | 83.5 | ... | 12.5 | ... | 0.2 | ... | 3.8 | 1047 |
| Pennsylvania | ... | 88.0 | ... | 11.2 | ... | ... | ... | 0.8 | 1146 |
| Louisiana | ... | 94.7 | ... | ... | ... | 0.4 | ... | 4.9 | 1066 |
| Coke-oven gas | 57.4 | 28.5 | 2.9 | ... | 5.1 | 1.4 | 0.5 | 4.2 | 536 |
| Carbureted water gas | 35.2 | 14.8 | 12.8 | ... | 33.9 | 1.5 | ... | 1.8 | 578 |
| Producer gas | 10.5 | 2.6 | 0.4 | ... | 22.0 | 5.7 | ... | 58.8 | 136 |
| Blast-furnace gas | 3.2 | ... | ... | ... | 26.2 | 13.0 | ... | 57.6 | 93 |

Thus large quantities of natural gas which were formerly wasted in the oil fields are now being burned usefully. The known reserves of natural gas, like the reserves of oil, are quite limited.

Table 4.6 shows the composition of a number of typical natural gases. They are composed principally of methane ($CH_4$) and ethane ($C_2H_6$) with a small amount of inert gases and consequently have a high heating value per cubic foot compared to manufactured gases.

*Coke-oven gas* is obtained from the by-product coke oven and is used in furnaces at the steel mills and for distribution as "city gas." It is obtained by collecting and cleaning the volatile matter that is distilled from coal in the process of making metallurgical coke.

*Producer gas* is a lean gas too low in heating value for economical distribution through city gas mains. It is made for use in industrial furnaces by passing air plus a small amount of steam through a thick fuel bed in which the fuel is burned incompletely. The resulting gases are high in hydrogen, nitrogen, and carbon monoxide.

*Blast-furnace gas* is obtained as a by-product from the blast furnace, which is in effect a large gas producer, and is an important fuel in steel manufacture.

### 4.11   Determination of the Heating Value of Fuels

The heating value or calorific value of a fuel is the amount of energy that is released by the complete combustion of a unit quantity of the fuel. For solid and liquid fuels, the heating value is expressed as Btu per pound of fuel in the English engineering system of units. For gaseous fuels, the heating value is expressed in Btu per cubic foot at some specified temperature and pressure.

The heating value of *solid* and *liquid fuels* is determined by means of a *bomb calorimeter*. The cross section of a bomb calorimeter is shown in Fig. 4.8. The bomb consists of a cup-shaped receptacle with a cover that can be screwed down to a gas-tight joint. A carefully weighed sample of fuel, about 1 gram of coal or 0.8 gram of a liquid fuel, is placed in the fuel pan. An electric fuse wire is installed between binding posts so that it dips into the fuel, and the bomb is then assembled and filled with oxygen at a pressure of about 300 psi. The bomb is immersed in a well-insulated container supplied with about 2000 grams of water at approximately room temperature. The water is kept in motion and at a uniform temperature by a motor-driven stirrer. The temperature of the water is measured by a high-grade thermometer equipped with a magnifying glass that permits the measurement of temperature to 0.005 F. The charge is fired electrically. The heating value is calculated from the amount of fuel present and the temperature rise of the jacket water after suitable corrections for radiation, energy supplied by the electric-fuse wire, etc. The calorimeter may be calibrated by burning carefully weighed samples of standardized materials whose heating values in Btu per pound are known.

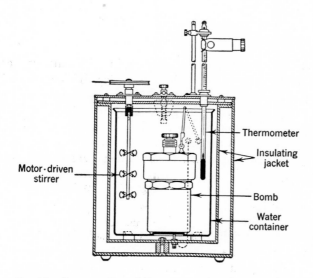

**Fig. 4.8.** Cross section of Parr oxygen-bomb calorimeter.

The heating value determined by the bomb calorimeter is called the *gross* or *higher heating value*, $Q_H$. When hydrogen is burned, 9 lb of water vapor are formed from 1 lb of hydrogen. In the bomb calorimeter, the products of combustion are cooled to within a few degrees of the initial temperature, and the water vapor formed from the combustion of the hydrogen in the fuel is condensed, thereby giving up its latent heat to the calorimeter. When fuel is burned, care must be exercised to prevent the condensation of this water vapor since practically all fuels contain sulphur. The products of combustion of sulphur will form sulphuric acid in the presence of water and cause serious corrosion. Since the latent heat given up by condensation of the water formed from the hydrogen cannot be utilized in commercial equipment, this latent heat may be deducted from the heating value as determined by the calorimeter in order to obtain a *net* or *lower heating value*, $Q_L$. In the United States the higher or gross heating value is normally used.

The higher heating value of gaseous fuels can be measured in a gas calorimeter. The volume of the gaseous fuel to be burned is accurately measured in a calibrated meter, and its pressure and temperature are noted. The gas is then burned in the calorimeter which is a specially constructed water heater in which the energy released by combustion may be computed from the quantity of water circulated through the calorimeter per cubic foot of fuel burned and the temperature rise of the water.

### 4.12  Combustion

Combustion is that chemical process in which oxygen combines rapidly with carbon, hydrogen, sulphur, and their compounds in solid, liquid, and gaseous fuels and results in the liberation of energy, the production of high-temperature gases, and the emission of light. The combustion process releases for the use of man some of the energy that was received upon the surface of the earth from the sun during the millions of years in which the fuel resources of the earth were being formed.

### 4.13  The Mole

The "molecular weight" of a given molecule is the mass of that molecule relative to the mass of the oxygen molecule, to which the number 32 is assigned. Thus the hydrogen molecule, which has a molecular weight of 2, has one sixteenth of the mass of the oxygen molecule. In this book, the term "molecular weight" will be used in order to conform to standard practice although the term "relative molecular mass" would be preferable.

The proportions in which the elements enter into the combustion reaction by mass are dependent upon the relative molecular weights, which are approximately as follows:

| Element | Symbol | Molecular Weight |
|---------|--------|------------------|
| Carbon | C | 12 |
| Sulphur | S | 32 |
| Hydrogen | $H_2$ | 2 |
| Oxygen | $O_2$ | 32 |
| Nitrogen | $N_2$ | 28 |

The symbols for the gases, hydrogen, oxygen, and nitrogen, are written above with the subscript 2 because they occur in nature as diatomic molecules: that is, as molecules composed of two atoms and not as individual atoms. Using these molecular weights, the molecular weights of various compounds can be determined as follows:

$CO_2$: $12 + 32 = 44$     $C_2H_6$:    $24 + 6 = 30$

$CO$: $12 + 16 = 28$     $SO_2$:    $32 + 32 = 64$

$H_2O$: $2 + 16 = 18$     $C_3H_8$:    $36 + 8 = 44$

$CH_4$: $12 + 4 = 16$     $C_{10}H_{22}$: $120 + 22 = 142$

Avogadro's law states that, at the same pressure and temperature, equal volumes of all perfect gases contain the *same number of molecules*. It follows, therefore, that the masses of equal volumes of all perfect gases at the same pressure and temperature are proportional to the molecular weights of the respective elements or compounds. Thus, at the same pressure and temperature, 2 lb of hydrogen, 28 lb of nitrogen, and 44 lb of carbon dioxide occupy equal volumes.

*The pound-mole, hereafter referred to as the mole, is that quantity of matter that has a mass in pounds numerically equal to the molecular weight of the substance.* Thus, 1 mole of hydrogen has a mass of 2 lb, 1 mole of nitrogen has a mass of 28 lb, and 1 mole of carbon dioxide has a mass of 44 lb. Also, 1 mole of any substance contains $2.7319 \times 10^{26}$ molecules.

In accordance with Avogadro's law, 1 mole of each of the various perfect gases occupies the same volume at equal pressures and temperatures. Therefore, for *perfect gases*, the mole is also a unit of volume. The volume of 1 mole of a perfect gas may be calculated as follows:

Let $\qquad\qquad\qquad M$ = molecular weight

For 1 lb of a perfect gas,

$$pv = RT$$

For $M$ lb of a perfect gas,

$$pMv = MRT$$

Let $\qquad\qquad\qquad V_m$ = the volume of 1 mole = $Mv$

Then, $\qquad\qquad V_m = MR\dfrac{T}{p}$

Since, for all perfect gases, $R = 1544/M$, then, $MR = 1544$, and

$$V_m = 1544\,\frac{T}{p} \qquad\qquad (4.6)$$

The volume of 1 mole of any perfect gas may be computed at any pressure and temperature by means of Equation 4.6. It is equal to 358 cu ft at 32 F and 14.7 psia. Therefore, at 32 F and 14.7 psia 2 lb of hydrogen, 28 lb of nitrogen, 44 lb of carbon dioxide, and 16 lb of methane ($CH_4$) each occupy 358 cu ft.

### 4.14   The Conversion of a Volumetric Analysis of a Gas to a Gravimetric Basis

The ultimate analysis of liquid and solid fuels such as oil and coal is expressed on a gravimetric (mass) basis. Gaseous fuels are analyzed volumetrically. By conversion of this volumetric analysis to a gravimetric ultimate analysis, it is possible to express the analysis of all kinds of fuels by one method and to develop a single method of calculation which will apply to the combustion of all kinds of fuels whether they be solids, liquids, or gases.

The products of combustion of all kinds of fuels are analyzed on a volumetric basis as discussed in Article 4.18. In order to perform combustion calculations, it is necessary to convert this volumetric analysis to a gravimetric basis so as to express the analyses of the fuel and its products of combustion in the same terms.

The conversion of a volumetric analysis of a gas to a gravimetric analysis can be made on the basis that a mole of gas is a unit of volume as well as being a quantity of matter having a mass numerically equal to the molecular weight of the element or compound. Thus, a volumetric analysis of a gas may be expressed as follows:

| Constituent | Volumetric Analysis, per cent | Cu Ft per 100 Cu Ft of Gas | Moles per 100 Moles of Gas |
|---|---|---|---|
| $H_2$ | 30 | 30 | 30 |
| CO | 30 | 30 | 30 |
| $CH_4$ | 20 | 20 | 20 |
| $C_2H_6$ | 10 | 10 | 10 |
| $O_2$ | 5 | 5 | 5 |
| $N_2$ | 5 | 5 | 5 |
| Total | 100 | 100 | 100 |

Taking 100 moles of gas as the basis upon which to make the computations, the mass of 100 moles can be found by multiplying the number of moles of each constituent by the molecular weight of that particular element or compound and adding the results. Thus, in Table 4.7, 10 moles of $C_2H_6$ (molecular weight $= 24 + 6 = 30$) have a mass of $10 \times 30 = 300$ lb. Also, since 1 mole of $C_2H_6$ contains 24 lb of carbon and 6 lb of hydrogen, the amount of carbon and hydrogen in 10 moles of $C_2H_6$ is the product of the number of moles of the compound and the molecular weight of each chemical element in the compound. The computations are shown in Table 4.7.

The specific volume, $v = RT/p$, of a gas having an analysis as given in Table 4.7 can be computed at a given pressure and temperature by first determining the value of the gas constant $R$ for this gas. Since 100 moles of this gas have a mass of 1820 lb as shown in column 4 of Table 4.7, the mass of 1 mole is 18.2 lb, or 18.2 is the molecular weight of a hypothetical gaseous compound having the same specific volume as the mixture of elements and compounds that constitute the gas whose analysis is shown in Table 4.7.

Then, since $R = 1544/M$, where $M$ is the molecular weight, $R = 1544/18.2 = 84.8$. At 100 F and 14.7 psia the specific volume of the gas would be

$$v = \frac{RT}{p} = \frac{84.8 \times 560}{14.7 \times 144} = 22.4 \text{ cu ft per lb}$$

## 4.15  Chemistry of Combustion

Except for special applications such as oxyacetylene welding, in which a high-temperature flame is necessary, the oxygen required for combustion is obtained from air. Air is a mechanical mixture of oxy-

**TABLE 4.7**

**Conversion of a Volumetric Analysis of a Gaseous Fuel to a Gravimetric Ultimate Analysis**

| (1) Constituent | (2) Volumetric Analysis, per cent | (3) Moles per 100 Moles | Pounds of Each Constituent per 100 Moles of Gas | | | | |
|---|---|---|---|---|---|---|---|
| | | | (4) Total | (5) Carbon | (6) Hydrogen | (7) Oxygen | (8) Nitrogen |
| $H_2$ | 30 | 30 | $30 \times 2 = 60$ | | $30 \times 2 = 60$ | | |
| CO | 30 | 30 | $30 \times 28 = 840$ | $30 \times 12 = 360$ | | $30 \times 16 = 480$ | |
| $CH_4$ | 20 | 20 | $20 \times 16 = 320$ | $20 \times 12 = 240$ | $20 \times 4 = 80$ | | |
| $C_2H_6$ | 10 | 10 | $10 \times 30 = 300$ | $10 \times 24 = 240$ | $10 \times 6 = 60$ | | |
| $O_2$ | 5 | 5 | $5 \times 32 = 160$ | | | $5 \times 32 = 160$ | |
| $N_2$ | 5 | 5 | $5 \times 28 = 140$ | | | | $5 \times 28 = 140$ |
| Total | 100 | 100 | 1820 | 840 | 200 | 640 | 140 |

Ultimate gravimetric analysis:

$$Carbon = \tfrac{840}{1820} \times 100 = 46.2\%$$
$$Hydrogen = \tfrac{200}{1820} \times 100 = 11.0\%$$
$$Oxygen = \tfrac{640}{1820} \times 100 = 35.1\%$$
$$Nitrogen = \tfrac{140}{1820} \times 100 = 7.7\%$$

gen and nitrogen, plus negligible amounts of other gases, and for engineering purposes may be considered to have the following composition:

$$1 \text{ lb of } O_2 + 3.31 \text{ lb of } N_2 \qquad = 4.31 \text{ lb of air}$$

$$1 \text{ cu ft of } O_2 + 3.77 \text{ cu ft of } N_2 = 4.77 \text{ cu ft of air}$$

$$1 \text{ mole of } O_2 + 3.77 \text{ moles of } N_2 = 4.77 \text{ moles of air}$$

When fuel is burned completely, all carbon is burned to carbon dioxide ($CO_2$), all hydrogen is burned to water vapor ($H_2O$), and all sulphur is burned to sulphur dioxide ($SO_2$).

The combustion reactions are written on the basis of the law of conservation of matter, which states that the quantity of matter entering into a reaction is equal to the quantity of the matter in the products of the reaction and that, although the chemical composition of the compounds may change, the amount of each element in the reaction remains the same.

The reaction for the complete combustion of carbon may be written as follows:

$$1C + 1O_2 = 1CO_2 + 14{,}600 \text{ Btu per lb of C} \qquad (4.7)$$

The numbers ahead of the symbols for the elements or compounds represent the number of molecules or moles entering into the reaction and are normally omitted if the number is 1. Equation 4.7 may be interpreted as follows:

$$1 \text{ molecule of C} + 1 \text{ molecule of } O_2 = 1 \text{ molecule of } CO_2$$

$$1 \text{ mole of C} + 1 \text{ mole of } O_2 = 1 \text{ mole of } CO_2$$

or, if the molecular weights are used,

$$12 \text{ lb of C} + 32 \text{ lb of } O_2 = 44 \text{ lb of } CO_2$$

$$1 \text{ lb of C} + 2\tfrac{2}{3} \text{ lb of } O_2 = 3\tfrac{2}{3} \text{ lb of } CO_2$$

Equation 4.7 indicates that the complete combustion of 1 lb of carbon releases 14,600 Btu.

In the absence of sufficient oxygen, the carbon may be burned incompletely in accordance with the following reaction:

$$2C + O_2 = 2CO + 4440 \text{ Btu per lb of C} \qquad (4.8)$$

This reaction releases only about 30 per cent of the energy released by complete combustion, and the escape of the CO thus formed is detri-

mental to efficient utilization of the fuel.  In the presence of sufficient
oxygen, the combustion of CO to $CO_2$ may be completed as follows:

$$2CO + O_2 = 2CO_2 + 10{,}160 \text{ Btu per lb of C} \tag{4.9}$$

Thus the total energy released is the same whether 1 lb of carbon is
burned directly to $CO_2$, or is burned to CO and the CO subsequently
burned to $CO_2$.

The complete combustion of hydrogen occurs as follows:

$$2H_2 + O_2 = 2H_2O + 62{,}000 \text{ Btu per lb of } H_2 \tag{4.10}$$

$$2 \text{ moles of } H_2 + 1 \text{ mole of } O_2 = 2 \text{ moles of } H_2O$$

$$4 \text{ lb of } H_2 + 32 \text{ lb of } O_2 = 36 \text{ lb of } H_2O$$

$$1 \text{ lb of } H_2 + 8 \text{ lb of } O_2 = 9 \text{ lb of } H_2O$$

Sulphur burns as follows:

$$S + O_2 = SO_2 + 4050 \text{ Btu per lb of S} \tag{4.11}$$

$$1 \text{ mole of } S + 1 \text{ mole of } O_2 = 1 \text{ mole of } SO_2$$

$$32 \text{ lb of } S + 32 \text{ lb of } O_2 = 64 \text{ lb of } SO_2$$

$$1 \text{ lb of } S + 1 \text{ lb of } O_2 = 2 \text{ lb of } SO_2$$

Much of the sulphur in fuel is in a chemical form in which it is partially oxidized or completely incombustible.  It adds little to the heating value of the fuel and is objectionable because of the corrosive
characteristics of the products of combustion.

The hydrocarbons or compounds of carbon and hydrogen, when
burned completely, are converted to $CO_2$ and $H_2O$ as in the following typical reactions:

$$2C_2H_6 + 7O_2 = 4CO_2 + 6H_2O \tag{4.12}$$

$$2 \text{ moles of } C_2H_6 + 7 \text{ moles of } O_2 = 4 \text{ moles of } CO_2 + 6 \text{ moles of } H_2O$$

$$60 \text{ lb of } C_2H_6 + 224 \text{ lb of } O_2 = 176 \text{ lb of } CO_2 + 108 \text{ lb of } H_2O$$

$$1 \text{ lb of } C_2H_6 + 3.733 \text{ lb of } O_2 = 2.933 \text{ lb of } CO_2 + 1.80 \text{ lb of } H_2O$$

$$2C_8H_{18} + 25O_2 = 16CO_2 + 18H_2O \tag{4.13}$$

$$2 \text{ moles of } C_8H_{18} + 25 \text{ moles of } O_2$$

$$= 16 \text{ moles of } CO_2 + 18 \text{ moles of } H_2O$$

$$228 \text{ lb of } C_8H_{18} + 800 \text{ lb of } O_2 = 704 \text{ lb of } CO_2 + 324 \text{ lb of } H_2O$$

$$1 \text{ lb of } C_8H_{18} + 3.509 \text{ lb of } O_2 = 3.088 \text{ lb of } CO_2 + 1.421 \text{ lb of } H_2O$$

Most fuels contain mixtures of complex hydrocarbon compounds which undergo ·a series of intermediate changes during combustion, resulting in the formation of many new species of compounds of carbon, hydrogen, and oxygen. However, the end products of complete combustion of these fuels will be $CO_2$, $H_2O$, and $SO_2$.

### 4.16 Theoretical Air

*Theoretical air* is the minimum amount of air that must be supplied to burn 1 lb of fuel completely. It may be calculated from the ultimate analysis of the fuel by using Equations 4.7, 4.10, and 4.11 to compute the mass of oxygen required to burn the carbon, hydrogen, and sulphur in 1 lb of fuel completely, than subtracting the oxygen in the fuel to obtain the oxygen required from the air, and calculating the amount of air from the known composition of air.

*Excess air* is air that is supplied for combustion in excess of the chemical requirements of the fuel. In practice, because of poor mixing and lack of time, excess air must be supplied if all the combustible constituents of the fuel are to be burned completely. The excess air is normally expressed as a percentage of the theoretical air required.

**Example 3.** Compute (*a*) the theoretical air required and (*b*) the air actually supplied if 25 per cent of excess air is supplied, when 1 lb of the Sullivan, Ind., coal having the ultimate analysis reported in Table 4.2 is burned completely.

*Solution:* The solution is worked out in tabular form in Table 4.8. The ultimate analysis is tabulated in column 1 of Table 4.8, the chemical reactions and combining proportions are given in column 2 for each combustible element, and the oxygen necessary to burn completely the carbon, hydrogen, and sulphur in 1 lb of fuel is shown in column 3. Since 1 lb of fuel contains 0.199 lb of oxygen, the oxygen in the fuel is subtracted from the total oxygen required to obtain a net oxygen requirement of 2.044 lb supplied as air. The nitrogen that accompanies this oxygen and the resulting air are computed from the known composition of air, giving a theoretical air requirement of 8.81 lb per lb of fuel. For 25 per cent excess air, the air supply is increased to 125 per cent of the theoretical requirement.

If a fuel is analyzed volumetrically (a gaseous fuel), the ultimate gravimetric analysis of the fuel can be determined as outlined in Table 4.7, after which the procedures used in Example 3 may be applied.

## TABLE 4.8

### Tabular Solution of Examples 3 and 4 for 1 Lb of Fuel

| Ultimate Analysis of Coal, per cent (1) | Chemical Reactions (2) | $O_2$ Required from Air, lb per lb of fuel (3) | Products of Complete Combustion, lb per lb of fuel | | | | |
|---|---|---|---|---|---|---|---|
| | | | $CO_2$ (4) | $H_2O$ (5) | $SO_2$ (6) | N (7) | $O_2$ (8) |
| C = 66.0 | $C + O_2 = CO_2$ $12 + 32 = 44$ | $\frac{32}{12} \times 0.66 = 1.760$ | $\frac{44}{12} \times 0.66 = 2.420$ | | | | |
| H = 5.9 | $2H_2 + O_2 = 2H_2O$ $4 + 32 = 36$ | $\frac{32}{4} \times 0.059 = 0.472$ | | $\frac{36}{4} \times 0.059 = 0.531$ | | | |
| S = 1.1 | $S + O_2 = SO_2$ $32 + 32 = 64$ | $\frac{32}{32} \times 0.011 = 0.011$ | | | $\frac{64}{32} \times 0.011 = 0.022$ | | |
| O = 19.9 N = 1.5 Ash = 5.6 | | Less $O_2$ in fuel = −0.199 | | | | 0.015 | |
| Total = 100.0 | | Theoretical $O_2$ required from air = 2.044 $N_2$ in air required to supply 2.044 lb of $O_2$ = 2.044 × 3.31 = 6.766 | | | | 6.766 | |
| | | Theoretical air = 2.044 × 4.31 = 8.810 | 2.420 | 0.531 | 0.022 | 6.781 | 0.0 |
| | | For 25% of excess air: Excess $O_2$ = 0.25 × 2.044 = 0.511 Excess $N_2$ = 0.25 × 6.766 = 1.692 | | | | 1.692 | 0.511 |
| | | Total air supplied, 25% excess = 11.013 | 2.420 | 0.531 | 0.022 | 8.473 | 0.511 |

Check: For 25% excess air:

Matter in = 1.0 lb of fuel + 11.013 lb of air = 12.013 lb

Matter out = 0.056 lb of ash + 2.420 lb of $CO_2$ + 0.531 lb of $H_2O$ + 0.022 lb of $SO_2$ + 8.473 lb of $N_2$ + 0.511 lb of $O_2$ = 12.013 lb

### 4.17   Products of Complete Combustion of a Fuel Analyzed Gravimetrically

When a fuel is burned completely, all the carbon, hydrogen, sulphur, and oxygen in the fuel appear in the products of combustion as $CO_2$, $H_2O$, and $SO_2$, respectively, together with all the oxygen in the theoretical air. The nitrogen in the fuel and the nitrogen in the theoretical air appear unchanged in the products of combustion. If excess air is supplied, the oxygen and nitrogen in the excess air appear unchanged in the products of combustion.

**Example 4.** Determine (a) the products of complete combustion without excess air and (b) the products of complete combustion with 25 per cent of excess air, when 1 lb of Sullivan, Ind., coal having the ultimate analysis reported in Table 4.2 is burned completely.

*Solution:* The solution is worked out in tabular form in Table 4.8. The masses of $CO_2$, $H_2O$, and $SO_2$ resulting from the complete combustion of 1 lb of fuel are calculated from the chemical equations and combining proportions as shown in columns 4, 5, and 6. Column 7 contains the nitrogen in the fuel and the nitrogen that accompanied the oxygen in the theoretical air. For the case of 25 per cent excess air, the excess oxygen and accompanying nitrogen appear in columns 8 and 7, respectively. A check on the accuracy of the computations for 25 per cent excess air is made by comparing the mass of matter entering into the process with the mass of matter resulting from the process.

### 4.18   Analysis of the Dry Gaseous Products of Combustion

If insufficient air is supplied, combustion will be incomplete, and part of the heating value of the fuel will be wasted through the escape of combustible gases such as CO and $H_2$. If more air is supplied than is necessary for complete combustion, the surplus air is heated from atmospheric temperature to the temperature at which the products of combustion are discharged to waste (usually 350 to 600 F), thus carrying away a considerable amount of energy. Practically, air in excess of that required theoretically to satisfy the chemical requirements of the fuel must be supplied because of inadequate mixing of air and combustible gases, lack of time to complete combustion, and other factors to be discussed in Chapter 6. At the same time, the quantity of this air must be kept to a minimum consistent with reasonably complete combustion. In order to determine the completeness of combustion and the quantity of excess air being

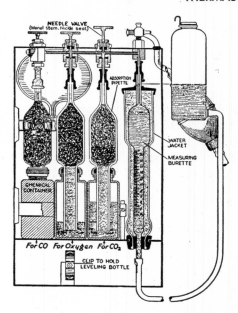

**Fig. 4.9.** Orsat apparatus.

supplied, the dry gaseous products of combustion may be analyzed volumetrically by an *Orsat* apparatus.

The Orsat apparatus is illustrated in Fig. 4.9. The volume of gas may be measured in a burette graduated to read volumes in per cent and water-jacketed to maintain a constant temperature. Three absorption pipettes contain suitable chemicals for the absorption, respectively, of $CO_2$, $O_2$, and CO. Each of the absorption pipettes may be connected to the measuring burette by a capillary tube and valve.

The measuring burette is filled at atmospheric pressure with a representative sample of the gas to be analyzed, after which the inlet valve is closed. By manipulation of a leveling bottle containing water, the gas can be transferred to the first pipette for absorption of $CO_2$. The gas is then drawn back into the measuring burette, and the reduction in volume, which is the $CO_2$ content of the original sample, may be read directly in per cent. The per cent by volume of $O_2$ and CO in the original sample may be determined successively by passing the sample into the respective absorption pipettes, withdrawing it into the measuring burette, and noting the further reduction in volume of the sample. The remainder of the original gas sample is assumed to be nitrogen. A typical Orsat analysis of the dry products of combustion of coal would be: $CO_2 = 14.2\%$, $O_2 = 5.5\%$, $CO = 0.1\%$, and (by difference) $N_2 = 80.2\%$.

The maximum percentage of $CO_2$ that is obtainable by the complete combustion without excess air of any given fuel depends upon the ultimate analysis of the fuel and varies from 6 to 25 per cent, as indicated in Fig. 4.10. Excess air dilutes the $CO_2$ with surplus $O_2$ and $N_2$ and thereby reduces the percentage of $CO_2$ in the gases as shown in Fig. 4.10. Therefore, for any given fuel, the $CO_2$ content measured in per cent by volume is an indication of the excess air supplied while the CO content is an indication of the incompleteness of combustion.

Carbon dioxide recorders have been on the market for many years for the purpose of continuously recording the $CO_2$ content of the products of combustion as an aid to the maintenance of optimum combustion conditions. Oxygen recorders are being used extensively as a guide to combustion control. Figure 4.11 shows the effect of excess air upon the $O_2$ content of the products of complete combustion of hydrogen, methane, and carbon. The curve for carbon in Fig. 4.11 corresponds closely to the curve for coke in Fig. 4.10, whereas the curve for methane in Fig. 4.11 corresponds closely to the curve for natural gas in Fig. 4.10. The curve for hydrogen would be a curve of zero per cent of $CO_2$ in Fig. 4.10. It will be noted that there is much less variation in $O_2$ content with the various types of fuel at a given per cent of excess air than the variation in $CO_2$ content for the same fuels. Also, the curve of $O_2$ content is fairly steep in the region

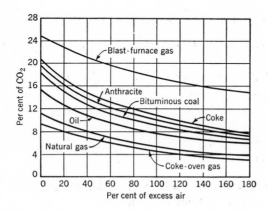

**Fig. 4.10.** Relation between excess air and $CO_2$ for representative fuels.

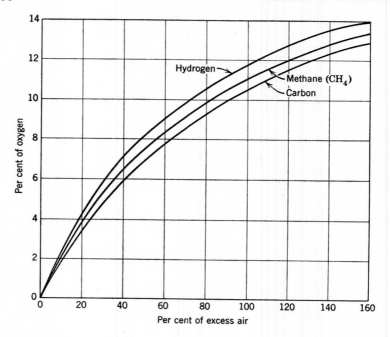

**Fig. 4.11.** Effect of excess air on the per cent of oxygen in the dry gaseous products of complete combustion.

from 10 to 40 per cent of excess air, which is the region of normal operation with good equipment.

### 4.19  Dry Gaseous Products of Combustion per Pound of Fuel

The dry gaseous products of combustion of 1 lb of fuel may be computed from the Orsat analysis of the dry products of combustion, the analysis of the fuel and refuse, and a carbon balance.

The dry products of combustion per *pound of carbon in the dry products of combustion* may be computed from the Orsat analysis by applying the methods outlined in Article 4.14.

Let $G_c$ = mass of dry products of combustion per lb of *carbon* in the dry products of combustion.  Then,

$$G_c = \frac{\left\{ \begin{array}{l} \text{lb of dry products of combustion per 100 moles of dry} \\ \quad \text{products of combustion} \end{array} \right\}}{\text{lb of C per 100 moles of dry products of combustion}}$$

The numerator and denominator in the above equation can be computed from the Orsat analysis of the dry products of combustion by using 100 moles of the gas as the basis of calculation as follows:

| Constit-uent | Per Cent by Volume | Moles per 100 Moles | Lb per 100 Moles | Lb of Carbon per 100 Moles |
|---|---|---|---|---|
| $CO_2$ | 10 | 10 | $10 \times 44 = 440$ | $10 \times 12 = 120$ |
| $O_2$ | 9 | 9 | $9 \times 32 = 288$ | |
| $CO$ | 1 | 1 | $1 \times 28 = 28$ | $1 \times 12 = 12$ |
| $N_2$ | 80 | 80 | $80 \times 28 = 2240$ | |
| Total | 100 | 100 | 2996 | 132 |

Then,

$$G_c = \frac{2996 \text{ lb of dry products of combustion per 100 moles of gas}}{132 \text{ lb of C per 100 moles of gas}}$$

$G_c = 22.7$ lb of dry gas per lb of *carbon* in the *gas*

The mass of the dry products of combustion *per pound of fuel* may be calculated as follows:

Let $G_f$ = lb of dry products of combustion per lb of fuel
$C_b$ = lb of carbon burned to CO and $CO_2$ per lb of fuel (lb of C in dry products of combustion per lb of fuel)

Then
$$G_f = G_c \times C_b \qquad (4.14)$$

or

$$G_f = \frac{\text{lb of dry products of combustion per 100 moles}}{\text{lb of C in dry products of combustion per 100 moles}}$$

$$\times \frac{\text{lb of C in dry products of combustion}}{\text{lb of fuel}}$$

$$= \frac{\text{lb of dry products of combustion}}{\text{lb of fuel}}$$

The ASME test code for steam-generating units specifies that $C_b$, the carbon burned to CO and $CO_2$ per pound of fuel, shall be determined as follows:

Let $C_f$ = carbon content of the fuel as reported in the ultimate analysis, expressed as a decimal

$m_f$ = mass of fuel burned during the test, lb

$m_r$ = mass of refuse collected during the test, lb

$Q_r$ = heating value of 1 lb of refuse as determined by the bomb calorimeter

Then $\dfrac{m_r}{m_f}$ = lb of refuse per lb of fuel burned

It is assumed that the combustible material in the refuse is carbon, which has a heating value of 14,600 Btu per lb.

Therefore
$$C_b = C_f - \left(\frac{m_r}{m_f} \times \frac{Q_r}{14,600}\right) \tag{4.15}$$

Figure 4.12 shows the carbon balance for the combustion of coal from which Equation 4.15 is derived.

When the fuel is oil or gas, the ash content is zero or negligible, $m_r$ is therefore zero, and $C_b = C_f$ = carbon content of the fuel as reported in the ultimate analysis.

The $SO_2$ formed from the combustion of sulphur is absorbed in the Orsat apparatus as $CO_2$. Since the molecular weights of carbon and sulphur are 12 and 32, respectively, and 1 mole of each produces 1 mole of gas (equal volumes), the sulphur may be treated as equivalent carbon by taking $1\frac{2}{32}$ or $\frac{3}{8}$ of its mass as reported in the ulti-

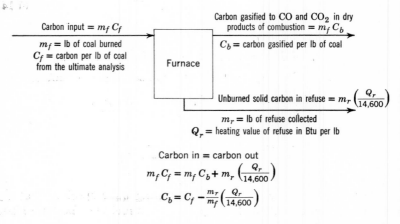

Fig. 4.12. Carbon balance for the combustion of coal.

mate analysis and adding this mass to that of the carbon in the fuel. The ASME test code for steam-generating units recommends, therefore, that the dry gaseous products of combustion be calculated as follows:

$$G_f = G_c(C_b + \tfrac{3}{8}S) + \tfrac{5}{8}S \qquad (4.16)$$

where $S$ = sulphur in the ultimate analysis, expressed as a decimal.

The calculation of the dry products of combustion per pound of fuel may be illustrated by an example.

**Example 5.** Determine the mass of the dry products of combustion per lb of coal if Sullivan, Ind., coal, the analysis of which is given in Table 4.2, was burned in a test in which the following data were collected:

> Coal burned = 10,000 lb $(m_f)$
> Ashpit refuse = 700 lb $(m_r)$
> Heating value of ashpit refuse = 2920 Btu/lb of refuse $(Q_r)$
> Orsat analysis in per cent: $CO_2$ = 14.1; $O_2$ = 5.1; $CO$ = 0.1;
> $N_2$ (by difference) = 80.7 per cent

*Solution:* From the ultimate analysis of the coal as given in Table 4.2, the carbon content is 66 per cent, or $C_f$ = 0.66 lb per lb of fuel.

$$C_b = C_f - \left( \frac{m_r}{m_f} \times \frac{Q_r}{14,600} \right)$$

$$= \frac{0.66 \text{ lb of carbon}}{\text{lb of fuel}} - \left( \frac{700 \text{ lb of refuse}}{10,000 \text{ lb of fuel}} \times \frac{2920 \dfrac{\text{Btu}}{\text{lb of refuse}}}{14,600 \dfrac{\text{Btu}}{\text{lb of carbon}}} \right)$$

$$= \frac{0.66 \text{ lb of carbon}}{\text{lb of fuel}} - \frac{0.014 \text{ lb of carbon}}{\text{lb of fuel}} = 0.646 \text{ lb of carbon}$$

per lb of fuel

The dry gaseous products of combustion per lb of *carbon*, $G_c$, may be calculated from the Orsat analysis as follows:

| Con-stituent | Analysis in Per Cent by Volume | Analysis in Moles per 100 Moles | Lb per 100 Moles | | Lb of Carbon per 100 Moles | |
|---|---|---|---|---|---|---|
| $CO_2$ | 14.1 | 14.1 | $14.1 \times 44 =$ | 620.4 | $14.1 \times 12 =$ | 169.2 |
| $O_2$ | 5.1 | 5.1 | $5.1 \times 32 =$ | 163.2 | | |
| $CO$ | 0.1 | 0.1 | $0.1 \times 28 =$ | 2.8 | $0.1 \times 12 =$ | 1.2 |
| $N_2$ (by difference) | 80.7 | 80.7 | $80.7 \times 28 =$ | 2259.6 | | |
| Total | 100 | 100 | | 3046.0 | | 170.4 |

$$G_c = \frac{3046 \text{ lb of dry gas per 100 moles of dry gas}}{170.4 \text{ lb of C per 100 moles of dry gas}}$$

$$= 17.87 \text{ lb of dry gas per lb of } \textit{carbon}$$

$$G_f = G_c(C_b + \tfrac{3}{8}S) + \tfrac{5}{8}S = 17.87[0.646 + \tfrac{3}{8}(0.011)] + \tfrac{5}{8}(0.011)$$

$$= 11.63 \text{ lb of dry gaseous products of combustion per lb of fuel}$$

If the sulphur is neglected and Equation 4.15 is used instead of Equation 4.16, the result is 11.55 lb of dry gaseous products of combustion per lb of fuel or a difference of about 0.5 per cent.

### 4.20   Carbon Burned to CO per Pound of Fuel

As shown by Equation 4.9, the incomplete combustion of carbon produces CO with the release of only about 30 per cent of the energy released by the complete combustion of carbon. It is important, therefore, to be able to compute the carbon burned incompletely to CO per pound of fuel so that the extent of this loss may be evaluated.

As shown in Article 4.19, the mass of carbon in 100 moles of dry products of combustion may be determined by multiplying the number of moles of $CO_2$ and CO in 100 moles of dry products of combustion by 12, which is the molecular weight of carbon or the mass of carbon in 1 mole of either $CO_2$ or CO. Thus, if the Orsat analysis shows 10 per cent $CO_2$ and 1 per cent CO, the mass of carbon in 10 moles of $CO_2$ equals $10 \times 12 = 120$ lb, and the mass of carbon in 1 mole of CO equals $1 \times 12 = 12$ lb. Therefore, 100 moles of dry products of combustion having 10 per cent of $CO_2$ and 1 per cent of CO contain $120 + 12 = 132$ lb of carbon, of which $12/132$ of the carbon burned is in the form of CO.

**Example 6.**   For the data in Example 5, compute the carbon burned to CO per lb of fuel.

*Solution:* The Orsat analysis in Example 5 shows 14.1 per cent of $CO_2$ and 0.1 per cent of CO in the dry gaseous products of combustion, or 14.1 moles of $CO_2$ and 0.1 mole of CO per 100 moles of dry gaseous products of combustion. Then

$$\text{Carbon in } CO_2 \text{ per 100 moles of gas} = 14.1 \text{ moles of } CO_2 \times \frac{12 \text{ lb of carbon}}{\text{mole of } CO_2}$$

$$= 169.2 \text{ lb of carbon}$$

$$\text{Carbon in CO per 100 moles of gas} = 0.1 \text{ mole of CO} \times \frac{12 \text{ lb of carbon}}{\text{mole of CO}}$$

$$= 1.2 \text{ lb of carbon}$$

Total carbon in 100 moles of dry products of combustion = 170.4 lb

Therefore:

Carbon burned to CO per lb of carbon

$$= \frac{1.2 \text{ lb of C in the form of CO per 100 moles of gas}}{(1.2 + 169.2) \text{ lb of C in the form of CO and CO}_2 \text{ per 100 moles of gas}}$$
$$= 0.00704 \text{ lb}$$

Carbon burned to CO per lb of fuel

$$= (0.00704 \text{ lb of C in CO per lb of C})(0.646 \text{ lb of C burned}$$
$$\text{to CO and CO}_2 \text{ per lb of fuel})$$
$$= 0.00455 \text{ lb}$$

### 4.21 Water Vapor in the Products of Combustion per Pound of Fuel

If the small amount of moisture in the air supplied for combustion is neglected, the moisture in the products of combustion of 1 lb of fuel comes from two sources: (1) the moisture formed from the combustion of hydrogen, and (2) the moisture in the fuel. Since 1 lb of hydrogen produces 9 lb of moisture when burned, the moisture in the products of combustion of 1 lb of fuel equals $(M + 9H)$, where $M$ and $H$ are the amounts of moisture and hydrogen in the ultimate analysis of the fuel, expressed as decimal parts.

**Example 7.** Compute the water vapor present in the products of combustion of 1 lb of Sullivan, Ind., coal having the analysis reported in Table 4.2.

*Solution:* From column $B$ of the solution to Example 2, the hydrogen and moisture per lb of fuel are found to be 4.4 and 13.5 per cent, respectively. Then the water vapor in the products of combustion of 1 lb of this fuel = $M + 9H = 0.135 + 9 \times 0.044 = 0.531$ lb.

The same result would have been obtained by using the analysis of column $A$, Example 2, since $9H = 9 \times 0.059 = 0.531$ lb. In column $A$, the moisture in the coal as given in the proximate analysis is reported as its chemical elements, hydrogen and oxygen, so that a separate item for moisture does not appear in the analysis. Care must be exercised to use the correct moisture and hydrogen values in computing total moisture in the products of combustion.

### 4.22 Air Supplied per Pound of Fuel

The ASME test code for steam-generating units specifies that the air actually supplied per pound of fuel burned shall be computed from a material balance as illustrated in Fig. 4.13. The ash in the coal appears in the refuse together with the unburned carbon. The moisture may appear in the ultimate analysis as a separate item, as in column $B$ of Example 2, in which case it appears unchanged in

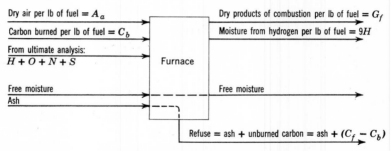

Matter entering into combustion reactions per lb of fuel $= A_a + C_b + H + O + N + S$
Matter resulting from combustion reactions per lb of fuel $= G_f + 9H$
Dry air supplied per lb of fuel $= A_a = (G_f + 9H) - (C_b + H + O + N + S)$

**Fig. 4.13.** Material balance for determining dry air supplied per pound of fuel.

composition in the products of combustion and may be disregarded in the material balance. If the moisture is reported as hydrogen and oxygen in the ultimate analysis, as in column $A$ of Example 2, then it appears in the products of combustion as "moisture from hydrogen." Consequently, the material balance may be written for 1 lb of fuel as follows:

Matter entering into the combustion reactions
$$= A_a \text{ lb of dry air} + C_b + H + O + N + S$$

where $H$, $O$, $N$, and $S$ represent the amounts of hydrogen, oxygen, nitrogen, and sulphur as reported in the ultimate analysis, expressed as decimals, and $A_a =$ lb of dry air supplied per lb of fuel.

Matter resulting from the combustion reactions
$$= G_f \text{ lb of dry products of combustion} + 9H \text{ lb of water vapor}$$

Therefore,   $A_a = G_f + 9H - (H + O + N + S + C_b)$       (4.17)

The excess air supplied for combustion is the actual air supplied, $A_a$, minus the theoretical air required, $A_t$, which may be calculated as outlined in Article 4.16. The percentage of excess air is the excess air divided by the theoretical air, expressed in per cent, and may be computed thus:

$$\text{Excess air, } \% = \frac{A_a - A_t}{A_t} \times 100 \qquad (4.18)$$

**Example 8.** Using the data and results from Examples 3, 5, and 7, compute the actual air supplied per lb of fuel and the excess air in per cent.

*Solution:* The material balance for 1 lb of fuel may be represented as follows:

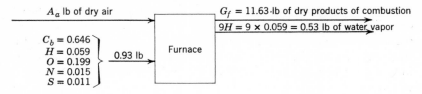

Then

$$A_a = G_f + 9H - (C_b + H + O + N + S) = 11.63 + 0.53 - (0.930)$$
$$= 11.23 \text{ lb of air actually supplied per lb of fuel}$$

From Example 3, the theoretical air required to burn completely 1 lb of this particular coal was found to be 8.81 lb $= A_t$. Therefore, excess air $= 11.23$ lb actually supplied $- 8.81$ lb required $= 2.42$ lb.

$$\text{Excess air, } \% = \frac{A_a - A_t}{A_t} \times 100 = \frac{2.24 \text{ lb}}{8.81 \text{ lb}} \times 100 = 25.4 \%$$

### 4.23  The Energy Balance

The energy balance is normally based upon 1 lb of fuel and is a tabulation of the amount of energy utilized and the amount of energy lost in various ways. The energy balance for 1 lb of coal may be written as follows:

| Item | Btu per Lb | Per Cent of Heating Value |
|---|---|---|
| I. Energy utilized | . . . | . . . |
| II. Losses | | |
|     1. Loss due to sensible heat in the dry gaseous products of combustion, $Q_1$ | . . . | . . . |
|     2. Loss due to CO in the dry gaseous products of combustion, $Q_2$ | . . . | . . . |
|     3. Loss due to carbon in the solid refuse, $Q_3$ | . . . | . . . |
|     4. Loss due to moisture in the fuel, $Q_4$ | . . . | . . . |
|     5. Loss due to water vapor formed from the hydrogen in the fuel, $Q_5$ | . . . | . . . |
|     6. Loss due to superheating the water vapor in the air supplied for combustion, loss due to unconsumed hydrogen and hydrocarbons, radiation, and unaccounted-for loss, $Q_6$ | . . . | . . . |
|     Total | Heating value of 1 lb of fuel | 100 |

For steam-generating units, the calculation of the energy absorbed or utilized is considered in detail in Article 8.12. This item, expressed as a percentage of the heating value of the fuel, is also the efficiency of the unit. The maximum efficiency with which the energy in fuel may be utilized is limited by the design of the equipment in which it is being used. The attainment of this maximum efficiency in daily operation of the equipment is dependent upon keeping the losses to a minimum. The computation of these losses and the intelligent utilization of the results of these computations are very important if maximum efficiency is to be maintained. The methods of computing the losses are considered in the following paragraphs and are illustrated by Example 9, which is based on the data and results of Example 5 and the following additional data: temperature of air $= 70\ F = t_a$; discharge temperature of gaseous products of combustion $= 500\ F = t_g$; temperature of fuel $= 60\ F = t_f$.

1. The loss due to sensible heat in the dry gaseous products of combustion results from the discharge to waste of the products of combustion at a temperature above that at which the air for combustion is supplied. This loss may be computed as follows:

$$Q_1 = G_f \times c_p \times (t_g - t_a) = 0.24 G_f (t_g - t_a) \qquad (4.19)$$

where $Q_1 =$ sensible heat loss, Btu per lb of fuel

$G_f =$ dry gaseous products of combustion, lb per lb of fuel

$c_p = 0.24 =$ mean specific heat of dry gaseous products of combustion, Btu per lb per F

$t_g =$ temperature at which the gaseous products of combustion are discharged to waste, F

$t_a =$ temperature of air, F

**Example 9a.** From Example 5,

$$G_f = 11.63 \text{ lb per lb of fuel}$$

$$Q_1 = 0.24 \times 11.63(500 - 70) = 1200 \text{ Btu per lb of fuel}$$

This is the largest loss that occurs in most fuel-burning installations. Inspection of formula 4.19 shows that this loss may be reduced by lowering the exit gas temperature, $t_g$, or by reducing the dry gaseous products of combustion of 1 lb of fuel, $G_f$. The minimum value of $t_g$ in an actual installation is generally fixed by the design of the equipment. The quantity of the dry gaseous products of combustion is dependent upon the excess air which cannot be reduced too much without increasing the losses due to incomplete combustion.

2. The loss due to the escape of CO in the dry gaseous products of combustion, $Q_2$, is the product of the carbon burned to CO per pound of fuel and the heating value of 1 lb of carbon in the form of CO. As pointed out in Article 4.15, 14,600 Btu are released when 1 lb of carbon is burned to $CO_2$, but only 4440 Btu are released when 1 lb of carbon is burned to CO. The undeveloped energy due to the incomplete combustion of 1 lb of carbon is therefore equal to 14,600 − 4440 or 10,160 Btu. Therefore,

$Q_2$ = carbon burned to CO per lb of fuel

$\times$ 10,160 Btu per lb of carbon burned to CO

**Example 9b.** In Example 6, the carbon burned to CO per lb of fuel was found to be 0.00455 lb. Therefore,

$Q_2$ = 0.00455 lb of C burned to CO per lb of fuel

$\times$ 10,160 Btu per lb of C

= 46 Btu per lb of fuel

3. The loss due to unburned carbon in the refuse, $Q_3$, may be computed from the heating value of the refuse, $Q_r$, Btu per lb of refuse, and the refuse collected per lb of fuel burned, $m_r/m_f$, or

$$Q_3 = \frac{m_r}{m_f} Q_r \qquad (4.20)$$

**Example 9c.** From Example 5, $m_f$ = 10,000 lb of coal, $m_r$ = 700 lb of refuse, and $Q_r$ = 2920 Btu per lb of refuse. Then

$$Q_3 = \frac{700 \text{ lb of refuse}}{10,000 \text{ lb of fuel}} \times \frac{2920 \text{ Btu}}{\text{lb of refuse}} = 204 \text{ Btu per lb of fuel}$$

4. The loss due to evaporating and superheating the moisture in 1 lb of fuel, $Q_4$, may be computed from the product of the moisture in 1 lb of fuel as given by the proximate analysis and the heat required to evaporate and superheat 1 lb of moisture. The following empirical equations are given in the ASME test code for steam-generating units:

$$Q_4 = M(1089 - t_f + 0.46t_g), \text{ where } t_g < 575 \text{ F} \qquad (4.21)$$

$$Q_4 = M(1066 - t_f + 0.50t_g), \text{ where } t_g > 575 \text{ F} \qquad (4.22)$$

where $t_f$ = the temperature of the fuel, F
and $M$ = the moisture in 1 lb of fuel as reported in the proximate analysis, expressed as a decimal

**Example 9d.** Since the proximate analysis of this fuel shows 13.5 per cent of moisture, $t_g = 500$ F and $t_f = 60$ F,

$$Q_4 = 0.135(1089 - 60 + 0.46 \times 500) = 170 \text{ Btu per lb of fuel}$$

5. The loss due to moisture formed from the combustion of hydrogen, $Q_5$, results from the fact that this moisture is condensed in the bomb calorimeter, and the energy given up by the condensation of this moisture in the calorimeter is included in the heating value of the fuel, whereas, when the fuel is burned in a commercial installation, the moisture is discharged to waste as a vapor at a temperature $t_g$. Since 1 lb of $H_2$ produces by its combustion 9 lb of $H_2O$, the loss due to the escape of this water vapor at $t_g$ may be computed thus:

$$Q_5 = 9H(1089 - t_f + 0.46t_g), \text{ where } t_g < 575 \text{ F} \qquad (4.23)$$

$$Q_5 = 9H(1066 - t_f + 0.50t_g), \text{ where } t_g > 575 \text{ F} \qquad (4.24)$$

In Equations 4.23 and 4.24, the symbol $H$ represents the hydrogen, expressed as a decimal, in the *ultimate analysis in which the moisture is included as a separate item of the ultimate analysis* as in column $B$ of Example 2. If the total hydrogen as reported in Table 4.2 is used in Equation 4.23 or 4.24, the loss thus calculated includes *both* the moisture loss, $Q_4$, and the loss due to water formed from the hydrogen, $Q_5$.

**Example 9e.** Since from column $B$ of Example 2 the hydrogen in 1 lb of fuel is found to be 4.4 per cent, then

$$Q_5 = (9 \times 0.044)(1089 - 60 + 0.46 \times 500) = 498 \text{ Btu per lb of fuel}$$

6. Item 6 of the energy balance includes the loss due to superheating the water vapor in the air supplied for combustion, the loss due to unburned hydrogen and hydrocarbons, radiation, and unaccounted-for losses. Some of these losses may be computed by methods beyond the scope of this book. Normally these losses are small and vary from 2 to 4 per cent of the heating value of the fuel. Where it is possible to measure the energy utilized, they are obtained by subtracting from the heating value of the fuel the sum of the energy utilized and the losses which may be computed.

An examination of formulas 4.21 to 4.24 shows that a moderate change in the exit temperature $t_g$ of the waste gases will not appreciably affect the numerical values of the moisture and hydrogen losses. Also, the moisture and hydrogen content of the fuel are not

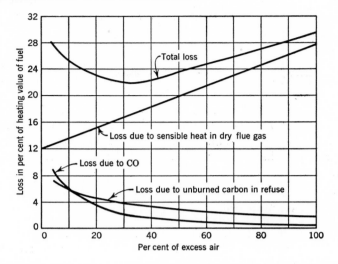

**Fig. 4.14.** Effect of excess air upon the losses due to CO, carbon in ash and refuse, and sensible heat in the dry flue gas.

under the control of the operator, and he therefore cannot do much to change these losses. However, the losses due to sensible heat in the hot waste gases and incomplete combustion are under control of the operator through regulation of the air supplied to burn the fuel. As indicated in Fig. 4.14, an increase in excess air reduces the losses due to incomplete combustion and increases the loss due to sensible heat in the hot waste gases. There is some air supply at which these losses will be a minimum, which for the particular curves of Fig. 4.14, is 30 per cent of excess air. The optimum air supply varies with the kind of fuel and the equipment for burning it and can best be determined by test. Efficient combustion is not attained unless the air supply is closely adjusted to requirements. Automatic $CO_2$ or $O_2$ recorders and elaborate systems of automatic-combustion control are often installed to make possible the efficient utilization of the fuel through close control of combustion.

The energy balance for combustion of oil is computed in the same way as the energy balance for combustion of coal, except that there is no solid carbon in the refuse and $C_b$, the carbon burned to CO and $CO_2$, is equal to the carbon in the fuel. When the fuel is a gas, the volumetric analysis can be converted to a gravimetric ultimate analysis as outlined in Article 4.14, after which the energy balance is computed as for oil; that is, there will be no solid refuse containing carbon when the fuel is a gas.

## NUCLEAR REACTIONS

The field of nuclear physics has developed very rapidly in recent years.   Since a balanced treatment of the subject is beyond the scope of this book, this chapter will be limited to a discussion of those basic principles in accordance with which mass is converted into energy.

### 4.24   Atomic Structure

Our solar system consists of a central sun in which most of the mass of the system is concentrated and a number of planets, of which our earth is one, each traveling in an elliptical orbit around the sun.   An atom is the smallest subdivision of matter which is unchanged during chemical reactions.   Each atom is constructed like our solar system.   Nearly all the mass of the atom is concentrated into a dense *nucleus*.   The nucleus in turn is composed of closely packed particles of two types: *protons* and *neutrons*.   The proton is a particle that carries a positive charge of electricity.   The neutron has approximately the same mass as the proton but is uncharged electrically.   The various species of atoms contain in their nuclei from 1 to approximately 100 protons and from zero to approximately 156 neutrons.

Traveling in orbits around the nucleus of the atom are one or more *electrons,* each of which carries a negative charge of electricity equal in magnitude to the positive charge on the proton.   Since the atom is electrically neutral, the number of orbital electrons equals the number of protons in the nucleus.   The mass of the electron is only $\frac{1}{1840}$ of the mass of the proton, while the distance from the center of the nucleus to the outermost electrons may be as much as 10,000 times the diameter of the nucleus.   Thus the mass of the atom is concentrated almost completely in the nucleus and the atom itself is mostly empty space as is our solar system.

During chemical reactions, atoms are regrouped without change in the structure of the atom to form new molecules.   The chemical properties of the atom are determined by the orbital electrons.   Physical properties are changed by alterations in the nucleus, and it is changes in the nucleus of the atom with which we are concerned in this chapter.

### 4.25 Atomic Nomenclature

The characteristics of an atom are determined by the number of protons and neutrons in the nucleus and the number of orbital electrons.

Let $Z$ = the atomic number of an element

then    $Z$ = the number of protons in the nucleus

and    $Z$ = the number of electrons

Also, let $A$ = the mass number of an element

then        $A$ = the sum of the numbers of protons and neutrons in the nucleus

and        $A$ = the "atomic weight" of an element to the nearest whole number

Furthermore, $A - Z$ = number of neutrons in the nucleus

Since a particular species of atom contains the same number of protons and electrons, the element is fully specified by indicating the number of protons (the atomic number) and the sum of the numbers of protons and neutrons in the nucleus (the mass number). The standard designation is as follows:

$$\text{Atomic number} [\text{Name of element}]^{\text{Mass number}} \quad \text{or} \quad {}_Z\text{X}^A$$

Thus, the atom ${}_8\text{O}^{16}$ is an oxygen atom which has 8 protons (also 8 electrons) and a nucleus consisting of 16 protons plus neutrons (therefore $16 - 8 = 8$ neutrons). Also, the atom ${}_{92}\text{U}^{238}$ is a uranium atom which has 92 protons (also 92 electrons) and a nucleus composed of 238 protons plus neutrons (therefore $238 - 92 = 146$ neutrons).

Table 4.9 contains data on the electron, proton, and neutron as well as some common elements and elements of particular interest in nuclear reactions. Attention is called to the following characteristics of the fundamental particles:

1. The simplest atom is the normal hydrogen atom, which consists of a nucleus of one proton with one orbital electron. Consequently, the proton is given the symbol ${}_1\text{H}^1$ or ${}_1p^1$, it being the nucleus of the normal hydrogen atom with one proton and a mass number of one.

### TABLE 4.9

#### Atomic Data

| Particle or Element | Symbol | Atomic Number $(Z)$ | Mass Number $(A)$ | Number of Neutrons | Mass, amu |
|---|---|---|---|---|---|
| Electron | $_{-1}e^0$ | 1 | 0 | 0 | 0.00055 |
| Proton | $_1p^1$ or $_1H^1$ | 1 | 1 | 0 | 1.00758 |
| Neutron | $_0n^1$ | 0 | 1 | 1 | 1.00894 |
| Hydrogen | $_1H^1$ | 1 | 1 | 0 | 1.00813 |
| Deuterium | $_1H^2$ | 1 | 2 | 1 | 2.01472 |
| Helium | $_2H^4$ | 2 | 4 | 2 | 4.00389 |
| Carbon | $_6C^{12}$ | 6 | 12 | 6 | 12.00386 |
| Carbon | $_6C^{13}$ | 6 | 13 | 7 | 13.00758 |
| Oxygen | $_8O^{16}$ | 8 | 16 | 8 | 16.00000 |
| Oxygen | $_8O^{17}$ | 8 | 17 | 9 | 17.00450 |
| Oxygen | $_8O^{18}$ | 8 | 18 | 10 | 18.00490 |
| Thorium | $_{90}Th^{232}$ | 90 | 232 | 142 | 232.11034 |
| Protactinium | $_{91}Pa^{233}$ | 91 | 233 | 142 | 233.11250 |
| Uranium | $_{92}U^{233}$ | 92 | 233 | 141 | 233.11193 |
| Uranium | $_{92}U^{235}$ | 92 | 235 | 143 | 235.11723 |
| Uranium | $_{92}U^{238}$ | 92 | 238 | 146 | 238.12514 |
| Neptunium | $_{93}Np^{239}$ | 93 | 239 | 146 | 239.12730 |
| Plutonium | $_{94}Pu^{239}$ | 94 | 239 | 145 | 239.12653 |

2. The neutron has the same mass number as the proton but its atomic number is zero in the absence of any protons. Therefore, it is designated by the symbol $_0n^1$.

3. The electron carries a negative charge equal in value to the positive charge of the proton but has a mass of only $\frac{1}{1840}$ of the mass of the proton. It is therefore designated by the symbol $_{-1}e^0$.

### 4.26   Isotopes

One hundred two different types of chemical elements are now known. They have from 1 to 102 electrons and an equal number of protons in the various known elements. However, most of the chemical elements may have several species, depending upon the number of neutrons present in the nucleus. These species of an element are called *isotopes*. For a given element, they will have the same chemical properties but different nuclear properties. Thus, ordinary oxygen is a mixture of three isotopes as follows: $_8O^{16}$ (99.76%), $_8O^{17}$ (0.04%), and $_8O^{18}$ (0.20%). Likewise, natural uranium is a mixture of three isotopes as follows: $_{92}U^{234}$ (0.006%), $_{92}U^{235}$ (0.712%), and $_{92}U^{238}$ (99.282%).

Since the atomic number of all the isotopes of a given element is the same, indicating the same number of protons and electrons, the difference lies in the number of neutrons in the nucleus. Accordingly, it is also common practice to designate the element by its symbol and mass number. Thus the three isotopes of uranium are frequently designated U-234, U-235, and U-238.

Over 300 isotopes are known to occur naturally among about 70 elements, in addition to which over 700 isotopes have been produced in recent years by nuclear reactions.

### 4.27 Atomic Mass

The atomic-mass unit, usually designated as *amu*, is based on the mass of the oxygen isotope, $_8O^{16}$ being assigned the number 16.00000.

Then $\quad$ 1 amu $= \dfrac{1}{16.00000}$ of the mass of one atom of $_8O^{16}$

$$= 1.66 \times 10^{-24} \text{ gram}$$

$$= 3.66 \times 10^{-27} \text{ lb}$$

The atomic masses in amu of a number of elements and some of their isotopes are listed in Table 4.9.

It should be noted that the atomic mass unit, amu, is based on the $_8O^{16}$ isotope of oxygen. Normal oxygen is composed of three isotopes as mentioned in the preceding article. The "atomic weight" of the chemist is based on the weighted average atomic mass of the three species of isotopes comprising normal oxygen and is therefore 0.0279 per cent larger than the amu.

### 4.28 Radioactive Decay

About 40 naturally occurring isotopes of elements heavier than lead (atomic number = 81) are unstable in varying degrees and subject to spontaneous change in an effort to reach stability. This phenomenon is called *radioactive disintegration* or *radioactive decay*.

The nucleus of the radioactive isotope expels a charged particle spontaneously, thereby leaving behind a so-called "daughter" isotope which has different properties from the parent isotope. The expelled radiation may be very dangerous to human beings and damaging to engineering materials. The elements lighter than lead as found in nature are usually stable although radioactive isotopes of all these elements have been produced in the physics laboratory. Three types of radiation are emitted during the processes of radioactive decay. They are (1) alpha ($\alpha$) particles, (2) beta ($\beta$) particles, and (3) gamma ($\gamma$) rays.

**Alpha particles ($\alpha$ or $_2He^4$).** Alpha particles are helium nuclei and consist of two protons and two neutrons with a mass number of 4. They are ejected with a velocity about one tenth of the speed of light, travel only a few inches in air before picking up two electrons and being converted into a helium atom, and are blocked by simple shielding such as a few sheets of paper. The radiation of the alpha particle causes the parent nucleus to lose two protons and two neutrons and thereby produces a new element. The process may be illustrated for Uranium-238 as follows:

$$_{92}U^{238} \rightarrow {}_2He^4 + {}_{90}Th^{234} \qquad (4.25)$$

The result is the conversion of Uranium-238 into a new element, Thorium-234, having an atomic number smaller by 2 and a mass number smaller by 4 because of the loss of two protons and two neutrons. It should be noted that in Equation 4.25, as in all equations for nuclear reactions, the sum of the mass and atomic numbers of the products is equal respectively to the sum of the mass and atomic numbers of the elements and particles entering into the reaction.

**Beta particles ($\beta$ or $_{-1}e^0$).** Beta particles are electrons that are ejected from the nucleus with a velocity approaching the speed of light. They are negatively charged, have very small mass (the mass of the electron is $\frac{1}{1840}$ of the mass of the proton), and will travel only a few feet in air. Since there are no electrons in the nucleus, the emission of an electron or beta particle results from the spontaneous conversion of a neutron to a proton and an electron and the release of a very small particle known as a neutrino.

Thus, neutron $\rightarrow$ proton + electron + neutrino. The neutrino is small compared to the electron, little is known about it, and it apparently has no engineering significance.

Since the radiation of a beta particle results from the conversion of a neutron to a proton, this process increases the atomic number by one and leaves the mass number unchanged. The process may be illustrated as follows:

$$_{90}\text{Th}^{234} \rightarrow {}_{-1}e^0 + {}_{91}\text{Pa}^{234} \tag{4.26}$$

**Gamma rays ($\gamma$).** When a nucleus emits an alpha or beta particle, the newly formed nucleus is in an excited state and emits energy in the form of gamma rays. Gamma rays are a form of electromagnetic radiation similar to X-rays except of shorter wave length. They possess a high energy level, are highly penetrating, and require considerable thickness of lead or other dense material to stop them. Because of their high penetrating power and damaging effect upon the human body, great care must be exercised to protect people from the dangers of gamma-ray radiation.

### 4.29   Rate of Radioactive Decay

Starting with a given mass of a radioactive element, radioactive decay will proceed spontaneously at a rate which is some function of the number of nuclei remaining at any subsequent time which are available for radioactive disintegration. The rate therefore decreases exponentially with time.

The *half-life* of a radioactive element is a constant that is used to measure the decay rate. It is the time required for a given number of nuclei of a radioactive isotope to decay to one half of the orig-

**Fig. 4.15.** Decay curve for radioactive elements.

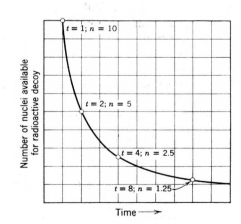

inal number. Thus, if the half-life of a given radioactive isotope is 20 minutes, then, at the end of 20 minutes, one half of the radioactive nuclei present at the beginning of the period have decayed. At the end of the second 20-minute period, one half of the radioactive nuclei present at the end of the first 20-minute period (or three quarters of the original material) have decayed, etc. Figure 4.15 shows the decay curve for a radioactive isotope. The half-life varies from a small fraction of a second to millions of years, depending upon the material. Radioactive waste products from nuclear reactors that have a long half-life therefore present a serious safety hazard for a long period of time, whereas those with a short half-life lose their hazard quickly.

### 4.30   Nuclear Reactions Induced by High-Energy Particles

Radioactive decay is a spontaneous process in which an unstable nucleus discharges radiation in an effort to attain stability. Another class of transmutations occurs when a high-speed nuclear particle such as a proton, neutron, or alpha particle, called the *incident particle,* strikes an atomic nucleus, called the *target nucleus.* The target nucleus and the incident particle combine to form a *compound nucleus,* which, because of the added energy, is unstable and disintegrates in an effort to become stable. In this manner, hundreds of isotopes have been made, some of which are of great scientific and industrial value.

The incident particles commonly used to initiate these nuclear reactions are the proton ($_1H^1$), alpha particle ($_2He^4$), neutron ($_0n^1$), gamma ray ($\gamma$), and deuteron ($_1H^2$), which is the nucleus of heavy hydrogen and differs from the nucleus of normal hydrogen by having one neutron in the nucleus in addition to the one proton in the normal hydrogen nucleus. Incident particles may be obtained from naturally radioactive elements or, in the case of charged particles, from particle accelerators such as the cyclotron, betatron, and synchrotron.

Some typical reactions are as follows:

| Incident Particle | Nuclear Reaction |
|---|---|
| Proton, $_1H^1$ | $_6C^{12} + {}_1H^1 \rightarrow {}_7N^{13} + \gamma$ |
| Neutron, $_0n^1$ | $_8O^{16} + {}_0n^1 \rightarrow {}_6C^{13} + {}_2H^4$ |
| Alpha particle, $_2H^4$ | $_7N^{14} + {}_2H^4 \rightarrow {}_8O^{17} + {}_1H^1$ |
| Deuteron, $_1H^2$ | $_8O^{16} + {}_1H^2 \rightarrow {}_7N^{14} + {}_2H^4$ |
| Gamma ray, $\gamma$ | $_4Be^9 + \gamma \rightarrow {}_4Be^8 + {}_0n^1$ |

### 4.31 The Fission Reaction

The fission reaction occurs when the nucleus of a large atom such as uranium is bombarded by a neutron under such conditions that it disintegrates to form two lighter daughter isotopes, releases several neutrons, and converts a substantial quantity of mass into energy. If one of the neutrons released in each fission reaction is able to strike and disintegrate another nucleus, the process becomes a self-propagating or *chain reaction*. The only element occurring in its natural state which is capable of supporting this chain reaction is one of the isotopes of uranium, U-235. U-235 constitutes only 0.7 per cent of natural uranium. Two other isotopes which can be manufactured are also capable of supporting the reaction and will be discussed in Article 4.36.

When a neutron strikes and is absorbed by the nucleus of an atom of U-235, the following reaction results:

$$_0n^1 + _{92}U^{235} \rightarrow _{92}U^{236} \tag{4.27}$$

The newly formed isotope of uranium is very unstable and has the unique property of immediately splitting into two smaller atoms and releasing on the average 2.5 neutrons per fission. The process may be described as follows:

$$_{92}U^{236} \rightarrow \text{two fission fragments} + 2.5_0n^1 \text{ (average)} \tag{4.28}$$

Figure 4.16 illustrates the four stages in the fission reaction. First, a nucleus of $_{92}U^{235}$ is bombarded by one neutron, resulting in the formation of the highly unstable isotope, $_{92}U^{236}$. The $_{92}U^{236}$ nucleus has the unique property of disintegrating almost instantly into two fission fragments with the release of an average of 2.5 neutrons per fission. In general, the masses of the newly formed fission fragments are unequal. Statistically, the smaller one will have a mass between 80 and 110 and the larger one will have a mass between 125 and 155. For the particular reaction illustrated in Fig. 4.16, the two fission fragments are $_{38}Sr^{94}$ and $_{54}Xe^{140}$. These isotopes are radioactive and undergo a series of transmutations which involve the emission of beta particles until the stable end products $_{40}Zr^{94}$ and $_{58}Ce^{140}$ are produced. The overall result of this sequence of reactions can be written as follows:

$$_{92}U^{235} + _0n^1 \rightarrow _{40}Zr^{94} + _{58}Ce^{140} + 2_0n^1 + 6_{-1}e^0 \tag{4.29}$$

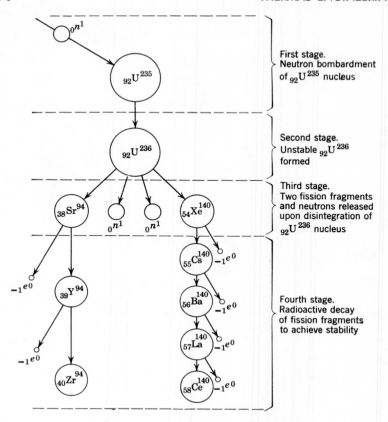

**Fig. 4.16.**  Typical fission reaction for U-235.

If exactly one neutron from each fission reaction creates the fissioning of another nucleus, this self-propagating *chain reaction* will continue at a constant rate.  This is the condition desired in a power reactor under steady-load conditions.  If considerably more than one neutron from each fission reaction creates the fissioning of another nucleus, the rate of reaction builds up at a fantastic rate.  If less than one neutron per fission creates another fission, then the reaction ceases to be a self-supporting chain reaction and is soon extinguished.

On the average, 40 per cent of the neutrons that are released by the fissioning of U-235 must be utilized to create the fissioning of additional nuclei of fissionable material if the chain reaction is to progress at a steady rate.  The remainder of the neutrons are absorbed in the system without producing fission or escape from the system.  One of the factors that determines the percentage of the total neutrons that escape from the system is the ratio of volume to external surface of the mass of fissionable material.  There is a minimum mass below which the loss of neutrons will prevent the support of a chain reaction.  This minimum mass is known as the *critical mass*.  Since the U-235 isotope in natural uranium is only 0.7 per cent of the total uranium present, the critical mass of natural uranium is large.  It may be reduced by enriching the uranium through the addition of U-235, which may be obtained by separation of U-235 from natural uranium or by the addition of other fissionable material that has been produced from fertile material as explained in Article 4.35.

### 4.32  Mass and Energy

Einstein has shown that there is an exact equivalence between mass and energy and that mass can be converted into energy in accordance with the following relation:

$$E = mC^2 \qquad\qquad (4.30)$$

where $E$ = energy produced, ergs
$m$ = mass that disappears, grams
$C$ = speed of light, $(3 \times 10^{10})$ cm per sec.

Conversely, energy may be converted into mass.

In chemical reactions such as the combustion of fossil fuels, the energy released is so small that less than one billionth of the mass involved is converted into energy and this cannot be measured.  However, in nuclear reactions, the reduction in mass is appreciable.

The units commonly used to express the relationship between mass and energy are as follows:

$$1 \text{ amu (atomic mass unit)} = \frac{1}{16.0000} \text{ of the mass of one atom of } _8O^{16}$$
$$= 1.66 \times 10^{-24} \text{ gram}$$

1 dyne = the force required to accelerate 1 gram at the rate of 1 cm

$$\text{per sec per sec} = 1.0 \frac{\text{gm cm}}{\text{sec}^2}$$

1 erg = work done when 1 dyne acts through a distance of 1 cm

$$= 1 \text{ dyne} \times 1 \text{ cm} = 1.0 \frac{\text{gm cm}}{\text{sec}^2} \times 1 \text{ cm} = 1.0 \frac{\text{gm cm}^2}{\text{sec}^2}$$

1 ev (electron volt) = energy acquired by a charged particle carrying
a unit electrical charge when it passes through
a potential difference of 1 volt
$$= 1.60 \times 10^{-12} \text{ erg}$$

1 mev (million electron volts) $= 1.60 \times 10^{-6}$ erg
$$= 1.52 \times 10^{-16} \text{ Btu}$$

Let $E$ per amu = energy produced in ergs when 1 amu of mass is converted into energy

$$E \text{ per amu} = mC^2 = \frac{1.66 \times 10^{-24} \text{ gm}}{\text{amu}} \times (3 \times 10^{10})^2 \frac{\text{cm}^2}{\text{sec}^2}$$

$$= \frac{1.49 \times 10^{-3} \text{ gm cm}^2}{\text{amu} \quad \text{sec}^2} = 1.49 \times 10^{-3} \text{ erg per amu}$$

$$(4.31)$$

$$E \text{ per amu} = \frac{1.49 \times 10^{-3} \dfrac{\text{erg}}{\text{amu}}}{1.60 \times 10^{-6} \dfrac{\text{erg}}{\text{mev}}} = 931 \text{ mev per amu} \qquad (4.32)$$

$$E \text{ per amu} = \left(931 \frac{\text{mev}}{\text{amu}}\right)\left(1.52 \times 10^{-16} \frac{\text{Btu}}{\text{mev}}\right) = 1.41$$

$$\times 10^{-13} \text{ Btu per amu} \qquad (4.33)$$

Equations 4.32 and 4.33 show that the energy released when one atomic mass unit of mass *disappears* is equal to 931 mev or $1.41 \times 10^{-13}$ Btu.

Equation 4.33 can be expressed in units of Btu released per lb of mass which disappears by means of the following conversion:

Let $E$ per lb = energy released per lb of mass which *disappears*

Then $\quad E$ per lb = $\dfrac{1.41 \times 10^{-13} \text{ Btu}}{\text{amu}} \quad \dfrac{1.0 \text{ amu}}{1.66 \times 10^{-24} \text{ gm}} \times 453.6 \dfrac{\text{gm}}{\text{lb}}$

$$= 38.7 \times 10^{12} \text{ Btu per lb} \qquad (4.34)$$

Equation 4.34 shows that when 1 lb of mass *disappears* in a nuclear reaction 38.7 trillion Btu appear. This is equivalent to the energy released by the combustion of about 3,000,000 lb of coal having a heating value of 13,000 Btu per lb.

### 4.33 Mass Defect

The mass of a given atom is less than the sum of the masses of the electrons, protons, and neutrons of which it is composed. This difference is known as the mass defect. This may be illustrated by comparing the mass of the $_8O^{16}$ atom, which by definition has a mass of 16.00000 amu, with the sum of the masses of its constituent particles as given in Table 4.9. Thus, for the atom $_8O^{16}$:

$$
\begin{aligned}
\text{Mass of 8 electrons} &= 8 \times 0.00055 = \quad 0.00440 \text{ amu} \\
\text{Mass of 8 protons} &= 8 \times 1.00758 = \quad 8.06064 \text{ amu} \\
\text{Mass of 8 neutrons} &= 8 \times 1.00894 = \quad 8.07152 \text{ amu} \\[6pt]
\hline
\text{Sum of masses of original particles} &= 16.13656 \text{ amu} \\
\text{Actual mass of the } _8O^{16} \text{ atom} &= 16.00000 \text{ amu} \\
\hline
\end{aligned}
$$

$$\text{Mass defect} = \quad 0.13656 \text{ amu}$$

Since the mass of the atom is less than that of its constituent particles, in this case by 0.13656 amu, energy in the amount of 0.13656 amu $\times$ 931 mev per amu or 127.1 mev was released in assembling the atom from its constituent parts. Likewise, an equal amount of energy must be supplied to take the atom apart and reduce it to its constituent particles. This is known as the *binding energy* of the atom.

When an atom of U-235 fissions, the mass of the fission products is less than the original mass of the U-235 atom and the neutron that produced fission. The loss in mass or mass defect is a measure of the energy released by the reaction.

Consider the fission reaction as illustrated in Fig. 4.16:

$$_{92}U^{235} + _0n^1 \rightarrow _{40}Zr^{94} + _{58}Ce^{140} + 2_0n^1 + 6_{-1}e^0 \qquad (4.29)$$

The following material balance may be written for this reaction:

| Original Mass, amu | | Final Mass, amu | |
| --- | --- | --- | --- |
| $_{92}U^{235}$ | 235.117 | $_{40}Zr^{94}$ | 93.936 |
| $_{0}^{1}n$ | 1.009 | $_{58}Ce^{140}$ | 139.949 |
| | | $2_{0}n^{1}$ | 2.018 |
| | | $6_{-}e^{0}$ | 0.003 |
| Total | 236.126 | | 235.906 |

Mass defect = 236.126 amu − 235.906 amu = 0.220 amu
Energy equivalence = 0.220 amu × 931 mev per amu = 204.8 mev

Thus, the energy released per fission is 204.8 mev. These calculations are approximate because of the emission of neutrinos, gamma rays, uncertainties in exact data of molecular masses, etc. Although the unstable $_{92}U^{235}$ isotope may fission into many different combinations of fragments, the energy released will be approximately 200 mev per atom fissioned. Also, on the average, 2.5 neutrons will be released per atom of U-235 undergoing fission.

### 4.34   Control of the Rate of Reaction

Some of the neutrons that are released by the fission of U-235 will escape from the system, some will be absorbed by the structure, coolant, and fission products, and some may be absorbed by uranium without producing fission. To maintain the fission reaction, at least one of the neutrons released per fission must create a fission reaction in another atom. In a practical power reactor, slightly more than one neutron per fission must be available to continue the chain reaction. Under such conditions, the output of the reactor would increase until the temperature resulting from the release of energy would destroy the reactor unless provisions were made to control the rate of reaction. Control of the rate of reaction is obtained by inserting or withdrawing control rods composed of boron, cadmium, or similar material that has the property of absorbing large numbers of neutrons. Automatic devices are available for moving these control rods in such a manner as to absorb the excess neutrons and thus regulate the rate of reaction to maintain a uniform rate of energy release or a gradually increasing or decreasing rate of energy release.

When U-235 undergoes fission, about 99.3 per cent of the neutrons that are ejected are released within $10^{-7}$ to $10^{-14}$ sec after fission. These are known as *prompt* neutrons. Fortunately, the rest of the neutrons that are released leave the fission products during a period of time extending up to minutes after fission. They are called *delayed* neutrons. These delayed neutrons provide a time element of sufficient magnitude that automatic controls can be devised to insert or withdraw the control rods so as to maintain the chain reaction at the desired rate.

### 4.35 Fertile Materials

Uranium-235, the only natural isotope that will support a chain reaction, constitutes only 0.7 per cent of natural uranium. Fortunately, two other materials, known as fertile materials, have the property of absorbing neutrons from the fission reaction and of being thereby converted into fissionable materials. These fertile materials are uranium-238 and thorium-232.

The reaction for U-238 is as follows:

$$_{92}U^{238} + _{0}n^{1} \rightarrow _{92}U^{239} + \gamma \tag{4.35}$$

in which neutrons from the fission of U-235 react with U-238 to produce U-239 plus gamma radiation. The U-239 is radioactive, has a half-life of 23.5 min, emits a beta particle and forms an isotope of element 93, neptunium:

$$_{92}U^{239} \xrightarrow{23.5 \text{ min}} _{-1}e^{0} + _{93}Np^{239} \tag{4.36}$$

Neptunium has a half-life of 2.3 days and decays by radiating beta particles to form plutonium-239:

$$_{93}Np^{239} \xrightarrow{2.3 \text{ days}} _{-1}e^{0} + _{94}Pu^{239} \tag{4.37}$$

Plutonium-239 may be separated from the fertile material and is a fissionable material like U-235. It is the fissionable material that is used in the atomic bomb.

Thorium-232 reacts in a similar manner when subject to bombardment by neutrons.

$$_{90}Th^{232} + _{0}n^{1} \longrightarrow _{90}Th^{233} + \gamma \tag{4.38}$$

$$_{90}Th^{233} \xrightarrow{23.5 \text{ min}} _{-1}e^{0} + _{91}Pa^{233} \tag{4.39}$$

$$_{91}Pa^{233} \xrightarrow{27.4 \text{ days}} _{-1}e^{0} + _{92}U^{233} \tag{4.40}$$

This series produces the intermediate element protactinium and finally another isotope of uranium, U-233, which, like U-235, is fissionable.

The conversion of fertile materials to fissionable materials requires neutrons from a fission reaction. However, the possibility exists of building a power reactor which not only supplies energy for steam generation but also converts fertile material to fissionable material at such a rate that the overall supply of fissionable material is actually increased during the process. The process of converting fertile material into fissionable material in a reactor under such conditions that more nuclear fuel is produced than is burned is called *breeding*. In order to produce breeding, in addition to the one neutron required per fission to maintain the chain reaction, there must be enough neutrons to provide for leakage from the system, absorption in the moderator (see Article 4.36), fission products, coolant, and structure, and still have some neutrons to combine with the nuclei of the fertile material. Since Pu-239 releases on the average 3.0 neutrons per fission whereas U-235 releases 2.5 neutrons per fission, there is an advantage in using Pu-239 as the nuclear fuel. The possibility exists of converting most of the U-238 into fissionable material, whereas in the natural state the fissionable U-235 is only 0.7 per cent of natural uranium. Experimental breeder reactors have been constructed and operated successfully, and breeder reactors are currently under construction for large central-station power plants.

### 4.36   The Moderator

The neutrons that are released during a fission reaction are ejected at a very high speed and have a kinetic energy of about 2 mev (2000 ev). They are called fast neutrons. Fast neutrons are capable of fissioning U-238 and Th-232 as well as U-233, U-235, and Pu-239. However, U-233, U-235, and Pu-239 are fissioned more readily by neutrons having a speed not greatly different from that of the nuclei with which they react in the fission reaction. Optimum results are obtained if the neutrons have a kinetic energy in the order of 0.025 ev. Such neutrons are known as slow or thermal neutrons.

If a fast neutron is allowed to undergo a series of elastic collisions with some light material that does not absorb neutrons readily, most of the kinetic energy of the neutron is transferred to this material and the kinetic energy of the neutron is thereby reduced to the thermal level. A material that has the property of slowing down fast neutrons to the thermal level by means of elastic collisions with a minimum of neutron absorption is known as a *moderator*. Atoms of low mass, such as hydrogen, carbon, and beryllium, and molecules of light water ($H_2O$) and heavy water ($D_2O$) are used in nuclear reactors as moderators. Carbon in the form of graphite and ordinary water are used most extensively as moderators. The nuclear fuel must be so dispersed in the moderator that the neutrons which are ejected during a nuclear fission are thermalized by the moderator before an appreciable portion of them can be absorbed in the system without producing fission.

U-238 and Th-232 are fertile materials that may be converted to Pu-239 and U-233 upon absorption of a neutron in accordance with reaction equations 4.35 to 4.40, or they may be fissioned by fast neutrons. The breeder reactor is intended to produce more nuclear fuel than it burns through the process of using neutrons not needed to maintain the chain fission reaction for the purpose of converting the relatively plentiful fertile materials into U-233 and Pu-239. This requires that neutrons be conserved and utilized as completely as possible. Since the best of moderators absorb part of the neutrons, breeder reactors are designed to operate on fast neutrons without a moderator. In general, other nuclear reactors operate on thermal neutrons and require a moderator.

## PROBLEMS

1. Determine (*a*) the proximate analysis on a moisture-free and a moisture-and-ash-free basis, (*b*) the ultimate analysis on a moisture-free and a moisture-and-ash-free basis, and (*c*) the theoretical air required for the complete combustion of 1 lb of coal, for each of the following coals, the analyses of which are reported in Table 4.2: (1) Cambria County, Pa., low-volatile bituminous coal; (2) Logan County, Ill., high-volatile *C* bituminous coal; (3) Coss County, Oreg., subbituminous *B* coal.

2. Compute the theoretical air required for the complete combustion of 1 lb of the following crude oils, the analyses of which are reported in Table 4.4: (1) Pennsylvania crude; (2) Ohio crude; (3) Oklahoma crude.

3. A coal has the following ultimate analysis: C = 75%; H = 5%; O = 7%; N = 1%; S = 2%; ash = 5%; free moisture = 5%. If this fuel is burned completely with 40 per cent of excess air, (*a*) how much air is supplied per lb of fuel; (*b*) what are the products of combustion?

4. Coal has the following ultimate analysis: C = 72%; H = 5%; O = 12%; N = 2%; S = 3%; ash = 6%. The proximate analysis is as follows:

fixed carbon = 53%; volatile matter = 32%; ash = 6%; free moisture = 9%. Compute the air required to burn 1 lb of this fuel completely with 50 per cent of excess air.

**5.** A fuel oil has the following ultimate analysis: $C = 81\%$; $H = 14\%$; $0 = 3\%$; $N = 2\%$. The specific gravity of the oil is 24 degrees API. Calculate the theoretical air required for the complete combustion of 1 lb of this fuel.

**6.** Compute the theoretical air required to burn 1 lb of each of the following fuels: (a) $CH_4$, (b) $C_2H_6$, (c) $C_3H_8$, (d) $CH_3OH$.

**7.** Determine the approximate heating value per lb and per gal of an uncracked fuel oil having a gravity of 32 degrees API.

**8.** A cracked fuel oil has a gravity of 26 degrees API and costs 14 cents per gal. At what price per ton will coal with a heating value of 12,000 Btu per lb have the same cost per million Btu?

**9.** A fuel gas has the following volumetric analysis: $CH_4 = 65\%$; $CO = 10\%$; $H_2 = 15\%$; $O_2 = 6\%$; $N_2 = 4\%$. Compute the lb of theoretical air required to burn 1 lb of this fuel.

**10.** Determine (a) the per cent of carbon in 1 lb of fuel gas, (b) the gas constant, $R$, (c) the specific volume at 15 psia and 80 F, and (d) the heating value per lb of fuel for the following gaseous fuels whose analyses are reported in Table 4.6: (a) Pennsylvania natural gas, (b) coke-oven gas, (c) producer gas.

**11.** Dry gas has the following volumetric analysis: $CO_2 = 14.2\%$; $O_2 = 5.5\%$; $CO = 0.2\%$; $N_2 = 80.1\%$. Compute (a) the per cent of carbon in 1 lb of this gas, (b) the per cent of nitrogen in 1 lb of this gas, (c) the specific volume of the gas at 400 F and 14.7 psia.

**12.** Calculate the products of complete combustion with 40 per cent of excess air from 1 lb of (a) Logan County, Ill., high-volatile $C$ bituminous coal, the analysis of which is given in Table 4.2; (b) Kern River, Calif., crude oil, the analysis of which is given in Table 4.4; (c) coke-oven gas, the analysis of which is given in Table 4.6.

**13.** The following data were collected from tests of coal-fired steam boilers:

| Item | | No. 1 | No. 2 | No. 3 |
|---|---|---|---|---|
| Ultimate analysis of coal as fired: | C | 65 | 57 | 64 |
| | H | 6 | 6 | 5 |
| | O | 14 | 16 | 8 |
| | N | 2 | 2 | 3 |
| | S | 3 | 4 | 2 |
| | Ash | 10 | 15 | 8 |
| | Moisture | ... | ... | 10 |
| Proximate analysis of coal as fired: | | | | |
| Fixed carbon | | 49 | 44 | 48 |
| Volatile matter | | 32 | 29 | 34 |
| Ash | | 10 | 15 | 8 |
| Free moisture | | 9 | 12 | 10 |

| Item | | No. 1 | No. 2 | No. 3 |
|---|---|---|---|---|
| Heating value of coal, Btu per lb | | 12,270 | 10,960 | 11,930 |
| Orsat analysis of flue gas: | $CO_2$ | 12.0 | 13.5 | 14.2 |
| | $O_2$ | 7.8 | 6.2 | 5.5 |
| | CO | 0.2 | 0.1 | 0.2 |
| Air temperature, F | | 70 | 80 | 75 |
| Flue-gas temperature, F | | 500 | 350 | 400 |
| Coal burned during test, lb | | 1,000 | 5,000 | 8,000 |
| Dry refuse collected, lb | | 130 | 940 | 850 |
| Heating value of dry refuse, Btu per lb | | 3,360 | 2,920 | 3,650 |

(a) Assuming a radiation and unaccounted-for loss of 3 per cent of the heating value of the fuel in each test, calculate a complete energy balance for each test, and express the results in Btu per lb of fuel and in per cent of the heating value of the fuel. (b) Compute the air actually supplied per lb of fuel and the per cent of excess air supplied.

**14.** The following data were collected from tests of oil-fired furnaces:

| Item | | No. 1 | No. 2 | No. 3 |
|---|---|---|---|---|
| Ultimate analysis of oil: | C | 88.1 | 85 | 84.6 |
| | H | 10.8 | 11 | 10.9 |
| | O | 0.3 | 2 | 2.6 |
| | N | 0.2 | 1 | 0.3 |
| | S | 0.6 | 1 | 1.6 |
| Heating value, Btu per lb | | 18,600 | 18,350 | 18,100 |
| Orsat analysis of flue gas: | $CO_2$ | 12.0 | 11.8 | 10.6 |
| | $O_2$ | 5.8 | 5.9 | 7.3 |
| | CO | 0.1 | 0.2 | 0.1 |
| Air temperature, F | | 80 | 60 | 70 |
| Flue-gas temperature, F | | 400 | 350 | 440 |

Assume in each test that the radiation and unaccounted-for losses are 3 per cent of the heating value of the fuel. (a) Calculate a complete energy balance for each test, expressing the results in Btu per lb of fuel and in per cent of the heating value of the fuel. (b) Calculate the air supplied per lb of fuel and the per cent of excess air supplied.

**15.** The following data were obtained from tests of gas-fired furnaces:

| Item | | No. 1 | No. 2 | No. 3 |
|---|---|---|---|---|
| Volumetric analysis, %: | $H_2$ | | 54.5 | 14 |
| | CO | | 11.9 | 27 |
| | $CH_4$ | 83.2 | 24.2 | 3 |
| | $C_2H_6$ | 15.6 | | |
| | $O_2$ | | 0.4 | 0.6 |
| | $N_2$ | 1.2 | 5.7 | 50.9 |
| | $CO_2$ | | 3.3 | 4.5 |
| Heating value, Btu per cu ft at 60 F and 14.7 psia | | 1025 | 530 | 160 |

| Item | No. 1 | No. 2 | No. 3 |
|---|---|---|---|
| Orsat analysis of products of combustion, % $CO_2$ | 9.5 | 9.8 | 10.1 |
| $O_2$ | 2.4 | 2.2 | 5.8 |
| CO | 0.1 | 0.2 | 0.1 |
| Air temperature, F | 75 | 90 | 85 |
| Flue-gas temperature F | 500 | 700 | 450 |

For each of these tests, calculate (a) the loss in Btu per lb due to sensible heat in dry gaseous products of combustion, (b) the loss due to the CO in the dry gaseous products of combustion, (c) the loss due to moisture formed from the hydrogen, (d) the heating value of the fuel in Btu per lb, (e) the air actually supplied per lb of fuel, and (f) the per cent of excess air supplied.

**16.** A Diesel engine has a thermal efficiency of 30 per cent when delivering 300 bhp (brake horsepower). The fuel has a heating value of 140,000 Btu per gal and a specific gravity of 32 degrees API. Calculate the fuel consumption in lb per hr.

**17.** Uranium-235 undergoes the following fission reaction:

$$_{92}U^{235} + {_0}n^1 \rightarrow {_{59}}Pr^{141} + {_{41}}Nb^{93} + 2{_0}n^1 + 8{_{-1}}e^0$$

The masses of $_{59}Pr^{141}$ and $_{41}Nb^{93}$ are 140.951 and 92.935 amu, respectively. Calculate the mass defect and the energy released in mev per nucleus of U-235 undergoing fission.

**18.** Plutonium-239 undergoes the following fission reaction:

$$_{94}Pu^{239} + {_0}n^1 \rightarrow {_{42}}Mo^{95} + {_{60}}Nd^{143} + 2{_0}n^1 + 8{_{-1}}e^0$$

Calculate the mass defect and the energy released per fission in mev if the masses of $_{42}Mo^{95}$ and $_{60}Nd^{143}$ are 94.936 and 142.954 amu, respectively.

**19.** Plutonium-239 undergoes the following fission reaction:

$$_{94}Pu^{239} + {_0}n^1 \rightarrow {_{54}}Xe^{132} + {_{46}}Pd^{105} + 3{_0}n^1 + 6{_{-1}}e^0$$

The masses of $_{54}Xe^{132}$ and $_{46}Pd^{105}$ are 131.946 and 104.938 amu, respectively. Calculate the mass defect and the energy released in mev per fission.

**20.** Uranium-233 undergoes the following fission reaction:

$$_{92}U^{233} + {_0}n^1 \rightarrow {_{39}}Y^{89} + {_{60}}Nd^{143} + 2{_0}n^1 + 7{_{-1}}e^0$$

The masses of $_{39}Y^{89}$ and $_{60}Nd^{143}$ are 88.934 and 142.954 amu, respectively. Calculate the mass defect and the energy released in mev per fission.

**21.** If 1 lb of Pu-239 is completely converted into $_{42}Mo^{95}$ and $_{60}Nd^{143}$ in accordance with the reaction given in Problem 18, how many Btu are released?

# CHAPTER 5

# Internal-Combustion Engines

## 5.1 Introduction

The desirability of supplying energy directly to the working fluid by combustion within the engine itself was recognized many years before a practical internal-combustion engine was devised. The steam engine, which came into use about 1700, was extremely uneconomical in its use of fuel. Its low efficiency was known to be largely the effect of the indirect application of energy, the fuel being burned under a boiler where water and steam absorbed a small portion of the energy liberated by combustion, and then an even smaller portion of this energy was converted into useful work in the engine. The combined losses incurred in generating the steam, conveying it to the engine, and converting its energy into work were large, much larger than in modern steam equipment. Many of those investigating the problem were convinced that it would be more economical as well as convenient to burn the fuel directly within the working cylinder of the engine. Although the early attempts to accomplish this were unsuccessful, they provided the background for later more successful attempts.

It was not until 1860 that an engine was developed to the extent that it could be manufactured and sold as a practical machine. This engine was designed by Lenoir in France and resembled structurally the steam engines of that period, differing basically only in that a combustible charge of fuel and air was supplied to the cylinder of the engine where it was ignited and burned, instead of steam under pressure being supplied from a boiler. This engine was inefficient in its use of fuel and developed very little power in proportion to its size and weight; nevertheless, it was more satisfactory for certain applications than the steam-power plant.

A few years later, another French scientist, Beau de Rochas, published a plan of operating procedure for an internal-combustion engine which differed fundamentally from that of Lenoir and was destined to become the basic principle of our modern engines. He did not build any engines which applied his theory of operation, but in 1876 Otto produced in Germany an engine of this type which operated successfully. The cycle of operation has since been called the Otto cycle, although Otto did not originate the idea.

Later developments include the work of Dugald Clerk in England, who succeeded in supplying the charge to the cylinder in such manner that the necessary processes of the Otto cycle could be completed in one revolution of the crankshaft instead of requiring two turns of the shaft as did the original Otto engine. Dr. Rudolf Diesel of Germany extended the range of operation of the internal-combustion engine and increased its efficiency by reducing to practice the compression-ignition principle that had been investigated theoretically and experimentally by several others who were not successful in carrying their ideas to completion. The adoption of many other advances has similarly been delayed, because the engineers who conceived them failed to correlate all the factors necessary for success.

## 5.2   The Lenoir Engine

A study of the working principles of the obsolete Lenoir engine is justified by the theoretical and practical evidence that it affords of the beneficial effect of compression upon the performance of internal-combustion engines. The fuel charge was ignited in the Lenoir engine at practically atmospheric pressure without being previously compressed, and it is therefore classed as a noncompression engine. Early experience with this engine convinced engineers that the compression process that is now a part of the cycle of all reciprocating engines is essential to the efficiency and power capacity of such engines. Starting with the Lenoir engine, which did not compress its charge prior to combustion, the subsequent development of the type of engine which inducts an inflammable fuel–air charge into its cylinder is seen to include a continuous increase in the extent to which the charge is compressed prior to combustion. The high-compression engine of today represents the present level of that trend.

In order to understand the operation of any engine, a knowledge must be acquired of the cycle upon which the engine operates. By definition, a cycle is a series of events which periodically returns the working parts to their original condition and which, when repeated over and over again, produces regular and continuous operation. Every mechanical device has a cycle which consists of certain movements that occur in regular succession and which, when completed, return all parts of the machine to their original positions ready to start another cycle. This is its mechanical or operating cycle and is distinct from the thermodynamic or theoretical cycle of the engine, which consists of the processes undergone by the working fluid or gases that are present in the cylinder during its operation.

Mechanically, the Lenoir engine differed little from the existing slide-valve steam engine. The intake valve opened as the piston started its downward stroke, and a combustible mixture of gas and air was inducted into the cylinder, occupying the space vacated by the descending piston. At about mid-stroke the intake valve was closed and the charge immediately ignited by an electric spark. Combustion of the charge increased the pressure in the cylinder and caused work to be done by the resulting force on the piston. The second half of the downward stroke was thus the only power-producing portion of the cycle, as power was expended in moving the piston during the rest of the cycle. The exhaust port was uncovered at the end of the power stroke, and the products of combustion were displaced from the cylinder by the return stroke of the piston. The cycle was thus completed in one turn of the crankshaft or two strokes of the piston, half a stroke for intake of charge, half a stroke for expansion and power, and a full stroke for exhaust.

The operating cycle of an engine is portrayed by an engine indicator in the form of a pressure–volume diagram which is a record of the pressure exerted by the gas and the corresponding volume changes of the working fluid within the cylinder as the strokes are executed. The engine indicator is considered in detail later. A pressure–volume diagram for the Lenoir engine, as traced by an engine indicator, would appear as shown in Fig. 5.1.

Horizontal distances on this diagram represent piston stroke and the corresponding volumes displaced by the piston. The straight line drawn at atmospheric pressure is the datum line from which pressures are measured vertically. The line 1–2 is drawn by the stylus during the first portion of the piston stroke and shows the depression caused by the pumping action of the piston. At the point 2 the valve closes and the charge is ignited, causing the pressure to rise to that of the point 3. Further movement of the piston results in a gradual decrease in pressure to the exhaust opening point 4, after which the pressure drops rapidly until the end of the stroke is reached at 5. The return

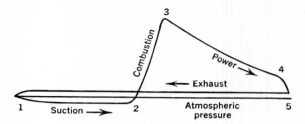

**Fig. 5.1.** Lenoir indicator diagram.

stroke 5–1 finds the pressure slightly above atmospheric pressure as the products of combustion are being forced from the cylinder by the returning piston. At the point 1 the exhaust closes, and the intake opens, returning all parts of the system to their original condition ready to begin another cycle.

The thermal efficiency of an engine relates the power developed to the energy expended. The amount of energy equivalent to the power developed, when expressed in per cent of the energy in the fuel burned, is the thermal efficiency. When based upon the net power actually delivered by the engine, it is called brake thermal efficiency. Tests of the Lenoir engine are said to have shown its brake thermal efficiency to be less than 4 per cent. This is much less than the efficiency attained by engines that compress the charge before combustion. Furthermore, the power developed was much less than that of engines of the same size and speed employing compression.

### 5.3  The Otto Engine

The Otto engine differed from the Lenoir engine essentially in that the entire outward stroke of the piston was used to fill the cylinder with charge. Since the piston was then at its greatest distance from the cylinder head, and burning the fuel at the maximum volume of the cycle could not cause any work to be done on the piston, it was necessary to return the piston to its starting point. This was the first possible opportunity that the force resulting from combustion would have to do work by pushing the piston outward. Compression of the charge during a return stroke of the piston was thus necessitated in order to place the piston in position to be acted upon by the energy released by combustion. Additional thermodynamic gains result from compression, which make it desirable as well as necessary.

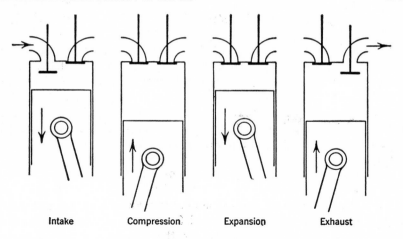

**Fig. 5.2.** Strokes of the four-stroke-cycle engine.

Combustion occurred, theoretically, while the piston paused at the end of the compression stroke, and the entire outward stroke was used for expansive work done by the products of combustion. The return of the piston the second time was with the exhaust valve open.

The differences between these two principles of operation may be summed up as the utilization of an entire stroke for intake of charge instead of half a stroke, compression of the charge by returning the piston to the position of maximum possible expansion after combustion, combustion at minimum cylinder volume, and expansion for an entire forward stroke instead of half a stroke. Since the exhaust strokes were common to both cycles, the additional half-stroke added to the intake period, the half-stroke increase in expansion, and the

**Fig. 5.3.** Four-stroke-cycle indicator diagram.

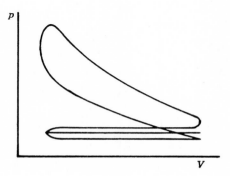

introduction of a compression stroke increased the number of strokes required to complete the cycle from two to four. The cycle is thus seen to be a four-stroke cycle that is completed every fourth stroke of the piston and every second revolution of the crankshaft. The four strokes are *intake, compression, expansion, exhaust,* repeated over and over in that order. Figure 5.2 shows a conventional cylinder and piston during the strokes of the four-stroke-cycle engine. An indicator diagram, showing the actual pressure–volume relationships during the four strokes, is reproduced in Fig. 5.3.

### 5.4   The Otto Cycle

Ideally, the $p$-$V$ diagram for the Otto cycle would appear as shown in Fig. 5.4. The processes 1–2 and 3–4 follow adiabatic paths for compression and expansion, whereas the indicator diagram shows an appreciable departure from the theoretical paths because of leakage, heat transfer between the gases and the engine parts, noninstantaneous combustion, and flow of charge to and from the cylinder. The constant-volume processes 2–3 and 4–1 assume that energy is supplied to the working fluid instantaneously when compressed to its minimum volume and that an instantaneous rejection of energy occurs at the maximum volume.

Only two strokes, compression and expansion or power, are required for the thermodynamic processes of the ideal cycle because it

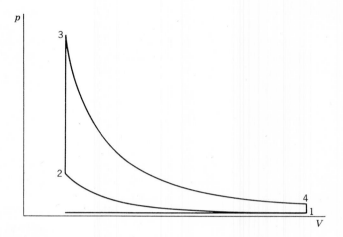

**Fig. 5.4.**   Ideal Otto cycle.

is assumed that the same quantity of gas is alternately heated, expanded, cooled, and recompressed in executing the cycle. Pumping the charge into and from the cylinder is not ideally necessary.

An ideal cycle is analyzed by evaluating the energy supplied during the cycle and either the net work done or the energy rejected during the cycle. The thermal efficiency of the ideal cycle is the ratio of the useful output to the energy input and may be expressed in equation form as

$$\eta_t = \frac{\text{net work}}{\text{energy supplied}} = \frac{Q_1 - Q_2}{Q_1} \tag{5.1}$$

where $Q_1$ is the number of Btu supplied by the fuel, and $Q_2$ is the number of Btu rejected and not used. These quantities may be evaluated for a specific cycle. An efficiency equation for that cycle can then be derived which contains the factors affecting the theoretical efficiency of any engine operating on a cycle comprising the assumed series of processes.

The ideal Otto cycle may be analyzed to find its thermal efficiency by evaluating the energy quantities supplied and rejected. Energy is supplied to the working fluid during the constant-volume process 2–3. If the cycle is assumed to operate with a mass of 1 lb of air, the energy supplied equals the change in internal energy, $c_v(T_3 - T_2)$. Rejection of energy after completion of the power stroke occurs during the constant-volume process 4–1 and is equal to $c_v(T_4 - T_1)$. The ideal thermal-efficiency expression may be written as

$$\eta_t = \frac{c_v(T_3 - T_2) - c_v(T_4 - T_1)}{c_v(T_3 - T_2)} \tag{5.2}$$

$c_v$ is present in each term and may therefore be eliminated. Dividing by $(T_3 - T_2)$ yields

$$\eta_t = 1 - \frac{T_4 - T_1}{T_3 - T_2}$$

This expression is not in a convenient form because only the initial temperature, $T_1$, is known. An equation in terms of volumes is most useful because the cycle operates between two volume limits that can be readily determined for an actual engine. For the frictionless adiabatic process, it is known that $\dfrac{T_1}{T_2} = \left(\dfrac{V_2}{V_1}\right)^{k-1}$. Applying this relationship to the two adiabatic processes of the cycle,

$$T_1 = T_2 \left(\frac{V_2}{V_1}\right)^{k-1} \quad \text{and} \quad T_4 = T_3 \left(\frac{V_3}{V_4}\right)^{k-1}$$

$$V_2 = V_3 \quad \text{and} \quad V_1 = V_4$$

Substituting in Equation 5.2,

$$\eta_t = 1 - \frac{T_3 \left(\frac{V_2}{V_1}\right)^{k-1} - T_2 \left(\frac{V_2}{V_1}\right)^{k-1}}{T_3 - T_2}$$

$$= 1 - \left(\frac{V_2}{V_1}\right)^{k-1}\left(\frac{T_3 - T_2}{T_3 - T_2}\right)$$

$$= 1 - \left(\frac{V_2}{V_1}\right)^{k-1}$$

The reciprocal of this volume ratio is the compression ratio at which the cycle operates. It is desirable, therefore, to express the equation as

$$\eta_t = 1 - \frac{1}{r^{k-1}} \tag{5.3}$$

where $r$ is the compression ratio of the cycle, $V_1/V_2$. This expression shows that the efficiency of the ideal Otto cycle depends upon the ratio by which the charge is compressed before energy is supplied.

The upper curve of Fig. 5.5 shows the theoretical thermal efficiencies of the Otto cycle corresponding to compression ratios over the combined ranges of spark-ignition and compression-ignition engines for $k = 1.4$. The lower curve shows the approximate brake thermal efficiencies of typical actual engines. Individual engines may differ slightly from this curve in their performance because of differences in design and operating conditions. It is evident from these curves that important gains in thermal efficiency accompanied increases in compression ratio since the early years of the development of the gasoline engine when compression ratios near 4 were used. Increases above the present automobile-engine compression ratio range of 8 to 10.5 will be less effective in improving the efficiency, as is evidenced by the decreased slopes of the curves. It is also apparent that there is little to be gained by the use of compression ratios in the Diesel engine greater than necessary to assure proper ignition of the fuel because the slope in that range is very slight. Since the days of the

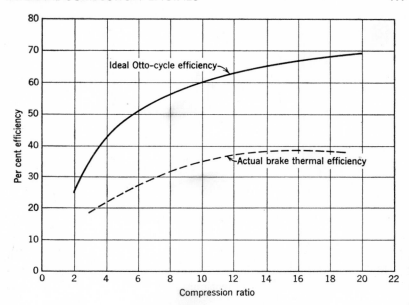

**Fig. 5.5.** Effect of compression ratio on efficiency of ideal Otto cycle and brake thermal efficiency of actual engines.

noncompression Lenoir engine, it has been recognized that the fundamental way to increase thermal efficiency is to increase the compression ratio. Other factors, principally mechanical efficiency, combustion efficiency, and control of heat losses, also affect the performance of engines and cause their thermal efficiencies to be above or below the values of the curve of Fig. 5.5.

**Example 1.** Find the thermal efficiency of an ideal Otto cycle with a compression ratio of 10, assuming air as the working fluid.

*Solution:* If we substitute in Equation 5.3, $r = 10$, and $k = 1.4$,

$$\eta_t = 1 - \frac{1}{10^{1.4-1}} = 1 - \frac{1}{10^{.4}} = 1 - \frac{1}{2.51} = 1 - 0.398 = 60.2\%$$

The compression ratio of the Otto-cycle engine cannot be increased at will in order to improve the thermal efficiency if it is of the spark-ignition type because the temperature and pressure at the end of the compression stroke will approach the ignition point of the fuel with the result that self-ignition of the charge will prevent proper control

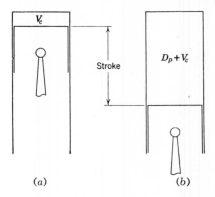

**Fig. 5.6.** Cylinder volumes.

of the combustion process. It is customary to design engines with the highest useful compression ratio, the maximum at which the engine will operate free from objectionable combustion knock without resorting to special fuels.

The compression ratio of an actual engine is established by the amount of space remaining between the piston and the cylinder head of the engine when the piston is at its point of closest approach to the head. Figure 5.6 shows the cylinder with the piston at each of the two extremes of its travel. In (a) the volume of the cylinder above the piston is the minimum or clearance volume, $V_c$. The volume above the piston in (b) is the maximum or total cylinder volume. The difference between these two extremes is the volume swept by the piston in its stroke and is called the piston displacement volume, $D_p$. It is equal to the product of the area of the piston and its stroke. The compression ratio is the total cylinder volume divided by the clearance volume:

$$r = \frac{\text{total volume}}{\text{clearance volume}} = \frac{D_p + V_c}{V_c} \tag{5.4}$$

**Example 2.** An engine of 4-in. bore and 3.25-in. stroke has a clearance volume of 5 cu in. Find the compression ratio.

*Solution:*

$$D_p = \frac{4 \times 4 \times 3.14 \times 3.25}{4} = 41 \text{ cu in.}$$

$$r = \frac{41 + 5}{5} = 9.2$$

### 5.5    The Clerk Engine

The Clerk engine operated on the same principle as the Otto engine except that its cycle was completed in two strokes of the piston. The intake and exhaust strokes were eliminated by providing a second cylinder and piston which functioned only as a pump to draw in the fuel–air mixture and place it under a slight pressure in a passage leading to the entrance of the power cylinder.  The charge was then transferred to the engine cylinder through inlet ports in the cylinder head that were uncovered by a slide valve near the end of the expansion stroke.  A second set of ports in the side of the cylinder wall leading to the exhaust disposal system had been uncovered by the piston an instant earlier, and the products of combustion were forced from the cylinder by the pressure of the incoming charge. Only the compression and expansion strokes were necessary for the completion of the cycle, the transfer of charge into the cylinder and exhaust of products of combustion occurring while the piston was at its most remote point from the cylinder head and occupying the last

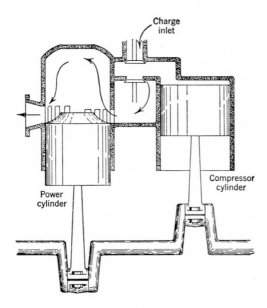

**Fig. 5.7.**   Two-stroke-cycle engine.

part of the expansion stroke and the first part of the compression stroke.

The two-stroke cycle as now widely used in Diesel-type engines is a development of the original Clerk idea. It has the advantage of permitting twice as many cycles to be completed, but requires some form of pump or blower to supply air. It is not so satisfactory for engines that induct a fuel–air mixture because some portion of the charge escapes through the exhaust ports during the transfer period. Figure 5.7 illustrates an early adaptation of the Clerk principle in which the air from the compressor enters through one set of piston-controlled cylinder ports, and the exhaust leaves through a second set.

Small modern two-stroke-cycle engines generally use the underside of the piston and the crankcase to precompress the air charge, whereas larger engines employ centrifugal or Roots-type rotary blowers to supply the charge under pressure. The ideal Otto cycle is followed by two-stroke-cycle engines, just as when four strokes are used.

### 5.6  The Diesel Engine

Ignition of the charge as originally effected by exposing it to a pilot flame or by an electric spark was equally applicable to all types of engines that have been discussed. The Diesel principle, which was commercialized by Dr. Rudolf Diesel just before the close of the nineteenth century, eliminates the need for an ignition system by utilizing the temperature rise which accompanies rapid compression to ignite the charge. If sufficiently compressed, the temperature of the air in the cylinder will exceed the ignition point of the fuel, and no spark is needed to ignite it. Practical Diesel compression ratios cause the air to attain a pressure of approximately 500 psi and a temperature of 1000 F. When this principle is employed, the charge taken into the cylinder on the intake stroke is air alone, and the fuel is sprayed in near the end of the compression stroke at the precise instant that will cause combustion to occur at the proper point in the cycle.

The early engines of the Diesel type were very large and heavy, running at low speeds and with fuel injection and combustion starting at top dead center and continuing during the first portion of the power stroke. The ideal cycle which approximates the indicator diagram of these engines was first described by Dr. Diesel in 1893

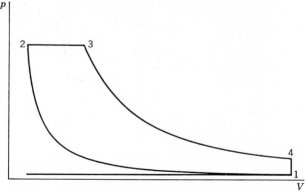

**Fig. 5.8.** Ideal Diesel cycle.

and differs from the Otto cycle only in the energy-addition process. Energy is supplied at constant pressure in the ideal Diesel cycle instead of at constant volume as in the Otto cycle. Figure 5.8 shows the $p$-$V$ diagram for the Diesel cycle.

Evaluating the quantities of energy in Equation 5.1 for this cycle yields the thermal efficiency expression

$$1 - \frac{1}{r^{k-1}} \left[ \frac{r_c{}^k - 1}{k(r_c - 1)} \right] \tag{5.5}$$

where $r$ is the compression ratio, $V_1/V_2$, of Fig. 5.8, $r_c$ is the cut-off ratio, $V_3/V_2$, which varies with load, and $k$ is the ratio of specific heats, $c_p/c_v$. Because $V_3$ is always greater than $V_2$, $r_c$ is greater than unity. An increase in the quantity of energy supplied increases $V_3$ and consequently $r_c$. The bracketed quantity of Equation 5.5, for any possible values of $r_c$, is greater than unity and increases with increased cut-off ratio or load. The subtractive portion of the efficiency equation, therefore, increases with the amount of energy supplied and is always greater than $1 - 1/r^{k-1}$, the corresponding quantity for an Otto cycle with the same compression ratio.

The efficiency of the Diesel cycle is shown by Equation 5.5 to depend not only upon the compression ratio, but also upon the amount of energy supplied, increasing with the compression ratio and decreasing with greater energy supply. As the amount of energy supplied approaches zero at light load, the efficiency approaches that of the Otto cycle with the same compression ratio, and becomes progressively less than that of the Otto cycle as the energy supplied increases. The efficiency of the Diesel cycle is lower principally because the expansion ratio decreases as the cut-off ratio increases.

**Example 3.** Find the thermal efficiency of a Diesel cycle with a compression ratio of 16 and a cut-off ratio of 4.

*Solution:* Substituting in Equation 5.5, $r = 16$, $r_c = 4$, and $k = 1.4$,

$$\eta_t = 1 - \frac{1}{16^{0.4}} \left[ \frac{4^{1.4} - 1}{1.4(4 - 1)} \right] = 1 - \frac{1}{3.03} \left( \frac{7 - 1}{4.2} \right)$$

$$= 1 - 0.472 = 52.8\%$$

The curve of Fig. 5.5 shows the efficiency of the ideal Otto cycle to be 67.5 per cent for a compression ratio of 16, an efficiency 28 per cent higher than that of the Diesel cycle at this cut-off ratio.

Combustion in the modern high-speed Diesel engine is very rapid and does not extend appreciably into the power stroke. The pressure rises considerably during combustion, and the process more nearly approximates constant volume than constant pressure. Only the extremely large Diesel engines, which run at very low speeds, and particularly those built some years ago when air injection was employed, can be considered to follow the theoretical Diesel cycle. All high-speed Diesels operate with peak cylinder pressures that greatly exceed compression pressures and actually approach the Otto cycle more nearly than does the spark-ignition engine. The thermal efficiency of the Diesel engine is higher than that of the spark-ignition engine because compression ratios of about 16 are employed, which are appreciably higher than those used with spark ignition.

### 5.7   Actual Thermal Efficiency

The thermal efficiency of the cycle, as expressed by the general Equation 5.1 and the special equations for each cycle, represents the portion of the energy consumed by a perfect engine that would be converted into work if there were no loss other than the inherent cycle loss caused by its rejection of the energy remaining in the working fluid at the end of the expansion stroke. No real engine could attain this thermal efficiency, because losses, such as imperfect combustion, heat transfer from the working fluid, leakage, and movement of the piston during combustion, are always present and reduce the work done below the theoretical amount. The actual thermal efficiency of a real engine is found by expressing the Btu equivalent of the net work done in per cent of the energy input.

The net work of the actual engine is evaluated by testing the engine in operation. The thermal efficiency is calculated by expressing the Btu equivalent of the work done in per cent of the higher heating value of the fuel consumed in the period of the test.

The power developed may be measured in the cylinder of the engine by means of the indicator diagram, in which case it is called the indicated horsepower, or it may be the power delivered by the shaft of the engine, which is called the shaft or brake horsepower. The engine indicator provides a record of the actual pressures existing in the cylinder at the corresponding volumes displaced by the moving piston. An area is enclosed by the lines traced by the indicator stylus (Fig. 5.3) which is proportional, according to the scale to which the diagram is drawn, to the work done on the piston per cycle.

### 5.8   The Engine Indicator

Engine indicators are of several types, each of which is suitable for a certain range of operating conditions. The displacement-type indicator, illustrated in Fig. 5.9, is the simplest device available for reproducing the cylinder pressure–volume record. It gives satisfactory results on engines of reasonably large cylinder dimensions operating at moderate speeds. Cylinder diagrams of steam engines, Diesel engines, gas engines, compressors, and pumps are usually made with displacement-type indicators. High-speed small-displacement engines require indicators that do not add an appreciable volume to the cylinder when connected to it and that are capable of responding to extremely rapid changes in cylinder pressure and volume without lag and distortion. The high-speed indicators are usually of the electric type, employing a pressure pickup which generates a small emf that is proportional to the pressure acting upon a small thin steel diaphragm exposed to the gases in the cylinder. A cathode-ray oscilloscope is used to portray the pressure changes, a permanent record of which may be made by photographing the screen.

The displacement-type indicator consists of two essential units, a drum and a cylinder fitted with a piston, spring, and stylus mechanism illustrated in Fig. 5.9. The drum (5) reproduces the volume changes within the cylinder. It is free to rotate slightly less than one full turn about a vertical axis when propelled by the application of tension to the cord (6) which is wrapped about the lower edge. This cord is connected to a reducing-motion mechanism attached to

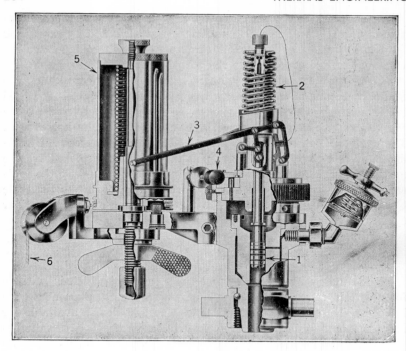

**Fig. 5.9.**  Sectional view of engine indicator.

the engine which reproduces exactly the motion of the piston scaled down to less than the circumference of the drum. A spring inside the drum returns it to a stop position and keeps the cord taut. A strip of paper, coated on one side with a metallic compound which causes a brass stylus to produce a black trace, is attached to the cylindrical surface of the drum by means of two spring clips. This strip of paper is customarily referred to as an indicator card.

The cylinder (1) has a normal cross-sectional area of $\frac{1}{2}$ sq in. but may be fitted with a smaller bushing rated in terms of the fraction that its area is of the normal half-inch. The lower face of the piston which operates in the cylinder is exposed to the cylinder pressure by a connection to the cylinder head, and its motion is restrained by a calibrated spring (2). A multiplying straight-line linkage connects the piston rod to the brass stylus on the end of arm (3) which can be made to bear against the surface of the card on the drum if the adjustable screw (4) is pressed against a stop. The vertical movement of the stylus will be proportional to the pressure acting on the piston, while the distance it moves will depend upon the scale of the indicator spring. Springs are rated in terms of the pressure in pounds per square inch required to displace the stylus 1 in. when used with a normal-size piston. When a smaller-than-normal piston is used, the scale is increased by dividing by the fractional rating of the piston. Thus, a 100-lb spring would permit a pressure of 100 psi to move the stylus 1 in. when used with a normal piston, but a pressure of 500 psi would be necessary to move the stylus 1 in. if the same spring were used with a $\frac{1}{5}$ normal piston.

In operation, the card moves horizontally in unison with the movement of the piston, while the stylus moves vertically in unison with the changes in cylinder pressure. Since the volume displaced by the piston of the engine is proportional to the linear movement of the piston, horizontal distances on the indicator card are also proportional to the volume displaced by the piston. When placed in contact with the card, the stylus makes a pressure–volume tracing of the cycle executed in the cylinder.

### 5.9 Mean Effective Pressure

It would be possible to determine the relationship between foot-pounds of work and unit areas on the diagram from the scale of the indicator spring and the scale to which cylinder volumes are reproduced as horizontal distances on the diagram. It is more convenient and useful, however, to evaluate the average pressure from the diagram than to find the work directly. A quantity known as the mean effective pressure is defined as that constant pressure which, if it acted on the piston for one stroke, would do the same amount of work as is done by the varying pressure during one cycle. It is apparent that this pressure is the average pressure during the return strokes subtracted from the average pressure during the forward strokes.

The mean effective pressure, or mep, is equal to the average ordinate of the area formed by the diagram multiplied by the spring scale of the indicator. The average ordinate may be measured directly by the use of a polar planimeter, or the enclosed area may be evaluated and divided by the length of the diagram parallel to the volume axis.

A better conception of the significance of the average ordinate of the indicator diagram and the corresponding mean effective pres-

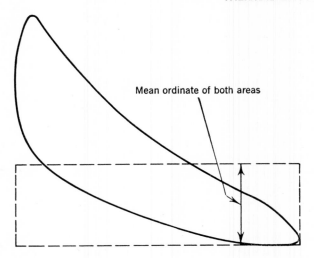

Mean ordinate of both areas

**Fig. 5.10.** Engine indicator diagram and equivalent rectangle.

sure may be gained from Fig. 5.10. The dotted line encloses a rectangle that is equal in area and length to the diagram traced by
the indicator stylus. This equivalent rectangle represents the indicator diagram that would have resulted, had the two strokes of the
piston been at the constant pressures corresponding to its upper and
lower horizontal boundaries, which are the pressures necessary to
do the same amount of work as was done by the varying pressures of
the actual cycle. The height of this rectangle is found by dividing
the area of the actual diagram by its length, and it obviously is equal
to the mean ordinate of the original indicator diagram. The difference between the maximum and the minimum pressures of this hypothetical cycle is the mean effective pressure indicated by the actual
diagram. Evaluation of the mean effective pressure is thus, in effect,
the substitution of a constant pressure that would do the same
amount of work during the power stroke of the piston for the actual
varying pressure of the cycle.

The mean effective pressure is a useful quantity for comparing the
performance of two engines, because it is proportional to the work
done per cycle per unit of cylinder size. The horsepower capacities
of different engines do not indicate how effectively the engines are
operating, because the factors of speed and piston displacement are
included. Two engines might conceivably develop the same horse-

power, but, if one were larger than the other and running at higher speed, its specific output would be less than that of the other. Their mean effective pressures would, in this case, be different and would reveal the extent to which the one was operating more effectively.

**Example 4.** Find the mep of an engine for which an indicator diagram drawn with a 200-lb spring has an area of 1.65 sq in. and a length of 3.5 in.

*Solution:* The mean ordinate of the diagram is 1.65/3.5 = 0.471 in. Each inch of mean ordinate corresponds to 200 psi; so 200 × 0.471 or 94.2 psi is the mep.

### 5.10   Indicated Horsepower

The indicated horsepower is the power developed in the cylinders at the faces of the pistons of an engine. It is calculated from the cylinder-pressure information revealed by the engine indicator and receives its name from that fact. The mean effective pressure (mep) of the cylinder in pounds per square inch multiplied by the area of the piston in square inches equals the average total force in pounds acting upon the piston during one stroke. When this force in pounds is multiplied by the length of the piston stroke in feet, the work in foot-pounds done on the piston by the gas during one power stroke is evaluated. One power stroke is completed in each cylinder of a four-stroke-cycle engine while the crankshaft turns through two revolutions and the engine completes one cycle. This product is thus the work done per cycle or per two revolutions of the crankshaft of a four-stroke-cycle engine. Since the two-stroke-cycle engine completes a cycle in one turn of the crankshaft, the work done per cycle is also the work done per revolution.

If the mean effective pressure in pounds per square inch is expressed by $P_i$, the length of stroke in feet by $L$, and the area of the piston in square inches by $A$, the work done per cycle is

$$W = \text{force} \times \text{distance} = P_i LA \text{ ft-lb per cycle} \qquad (5.6)$$

If $N$ cycles are completed per minute, the work done per minute is $P_i LAN$ ft-lb per min and, by the definition of a horsepower,

$$\text{Ihp} = \frac{P_i LAN}{33,000} \qquad (5.7)$$

$N$ is evaluated by measuring the rpm of the engine and taking into account the cycles completed per revolution. When there are several cylinders in the engine, the average mep of all cylinders may be used for $P_i$, and $N$ will then be the number of cycles completed per minute by all the cylinders in the engine. It is also possible to calculate the power developed in each cylinder separately and sum up the total.

**Example 5.** Find the ihp of a four-cylinder four-stroke-cycle engine of 3-in. bore and 5-in. stroke when it is running at 1500 rpm with a mep of 90 psi.
*Solution:*

$$P_i = 90, \quad L = \tfrac{5}{12}, \quad A = 3.14 \times 1.5 \times 1.5, \quad \text{and} \quad N = 1500 \times \tfrac{4}{2}$$

When substituted in Equation 5.7,

$$\text{Ihp} = \frac{90 \times 5 \times 3.14 \times 1.5 \times 1.5 \times 1500 \times 4}{12 \times 33,000 \times 2} = 24$$

### 5.11  Brake Horsepower

The brake horsepower (bhp) is the useful power delivered by the engine at the crankshaft coupling. The name comes from the method of loading an engine for test by means of a prony brake or any equivalent device arranged to brake or resist the crankshaft rotation to a controllable and measurable extent. The action of all power-absorbing devices used for testing engines resulting in converting the rotational tendency of the crankshaft into a tangential force acting at some established distance from the center of rotation.

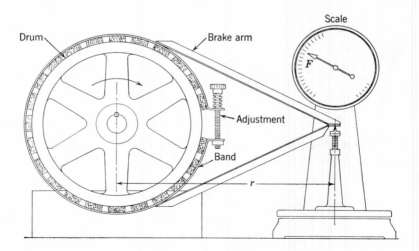

**Fig. 5.11.**  Prony brake.

Figure 5.11 shows a type of prony brake in which the power developed by the engine is converted into heat by friction between the brake shoes and a drum that is rotated by the crankshaft of the engine. The heat generated is absorbed by cooling water circulated through the drum. The torque arm of the brake terminates in a knife-edge located a precise distance $r$ ft from the center of the rotating brake drum. A force $F$ acts upon a pedestal resting on a scale platform, permitting its measurement in pounds. The product $Fr$ is the torque or turning movement in pound-feet which, in conjunction with the rotational speed $n$, determines the horsepower absorbed by the brake.

The force $F$ acts tangentially to a circle of radius $r$ and, in one turn of the brake drum, acts through a distance equal to the circumference of the circle, $2\pi r$ ft. Work is done in the amount,

$$W = 2\pi rF \text{ ft-lb per revolution}$$

$$= 2\pi rFn \text{ ft-lb per min}$$

where $n$ is the rpm of the engine shaft.

$$\text{Bhp} = \frac{2\pi rFn}{33,000} \tag{5.8}$$

Since $rF = t$, where $t =$ torque in pound-feet,

$$\text{Bhp} = \frac{2\pi tn}{33,000} = \frac{tn}{5252} \tag{5.9}$$

**Example 6.** An engine is loaded by a prony brake with a 48-in. arm. When it was running at 300 rpm, the scale reading was 60 lb. Find the bhp of the engine loaded by this brake.

*Solution:*

$$\text{Bhp} = \frac{2\pi rFn}{33,000} \text{ from Equation 5.8}$$

$$r = 4, \quad F = 60, \quad \text{and} \quad n = 300$$

$$\text{Bhp} = \frac{2 \times 3.14 \times 4 \times 60 \times 300}{33,000} = 13.7$$

$$\text{Torque} = \frac{\text{bhp} \times 5252}{n} \text{ from Equation 5.9}$$

$$= \frac{13.7 \times 5252}{300} = 239.8 \text{ lb-ft}$$

The prony brake is limited to low operating speeds. Its low cost and simplicity make it desirable for testing engines within its speed range, but it is unsatisfactory for higher speeds because of vibration and difficulty in adjusting and maintaining loads. At increased speeds, its convenience of operation and accuracy of power measurement decrease rapidly.

Electric dynamometers apply the same fundamental principle as the prony brake. Magnetic linkage between the stator and the rotor replaces the mechanical friction of the prony brake as the means of creating the desired turning effort or load. The torque thus applied to the stator by the rotation of the rotor is transmitted, by means of a horizontal arm attached to one side of the stator, to a sensitive type of weighing scales. The stator is cradled on pedestals and is free to rotate through the small angle necessary to allow the arm to act upon the scale linkage. The rotor or armature is carried on ball bearings mounted in the ends of the stator, which, in turn, are carried on bearings in the supporting pedestals. Thus, any bearing friction between the rotor and the stator is included in the force applied to the scales, and need for tare correction in the scale reading is eliminated. The point of contact between the dynamometer arm and the scale linkage tends to rotate in a circle of radius $r$, which is the distance from the point to the center of rotation, just as in the prony brake.

Electric dynamometers are more generally used for high speeds than other types of brakes. They are of two general types, the eddy current and the motor–generator. The eddy-current dynamometer is, in effect, a magnetic brake in which a toothed steel rotor turns between the poles of an electromagnet attached to a trunnioned stator. The resistance to rotation is controlled by varying the current through the coils and hence the strength of the magnetic field. The flux tends to follow the smaller air gaps at the ends of the rotor teeth, and eddy currents are set up within the metal of the pole pieces, resulting in heating the stator. This energy is removed by circulating cooling water through a jacket formed in the stator. Figure 5.12 illustrates a Midwest Dynamatic Dynamometer of the eddy-current type which has a capacity of 1100 hp and is capable of operating up to 8000 rpm.

The motor–generator-type dynamometer consists of a cradled generator that is operated as a generator for loading an engine and as a motor for cranking or measuring friction horsepower. Torque reaction is applied to scales through a reversible linkage which causes

**Fig. 5.12.** Dynamatic eddy-current dynamometer.

the force at the scales to act in the proper direction, regardless of the direction of rotation of the armature or whether the dynamometer is loading or motoring the engine. Load and speed are controllable both by varying the resistance of the circuit to which the generator terminals are connected and by varying the field strength.

The horsepower absorbed by a dynamometer is calculated in the same manner as when an engine is loaded by a prony brake. The linkage connecting the stator to the scales may be complicated, but its effective length may be determined and substituted in Equation 5.8. It is customary to calculate a constant for a dynamometer such that the horsepower absorbed is equal to the product of the constant, the speed, and the scale reading.

Since the brake horsepower is equal to $2\pi rFn/33{,}000$, calculations of horsepower will be simplified if, in the design of the dynamometer, the arm is given a length $r$ such that $2\pi r$ will equal 33 ft or some simple fraction thereof. If the length of the arm were made equal to 63.02 in. or 5.252 ft, $2\pi r$ would equal 33 ft, and the horsepower expression would become $Fn/1000$. This length, however, is too great for dynamometers of usual size, and so the typical lengths employed are 21 in. or 15.75 in., and the denominator of the above expression becomes 3000 or 4000. It is customary to express the horsepower determination as

$$\text{Bhp} = CFn \qquad (5.10)$$

where $C$ is $\frac{1}{3000}$ for a 21-in. arm and $\frac{1}{4000}$ for a 15.75-in. arm. Either the value of $C$ or the length of the arm will be found on the rating plate attached to the dynamometer by the manufacturer.

**Example 7.** An automobile engine is attached to an electric dynamometer which has a 15.75-in. arm. When it is running at 3500 rpm, the scale reading is 125 lb. Find the bhp developed by the engine.
*Solution:*

$$\text{Bhp} = CFn$$

$$C = \tfrac{1}{4000}, \quad F = 125, \quad \text{and} \quad n = 3500$$

$$\text{Bhp} = \frac{125 \times 3500}{4000} = 109.4$$

### 5.12 Friction Horsepower

The indicated horsepower is invariably considerably greater than the brake horsepower. The difference between these two quantities represents the portion of the power developed in the cylinders that was absorbed by the engine and not delivered by its shaft. It includes the power required to overcome friction in the bearings of the engine and between the piston and cylinder walls, the power required to operate such auxiliaries as camshafts, magnetos, oil and water pumps, and fans, in addition to the pumping work done in drawing in the charge and expelling the exhaust gases. These power losses are summed up as the friction horsepower of the engine. The indicated horsepower of slow-speed engines can be accurately determined from the indicator diagrams, and the friction horsepower found as the difference between the indicated horsepower and the brake horsepower. It is customary to evaluate the friction horsepower of high-speed engines by motoring them with the fuel supply

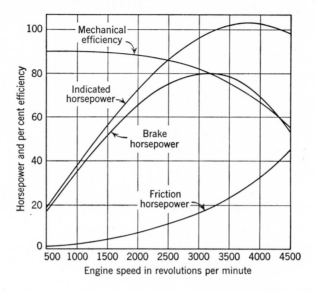

**Fig. 5.13.** Effect of speed on friction losses of an automobile engine.

cut off and all operating conditions kept as nearly as possible the same as when the engines are running. The horsepower developed by an electric dynamometer in motoring an engine is a close approximation of its friction horsepower.

The friction horsepower of an internal-combustion engine varies but little with changes in load at constant speed and is essentially a function of the speed of the engine, increasing gradually at low speeds and more rapidly at higher speeds. Figure 5.13 shows a typical relationship of indicated horsepower, brake horsepower, and friction horsepower for an automobile engine.

### 5.13 Mechanical Efficiency

Mechanical perfection is approached as the friction horsepower decreases in proportion to the indicated horsepower. The mechanical efficiency of an engine is evaluated as the brake horsepower in per cent of the indicated horsepower. It is a measure of the mechanical perfection of an engine or its ability to transmit the power developed in the cylinders to the driveshaft. Mechanical efficiency is expressed by the equation,

$$\eta_m = \frac{bhp}{ihp} \times 100 \qquad (511).$$

The mechanical efficiency of a constant-speed engine, such as a stationary power-plant engine, increases with load, because the approximately constant friction horsepower becomes a smaller per cent of the increasing indicated horsepower. The mechanical efficiency of a variable-speed engine decreases with increased speed, because of the rapid increase in friction horsepower. Mechanical efficiencies of internal-combustion engines usually range from 70 to 90 per cent.

### 5.14   Brake Mean Effective Pressure

It is more convenient and accurate to determine the brake horsepower of an internal-combustion engine than the indicated horsepower. Because of the rapidity with which the cylinder pressure changes in high-speed engines, very elaborate engine indicators are necessary for accurate evaluation of mean effective pressures. The friction horsepower can be measured with fair accuracy by motoring an engine with a motor–generator-type dynamometer and calculating the power required to rotate the crankshaft at constant speed from the reaction on the scales. The indicated horsepower can be found as the sum of the brake horsepower and the friction horsepower, instead of by calculating it from the mean effective pressure.

A hypothetical quantity known as the brake mean effective pressure is a widely used index to the pressure developed in the cylinder and, consequently, to the power capacity of the engine per unit of engine size. It may be defined as the constant pressure in pounds per square inch gage which, if acting on the piston through each power stroke, would develop in the cylinder an amount of power equal to the brake horsepower of the engine. It will be seen that the bmep is less than the imep by the ratio of the bhp to the ihp, making it equal to the imep multiplied by the mechanical efficiency. The bmep is found by calculating the bhp in the usual way and then equating it to

$$P_b LAN/33,000$$

Thus,
$$P_b = \frac{bhp \times 33,000}{LAN} \qquad (5.12)$$

where $P_b$ is the bmep in pounds per square inch gage, $L$ is the length of piston stroke in feet, $A$ is the piston area in square inches, and $N$ is the cycles per minute for the entire engine.

**Example 8.** A six-cylinder engine of 4-in. bore and 3.5-in. stroke develops 165 bhp at 4000 rpm. Find the brake mean effective pressure.

*Solution:*

$$L = \frac{3.5 \text{ in.}}{12 \frac{\text{in.}}{\text{ft}}} = 0.291 \text{ ft}, \ A = 3.14 \times 2 \text{ in.} \times 2 \text{ in.} = 12.56 \text{ sq in.}$$

$$N = \frac{4000 \text{ rpm} \times 6 \text{ cyl.}}{2 \frac{\text{strokes}}{\text{rev.}}} = 12,000 \text{ cycles per min}$$

$$P_b = \frac{\text{bhp} \times 33,000}{LAN} = \frac{165 \times 33,000}{0.291 \times 12.56 \times 12,000} = 124.3 \text{ psi}$$

### 5.15   Brake Thermal Efficiency

The actual thermal efficiency involves the weight of fuel consumed by the engine in an actual test. The fuel burned by the engine during the test period is expressed in pounds per hour and converted into a unit quantity by dividing the total pounds of fuel per hour by the horsepower developed. This quotient is called the brake specific fuel consumption, bsfc, or indicated specific fuel consumption, isfc, depending upon whether the bhp or ihp is used as the divisor.

The thermal efficiency of an engine is the energy equivalent of the horsepower developed in per cent of the energy supplied to the engine during a test. Since 1 hp delivered for 1 hr is equivalent to 2545 Btu, the product of the specific fuel consumption and the higher heating value of the fuel will be the energy input corresponding to an engine output of 2545 Btu. In equation form,

$$\text{Thermal efficiency, } \eta_t = \frac{\text{work output, Btu per hr}}{\text{energy input, Btu per hr}}$$

$$(5.13)$$

$$\text{Brake thermal efficiency, } \eta_{bt} = \frac{\text{bhp} \times 2545}{m_f \times Q_H}$$

$$(5.14)$$

$$\text{Indicated thermal efficiency, } \eta_{it} = \frac{\text{ihp} \times 2545}{m_f \times Q_H}$$

$$(5.15)$$

where $m_f$ is pounds of fuel per hour and $Q_H$ is the higher heating value of the fuel in Btu per pound.

**Example 9.** An engine loaded by a dynamometer with a 21-in. arm is operated at 500 rpm. When it was tested under load, the scale reading was 120 lb, and, when motored, the reading was 24 lb. It burned 2.54 lb of fuel for which $Q_H = 20,000$ Btu per lb during a 15-min test. The compression ratio of the engine is 5.65. Find the efficiencies determined by the test.

*Solution:*

$$\text{Bhp} = \frac{2\pi r F n}{33,000}$$

$$\text{Bhp} = \frac{2 \times 3.14 \times 1.75 \times 120 \times 500}{33,000} = 20$$

$$\text{Fhp} = \frac{2 \times 3.14 \times 1.75 \times 24 \times 500}{33,000} = 4$$

$$\text{Ihp} = \text{bhp} + \text{fhp} = 20 + 4 = 24$$

$$\text{Fuel consumption} = \frac{2.54 \times 60}{15} = 10.16 \text{ lb per hr}$$

$$\text{Bsfc} = \frac{10.16}{20} = 0.51 \text{ lb per bhp per hr}$$

$$\text{Isfc} = \frac{10.16}{24} = 0.42 \text{ lb per ihp per hr}$$

$$\eta_m = \tfrac{20}{24} \times 100 = 83.3\%$$

$$\eta_{bt} = \frac{20 \times 2545}{10.16 \times 20,000} = 25\%$$

$$\eta_{it} = \frac{24 \times 2545}{10.16 \times 20,000} = 30\%$$

$$\text{Ideal cycle efficiency} = 1 - \frac{1}{r^{k-1}}$$

$$\eta_c = 1 - \frac{1}{5.65^{0.4}} = 1 - \frac{1}{0.2} = 50\%$$

Figure 5.14 is a flow chart that shows diagrammatically the destination of the energy supplied to the engine of Example 9 and makes it possible to visualize the various efficiencies that have been discussed. Energy enters the system at the left in the form of fuel. An assumed 100 Btu or 100 per cent of the energy supplied is represented by the width of the stream. The cycle on which the engine operates

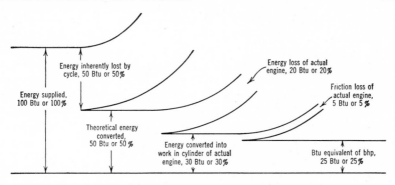

**Fig. 5.14.**  Flow diagram showing destination of energy supplied to an engine.

is shown to be 50 per cent efficient, and the stream, consequently, divides equally between the energy theoretically converted into work and that rejected because it is theoretically unavailable for conversion when following that cycle.  The 50 Btu that can be theoretically converted into work suffer a loss of 20 Btu, representing the imperfections of the practical engine as an energy converter.  These energy losses are largely the effects of incomplete combustion and heat transfer from the combustion chamber.  It should be noted that these losses are considered to be at least partially avoidable, whereas the 50-Btu inherent cycle loss is unavoidable.  The remaining energy, 30 Btu, is actually converted into work in the cylinder of the engine and represents the indicated horsepower of the engine.  Of these 30 Btu, 5 Btu are lost mechanically by the engine, and the work equivalent of only 25 Btu finally arrives at the flywheel.  These last losses are occasioned by the rubbing together of parts, pumping action of the pistons, and such mechanical operations as opening the valves and driving fans, pumps, and other needed accessories.

The various efficiencies indicated by this chart take the form of ratios of energy quantities and, consequently, widths of streams. The theoretical cycle efficiency is the theoretical work in per cent of the energy supplied, 50 Btu/100 Btu or 50 per cent.  The indicated thermal efficiency is the indicated work in per cent of the energy supplied, 30 Btu/100 Btu or 30 per cent.  The brake thermal efficiency is the brake or delivered work in per cent of the energy supplied, 25 Btu/100 Btu or 25 per cent.  The mechanical efficiency is the Btu equivalent of the brake horsepower in per cent of the Btu equivalent of the indicated horsepower, 25 Btu/30 Btu or 83.3 per cent.

### 5.16  Fuels

Consideration must be given early in the study of internal-combustion engines to the properties of the fuels burned in them. The development of these engines has to a large extent paralleled the progress made in producing fuels suitable for internal-combustion-engine use. Nothing useful is accomplished by designing an engine that is ideally suited to the combustion of a hypothetical but nonexistent fuel. The efforts of the engine designers have, accordingly, been directed toward the creation of engines that would operate as satisfactorily as possible on the fuels that were available on the market at that time. The gasoline engines of 25 years ago represent the best that the engineers of that period knew how to design for operation on the gasolines marketed at that time. The best-performing engines of today would have been useless had they existed then, because they could not have been run on the fuels that were available.

Several important differences between the engines of a generation ago and the best gasoline engines of today are improvements made possible by the acquisition of a better understanding of fuel properties and by the changes in those properties which were the result of technological advances in petroleum production. The fuels engineer strives to produce fuels that will best meet the increasingly severe requirements of the improved engines, while the engine designer modifies his designs to utilize the better characteristics of the improved fuels as fully as possible, both keeping within the limitations imposed by economic and practical considerations.

Since there will probably always be several grades of each of the types of manufactured fuels, there should be engines whose requirements match as closely as practicable the characteristics of each grade and type. Many of the imperfections of the early internal-combustion engines were the result of the designer's lack of knowledge of fuel characteristics, and unsatisfactory performance has frequently been traceable to the operator's lack of knowledge of the fuel requirements of the engine. Effective use of both fuels and engines requires knowledge of the fuels that are available. Although it is true that internal-combustion engines may operate on all types of fuels if the necessary provisions are made, only gaseous and liquid fuels have practical significance.

### 5.17    Gaseous Fuels

Gases are very desirable fuels when available at reasonable cost per unit of heating value. Compared with liquid fuels, gases have both good and bad qualities. Aside from the need for removing dirt and other foreign matter, gases need no preparation for combustion other than mixing in appropriate proportions with air. Liquid fuels must be vaporized before they can be burned. Problems also arise in carbureting a liquid-fuel–air charge and in the combustion of partially vaporized mixtures. The advantage of easily preparing a uniform charge of gas and air that is immediately ready to burn is offset to a considerable extent by the disadvantage of its high specific volume which limits the use of gases that cannot be readily liquefied to stationary power plants. The high self-ignition temperatures of gases improve their anti-knock quality for spark-ignition engines and permit higher compression ratios than are practical with liquid fuels, but require pilot injection of liquid fuel to attain ignition in Diesel engines. Natural gas is used extensively, in regions where it is available, for both spark-ignition and Diesel stationary engines. By-product gases from certain industries are also burned to limited extents where circumstances justify their use.

A mixture of liquefied petroleum gases, usually called LP-gases or LPG, consists largely of propane with small amounts of butane and unsaturates. Originally sold for domestic heating and cooking, with some commercial applications, LPG has proved desirable as an internal-combustion engine fuel because of its high anti-knock quality, clean burning, and low cost. The most important power users of LPG are city buses, delivery trucks, highway trucks and buses, and farm tractors. Because of its high knock resistance, higher compression ratios can be used than with gasoline, and its cleaner burning characteristics reduce maintenance requirements and increase engine life.

Generally speaking, gaseous fuels cost less per Btu and can be burned more efficiently than liquid fuels.

### 5.18    Liquid Fuels

A large percentage of all internal-combustion engines burn liquid fuels, most of which are refined from crude petroleum. Fuels processed from certain vegetable materials are quite satisfactory, but cannot compete with petroleum fuels on the present market. Petroleum is a mixture of many hydrocarbon compounds and comes from the oil wells as a brown or black liquid with a peculiar pungent odor. It can be burned in some large slow-speed Diesel engines by merely being centrifuged to remove the sand and water. All other petroleum fuels are refined from the crude oil by processes ranging from those which merely segregate certain groups of hydrocarbons that are already present in the crude oil to those which break down the chemical structure of some of the compounds and form new molecules.

The basis of the separation of the crude oil into the various products that are derived from it is the difference in the boiling points of its hydrocarbon compounds. The lightest constituents of the crude oil boil at temperatures well below zero, whereas the heavy ends are waxes and tarry compounds that are practically nonvolatile. The mixture can be separated into groups containing only compounds whose boiling points lie within a desired temperature range. This process is called fractional distillation and consists essentially in progressively boiling off and condensing fractions that boil below certain temperatures. Highly volatile petroleum ethers are evolved by a slight warming of the crude oil; then other ingredients too volatile for use in gasolines are driven off, followed, at progressively higher temperatures, by gasoline, naphtha, kerosene, light fuel oils or distillates, and lubricating oils. The residue contains the heavy fuel oils and solids.

In the early years of the petroleum industry, kerosene was the product for which the greatest demand existed; gasoline and heavier fuels, having little market value, were sometimes destroyed. Increased use of internal-combustion engines made gasoline the essential product, and refinery practice has progressed in the direction of getting more and better gasoline from the crude petroleum. More recently, the demand for Diesel and oil-burner fuels has necessitated increased yields of these less volatile fuels.

### 5.19   Hydrocarbons

Petroleum fuels, whether liquids or gases, are composed of compounds of hydrogen and carbon known as hydrocarbons. Each commercial fuel, such as gasoline, kerosene, and fuel oil, is a mixture of many hydrocarbons whose boiling points lie within the volatility range appropriate for that fuel. These hydrocarbon compounds vary widely in molecular weight and structure. They are classified according to their molecular structure into several series or families, whose members have the same general formula but different molecular weights. Those present in petroleum fuels are paraffins, olefins, diolefins, naphthenes, and aromatics. The important significance of these series lies in their different combustion characteristics and their consequent effect upon engine performance. The eastern crude oils consist largely of paraffins, whereas the western crudes have large aromatic and naphthenic content. Olefins and diolefins are not common in natural petroleums, but are formed at the high temperatures used in some refining operations.

The paraffins are chain compounds, either straight or branched, and range from $CH_4$ containing only one carbon atom to the waxes containing about 30 carbon atoms. They are classified as saturated compounds, because no additional hydrogen atoms can be attached to the carbon atoms. Normal paraffins have all the carbon atoms linked together in a straight chain with the hydrogen atoms attached to each side and at each end of the carbon chain. Typical of these is normal heptane which has its seven carbon atoms linked together by single bonds and the hydrogen atoms attached so as to satisfy the valence of each. Since carbon has a valence of four and hydrogen of one, there must be 16 hydrogen atoms in the heptane molecule. The normal heptane molecule may be shown graphically in the following manner:

$$\begin{array}{ccccccc}
\text{H} & \text{H} & \text{H} & \text{H} & \text{H} & \text{H} & \text{H} \\
| & | & | & | & | & | & | \\
\text{H—C—C—C—C—C—C—C—H} \\
| & | & | & | & | & | & | \\
\text{H} & \text{H} & \text{H} & \text{H} & \text{H} & \text{H} & \text{H}
\end{array}$$

The chemical formula for heptane is seen to be $C_7H_{16}$, conforming to the general paraffin formula, $C_nH_{2n+2}$.

A normal straight-chain paraffin molecule may be so rearranged that it has a branched-chain structure. The new compound is called an isomer of the original one or an isoparaffin. Normal heptane thus becomes isoheptane. The branches formed by removing carbon atoms from the chain and attaching them to its sides are called radicals and usually consist of one carbon atom and three hydrogen atoms, the methyl radical. Paraffins with less than four carbon atoms cannot form isomers, and the number of possible rearrangements increases with number of carbon atoms present. The important change in the characteristics of the compound resulting from isomerization of a normal paraffin is an appreciable improvement in anti-knock quality accompanying each transfer of a carbon atom from the chain to a methyl-radical branch.

The progressive improvement in anti-knock quality as the heptane molecule is shortened and compacted by reforming into isoheptanes is illustrated in the following simplified structural diagrams. The hydrogen atoms are omitted for simplicity, but their bonds to the carbon atoms are shown. The corresponding octane numbers indicate their relative anti-knock quality. The name of each compound is based upon the number of methyl radicals and the number of carbon atoms in the chain, but they are all heptanes because each has a total of seven carbon atoms.

Normal heptane                 0 octane number

Methyl hexane                 60 octane number

Dimethyl pentane             90 octane number

Trimethyl butane (triptane)   125 octane number

One of the iso-octanes, 2,2,4-trimethyl pentane, is the compound that is blended with normal heptane in the reference fuels used to establish the octane number of a gasoline. This iso-octane was arbitrarily assigned an anti-knock rating of 100 octane, and normal heptane was assigned an anti-knock rating of zero octane. The iso-octane used in knock rating has the following molecular structure:

Olefins differ from paraffins in that they are unsaturated because each molecule lacks one hydrogen atom, whereas the diolefin molecule lacks two hydrogen atoms. Both are formed when compounds of higher molecular weight are split into two or more smaller molecules by the cracking process that is extensively used in refining petroleum crudes as a means of increasing gasoline yield. Olefinic compounds are quite generally considered to be the most undesirable constituents of gasoline because they form gums and also decrease anti-knock quality.

Naphthenes or cycloparaffins are saturated ring-structure compounds, structurally equivalent to normal paraffins that have been formed into rings by joining their ends together. The molecules are thus centralized and the anti-knock quality improved.

Aromatics have the ring structure of the naphthenes, but they are unsaturated because each carbon atom carries only one hydrogen atom. They differ further in that there are always six carbon atoms in the rings. Naphthalenes are double-ring aromatics with ten carbon atoms and benzenes are single-ring compounds. Both may have side chains composed of methyl or larger radicals.

From the standpoint of their heating value, volatility, anti-knock quality, and preignition resistance, branched-chain paraffins are considered the most desirable gasoline compounds, and cyclic paraffins are next, with olefins and aromatics least desirable.

The characteristics of the hydrocarbons which affect their suitability as engine fuels depend upon both their molecular structure and their molecular weight. Volatility, as evidenced by decreased boiling point, increases as molecular weight decreases. Ignition temperature increases with decreased molecular weight for any one series of compounds. Ignition temperature also increases as the compactness or complexity of the molecule increases. Volatility and ignition temperature are the principal factors that determine whether a fuel is suitable for a spark-ignition or a compression-ignition engine. Most spark-ignition engines induct a combustible mixture of fuel and air that is prepared outside the cylinder of the engine. The fuel should be sufficiently volatile to permit it to be substantially vaporized during the intake stroke. Gasoline, with a boiling-point range of about 100 to 400 F, meets the volatility requirements of spark-ignition engines of the carburetor type. Some spark-ignition engines have mechanical-injection systems instead of carburetors and can use fuels of lower volatility.

### 5.20   Spark-Ignition Combustion

In the spark-ignition engine, combustion originates at the spark-plug points, where a small quantity of fuel is heated above its ignition point by the high temperature of the electric spark.   The surface of the particle of fuel that was ignited by the spark presents a tiny incandescent flame front to the particles of fuel adjacent to it, causing them to ignite and, in turn, present a flame front of larger area to the fuel particles surrounding them.   This flame front travels through the unburned charge radially across the combustion chamber and, if undisturbed, progressively burns all the fuel.   This process is illustrated in $(a)$, Fig. 5.15.   As each particle of fuel is burned, its chemical energy is released, causing its products of combustion to attain a very high temperature.   The resulting expansion of the burned portion compresses the unburned particles by crowding them into a progressively smaller specific volume, and the accompanying temperature rise may cause the ignition point of the fuel to be attained ahead of the flame front.   If this happens, all the remaining unburned fuel will ignite spontaneously, and, instead of burning progressively, it will burn almost instantaneously with a rapid pressure rise which sets up a high-frequency pressure wave and causes an audible knock that has been called detonation but now is known as normal knock or simply knock.   Knocking combustion is shown in $(b)$ of Fig. 5.15, whereas nonknocking combustion proceeds entirely across the combustion chamber as in $(a)$.   It should be noted that this combustion irregularity occurs only after burning has progressed considerably, and it involves the last portion of the charge to burn. The higher the ignition point of the fuel, the less possibility there will

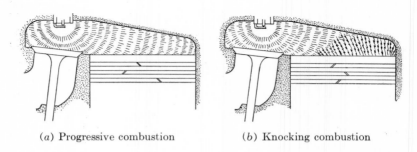

$(a)$ Progressive combustion            $(b)$ Knocking combustion

**Fig. 5.15.**  Combustion in the spark-ignition engine.

be for this spontaneous combustion to occur under given combustion-chamber conditions, and the higher its anti-knock quality.

Another mode of uncontrolled ignition is known as surface ignition. This occurs when the fuel–air charge ignites either before the spark or before the flame front set up by the spark reaches the point of surface ignition. Ignition is caused by contact with a hot surface, such as overheated spark-plug electrodes, a protruding particle of chamber deposit material, or an exhaust valve. Temperatures endangering the piston crown may be attained if surface ignition occurs appreciably before the spark. This irregularity is called preignition. In extremely high-compression engines a low-pitched thudding knock, called rumble, results when surface ignition causes the rate of pressure rise to become excessive. Paraffins are more resistant to surface ignition than other gasoline compounds, and olefins are the most susceptible.

### 5.21   Octane Rating

The compression ratio of the engine determines the temperature and pressure attained by the charge at the end of the compression stroke for given initial conditions. The more nearly the charge approaches self-ignition conditions before being ignited by the flame front created by the spark and the longer it is exposed to these conditions, the greater the possibility of knock. The factors that control self-ignition include not only temperature and time but also the density and burning rate of the charge. Consequently, the higher the volumetric efficiency of the engine, the higher its compression ratio, and the more powerful the fuel–air ratio of the mixture in the cylinder, the more knock-resistant the fuel must be. Because the speed of the engine affects the time of exposure to self-ignition temperature, operation at low speed encourages knock. It naturally follows that, the higher the compression ratio, other factors being equal, the greater the tendency for knock. The compression ratio at which a fuel can be used without knock is thus a measure of its anti-knock quality. This fact is utilized in rating the anti-knock quality of gasolines by the octane scale.

The octane method of rating fuels requires an actual test of the fuel in a special knock-rating engine that is so constructed that the compression ratio can be varied while all other operational factors are held constant. The tendency of the fuel to knock can be measured by varying the compression ratio of the test engine until a certain intensity of knock is indicated by a knock meter which measures the rate of pressure rise during the combustion period. This knocking tendency is then matched by preparing a reference fuel composed of iso-octane, a branched-chain paraffin of high anti-knock quality, and normal heptane, a straight-chain paraffin of high knocking tendency, in such proportions that the mixture behaves the same in the engine as the fuel being rated. The higher the percentage of iso-octane in the reference fuel that has the same knocking tendency as the fuel being rated, the higher the anti-knock quality that is indicated.

The octane number of the fuel is expressed as the per cent by volume of iso-octane in a reference blend with normal heptane that has the same anti-knock quality as the fuel. The term 100 octane means a high anti-knock quality equal to that of pure iso-octane, while zero octane means a very low anti-knock quality equal to that of pure normal heptane. If the fuel has higher anti-knock quality than pure iso-octane, as some aviation gasolines now have, this method of testing cannot be used, but approximate octane numbers can be assigned by extrapolation. It should be noted that the octane rating is merely a comparison between the two fuels and does not mean that there is any *iso*-octane actually present in the gasoline.

### 5.22   Volatility Rating

The volatility of a liquid fuel is determined by a standard distillation test designed to find the lowest or initial boiling point at which the first drop of the fuel distils and the percentage of the original amount of fuel distilled at each chosen increment of temperature rise up to the end point or maximum boiling point. The standard apparatus for this test is illustrated in Fig. 5.16. The fuel sample is placed in the glass flask and heated by the gas burner so that it is boiled off at a prescribed rate. The thermometer inserted in the flask indicates the temperature of the vapor leaving the flask at each interval. The vapor enters the condenser and is cooled to a liquid which drips into the graduated cylinder. The quantity distilled is recorded with the corresponding temperature at each stage.

A volatility curve is plotted from these data which shows, not only the boiling-point range of the fuel, but also the temperature necessary to boil each fraction. Two fuels may have the same boiling range, but differ widely in overall volatility, because the amounts that vaporize when the fuel is heated to each temperature within that range are greatly different. Starting quality, rapidity of warming

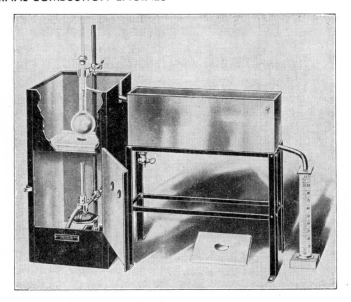

**Fig. 5.16.** Tag ASTM distillation apparatus.

up, acceleration, power, and completeness of combustion are related
to the volatility or distillation curve of a gasoline. Figure 5.17 shows
typical distillation curves for a gasoline and a Diesel fuel. Note
that the least volatile portion of the gasoline boils at the boiling point
of the most volatile portion of the Diesel fuel. These are the boiling
points at atmospheric pressure and should not be confused with the
evaporation temperatures when atomized and mixed with air. The
vapor pressure is then much lower than the total pressure of the mixture, and the temperature of vaporization is correspondingly lower.

### 5.23 Compression-Ignition Combustion

Combustion in the compression-ignition or Diesel-type engine differs extremely from that in the spark-ignition engine. The compression ratio must be sufficiently high to raise the temperature of
the compressed-air charge well above the ignition point of the fuel.
The fuel is sprayed into the combustion chamber as the end of the
compression stroke is approached and is distributed in the form of
tiny droplets throughout the air charge. The particles of fuel receive heat by transfer from the air, and their temperature rises to the

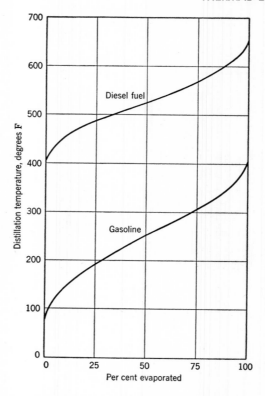

**Fig. 5.17.** Distillation curves of gasoline and Diesel fuel.

boiling point of some portion which vaporizes and absorbs additional heat until the ignition point of the most easily ignited particle is reached. Ignition is thus originated at many points throughout the combustion chamber after a certain delay during which the vaporization, heating, and precombustion reactions occur. The length of this delay or ignition lag, as it is called, is not the same for all fuels, and it is a measure of the ignition quality of the fuel. It determines whether or not the engine will run smoothly, because the rate of pressure rise in the cylinder accompanying ignition of the fuel depends upon the amount of fuel that has accumulated in the combustion chamber before ignition.

If the ignition lag is long, a large portion or perhaps all of the fuel will have been injected before any of it ignites. This accumulated fuel will then burn almost instantaneously, and a large amount of energy will be released during the first instant of combustion with a resulting overly rapid rise in cylinder pressure. If this initial burning is too rapid, there will be a combustion knock, and the engine will be rough running and noisy. If the fuel ignites quickly, only a small amount of fuel will be involved in the initial burning, and the remainder will burn progressively as it is sprayed in. Combustion will then start without any audible knock, and the engine will run smoothly. It should then be noted that the knock in a Diesel-type engine occurs at the beginning of combustion instead of at the end as in the spark-ignition type.

### 5.24  Cetane Rating

The ignition quality of a Diesel fuel is rated in the form of a cetane number by a comparison method analogous to the octane rating system. A special variable-compression test engine is used, and the compression ratio is established which causes ignition a certain standard delay period after the injection of the fuel. A reference blend of cetane, a long straight-chain easily ignited paraffin, and α-methyl naphthalene, a closely knit ring-type naphthalene, is then prepared which matches the ignition behavior of the fuel being rated. The per cent by volume of cetane in this reference blend is the cetane number assigned to the fuel. A fuel of high ignition quality will have a high-cetane number. In order to have a high-cetane number, the molecular structure of a hydrocarbon must be simple and its molecular weight high. This latter requirement limits the volatility range of Diesel fuels to temperatures above those of gasolines.

It is fortunate that the requirements, as to both volatility and ignition temperature of gasoline and Diesel fuel, are directly opposite, because each uses a different group of the hydrocarbons present in petroleum. The high volatility and high ignition temperature required of gasoline are associated in the same hydrocarbons, whereas the low volatility and low self-ignition temperature required of Diesel fuels are found in other heavier compounds.

### 5.25  Refining Methods

Early refining methods consisted simply in fractional distillation of the crude oil at desired temperature increments which bracketed together the ingredients appropriate to the different products. It was later found that the yield of gasoline from this process was insufficient to meet the demands, as less than 20 per cent of the crude oil was convertible to gasoline. The thermal-cracking process was then devised, whereby the larger and simpler molecules were broken down by heating to high temperatures into lighter molecules in the gasoline range of volatility. More stable molecular structures were also attained by cracking, and gasolines of higher-octane rating were produced besides increasing the yield to nearly 50 per cent of the crude oil. Olefins, diolefins, and aromatics are formed by high-temperature cracking, whereas isomers of the paraffins are formed at lower temperatures.

More recently developed cracking processes employ catalysts which permit accurate control of the compounds formed and produce gasoline of higher-octane number than thermal cracking. Olefins and diolefins, which are unstable in storage and cause gum formation, are not present in catalytically cracked fuels.

Polymerization is a process that is chemically the opposite of cracking and is used extensively to convert hydrocarbons that are too light and volatile for gasoline into heavier compounds of the desired volatility. Numerous other processes for synthesizing desirable molecules from those in the portion of the crude oil that is not suitable for gasoline have been developed by petroleum technologists, and further increases in yield and octane number have resulted.

Improvement in molecular structure and selection of components alone did not meet the octane requirements of continuously increasing compression ratios. Most gasolines now contain up to three milliliters of tetraethyl lead per gallon. This additive retards the precombustion reactions that lead to self-ignition and cause knock. The effect of tetraethyl lead addition on the octane number of a gasoline depends upon the refining processes used in its production because some hydrocarbons are more susceptible to lead than others. Phosphorous, bromine, and chlorine compounds are included in the anti-knock additive to decrease the spark-plug fouling and surface-ignition tendencies of the products formed by combustion of the lead compound.

Injection of water and methyl alcohol into the inlet manifold has proved an effective means of reducing the octane requirement of high-compression engines. The function of the water-alcohol mixture in improving the anti-knock performance of engines is largely as an internal coolant. The high latent heat of vaporization of the liquid injected causes it to act as a refrigerant to reduce the temperature rise of the charge before ignition. The take-off power of airplane engines is increased by this means because higher supercharge pressure may be used without causing knock.

## 5.26 Nomenclature

In order to read intelligently and understandingly discussions of internal-combustion engines, it is necessary to learn the nomenclature by which the many components of an engine are designated. Some of these terms are descriptive of the parts or their functions, but others are not so readily associated with the devices to which they have been applied. Figure 5.18 illustrates those parts of an automobile engine that would be made visible if it were cut across through

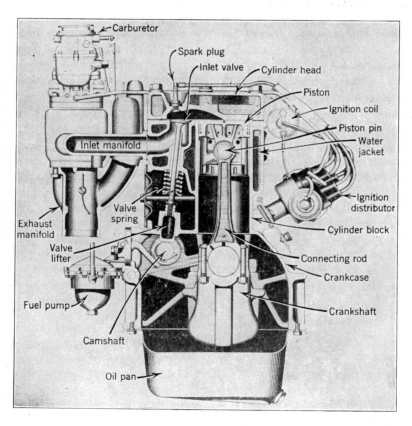

**Fig. 5.18.** Cross section of automobile engine.

one of its cylinders. The more important parts exposed to view by sectioning the engine in this manner are indicated. The following discussion is intended to define the terms that are used to designate the important parts of an engine and to explain their functions.

**Cylinder Block.**  The cylinder block forms the structural framework of the engine except in certain large units that are built up from a foundation or bedplate and in aviation engines with air-cooled cylinders.  In the automotive-type engine, the cylinder block is an alloy-iron casting extending ordinarily from the crankshaft vertically to the top of the cylinder bores.  Machined surfaces are provided at appropriate places for attaching the bearings that support the crankshaft and for connecting adjacent units of the engine to the block. The upper surface is fitted to retain the cylinder head in position, and the lower extremity is machined for attaching an oil pan which completes the enclosure for the crankshaft and acts as an oil sump for the lubricating system.  The extension of the cylinder block below the cylinders is called the crankcase.

The cylinders proper consist of cylindrical passages extending vertically through the block, bored to accurate dimensions, and honed to smooth bearing surfaces.  The cylinders are frequently fitted with thin hardened liners which form the cylinder walls instead of these being machined in the block itself.  The camshaft which operates the valves is fitted into a passage bored lengthwise of the cylinder block, and suitable guides for the valve lifters are formed above it.  The cylinder block illustrated in Fig. 5.18 is of the L-head type and carries the valve passages and valve seats beside the upper extremities of the cylinder bores.  Machined faces for attaching the inlet and exhaust manifolds are provided where the valve ports terminate on the side of the block.  Water passages or jackets are cored into the block around the cylinders and valves for cooling.  The supports by which the engine is mounted are usually attached to the front cover which encloses the gears or chains by which the camshaft is driven from the crankshaft and to the bell housing which encloses the flywheel at the rear of the engine.

**Cylinder Head.**  The gas-tight compartments which serve as the combustion chambers are formed between the cylinder head and the pistons.  With the valves in the cylinder block, as shown, the cylinder head is merely a water-jacketed cover, the underside of which is profiled to form spaces of the exact size and shape desired above each cylinder bore.  Threaded openings for the spark plugs are provided at the proper locations.  The joint between the cylinder head and the cylinder block is made gas-tight by a composition gasket.  The cylinder head of the valve-in-head engine is complicated by the presence of the valve ports, guides, and seats, which are in the cylinder block of the L-head engine. The supports for the rocker arms which operate the valves must also be provided on top if the valves are in the head.

**Pistons.**  The pistons are cylindrical castings of iron, aluminum alloy, or steel, with one end closed by a crown which forms the lower surface of the combustion chamber.  The outer periphery of each piston is accurately machined to a running fit in the cylinder bore and is provided with several grooves into which piston rings are fitted. The piston rings are elastic rings of high-grade cast iron which are so made that they exert a radial force against the cylinder wall and tend to prevent gas leakage through the joint between the piston and the cylinder wall.  The upper rings are plain rings, called compression rings, which serve to prevent escape of gases from the combustion chamber.  The rings in the lower grooves are vented oil rings that are especially designed to distribute the oil uniformly around the cylinder in a film of proper thickness.  The oil rings also scrape excess lubricating oil from the cylinder wall and dispose of it through drain holes leading from the ring grooves to the piston interior.

The pistons are attached to the connecting rods by hardened steel piston pins fitted into bosses cast in opposite sides of the piston. Piston pins are said to be floating when they are free to turn in bearings both in the upper ends of the connecting rods and in the piston bosses but are restrained by suitable retainers from contacting the cylinder wall.  In some installations the pin is secured in either the connecting rod or the piston bosses and is not free to turn in both. The portion of the piston below the pin bosses is called the skirt and serves as a bearing surface for the piston.

**Connecting Rods.**  The reciprocating motion of the pistons is adapted to the rotary motion of the crankshaft by forged steel rods which connect them together.  The upper ends of the rods are formed into eyes and bushed with bronze or similar bearing material unless the design requires that the piston pins be clamped in the rods.  The lower ends are fitted with removable caps that are retained by two or more bolts, and the bored openings are lined with babbitt or similar bearing metal. This lining may be cast in place, or it may be in the form of a precision insert that fits closely within the bore. The rods are usually of I section and may be rifle-drilled to convey lubricant from a passage in the crankshaft to the piston pins.

**Crankshaft.** The crankshaft is one of the most important parts of the engine and serves to convert the forces applied by the connecting rods into a rotational force. It controls the motion of the pistons and must be designed to cause them to reciprocate in proper sequence. Alloy-steel forgings are usually employed in the crankshaft, although cast-alloy-iron shafts are in common use.

The crankshaft is made up of a throw for each cylinder, or each pair of cylinders if the engine is of the V-type, and a suitable number of main bearings located at the two ends and between adjacent throws. The throws are arranged in one or more planes passing through the center of rotation that is common to all the main bearings. The angles between the throws depend upon the number of cylinders served by the shaft, which determines the firing interval between the cylinders. The location of the throws in the various radial directions depends upon the firing order of the cylinders. The firing order is chosen to distribute the power impulses along the length of the engine, and in the six-cylinder engine the standard sequence is 1–5–3–6–2–4.

Since the crankshaft is made up of rotating masses that are displaced from the center of rotation, centrifugal forces are set up that must be balanced by counterweights opposite the throws. The bearing areas on the crank throws to which the connecting rods are attached are called crankpins, and the radial arms are called cheeks. The distance from the crankpin centers to the main bearing center establishes the stroke of the pistons. The crankshaft is usually drilled to permit oil to be circulated from entrances in the main bearing areas to the crankpins. The main bearing journals operate in soft metal liners similar to those used in the lower ends of the connecting rods.

**Valve Mechanism.** The camshaft originates the motion that is imparted to the valves and is driven from the crankshaft by a chain or gears. In four-stroke-cycle engines the entire group of valves operates once in two turns of the crankshaft. The camshaft operates each of the valves in one turn and must therefore turn at half the speed of the crankshaft. Each valve requires a cam to open and close it at the proper point in the engine cycle. The complete cam-shaft for a multicylinder engine consists of a shaft extending the length of the engine with a cam formed adjacent to each valve. The sequence in which the lifting surfaces of the several cams present themselves as the shaft rotates is established by the firing order of the cylinders chosen by the designer. The shape of the cams is established by the valve timing and the lift imparted to the valves.

Associated with the camshaft is a set of cam followers or valve lifters which are reciprocated by contact with the cam surfaces. The valve lifters are interposed between the camshaft and the valves to take the side thrust caused by the rotation of the cams and to provide an adjustment by which the lash in the mechanism can be controlled. Push rods and rocker arms are also included in the camshaft mechanism for valve-in-head engines, because the motion of the lifter must be conveyed to the cylinder head and reversed in direction. The camshaft can be carried on the head above the valves to eliminate this extra linkage.

The valves are mushroom-shaped and are called poppet valves. The bevel-seated heads seal against matching seats in the combustion-chamber wall. The valve stems are closely fitted in guides with which they form gas-tight joints to prevent manifold leakage. Helical springs apply the necessary closing force through spring seats that are attached to the ends of the valve stems by retainers.

**Manifolds.** The inlet-valve passages in the cylinder head of the valve-in-head engine or in the cylinder block of the L-head engine must be connected to the outlet of the carburetor. The inlet manifold is simply a tube extending along the length of the cylinder block with a branch at each inlet valve and a central inlet at the top to which the carburetor is attached. The principal objectives in the design of the manifold are to provide equal distribution of fuel and air to the several cylinders and to deliver as large a charge as possible when the engine is not throttled.

The exhaust manifold gathers together the streams of products of combustion leaving the exhaust passages of the cylinders and conveys them to the muffler. Minimum possible restriction to flow is desirable.

**Fuel Pump.** The fuel-supply tank is ordinarily at a lower level than the carburetor, necessitating a pump to supply the fuel to the engine. The usual type of pump employs a diaphragm of impregnated fabric that is displaced on its suction stroke by a lever which bears against one of the cams on the camshaft. It is propelled on its return or delivery stroke by a spring which exerts only sufficient force to push the diaphragm against a fuel discharge pressure of about 4 psi. The pump executes delivery strokes only in accordance with the amount of fuel needed by the carburetor, because the spring can return the diaphragm only as rapidly as the carburetor accepts the fuel.

### 5.27   Classification

The internal-combustion engine has attained a wide variety of forms in order to meet the requirements of the many widely differing applications to which it has been adapted. Although basically the same in fundamental principle, these engines differ appreciably in structural design, arrangement of components, accessory equipment, and operating characteristics. Any attempt to classify them into clearly defined groups is complicated by the fact that some of the distinguishing features of each type engine are found in varied combinations with certain other features which are distinctive of another type engine. It is necessary to specify a number of things about any engine in order to define its field of application, fuel limitations, and manner of functioning. A knowledge of some of the more important functional and constructional differences between these widely diversified engines is essential to the acquirement of a proper background for the study of internal-combustion engines. Some of the more important methods of classifying engines are discussed in the following paragraphs.

**Number of Strokes per Cycle.** One of the most important differences that may be used as a basis in grouping together certain engines is whether two or four strokes of the pistons are required to complete the operating cycle of the engine. Regardless of their classification on any other basis, engines of all types and for all purposes may be of either the two-stroke-cycle or the four-stroke-cycle type. Actually, the two-stroke engine has not been applied to a wide field because of difficulties in charging and scavenging the cylinders under certain conditions of operation. All high-speed engines are of the four-stroke type with the exception of small outboard marine engines and others of similar structural type intended for different usage. Other than these small gasoline engines, practically all the two-stroke engines are now of the Diesel or compression-ignition type. The fact that the incoming charge is used to push out the products of combustion remaining in the cylinder at the completion of the power stroke prevents efficient operation on the two-stroke cycle when the incoming charge is a combustible mixture of fuel and air. Some portion of the entering charge invariably escapes with the exhaust gases. This loss of charge to the exhaust is of small consequence if the fuel enters after scavenging is completed, as in the Diesel, but constitutes a serious loss if the fuel has already been mixed with the entering air. The distinguishing feature of the two-stroke engine is the inclusion of some form of pump or blower to place the incoming charge under pressure. This is not necessary in the four-stroke engine unless it is supercharged.

**Method of Igniting Charge.** All engines may be divided into two groups according to the method of raising the fuel temperature to the ignition point. Those employing compression ratios of between about 14 and 18 cause the air charge to attain a temperature sufficient to ignite ordinary fuels without any supplementary ignition device. Such engines operate on the Diesel principle of compression ignition which requires that the charge inducted into the cylinders be noninflammable and that the fuel be injected late in the compression stroke at the precise instant that will properly time combustion. The second group employs compression ratios that are limited to a maximum of about seven to avoid self-ignition. The fuel and air must be combined to form an inflammable mixture before the combustion period. Ignition is by means of an electric spark.

**Fuel Systems.** Engines have been designed for operation with three general types of fuels: solid, liquid, and gaseous. Solid fuel, in the form of pulverized coal, has been burned with some success, but the abrasive nature of the ash presents a problem that has not been mastered. Liquid fuels are burned by either carburetion or injection. Gases require no conversion preparatory to combustion and permit the use of a simple mixing valve.

Fuel systems are necessarily adapted, not only to the type of fuel used, but also to the operating cycle of the engine. Volatile fuels may be aspirated into the air stream as it flows through a carburetor, or they may be injected at moderate pressure into the air as it enters each cylinder or directly into the cylinders after the air has entered. Nonvolatile fuels, whether burned by compression ignition or spark ignition, are injected into the engine cylinder. Compression ignition requires a more elaborate injection system, because extremely high fuel pressures are necessary to atomize the fuel, and accurate timing must be maintained.

**Number and Arrangements of Cylinders.** Whether an engine has but one cylinder or several depends to a large extent upon the purpose for which the engine is intended. When a relatively small power output is required at slow speed under stationary power-plant conditions, a single-cylinder engine of quite large dimensions may be appropriate. When flexibility and smoothness of operation are important, six or eight small cylinders may be a more logical choice, even for power requirements that could easily be met with fewer and larger cylinders. Extremely large power capacity necessitates an increase in the number of cylinders, because the power that can be developed in each cylinder is limited by practical considerations. Engines having as many as 24 cylinders are produced for aviation use.

Single-cylinder engines may be built with the axis of the cylinder horizontal, but usually the cylinder is mounted vertically. Multicylinder designs permit a wide latitude in cylinder arrangement. The inline type has all cylinders in a single row, and they may be vertical, as in many automobile engines, or horizontal, as in some truck and bus engines that are disposed beneath the body to permit full space utilization above. V-type engines have half the cylinders in one line or bank and the other half in a second bank, with a common crankshaft at the intersection of the planes of the axes of the two rows of cylinders. Two connecting rods are attached to each throw of the crankshaft, and each cylinder in one bank is thus paired with the opposite cylinder in the other bank. The crankshaft has as many throws as there are cylinders in one bank. The angle between the two banks is determined by the application of the engine and the number of cylinders. It may be as small as 45 degrees or it may be 180 degrees, in which case it is classed as an opposed cylinder engine rather than a V-type.

Figure 5.19 shows a cross section of an eight-cylinder 90-degree V-type Studebaker engine of the valve-in-head design. This engine is representative of the compact short-stroke design that has been found desirable in providing the greater rigidity necessitated by the higher pressures encountered in engines of high compression ratio.

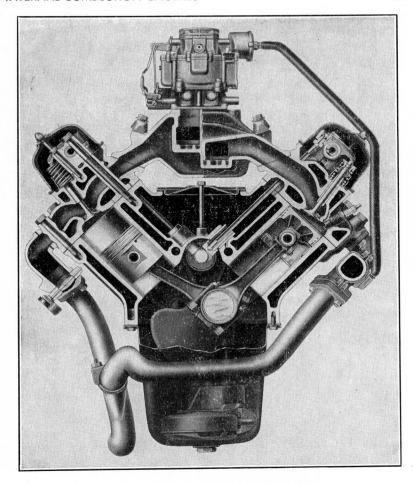

**Fig. 5.19.** Cross-sectional view of Studebaker V-8 engine.

Several arrangements of cylinders are used in combining more than ordinary numbers of cylinders into a single engine. Engines with four banks of cylinders and only one crankshaft are called X-type engines. Four banks of cylinders are also combined with two crankshafts in the form of a double V. Multicylinder engines with all the connecting rods attached to a single crank throw are called radial engines. They may have two such rows with as many as nine cylinders in each row.

**Valve Type and Location.** Most engines now employ poppet or mushroom valves that are held in the closed position by helical springs and are operated by a camshaft and associated linkage. The seat of the poppet valve is usually beveled at an angle of either 30 or 45 degrees with the plane of the valve head. Double-sleeve-valve engines, in which two reciprocating cylindrical sleeves containing inlet and exhaust ports were interposed between the piston and the cylinder wall, were quite widely used several years ago but are seldom seen now. Single-sleeve-valve types in which the sleeve has a combined reciprocating and oscillating motion are used by a few foreign manufacturers.

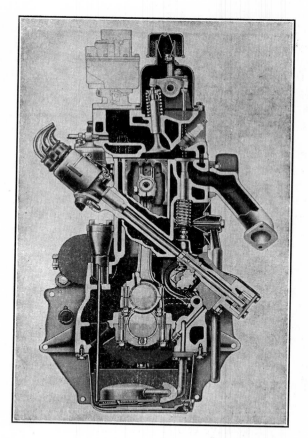

**Fig. 5.20.** Cross-sectional view of F-head Willys-Overland engine.

The inlet and exhaust valves of the four-stroke engine may both be inverted over the top of the piston in the widely used valve-in-head construction or I-head type. When located in the head, the valve stems may be parallel with the cylinder axis or inclined so that the heads lie in the sloping upper walls of a domed combustion chamber. All four-stroke Diesel engines are of the valve-in-head type, and two-stroke Diesels that use exhaust valves have them located in the cylinder head, because the resulting clearance space is more compact.

The valves of the four-stroke engine may be located in the cylinder block and open upward into pockets formed by extending the clearance space in the cylinder head so that it projects beyond the cylinder bore. If both inlet and exhaust valves are located at one side of the cylinder bore, the engine is classed as an L-head type, but, if they are on opposite sides of the cylinder, it is a T-head engine. Some engines have one valve in the head and one on the side in an arrangement known as the F-head type. Figure 5.20 is a cross section of the F-head Willys-Overland engine. In this design the inlet valve is located in the cylinder head and the exhaust valve in the block. This arrangement permits the use of a large inlet valve with an unrestricted passage for charge flow into the cylinder. The combustion chamber is compact and the spark plug favorably located.

**Application.** The purpose for which an engine is intended probably has more effect upon its design than any other factor. Weight and space limitations dictate the speed at which the engine must operate in order to develop sufficient power per unit of weight and size. Cost limitations are also reflected in the speed at which the engine is designed to operate, because low-speed engines cost much more to build per horsepower output. Low-speed engines are usually large heavy units that are governed to maintain a relatively constant speed. They are largely of the Diesel type, although most gas engines are also in this classification. Large marine and stationary power applications are the principal fields of the low-speed type.

Automotive applications require the light weight and flexibility of the high-speed multicylinder type. High-speed gasoline engines, such as the automobile type, commonly operate over a wide range of speeds with top speeds well above 4000 rpm. High-speed Diesel engines intended for automotive applications operate at speeds that would be medium for gasoline engines, and few go higher than 2500 rpm. Diesels of this type are extensively used for truck and tractor propulsion. Figure 5.21 illustrates a Cummins high-speed Diesel

**Fig. 5.21.** Cut-away view of Cummins Diesel engine.

**Fig. 5.21 (Continued).** Some of the working parts of Cummins Diesel engine.

engine. It is representative of the type that is applicable to heavy-duty automotive, portable, stationary, and marine uses. It differs from the automobile engine of Fig. 5.18 in that it is of the value-in-head type, has a Diesel fuel-injection system instead of a carburetor, has removable cylinder liners, and is of heavier more rugged construction throughout.

Aviation engines operate under conditions that impose very special requirements. The need for large horsepower capacity, light weight, and absolute reliability justifies construction costs that would be considered prohibitive for other uses. High-strength alloys are used for highly stressed parts and lightweight alloys contribute further weight savings at other points. Superchargers are necessary to enable the engine to obtain sufficient weight of charge at high altitudes where the specific weight of the air is low. Aviation engines are usually of the V-type or the radial type and may be either air- or liquid-cooled.

### 5.28   Cylinder Charging

The first step that must be taken in order to burn a fuel–air mixture within the cylinder of an engine is to supply the charge to the cylinder. Air will not flow into an engine cylinder of its own accord. It must be pumped in quantities of 100 or more cu ft per hr for each horsepower developed by the average engine. The first function of the engine is therefore that of an air pump to supply oxygen to the cylinders to react with the fuel. This function includes forcing from the cylinder the products of combustion which must be removed to make room for the next air charge, either by direct pumping action of the piston, as in the four-stroke cycle, or by requiring the entering air charge to push out the exhaust gases under the influence of the air-supply pump, as in the two-stroke cycle.

The pumping mechanism of the four-stroke-cycle engine consists of a system of inlet and exhaust valves that are mechanically actuated in phase with the movement of the piston. The exhaust valve must be held open during the inward stroke of the piston following the power stroke, and the inlet valve must be open during the next outward stroke.

Theoretically, the opening and closing events of the valves would occur at the crankshaft dead-center positions coincident with the beginning and ending of the piston strokes, but the time necessary for opening and closing the valves and the effects of high speeds upon the flow of the gases require that these events be displaced somewhat from the dead-center points. The intake and exhaust periods of the cycle are therefore not truly intake and exhaust strokes, since each includes 50 or more degrees of crank rotation in excess of a full piston stroke, and the two periods actually overlap to some extent. The most desirable timing of the valve operation is determined experimentally for each engine. A typical automobile-engine valve-timing spiral is shown in Fig. 5.22a.

This diagram shows all four valve events, inlet opening, inlet closure, exhaust opening, and exhaust closure, to be displaced appreciably from the top and bottom dead-center points, where they would theoretically occur. Opening events are earlier and closing events later to allow for time required to open and close the valves without excessive inertia loading or slamming. Ramps are provided on each cam to start the valve lift gradually and to slow down the valve return motion slightly before the seat is contacted. Of still greater significance, however, are the advance in opening timing and delay in closure for the purpose of utilizing the inertia of the flowing gases and for other practical considerations.

The exhaust valve is shown to open about 45 degrees before the piston reaches its bottom dead center. This so-called release point occurs well before the end of the power stroke in order to provide a blow-down period, during which the pressure in the cylinder is re-

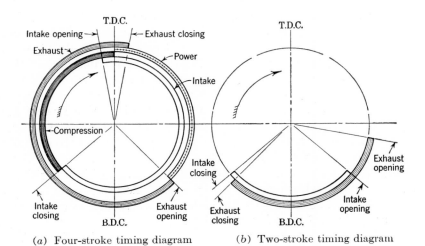

(a) Four-stroke timing diagram    (b) Two-stroke timing diagram

**Fig. 5.22.** Timing of intake and exhaust events.

duced by outflow of products of combustion to nearly that of the exhaust manifold by the end of the piston stroke. This results in some loss of work done on the piston by the expanding gases, but effects a greater reduction in the negative work that the piston must do in being pushed up against the back pressure of the gases during the early part of the exhaust stroke. The optimum exhaust-opening point is that at which the net work of the cycle is a maximum.

The exhaust valve is allowed to remain open some 10 to 20 degrees after the piston has reached its top dead-center position in the typical high-speed engine. The exhaust gas flows rapidly from the cylinder during the entire exhaust stroke, and the moving column has considerable kinetic energy because of its high velocity. Flow will thus continue outward after the piston is no longer pushing the gases. If the valve is still open, this inertia effect will reduce the pressure in the combustion chamber appreciably below exhaust-manifold pressure, thereby decreasing the weight of spent gases retained in the clearance space to dilute the incoming charge. This inertia effect of the outflowing exhaust gases is also the reason for opening the inlet valve before the end of the exhaust stroke. A partial vacuum is formed in the cylinder behind the moving column of exhaust gases, and this vacuum can be used to start the flow of fresh charge into the cylinder by opening the inlet valve at the point in the cycle at which the cylinder pressure is reduced to that of the inlet manifold. This valve-overlap period results in improved scavenging of the clearance space and more complete filling of the cylinder with fresh charge.

The inlet valve is allowed to remain open 50 degrees or more after bottom dead center to utilize the ram effect resulting from the kinetic energy of the column of rapidly moving charge in the inlet manifold. The flow of fresh charge persists after the piston has started upward on the compression stroke, and at some point, depending upon the engine speed and the flow characteristics of the induction system, a maximum weight of charge will have entered the cylinder. The inlet valve should close late enough to utilize the ram effect of the charge column in the manifold, but not so late as to permit some of the charge already inducted to be returned to the manifold by the rising piston. In general, the higher the speed at which the power of the engine is desired to reach its peak, the farther all the valve events will depart from the dead-center points.

The two-stroke-cycle engine requires a pump external to the working cylinder to force the air into the cylinder and the exhaust gases out. This pump may consist of the underside of the power piston

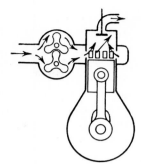

**Fig. 5.23.** Uniflow scavenged two-stroke engine.

and a closed crankcase, a second cylinder with a pump piston driven from the engine crankshaft, or a rotary pump of either the centrifugal or positive-displacement type. This pump or compressor places the air charge under sufficient pressure to force the required amount of air through the inlet ports in the cylinder wall against the back pressure of the outgoing exhaust gases within the allotted portion of the cycle. Figure 5.23 illustrates the scavenging action of a modern uniflow Diesel engine with air supplied by a Roots blower. Instead of more than 200 degrees of crank angle available for pumping in air and a similar period for pumping out exhaust products, as in the four-stroke cycle, the entire transfer of air and gases must be completed in about 130 degrees in the two-stroke-cycle engine. The timing circle for a two-stroke cycle engine is shown in Fig. 5.22b.

The exhaust valve or port is shown to open somewhat earlier than the release point of the four-stroke cycle because the blow-down period must now be completed before the inlet-port opening point. Unless the pressure in the cylinder is reduced to that maintained on the fresh charge by the scavenging blower, back flow of exhaust gases into the air box surrounding the inlet ports will result. This necessarily early release is a defect of the two-stroke cycle because it shortens the effective power stroke and decreases the mean effective pressure. Both inlet and exhaust passages are open for nearly 100 degrees, and the exhaust valve is then closed a few degrees before the inlet ports are again covered by the rising piston. This period during which the inlet remains open after the exhaust closure is referred to as a supercharging period, but it is doubtful that the high-speed two-stroke-cycle engine can be supercharged to any great extent by additional charge flow during so short an interval of time unless exceedingly high blower pressure is employed.

Engines of the back-flow-scavenged type differ from the uniflow type of Fig. 5.23 in that two sets of cylinder ports are provided at opposite sides of the cylinder bore, one serving as the inlet and the other as the exhaust passage, both opened and closed by the piston motion. The fresh air entering one of the ports is deflected sharply upward toward the cylinder head by a baffle formed on the piston crown, and the combustion products from the previous cycle flow downward on the opposite side of the cylinder bore, leaving via the exhaust ports. The inlet and exhaust timing of these engines is symmetrical in that the exhaust ports are covered by the piston at the same point on the compression stroke at which they were uncovered on the power stroke. This is also true for the inlet ports, and, since a blow-down period is necessary before the inlet opens, the exhaust must open before the inlet and close later than the inlet. This late exhaust closure is a defect of this type engine because some air charge will be lost from the cylinder after the inlet closes.

The opposed-piston-type two-stroke-cycle engine of Fig. 5.24 has two pistons in the same cylinder, and each controls a set of ports extending entirely around its circumference. By advancing one of the crankshafts with respect to the other and properly locating the

**Fig. 5.24.** Fairbanks-Morse opposed-piston Diesel engine.

ports in the cylinder walls, the timing of events is made similar to that of Fig. 5.22b. The large port areas made possible by this construction, the compact combustion chamber resulting from the high stroke–bore ratio, and the nearly perfect balance of reciprocating parts are advantages of the opposed-piston-type Diesel engine. The flow of gases during scavenging and charging of the cylinder is in one direction, resulting in less mixing of fresh air with exhaust gases than in engines wherein the two gases flow in opposite directions. The use of two crankshafts and the need for a means of connecting them together make the opposed-piston engine somewhat complicated mechanically. The engine illustrated has bevel gears on the front ends of the two crankshafts which mesh with gears on a vertical connecting shaft.

A comparison of Fig. 5.22a and Fig. 5.22b shows that, although the pumping operations of the two cycles are aimed at the same accomplishment, that of replacing the products of combustion in the cylinder by a fresh charge, the portions of the two cycles devoted to this exchange are quite different, as are the methods and mechanisms employed.

### 5.29  Volumetric Efficiency

The time during which air flows into the cylinder of an engine is not ordinarily sufficient to permit complete filling. Since the extent to which the cylinder is filled with air limits the amount of fuel that can be burned in the engine, a measure of the effectiveness of the engine as an air pump in charging itself will serve as an index to its possible power output. The weight of air inducted into the cylinder expressed in per cent of the weight required to completely fill the piston-displacement volume at outside temperature and pressure is known as the volumetric efficiency of a four-stroke-cycle engine. When the air entering the inlet system is measured volumetrically at atmospheric conditions, the volumetric efficiency is this measured volume in per cent of the volume displaced by the pistons during the same interval of time. High-speed engines usually show an increase in volumetric efficiency with increased speed to a peak value of about 80 per cent at about half speed and then a decrease at an increasing rate at higher speeds. Improvements in valve timing, intake-manifold design, valve design, carburetor design, and engine cooling are responsible for much higher volumetric efficiencies at high speeds than were formerly possible.

**Example 10.** A four-stroke engine has six cylinders of 3-in. bore and 4-in. stroke and runs at 3000 rpm. The air entering the carburetor was metered at 110 cu ft per min at outside temperature and pressure. Find the volumetric efficiency.

*Solution:* The piston-displacement volume of the engine is $1.5 \times 1.5 \times 3.14 \times 4 \times 6 = 169.5$ cu in. The engine completes a cycle in two revolutions of the crankshaft, so that in 1 min $169.5 \times 3000/2$ or 254,000 cu in. are required to fill the cylinders completely.

$$110 \times 1728 = 190,000 \text{ cu in. supplied}$$

$$\eta_v = \frac{190,000}{254,000} = 74.8\%$$

The volumetric-efficiency curve of Fig. 5.25 is typical of modern high-speed automobile engines. The weight of air supplied to the cylinders per cycle at full throttle when running at 1800 rpm is shown to exceed considerably that at 4000 rpm. Since the fuel burned per cycle depends upon the amount of air inducted, the energy released during each combustion period decreases as the engine speed increases above that at which the volumetric-efficiency curve peaks. The mean-effective-pressure curve consequently follows the descent of the volu-

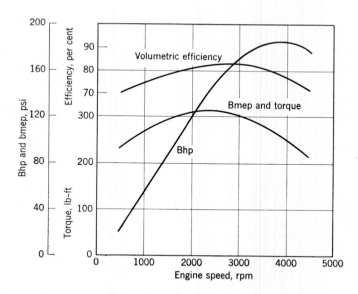

**Fig. 5.25.** Effect of volumetric efficiency on automobile-engine performance.

metric-efficiency curve because the decreased weight of charge burned results in lower cylinder pressures during each power stroke. Since the mechanical efficiency of an engine decreases as the speed increases, the brake-mean-effective-pressure curve peaks at a lower speed than that at which the volumetric-efficiency curve peaks and then decreases more rapidly than the indicated mean-effective-pressure curve. Torque is mathematically related to brake mean effective pressure; therefore the same curve can be used for both with suitable ordinate scales, as in Fig. 5.25.

The interdependency of torque and bmep is apparent from the equation for brake horsepower in terms of each:

$$\text{Bhp} = \frac{P_b LAN}{33,000} = \frac{tn}{5252}$$

This equation also indicates that the horsepower delivered would increase linearly with speed if the torque remained constant. Because the torque reaches a peak at a speed slightly below the peak of volumetric efficiency, the horsepower increases less rapidly than it does at the lower speeds where the torque is rising. When the speed is reached at which the rate of decrease in torque equals the rate of increase in speed, their product no longer increases, and the maximum horsepower of the engine is reached. At speeds above this peak, the torque is decreasing faster than the speed is increasing, and the horsepower falls off. The shape of the full-throttle horsepower curve is thus seen to depend upon how well the volumetric efficiency is maintained at increasing speed, and, consequently, the speed at which maximum horsepower is developed will be higher if the high-speed pumping effectiveness of the engine is improved.

### 5.30 Supercharging

In certain-type engines it has been found desirable to incorporate an auxiliary pump or blower to supplement the pumping action of the cylinders. A notable example of the application of this idea is in supercharging the cylinders of an aviation engine in order to increase its output per unit of size and weight and particularly to offset the loss in power at altitude caused by the decreased specific weight of the air. Diesel engines for marine, locomotive, and truck propulsion are sometimes supercharged to permit more horsepower output from a given space and weight limitation.

Supercharging appears to have greater advantages when used with compression ignition than with spark ignition, because it causes higher temperatures and pressures at the end of compression which increase the tendency of the spark-ignition engine to detonate and raise its gasoline octane requirement, whereas ignition is improved in the compression-ignition engine and its cetane requirement decreased by these same factors.

The supercharger merely raises the intake-manifold pressure, making it possible to fill the cylinders at a greater charge density. Since the density of the charge is intentionally reduced to regulate the input to the engine at partial loads, supercharging is necessary only when the engine operates at full load. The possibility of driving the supercharger blower by an exhaust turbine makes the idea even more attractive, incorporating as it does the utilization of otherwise wasted heat.

Supercharged engines differ from those that are naturally aspirated in that a pump or blower discharging into the intake manifold is incorporated. Blowers of the Roots type (Fig. 5.23) are usually employed for supercharging relatively small Diesel engines where the blower is geared to and driven by the crankshaft of the engine. The centrifugal-type blower is often geared to the crankshafts of aviation engines and, to a lesser extent, engines for other applications where the speeds are fairly constant and high manifold pressures are maintained.

Turbocharging, in which a centrifugal blower is driven by a gas turbine operated by the exhaust gas from the engine cylinders, is rapidly increasing in use. A majority of the large Diesel engines of the four-stroke-cycle type now manufactured are turbocharged, and their specific output is thus made to compare favorably with the power capacity of two-stroke-cycle Diesels. Locomotive Diesels of the four-stroke-cycle type are all turbocharged, in one instance the horsepower being doubled by increasing the inlet pressure to 22 psig pressure. When the blower is geared to the crankshaft, the friction horsepower of the engine is considerably increased, and it is not likely that the fuel economy of the engine will be improved by supercharging. The turbocharged engine, however, operates at improved mechanical efficiency and lower brake specific fuel consumption. It is conceivable that all four-stroke-cycle Diesel engines will eventually be supercharged as a means of decreasing the weight and cost per horsepower.

### 5.31   Combustion Chambers

Two important developments made possible the large increases in knock-free compression ratios for spark-ignition engines. First, H. R. Ricardo, in England, found that combustion knock can be controlled to a considerable extent by combustion-chamber design. Shortly after World War I he introduced a basic design for L-head engines in which the clearance space, instead of comprising a flattened dome of uniform depth over the cylinder bore and valve pocket, was concentrated in a deepened region over the valves with the portion over the far side of the piston thinned out to the thickness of the head gasket. The spark plug was located in the center of the deep section. Figure 5.18 shows a later modification of Ricardo's combustion chamber.

The second factor that made higher compression ratios possible is the improved anti-knock quality of gasoline. Mutual adaptation of engines and fuels, the combined efforts of the automobile and petroleum industries, gradually evolved the modern spark-ignition engine. The best gasolines produced in 1920 would knock severely in these engines.

In order to approach the ideal Otto cycle and its higher efficiency, the charge should be burned as rapidly as is possible without causing so high a rate of pressure rise that the engine is rough running and heat transfer is excessive. All forms of uncontrolled ignition must be avoided also. The portion of the charge farthest from the spark plug is the most likely to cause knock, either by self-ignition or by normal combustion at too high a mass rate. An important aspect of combustion-chamber design is thus control of the burning of the end gas remote from the spark plug. Other considerations are heat transfer to and from the charge and the effect of valve size and location on volumetric efficiency.

The rate of pressure rise in the cylinder depends upon the mass rate of combustion. The volume of charge burned during each increment of flame travel is a function of flame speed and the area of the flame front. The mass rate of burning depends upon these two factors together with the density of the charge as it is ignited. The density of the charge ahead of the flame front increases progressively during combustion because the burned products attain a high temperature and expand, pushing the flame front forward and compressing the unburned charge ahead of it. Successive layers of charge are ignited at progressively higher pressure and density, the mass rate of combustion increases, and the rate of pressure rise tends to become excessive in the end zone.

This tendency can be opposed by so shaping the combustion chamber that the cross-sectional area normal to the flame front decreases as the flame front advances. Peaks in the rate of pressure rise will be avoided if the chamber cross section approaches a triangle and the combustion space is similar to a cone with the spark plug located in the center of the base. The flame front, radiating hemispherically from the spark, enlarges rapidly at first to burn a large volume of charge while its density is low and then decreases as the chamber walls converge. The decrease in flame-front area compensates for the increase in charge density as burning progresses, and a uniform rate of pressure rise may be maintained. This effect is approached by the wedge-type chamber that is widely used in valve-in-head engines, as illustrated in Fig. 5.19, in which a high initial volume rate of burning tapers off to the quench area.

The temperature of the unburned charge is increased by the expanding combustion products as well as the density of the charge. If the last portion of the charge is at a high temperature long enough for precombustion reactions to occur, it may self-ignite and burn almost instantaneously. The lower the end-gas temperature and the shorter its time of exposure to igniting temperatures, the greater the probability that the flame front will reach and burn the end gas before it has time to self-ignite and cause knock. Burning time is reduced by a compact combustion chamber with the spark plug centrally located to decrease flame travel, and by creating turbulence to increase flame speed. Special provisions for cooling the end zone will keep down the end-gas temperature. Ricardo's chamber design satisfies all these requirements. He referred to the close-clearance region over the piston as squish area because of its turbulent effect in squeezing out the charge from the far side of the chamber. It is also called quench area because the high surface-volume ratio keeps the end gas cooler by increasing heat transfer to the piston crown and water-jacketed cylinder head. Shortened flame travel, charge motion toward the spark plug, and cooling the end gas combine to prevent knock by self-ignition in all modern high-compression engines.

Departures from the practice of forming the combustion chambers as cavities cast in the cylinder head have recently been made in designing wedge-type chambers. In these engines the cylinder head is machined flat in the plane of the valve heads and tilted at an angle

QUENCH and
SQUISH AREA

INTAKE
VALVE

FULLY
MACHINED
COMBUSTION
CHAMBER

SQUISH
SURFACE

**Fig. 5.26.** Chevrolet V-8 combustion chamber and pent-roof piston.

of 10 or more degrees from a plane normal to the cylinder axes. A part of the piston crown is machined flat at an angle which matches the tilt of the cylinder head, thus forming a quench area as shown in Fig. 5.26. The combustion chamber is deep over one edge of the piston and tapers to the shallow quench area. All surfaces of the combustion chamber are machined, permitting the accurate dimensions and smooth surfaces that are increasingly necessary as the combustion-chamber volume becomes smaller with increased compression ratio.

### 5.32  Firing Orders

The throws of the crankshaft of a multicylinder four-stroke-cycle engine are arranged to permit the desired firing sequence of the cylinders and to establish the intervals between impulses. The firing order of the cylinders is chosen to distribute the power impulses along the engine and to help the intake manifold distribute the charge uniformly to the cylinders. The average firing interval for a four-stroke engine is found by dividing 720, the number of degrees through which the crankshaft turns in one cycle, by the number of cylinders. A six-cylinder engine will accordingly fire at 120-degree intervals and an eight-cylinder engine every 90 degrees, if the intervals are equal.

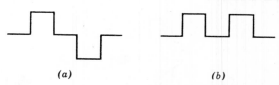

(a)                              (b)

**Fig. 5.27.** Two-cylinder four-stroke-cycle crankshafts.

Analysis of a two-cylinder in-line engine brings out a fundamental requirement for the throw arrangement of a four-stroke crankshaft. If the two throws are arranged as in Fig. 5.27a, the crankshaft has static balance because the eccentric masses are diametrically opposed. The pistons of the two cylinders arrive at the top-dead-center position at alternate half-turns of the crankshaft. Cylinder No. 1 fires when its crank throw points upward and the No. 2 throw is up 180 degrees later with the cylinder either in firing position or ending its exhaust stroke. Assuming that No. 2 now fires, then 180 degrees later No. 1 is again at top dead center, this time on its exhaust stroke. A full additional turn or 360 degrees must now be made before No. 1 is ready to fire a second time, and No. 2 follows 180 degrees later. The firing intervals are 180, 540, 180, 540 degrees, with both cylinders firing during one turn of the crankshaft and neither firing during the next turn. This is not desirable because the torque is not uniform and the speed tends to fluctuate cyclically.

In Fig. 5.27b the two throws are in alignment and the pistons move up and down in unison. This is obviously not as good as the previous arrangement from the standpoint of mechanical balance. However, No. 2 is ready to fire 360 degrees after No. 1, and No. 1 fires again 360 degrees later. One cylinder fires during each crankshaft turn and firing intervals are equal. It is now apparent that equal firing intervals for a four-stroke-cycle engine are possible only if another piston moves in unison with each one so that there is another cylinder in firing position exactly 360 degrees after each cylinder fires. Half of the cylinders must fire in each turn of the crankshaft. This necessitates that in-line engines have even numbers of cylinders so that there can be pairs of cylinders whose pistons are in phase that fire in alternate turns of the crankshaft.

Crankshafts for multicylinder, four-stroke engines may now be laid out following the general plan that the first and last throws be in alignment, the second and next to the last, the third and third from the last, and so on, similarly aligned. Pairing the throws equidistant from the center of the shaft eliminates rotating couples that tend to rock the engine longitudinally.

The sequence of the radial planes in which these pairs of throws lie is so established that the permissible firing orders will best distribute the power impulses along the length of the engine. Figure 5.28 shows diagrammatically the conventional crank-throw arrangements for four-, six-, and eight-cylinder automobile engines viewed from the front of the engine. The numbers indicate the cylinders whose crankpins are colinear. The crankshafts rotate clockwise and two pistons reach their upper limits of travel each time that the shaft turns through an angle equal to the firing interval. Since one of these two is in firing position and the other just completing its exhaust stroke, firing orders can be formulated assuming either of the two to be firing. The number of possible firing orders thus established is only one for a two-cylinder engine and doubles with each pair of additional cylinders. The two possible firing orders for a four-cylinder engine are 1–2–4–3 and 1–3–4–2, neither of which offers any advantage over the other because the front two and the rear two cylinders always fire in succession. Of the four possible firing orders for the six-cylinder engine, only 1–5–3–6–2–4 is used because the others do not distribute impulses desirably. Of the eight possible firing orders for the eight-cylinder, in-line engine, only 1–6–2–5–8–3–7–4 avoids successive firing of adjacent cylinders.

The V-8 crankshaft has only four throws because the connecting rods of opposite cylinders are attached to the same throw. Two-plane shafts are used with 90-degree angles between the throws. Equal firing intervals are possible only with a 90-degree angle between the cylinder banks. The banks of cylinders are usually staggered to permit the two connecting rods on each crankpin to be side by side. The front cylinder in the bank that is offset forward is thus the No. 1 cylinder, and, if it is in the right bank as viewed from the

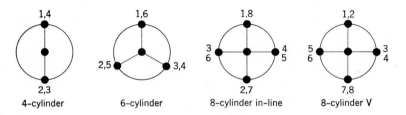

**Fig. 5.28.** Crank-throw arrangements for automobile engines.

front, the firing order is usually 1–8–4–3–6–5–7–2. Ideally, the power impulses should alternate between the two banks, but this is not possible with the two-plane crankshaft.

It should be noted that the arrangement of the crank throws for a four-stroke-cycle engine merely establishes the possible firing orders for the cylinders. The firing order that is selected for the engine is incorporated in the design of the camshaft. The crankshaft for the two-stroke-cycle engine has all its throws at firing-interval angles with each other, and their sequence is dictated by the desired firing order for the cylinders.

### 5.33   Mixture Making

In carburetor-type engines the fuel–air mixture necessary for combustion is created at the entrance to the intake manifold through which the charge is led to the cylinders. In injection-type spark-ignition engines the mixing of fuel and air is started either at the inlet valve as the air flows in or later within the cylinder. It is desirable that the fuel and air be brought into contact as early as possible in the cycle of the spark-ignition engine so that the liquid particles of fuel may have opportunity to vaporize and mix with the air. The carburetor excels other fuel-feeding devices in this respect but is inferior to injection systems in other respects. Injection of fuel directly into the combustion chamber immediately before the spark occurs offers the advantage that the fuel is exposed to high temperatures for a shorter time before its combustion and is, consequently, less likely to attain its self-ignition temperature and knock. It is also possible to stratify the charge in the combustion chamber by direct injection of the fuel as a means of controlling knock and operating on lean mixtures.

### 5.34   Mixture Requirements

The chemical reactions that occur when a hydrocarbon fuel is burned can be expressed in the form of combining-weight equations which make possible determinations of the relative amounts of fuel and air that are theoretically necessary for combustion. The most important internal-combustion-engine fuels are mixtures of many hydrocarbons, and the equation for the combustion of each compound would be necessary for an accurate analysis of their combustion requirements. It is customary, since the identity and proportions of the various ingredients are never known for fuels such as gasoline, to assume that the mixture is a single compound having the approximate molecular weight and carbon–hydrogen ratio of the mixture. Gasoline may, without great error, be considered to be octane, $C_8H_{18}$, and its combustion may be expressed by the equation:

$$C_8H_{18} + 12.5O_2 + 47.5N_2 = 9H_2O + 8CO_2 + 47.5N_2$$

If we multiply by molecular weights,

$$114 \text{ lb} + 400 \text{ lb} + 1330 \text{ lb} = 162 \text{ lb} + 352 \text{ lb} + 1330 \text{ lb}$$

When we divide by 114,

$$1 \text{ lb } C_8H_{18} + 3.51 \text{ lb } O_2 + 11.66 \text{ lb } N_2$$
$$= 1.42 \text{ lb } H_2O + 3.09 \text{ lb } CO_2 + 11.66 \text{ lb } N_2$$

$$1 \text{ lb } C_8H_{18} + 15.17 \text{ lb air} = 16.17 \text{ lb products of combustion}$$

These equations show that 1 lb of gasoline requires slightly over 15 lb of air to form a mixture of chemically correct proportions. If such a mixture were completely burned, there would be no oxygen or fuel in the products of combustion. Such perfect combustion is not possible under the conditions existing in an engine cylinder where mixing is not sufficiently complete to bring each fuel particle into contact with its quota of the air, and the time available for combustion is not always great enough to allow the reaction to go to completion.

The exhaust gases from the internal-combustion engine always contain fuel in the form of CO, $CH_4$, and $H_2$, and some oxygen is also present. The amount of unburned fuel discharged from the engine may be decreased by supplying additional air in excess of the chemically correct amount, and it has been found experimentally that air–fuel ratios of around 17 lb of air to 1 of fuel are most economical to burn except at light loads. Greater economy is possible when so-called lean mixtures are burned, because combustion is more complete if excess air is supplied, and mixing need be less perfect. It has been found experimentally that the maximum horsepower that an engine is capable of delivering is reduced when the air–fuel mixture is lean. This is explainable because the total volume of charge that can be inducted by natural means into the cylinder is fixed by the cylinder dimensions, and the space occupied by the excess air reduces that remaining for fuel–air mixture. Any increase in the amount of air supplied will result in a decrease in the amount of fuel that can be supplied. Consequently, the amount of energy liberated by combustion will decrease in spite of more complete combustion. This same effect makes it possible to increase the amount of energy released by combustion of rich mixtures containing a deficiency of air, in spite of the lowered efficiency of such combustion. Experiment has proved that air–gasoline mixtures of about 12 to 1 ratio produce the most horsepower from an engine. This difference in air–fuel ratios necessary for maximum economy and maximum power makes it impossible for full power to be developed by a spark-ignition engine without a sacrifice in fuel economy and prevents full-load operation at maximum economy. Figure 5.29 shows the effect of air–fuel ratio on the power and economy of an engine.

Because the charge in a spark-ignition engine is burned by being ignited at one point and sending a flame front across the combustion chamber, the charge must consist of fuel and air in such proportions as will support combustion. There are definite limits of inflammability for fuel–air mixtures which impose both a rich limit and a lean limit beyond which the proportions of the charge cannot go or it will not burn under cylinder conditions. The clearance space in the cylinder is seldom scavenged of exhaust gases to any extent, and the inert material carried over tends to dilute the next charge. At light-load conditions the proportionate effect of this neutral dilution is much greater than at heavier loads and necessitates enriching the mixture

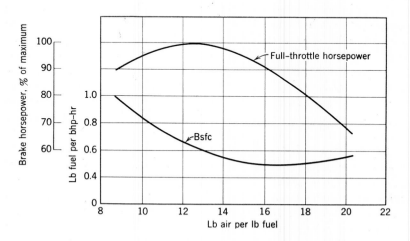

**Fig. 5.29.** Effect of air–fuel ratio on power and fuel economy.

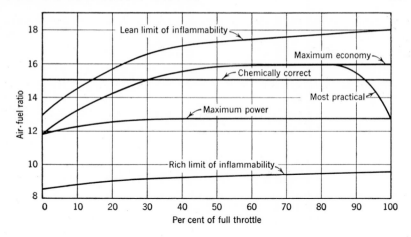

**Fig. 5.30.** Engine mixture requirements.

at light loads. Thus, the maximum-power, maximum-economy, lean-limit, and rich-limit mixtures are all richer at light loads than at heavy loads to compensate for this dilution.

These factors all combine to require that the charge supplied to a spark-ignition engine contain fuel and air in precise proportions which are different for different loads and operating conditions if optimum performance is to be had. Compression-ignition engines are not limited in this respect and do not require inflammable mixture ratios for their operation. Very high air–fuel ratios result at light loads in such engines, because the amount of air supplied is not varied with the fuel supplied at different loads.

Individual spark-ignition engines have different mixture requirements because of the effects of differences in the design of various functional parts. Figure 5.30 shows a typical range of mixture ratios.

## 5.35  Carburetion

The function of the carburetor is to atomize and meter the liquid fuel and mix it with the air as it enters the induction system of the engine, maintaining under all conditions of operation fuel–air proportions appropriate to those conditions. The process of carburetion is complicated by the wide range of fuel–air ratios required for best performance under different load and speed conditions. Automatic metering of fuel and air over a wide range of charge quantities by devices based upon hydraulic-flow principles is not easily accomplished when one of the fluids is a liquid and the other a compressible gas. Much of the difficulty in carburetion arises from the failure of the liquid gasoline to behave the same as the air under the changing pressures produced in the carburetor as the quantity of charge supplied varies.

All modern carburetors are based upon Bernoulli's theorem, which leads to the equation,

$$V^2 = 2gh \qquad (5.16)$$

where $V$ is the velocity in feet per second at which a fluid flows, $g$ the acceleration of gravity in feet per second per second, and $h$ is the head causing the flow expressed in feet of height of a column of the fluid.

The quantity in cubic feet per second of fluid flowing in a stream having a cross-sectional area of $A$ sq ft and flowing at a velocity of $V$ fps equals the product of $A$ and $V$. Multiplying this volume rate of flow by $\rho$, the density of the fluid in pounds per cubic foot, converts the quantity to pounds mass per second. Substitution of $\sqrt{2gh}$ for $V$ then makes the equation for mass rate of flow:

$$m = \rho A \sqrt{2gh} \qquad (5.17)$$

Figure 5.31 shows diagrammatically a simple elementary carburetor. $A$ is the float chamber in which a constant gasoline level is maintained by the float valve $B$, and $C$ is the jet from which the gasoline is sprayed into the air stream as it enters the carburetor at the inlet $G$ and passes through the throat or Venturi $E$. The gasoline level is slightly below the outlet of the jet when the carburetor is inoperative. All modern carburetors are basically of this type. Functioning is automatic in that no gasoline is discharged except when air is flowing through the carburetor. The gasoline and air mixture leaves the carburetor through the throttle valve $F$ and enters the inlet manifold at $D$.

Air flowing through the carburetor air passage is accelerated by the constriction at the throat $E$, with a resulting pressure drop caused by conversion of a portion of its pressure head to kinetic energy. The total pressure exerted by the air equals its static pressure plus its velocity head. The total pressure at the Venturi throat equals the total pressure at the entrance, but the velocity head is increased at the throat because acceleration is necessary when the same amount of air passes through the smaller throat area. The static pressure

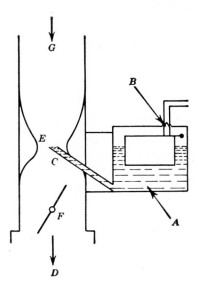

**Fig. 5.31.** Elementary carburetor.

at the throat is correspondingly decreased as the velocity increases; the kinetic energy increases while the total energy remains unchanged. This energy conversion reduces the pressure at the throat section $E$ below that at the entrance $G$ by an amount related to the rate of air flow, in accordance with Equation 5.17. Since the fuel jet has its outlet $C$ at the point where the air pressure is reduced by the air flow, and the surface of the fuel in the float chamber $A$ is under atmospheric pressure, a head, $h$ of Equation 5.17, is applied to the fuel, causing it to flow from the jet into the air stream and to form a mixture with the air as it enters the inlet manifold at $D$. The throat pressure, being a function of the velocity at which the air is flowing, the head on the fuel, and the consequent flow of fuel, will depend upon the rate of flow of air. The amount of charge supplied to the engine is regulated in accordance with the needs of the engine by varying the position of the throttle valve $F$.

Equation 5.17 applies to both the flow of air through the throat of the carburetor and the flow of fuel through the jet. In the case of the air flow, the quantity of air is drawn through the carburetor by the pumping action of the pistons in the engine cylinders. In flowing through the smaller area at the carburetor throat, the air pressure is decreased by an amount $h$ below the pressure at the entrance.

The air–fuel ratio on the mass or weight basis of the mixture formed by the merging streams of fuel and air in the carburetor may be written as

$$\frac{m_a}{m_f} = \frac{\rho_a A_a \sqrt{2gh_a}}{\rho_f A_f \sqrt{2gh_f}}$$

where $A_a$ is the cross-sectional area of the air stream at the carburetor throat and $A_f$ is that of the fuel stream at the jet. $\rho_a$ and $\rho_f$ are the densities of the air and fuel at the carburetor throat. $m_a$ and $m_f$ are the pounds of air and fuel flowing per second. If the areas are fixed, the air–fuel ratio will depend upon the constancy of the densities. Because gasoline is not compressible at the pressures encountered in the carburetor, whereas the density of air is appreciably decreased by its expansion as it is accelerated through the carburetor throat, the rate at which the mass rate of gasoline flow increases in response to opened throttle exceeds that at which the mass rate of air flow increases.

The elementary carburetor proportions the volumes of air and fuel according to the relative sizes of the air orifice $E$ and the fuel jet $C$ and serves as an accurate volumetric metering device. Because the density of the air at the throat of the carburetor becomes progressively less as the pressure is decreased by increasing rates of flow, the mass ratio of fuel to air increases, and the mixture supplied becomes richer in fuel. Fuel-jet and air-orifice sizes that provide a fuel–air mass ratio which meets the needs of the engine at one rate of charge flow will not maintain that ratio at other rates of flow, although the volume ratio will be unchanged.

The curves of Fig. 5.32 illustrate the effect of the compressibility of air upon the metering accuracy of the simple elementary carburetor. Air enters the carburetor at 0.075 lb per cu ft density and expands to progressively lower density as the rate of flow increases. At a flow rate that causes a throat pressure 60 in. of water below atmospheric pressure the air weighs only 0.0615 lb per cu ft. The actual pounds of air flowing per unit of time is indicated by the solid line, and the dotted curve shows the pounds that would flow if the air did not expand under the reduced throat pressure. This latter quantity of air is the amount necessary to maintain the air–fuel ratio at the initial value, and the difference between these two curves represents the deficiency in air supply which causes the mixture delivered by an uncompensated carburetor to become richer in fuel at higher rates of

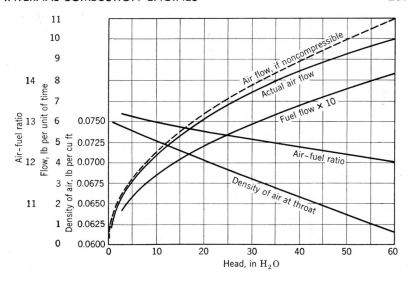

**Fig. 5.32.** Effect of compressibility of air on air–fuel ratio of charge delivered by simple elementary carburetor.

operation. No fuel is delivered to the air until sufficient air flow occurs to raise the fuel level to the discharge outlet and overcome the surface tension of the liquid. For this reason the fuel flow and air–fuel ratio curves do not start at the zero of air flow. The enriching effect accompanying increased charge flow is indicated by the decrease in air–fuel ratio from 13.2 to 12 in the range covered by the curves.

This tendency of the simple elementary carburetor to supply an increasingly rich mixture at increased rates of charge flow is a fundamental defect which makes it unsatisfactory for use with an engine that requires varying quantities of charge as loads and speeds are changed. Since few engines operate at constant input and output, the designer has had to conceive compensating devices to eliminate the need for readjustment of the size of the air or fuel passage whenever changes in engine output occur. The problem is further complicated by the fact that a constant air–fuel ratio would not be satisfactory for all operating conditions, even if a carburetor capable of providing a constant ratio were obtainable, because of the factors explained in Article 5.34.

When operating at constant loads and speeds within the middle portion of its total operating range, the engine should receive a mixture having as nearly as possible the maximum-economy ratio of air to fuel. At heavy loads, the mixture should be enriched to provide a maximum-power ratio, and at light loads it should again be enriched to compensate for greater neutral dilution. At idling conditions, the velocity at which the reduced amount of air flows through the carburetor is not sufficient to cause fuel to issue from the fuel jet, and it is necessary to equip the carburetor with an additional idling jet which functions only when the throttle is nearly closed. It must also be equipped to provide a greatly enriched mixture for starting a cold engine when the low temperature will not vaporize the heavier portions of the fuel, and, consequently, they will not enter into the mixture that has to be ignited in the cylinder. Rapid acceleration, as demanded of an automobile engine, also requires a momentarily enriched and more powerful mixture.

The inlet manifold, which connects the carburetor outlet to the inlet ports of the several cylinders of a multicylinder engine, introduces additional carburetion problems. Increased throttle opening raises the absolute pressure in the manifold, whereas closing the throttle produces a higher manifold vacuum. These changes in pressure occur because the restriction to flow into the manifold varies with the throttle position, while the vacuum-creating pumping action of the pistons does not change until the engine speed changes. The boiling points of the various fuel constituents are increased with the pressure rise that accompanies greater throttle opening. The normal operating temperature maintained in the inlet manifold causes slightly more than half of the gasoline in the charge to vaporize before it reaches the cylinders. Any increase in manifold pressure resulting from throttle manipulation will cause condensation of such fuel constituents as have boiling points that were only slightly exceeded at the previous pressure. The mixture arriving at the cylinders immediately after the throttle is opened will thus be leaner than it was when formed in the carburetor, some of its fuel remaining in the manifold to establish equilibrium at the higher pressure. Closing the throttle has an opposite effect, and the charge gains fuel in passing through the manifold. During acceleration, additional fuel must be supplied by the carburetor to compensate for the momentary loss of fuel from the charge in the manifold in order that the mixture arriving at the cylinders will not be so lean as to cause misfiring or so slow burning as to cause popping back through the carburetor.

The major differences between commercial carburetors as they are now produced are in the mechanisms provided for compensating for the inherent enriching tendency of increased flow and for varying the mixture proportions to meet the special needs of operation under varying loads, speeds, and temperatures.

### 5.36  Carburetor Systems

The following illustrations are sectional views of a Stromberg carburetor, a typical automobile engine device.   Figure 5.33 depicts the float mechanism which maintains the fuel at the proper level, just slightly below the main discharge jet, by means of a needle valve which is pressed horizontally against its seat when the float rises to a predetermined height.   A passageway formed in the cover casting connects the space above the fuel in the float chamber to a small tube, shown at the air entrance, which serves as a vent to maintain the same pressure on the fuel and the air entering the carburetor.

The throttle is shown in the closed or idle position.   The velocity of the air through the Venturi is now too low to produce enough head to raise the fuel to where it will flow from the main discharge jet.   The idle system, shown in Fig. 5.33, controls the mixture under idling and slow-speed conditions until the throttle is opened sufficiently to cause the main metering system to function.   A passage, terminating at a point below the throttle plate, extends vertically to the air entrance above the Venturi.   A horizontal opening near the upper end of this

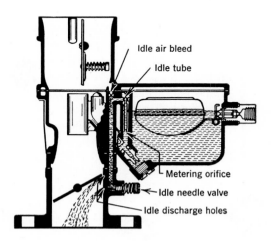

**Fig. 5.33.**  Idle system (second stage).

passage connects it to a second vertical tube in the float chamber, called the idle tube, which extends down below the fuel level to a point above the outlet of the main metering jet. This tube is kept filled with gasoline by the passageway through the main metering jet and the metering orifice in the idle tube. The air opening at the top is partially restricted by the inserted idle bleed, while the outlet below the throttle plate is provided with an adjustable needle valve as a means of so regulating the flow of fuel that a proper mixture ratio can be established. Two discharge holes leading to the main air tube are provided, one above and one below the throttle plate when in its closed position.

The idle system functions only when the throttle is closed, as in idling, or when operating at very low speed and light load. Under these conditions, the pressure in the inlet manifold and in the region below the throttle plate is very low, a vacuum of about 18 in. of Hg commonly existing when idling. This low outlet pressure causes the fuel to be forced through the idle system by the atmospheric pressure acting in the float chamber and to flow from its lower discharge hole. When the throttle is slightly opened, as in Fig. 5.33, the region of high vacuum is extended upward to include the upper discharge hole, and fuel then flows from both holes in increased amount to compensate for the greater amount of air now flowing. When it is opened beyond a certain point, the pressure at the throttle will have risen until there is not enough head to operate the idle system and fuel discharge will cease.

The main metering system controls the flow of fuel during the intermediate or part-throttle range. Fuel flow, in this system, is from the float chamber through the main metering jet and up through the inclined passage and inserted jet tube to the main discharge jet. An enlarged section of this jet tube and associated parts is shown in Fig. 5.34. The annular space around this tube $A$ communicates with the interior of the tube through a row of small holes arranged along its length and also opens vertically to a high-speed bleeder $B$ at the Venturi entrance. The main discharge jet $C$ is located in a small boost Venturi, which is so mounted in the center of the main Venturi that its mouth or exit is at the throat of the main Venturi. The pressure at the throat of the boost Venturi will thus drop more rapidly with increased air flow than that at the throat of the main Venturi. The restriction to air flow of this double Venturi is less than that of a single Venturi of the size that would produce the same throat pressure at the same rate of air flow.

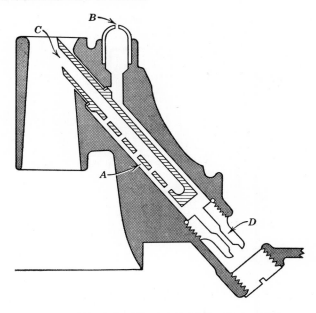

**Fig. 5.34.** Enlarged section of air-bled carburetor jet.

As the rate of air flow through the carburetor increases with increased throttle opening, the pressure at the main discharge jet becomes progressively lower. The pressure in the passage leading to the main discharge jet will decrease correspondingly, and the level of the gasoline in the annular space surrounding the jet tube will be depressed by the pressure of the air entering at the high-speed bleeder. At a certain rate of air flow through the carburetor throat, the fuel level outside the jet tube will drop below the highest of the holes leading into the tube and air will then flow from the high-speed bleeder, through this hole into the jet tube, and accompany the fuel through the main discharge jet. At further increases in the rate of operation, the level will continue to drop, and the other holes will be uncovered in sequence, admitting additional air to flow with the fuel. The effect of this air entrance to the fuel stream will be to partially break the vacuum in the tube and prevent the pressure at the exit from the metering jet $D$ from dropping as rapidly as it drops at the main discharge jet $C$ in response to increased rates of air flow through the carburetor throat. The head across the metering jet will, consequently, not increase so much for a given increase in air flow as it

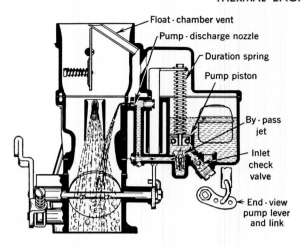

**Fig. 5.35.** Accelerating system.

otherwise would and, since the gasoline flow varies with the head across the metering jet, the mixture will not become so rich as it does in the simple elementary carburetor with no air bleed. This air-bleeding system is, therefore, a compensating device which automatically maintains a reasonably constant air–fuel ratio throughout the intermediate range of operation. The rate of fuel flow in this range can be adjusted to meet the requirements of the engine by substitution of different-size main and air-bleeder jets.

For smooth and rapid acceleration it is necessary to supply momentarily an extra quantity of fuel when the throttle is suddenly opened. In most designs, the accelerating pump (Fig. 5.35) is directly connected to the throttle, so that, when the throttle is closed, the pump piston moves up, taking in a supply of fuel from the float chamber through the inlet check valve into the pump cylinder. When the throttle valve is opened, the piston on its down stroke has a tendency to compress the fuel in the cylinder, which action closes the inlet check valve, forces open the by-pass jet, and discharges a metered quantity of fuel through the pump-discharge nozzle. This occurs only momentarily during the accelerating period. The pump duration spring provides a follow-up action so that the discharge carries out over a period of time.

The stroke of the accelerating pump plunger is made adjustable by providing three holes in the pump lever on the throttle shaft. When the pump link is placed in the center hole, as shown, the plunger has a medium stroke.  Should conditions require, as in cold weather, the link may be moved to the outer hole and the stroke increased, whereas, in warm weather, the inner hole may provide sufficient accelerating enrichment.

For maximum power operation, a richer mixture is required than that necessary for part-throttle operation at lighter loads.  A vacuum-controlled piston (Fig. 5.36) automatically operates the power by-pass jet in accordance with the engine load condition.  At light loads, the engine speed is high in proportion to the throttle opening, a high manifold vacuum is present, and the vacuum piston is moved to its "up" position against the tension of the spring.  When the load on the engine is so heavy that the throttle must be opened beyond a certain point to maintain the engine speed, the manifold vacuum decreases sufficiently so that the spring on the piston assembly moves the piston down and thereby opens the power by-pass jet to feed additional fuel into the main metering system.  When the manifold vacuum is high, fuel flows only from the main metering jet, and a maximum-economy air–fuel ratio is supplied.  When the manifold vacuum is low, fuel flows from the by-pass jet and the main jet, their combined flow enriching the mixture to the maximum-power ratio.

The butterfly valve shown in the air entrance at the top of the carburetor serves as a choke valve to supply the greatly enriched

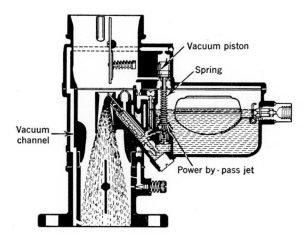

**Fig. 5.36.**  Power system.

mixture necessary when starting a cold engine. When the choke is fully closed, a mixture of approximately equal weights of fuel and air will leave the carburetor. Only a small fraction of the gasoline entering the cyinders will vaporize at low starting temperatures. If as little as 8 per cent of the liquid fuel vaporizes when the carburetor is fully choked, an ignitable air–vapor mixture will be formed, and the engine will start. The spring-loaded poppet valve shown in the lower portion of the choke valve will open automatically when the engine starts, thus making the control of the choke less sensitive.

### 5.37   Multiple Carburetors

Engines with eight or more cylinders present a problem in distributing the fuel–air mixture uniformly to all cylinders. This difficulty is somewhat alleviated by dividing the intake manifold so that half of the cylinders receive charge through one throat and the other half through a second throat. Each of these throats is half as large as the single throat that they replace. The two throats are combined in a single carburetor and receive fuel from the same float chamber. The two throttle plates of this dual carburetor are mounted on the same shaft and operate in unison.

The speed at which the horsepower of an automobile engine peaks is limited by the decrease in mean effective pressure resulting from the difficulty in charging the cylinders as the speed increases. Enlarging and improving the design of the intake system raise the peaking speed by maintaining the charging effectiveness at higher speeds. The restriction to flow imposed by the carburetor Venturi then becomes a limiting factor, unless a carburetor large enough to avoid excessive throat velocity and pressure drop at maximum capacity is used. If this oversize carburetor is used, however, the throat velocity at low-speed light-load conditions will be so low that fuel flow through the main jets will be erratic or will cease entirely. There is a minimum throat velocity below which a carburetor ceases to function properly, so a small, undersize carburetor is desirable at light loads.

The capacity range of a carburetor can be extended to meet the needs of the modern high-speed, high-output automobile engine by employing two small throats whose throttles open in series. The throttle linkage is so arranged that only one throat is opened during the first half of the accelerator travel and the second half of the carburetor is brought into operation by further depressing the accelerator. Since air flows through only one throat at loads up to half throttle and through both throats in parallel at more than half throttle, the effect is that of a small carburetor at low operating rates and a large carburetor at high rates. The total capacity can thus be made large without impairing the performance at low rates of charge flow. This type carburetor is made only for engines with divided manifolds employing dual carburetors. Two smaller throats thus replace each of the two throats of the dual carburetor, making a total of four throats or barrels as they are commonly called. Each of the two front throats of the four-barrel carburetor serves half of the cylinders of the engine up to half-throttle operation. The output of the two rear barrels is added as the throttle is opened further and the charge capacity is much greater than that of a dual carburetor of appropriate size for best all-round performance.

Installations totaling six barrels are also used for extreme power and flexibility. Three small dual carburetors are placed in a row and their throttles connected by a linkage that first opens the throttle of the center carburetor alone and then the front and rear throttles in unison. This arrangement is less complicated than the single four-barrel unit and manifold flow is better balanced.

### 5.38 Fuel Injection

The fuel charge is injected into the cylinder of the compression-ignition engine after the air has been inducted into the cylinder and compressed. The time at which fuel is injected is so chosen that the pressure rise accompanying combustion will occur approximately at the piston upper dead-center position. Allowance for the ignition lag and the initial pressure rise to occur before dead center usually makes the injection period start about 25 degrees before top dead center in high-speed engines and somewhat later in slow-speed types. The air has been compressed to at least 500 psi at this point in the compression stroke in order that its temperature exceed the ignition point of the fuel. Basically, all Diesel injection systems work on the same principle, that of placing the liquid fuel under an extremely high pressure, 30,000 psi in some systems, and forcing it through a spray nozzle having very small openings leading to the combustion chamber. The high velocity at which it flows from the spray nozzle breaks up the particles of oil and distributes them in a pattern dependent upon the number and type of openings in the nozzle.

The most commonly used injection systems employ a small

plunger-type pump for each cylinder of the engine, either located separately adjacent to each cylinder or combined in a single multi-pump unit and driven from the crankshaft. The injection pump is small in comparison with the engine cylinder and usually has a displacement of about 1/10,000 of the piston displacement of the power cylinder. Because of the high pressure and the need for extreme accuracy in metering the fuel, the injection pump must be made of high-grade materials, and the standards of workmanship and tolerances in fitting parts are much more exacting than in other units of the internal-combustion engine.

The metering function of the injection pump imposes one of its most difficult requirements. The maximum amount of fuel injected per cycle is usually less than 1 cu mm per cu in. of piston displacement, and at light loads it is proportionately less. This minute quantity of fuel must not only be accurately measured and delivered under very high pressure, but the start of injection must be precisely timed so that ignition will occur at the optimum point in the cycle. Early injection timing lengthens ignition lag and causes rough running; late injection timing causes loss of power and reduced fuel economy. The rate at which the fuel is delivered must also match the requirements of the engine so that the duration of injection will be proper. Metering, timing, and control of rate of injection are responsibilities of the pump in direct-pump injection systems. Another important requirement, dispersion of the fuel throughout the air in the combustion chamber, is a function of the spray nozzle.

Fuel-injection pumps differ from other types in that they place the fuel under pressure only spasmodically and deliver it in a single spurt of extremely short duration during each cycle. For this reason they are commonly called jerk pumps. An individual jerk pump is provided for each cylinder in all modern injection systems except the distributor type and the common-rail system. Distributor systems employ some form of distributor to index the several spray nozzles of a multicylinder engine in proper sequence with a single jerk pump that executes as many cycles as there are cylinders. Common-rail systems maintain fuel under injection pressure continuously in an accumulator, from which delivery lines lead to the spray nozzles via timed valves that permit flow during the injection period of each cylinder.

Metering provisions incorporated in jerk pumps range from mechanisms for varying the actual stroke of the pump plunger so that its displacement is changed in accordance with the load on the engine to valve and port arrangements for varying the effective stroke of the plunger by preventing delivery during the initial or final or both portions of the actual stroke. The trend is toward the use of a spirally recessed plunger that covers a port through the wall of the pump barrel during a portion of its stroke which is varied by rotating the plunger. During the initial and final portions of the stroke, the fuel in the pump barrel flows out through the port, which is then not

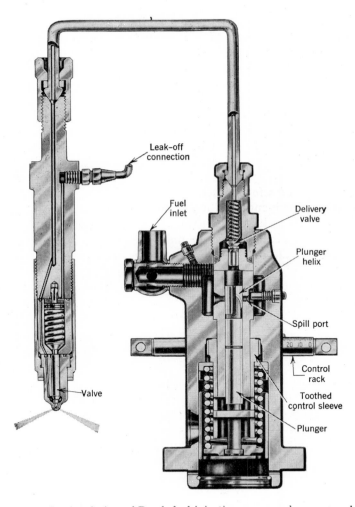

**Fig. 5.37.** Sectional view of Bosch fuel-injection pump and spray nozzle.

covered by the plunger, and returns to the supply side of the pump instead of being delivered to the spray nozzle. The sculptured plunger thus functions as a by-pass valve that is closed only while the unrecessed part of the plunger is passing the port. The length of the stroke completed while the port is covered changes with the angular position of the plunger because of the helical control edge formed by the plunger relief.

Figure 5.37 shows a sectioned Bosch plunger-controlled-port pump of the type that has each pump barrel and plunger housed in a separate pump body. There are a helical spring that retracts the plunger during the filling stroke and a rack and pinion that position the plunger angularly to control its effective stroke. This unit is mounted adjacent to the camshaft of the engine and the plunger is actuated by an additional cam formed on the camshaft. The racks of all the pumps of a multicylinder engine are joined by adjustable reach rods so that they are all in phase and moved in unison by the governor.

The same plunger and barrel units are used in multiple pumps containing as many of these components as there are cylinders to be served. This type pump is a self-contained unit, including the camshaft, and is driven by gears from the crankshaft. Figure 5.38 is a schematic view of a four-cylinder Bosch pump of this type, showing the additional components of a complete Diesel fuel system.

Figure 5.39 is a schematic diagram of a common-rail fuel-injection system. Two such four-cylinder units are employed for each bank of cylinders on the sixteen-cylinder Cooper-Bessemer Diesel engine of Fig. 5.46. The supply pump is a pressure-regulating type that maintains a pressure of several thousand pounds per square inch on the fuel in the accumulator. The injector valve unit comprises a triple-disc valve for each cylinder served. Each valve is raised from its seat at the proper instant for injection by a cam-actuated push rod. The amount of fuel delivered to the spray nozzles per cycle depends upon the length of time the valves are open, which is varied by increasing or decreasing their lift. The spray nozzles receive fuel in timed sequence through high-pressure delivery lines.

The Cummins P-T injection system uses a common-rail arrangement supplying fuel at low pressure to unit injectors which place the fuel under injection pressure. The unit injector has the pump plunger and barrel combined with the spray-nozzle body and is operated by a push rod and rocker arm from a third cam on the camshaft, as in Fig. 5.21. Fuel is metered by flowing through a timed orifice into the injector and is controlled by varying the rail pressure.

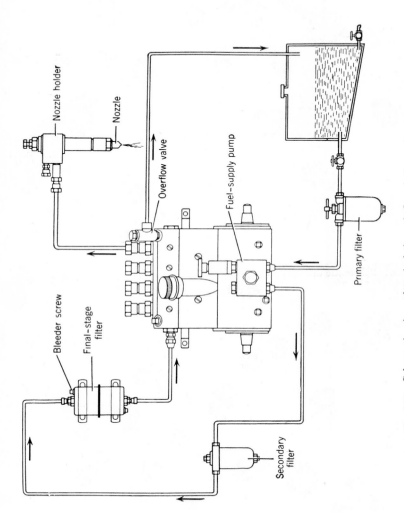

**Fig. 5.38.** Schematic view of typical through-flow fuel-supply system.

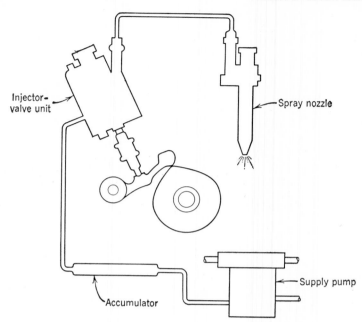

**Fig. 5.39.** Schematic diagram of Cooper-Bessemer common-rail fuel-injection system.

Spray nozzles are frequently of the multihole type, in which orifices of from 0.005-in. diameter upward are drilled in such manner that about six jets of fuel are discharged along ordinates of a cone. Conical spray nozzles of the pintle type are also widely used, in which the orifice is an annular one formed by a pin centered in a hole about 0.001 in. larger in diameter than the pin. This construction provides a nozzle that is easier to produce and maintain than those with several very small openings. Figure 5.37 illustrates a nozzle of this type. Fuel delivery is prevented until sufficient pressure is built up by the pump to force the spring-loaded valve from its seat. A predetermined opening pressure is established in the design of the valve, and its closure when the pressure drops serves to produce a sharp end to the injection period, preventing dribbling fuel into the cylinder between injections. The fuel is conveyed to the spray nozzle from the injection pump by a heavy-walled steel tube of small internal diameter which must be rigid to resist expansion under the high fuel pressure

### 5.39   Gasoline Injection

The failure of the carburetor to meter the gasoline accurately into the air stream and the difficulty in delivering equal amounts of charge of uniform fuel and air proportions to the several cylinders of a multi-cylinder engine cause one of the most serious imperfections of the gasoline engine.  By injecting the fuel, either directly into the engine cylinders or into the air stream at the entrance to each cylinder, the distribution of the charge can be greatly improved, much greater accuracy in metering can be maintained, and the atomization of the fuel can be improved.  These accomplishments tend to increase the power output of the engine and to lower its specific fuel consumption, the two important objectives of the engine designer.

Compared with the carburetor, the gasoline-injection system is more complicated and expensive to produce, and this factor has delayed its application to any except highly refined types of engines, such as military aviation engines.  Improved designs permitting lower production cost and simplification to reduce maintenance difficulties have extended the use of gasoline injection to automobiles, where it has excellent future prospects.

The fundamental principle of the gasoline injector is the same as that of the Diesel injector, and the general construction features are the same.  The lower back pressures against which the fuel is injected, the greater ease of atomizing and vaporizing the fuel, and the elimination of the need for accurate timing and control of the rate of injection simplify the requirements appreciably, but lubrication of the injection pump by the fuel itself, as is done in Diesel systems, presents a difficulty arising from the gasoline's lack of lubricating body.  A further complication results from the provision for controlling the ratio of fuel to air that is necessary for spark-ignition combustion.

Figure 5.40 illustrates the Ex-Cell-O gasoline-injection system.  The injection-pump unit comprises three essential parts, the supply pump, the fuel-metering valve, and the combined pumping and distributing plunger.  The gear-type supply pump takes the fuel from the storage tank and delivers it through the metering orifice to the barrel in which the plunger operates.  The metering orifice varies in size with movement of the throttle valve in the air inlet, thus proportioning the amount of fuel delivered to the amount of air flowing to the cylinders.  The cams formed on the upper face of the plunger gear cause it to reciprocate as it rotates and execute a delivery stroke

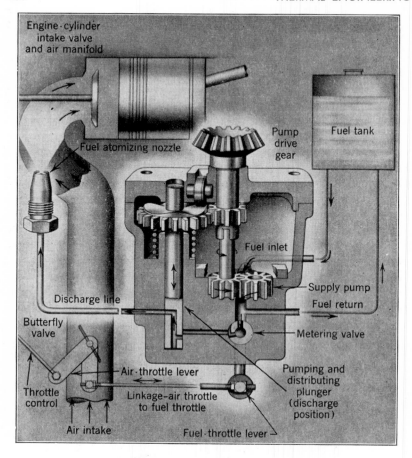

**Fig. 5.40.** Flow diagram and section view of Ex-Cell-O gasoline-injection system.

in phase with each suction stroke of the engine. Ports entering the distributor section of the pump barrel are alternately indexed as the plunger rotates. The pump illustrated is a four-cylinder unit, but only one discharge port is shown. Each of the ports connects to a discharge line leading to a spray nozzle located outside the inlet valve of the engine cylinder which atomizes the fuel charge into the air stream during the intake stroke. The fuel charges are thus metered separately to the individual cylinders, largely eliminating errors in distribution.

### 5.40   Ignition

Two forms of ignition systems are now in use, the magneto and the battery and distributor systems.  The magneto is a self-contained device which generates its own electric power, times the occurrence of the sparks, and distributes them to the spark plugs in proper sequence. It requires only the spark plugs and the wires connecting them to the magneto to complete the system.  The magneto is the same in principle as the battery and distributor system except for the source of electric energy and the arrangement of parts.  Both types provide all the electrical devices necessary to produce a voltage sufficient to cause a spark to pass across a gap of 0.025 to 0.040 in. against the resistance of the compressed charge in each cylinder at the precise instant when needed.

Figure 5.41 shows the component units of a four-cylinder battery and distributor system and the electric circuit.  One terminal of the battery is grounded to the frame of the engine, and the other is connected through the ignition switch to one primary terminal of the in-

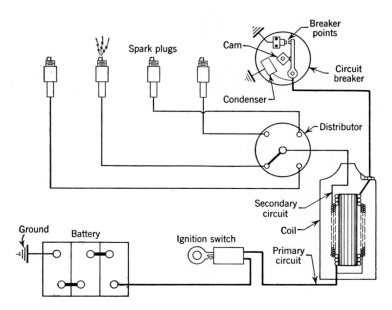

**Fig. 5.41.** Ignition system.

duction coil. The other primary terminal is connected to one of the contact points of the circuit breaker and through the closed points to the ground. The primary circuit of the coil is thus completed when the contact points of the circuit breaker are together and the switch is closed. The secondary terminal of the coil is connected to the central contact of the distributor and thence to the distributor rotor. Each spark plug is connected to one of the outer contacts of the distributor to be contacted in proper sequence by the turning rotor. The other end of the secondary winding is connected within the coil to the primary winding.

The battery is the source of electric energy for the ignition system in conjunction with the generator which maintains it in a charged state. The storage battery also serves to operate the starting motor and, in many installations, supplies electricity for lights and accessories. Batteries of six or twelve volts are ordinarily used for this purpose.

The distributor assembly is a double unit, comprising the distributor proper, which is a rotary switch whose function is to connect the secondary of the coil alternately to each of the several spark plugs in their firing sequence, and the circuit breaker by which the current in the primary winding of the coil is controlled. The two devices are not electrically connected within the distributor assembly and are combined only because both must be driven at the same speed by a shaft connection to the engine crankshaft. The distributor is located on the outer end of the spindle which passes through the assembly and is driven by gears from the camshaft or crankshaft. A cam mounted just beneath the distributor is formed with as many lobes as there are cylinders in the engine and operates the primary circuit breaker, the second unit of the distributor assembly. The breaker points are held in contact by a spring except when forced apart by the lobes of the cam. The cam rotates at half crankshaft speed on four-stroke-cycle engines and breaks the primary circuit once for each cylinder in the engine during one complete cycle of the engine. The distributor assembly is shown in Fig. 5.42.

The ignition coil is merely a special form of step-up transformer to raise the battery voltage, which is far too low to jump the spark-plug gap, to the necessary 8000 to 25,000 volts. It consists of a primary winding of a relatively few turns of copper wire wound about a soft-iron core and a secondary winding of a great many turns of very fine wire wound over the primary winding. Both ends of the primary winding are brought out to exterior terminals. One end of the secondary winding is connected to the primary winding, and the other is brought out to the high-voltage terminal of the coil.

When the primary circuit is closed, the current flows through the primary winding of the coil and creates a magnetic field through the core. When the primary circuit is opened by the action of the circuit

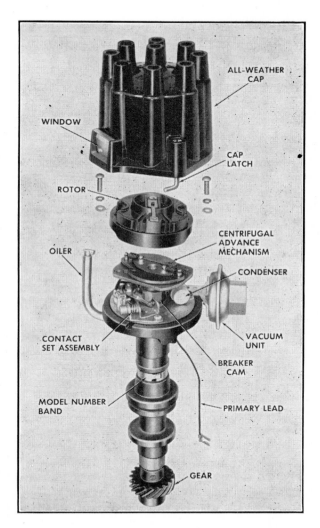

**Fig. 5.42.** Exploded view of Delco-Remy 8-cylinder ignition distributor and circuit breaker.

breaker, the magnetic field rapidly collapses, and a high voltage is induced in the secondary winding because of the high rate of change in the magnetic flux through the coil. A high voltage is thus built up across the secondary winding at each instant that the circuit-breaker points separate.

The distributor rotor is mounted in phase with the circuit-breaker cam so that it has established the secondary circuit to one of the spark plugs at the instant that the breaker points open. When the next lobe of the cam opens the breaker points, the distributor rotor has advanced to the contact leading to the next spark plug.

A considerable voltage is built up by self-inductance when the primary circuit is broken, and a spark results at the circuit-breaker points which tends to burn and pit the contact surfaces. A capacitor, commonly called a condenser, connected across the contact points quenches this spark by absorbing the surge which otherwise would jump between the contacts at the instant of separation. The charge received by the capacitor immediately discharges back through the primary winding of the coil in the opposite direction from that in which the current flows while the points are closed. The capacitor and the inductive primary-coil winding are now in series, completing a resonating circuit with a high-frequency characteristic. An oscillating current of damped sine-wave form flows until its amplitude drops to zero when the stored energy in the capacitor is dissipated. Without the capacitor, the contact points would be short lived, and less energy would be delivered to the spark plugs because of the energy drained from the coil to maintain the arc at the contact points.

The spark plug is essentially a pair of electrodes insulated from each other by an insert of fused aluminum oxide or similar heat-resistant material of high electrical resistance. One electrode is attached to the metal body of the plug which is threaded to screw into the cylinder of the engine whereby it is electrically grounded. The central electrode extends outside the plug through the insulator and provides a connecting terminal for the high-voltage lead from the distributor.

In operation, the coil is energized when the breaker contacts close, and a magnetic field is established. This field collapses when the contacts open, and the resulting high voltage sends a current to the distributor which selects the spark plug whose turn it is to fire. Very little current flows through the secondary circuit during this discharge, but considerable energy is expended at the spark-plug gap because of the high voltage.

The spark must be timed precisely with the cycle of the engine, and this is accomplished by adjusting the relationship between the circuit-breaker cam and the engine crankshaft. Changes in engine speed and load change the optimum point in the cycle at which the spark should occur. Since an earlier timing of spark is needed at higher speed, a simple fly-weight mechanism is incorporated in the distributor assembly beneath the breaker mechanism to advance the relative position of the cam with respect to the driving shaft as the speed increases. The charge in the cylinder burns more rapidly at heavy load conditions than at light load because of the higher pressure and temperature and relatively less dilution with exhaust gas from previous cycles. It is desirable, therefore, to retard the timing of the spark as the load increases. Since the vacuum in the intake manifold is higher with the throttle partly closed than at full throttle, this vacuum can be used to advance the spark timing at light loads. This is accomplished by a diaphragm attached to the breaker plate so as to rotate the contact points relative to the cam and cause the points to separate earlier as the manifold vacuum increases at lighter loads. These control devices are shown in Fig. 5.42.

### 5.41 Internal Temperature Control

Provision must be made in all internal-combustion engines to remove a portion of the energy released within the combustion chamber and dissipate it to the surroundings. Although the objective of the combustion system is to liberate as much energy as is possible from the fuel that is burned, only a fraction of the energy supplied by the fuel can be converted into useful work. That fraction, represented by the thermal efficiency of the engine, seldom exceeds 30 per cent; the remainder is waste energy and must be removed or the engine will overheat.

The instantaneous temperature attained by the burning gases in the combustion chamber probably exceeds 4000 F when an engine operates under load. If the unused portion of the energy in the cylinder is not removed from the engine, the temperature of the metal parts will rise rapidly and approach that of the gases. The major portion of the rejected energy is carried away by the exhaust gases, radiated, or unaccounted for, but a considerable amount, usually 20 to 30 per cent of the energy input to the engine, must be removed from the metal of the cylinders by a cooling system.

The amount of heat transfer to the cooling system should be the minimum that will prevent the temperature of the engine from becoming too high for proper functioning. The limiting temperature at which the combustion-chamber walls of a well-designed spark-ignition engine will permit satisfactory operation is that at which preignition and knock are avoided. Compression-ignition engine temperatures are limited only by the ability of the parts to function properly, since the normal combustion process is aided rather than interfered with by high cylinder temperature. In any case, cooling must be sufficient to prevent excessive loss of strength, erosion, warpage, overexpansion, and lubrication difficulties which accompany high temperatures. The parts that are most likely to suffer from overheating are exhaust valves, spark-plug electrodes, and piston crowns. Cooling is a matter of equalization of internal temperatures to prevent local overheating as well as one of removing sufficient energy to maintain a practical overall working temperature.

There are two systems of cooling in use: liquid and air. Each has certain advantages, and each is especially adapted to engines intended for certain applications. Marine engines are logically liquid-cooled, and aviation engines are very successfully air-cooled. The principal problem in liquid cooling is that of finding a coolant that has a high boiling point and a low freezing point and is chemically inert. Although water is largely used for this purpose, its low boiling point requires the engine to operate at lower than optimum temperatures, and its high freezing point necessitates the addition of anti-freeze agents in cold weather. Ethylene glycol has a much wider temperature range in the liquid phase than water, but is to some extent chemically active and is quite expensive.

Passages called cooling jackets are cored in the cylinder and cylinder-head castings of liquid-cooled engines. These jackets are so arranged that cool water can be forced in through an entrance at the lower extremity, pass around the cylinder walls, and flow upward into the cylinder-head jackets, and the heated water can be discharged through an outlet at the top. A radiator is commonly used to transfer the heat absorbed by the coolant to air that is circulated through the openings in the radiator core. A pump circulates the coolant at a rate sufficient to effect the necessary heat transfer. A thermostat is usually placed at the water outlet from the engine to regulate the flow and maintain the jacket temperature constant, regardless of the load and speed of the engine and the temperature of the surrounding air.

The principal problem in air cooling is to expose sufficient cooling surface to the air and to circulate the necessary amount of air without the expenditure of large amounts of power. The area of the cooling surface of the engine cylinder and cylinder head is increased by forming thin fins, either integrally by machining them on the outer walls of the parts or by attaching separate fins to them. Except in airplanes where the high speed and the propeller wash provide the necessary cooling air, the power required to circulate the air consumes an appreciable portion of the power developed by the engine.

### 5.42 Lubrication

Lubrication of the parts of an engine that have relative motion is necessary in order to reduce friction and wear that would quickly destroy the parts. The importance of proper lubrication of any mechanical device can hardly be overestimated, because successful operation without lubrication is impossible.

The theory of lubrication is to interpose a fluid film between the moving surfaces, such that metallic contact between the parts is prevented. Fluid friction between the many microscopically thin layers of oil that may be considered to make up the oil film is substituted for the rubbing friction between the metal parts. The forces pressing the surfaces together tend to squeeze out the lubricant and may prevent the maintenance of an oil film of sufficient thickness to eliminate metallic contact entirely. The resulting partial film will not eliminate friction so fully as a complete film but will permit the parts to function with only moderate heating and wear. Depending upon the load carried by the bearing, its lubrication may be so perfect that the moving part is entirely floated on an oil film, it may operate under a boundary condition with a film so thin that there is partial metallic contact, or the load may be so great that extreme-pressure lubricants are necessary which react with the metal to plate it with a slippery compound and do not interpose an oil film. Friction is never entirely eliminated, but is reduced to a very much smaller amount varying with the thickness of the oil film, the characteristics of the bearing surface, and the fluid friction of the lubricant.

The principal objective of the lubricating system of an internal-combustion engine is to restore continuously the oil film on all the bearing surfaces of the engine so that friction and wear are reduced to the minimum permitted by operating conditions. Several other purposes are served by the lubricating system, however. When the usual circulating system is employed, the lubricant serves as a coolant and picks up heat from the metal surfaces and permits its dissipation to the surroundings. Particles of metal, resulting from wear, and other contaminants, such as carbon that may form on cylinder walls, are flushed from the bearing surfaces and carried away. Leakage between the piston and cylinder walls is more effectively blocked, and noise resulting from contact between moving parts is reduced by the cushioning action of the oil film.

Lubricants suitable for engine use are refined from petroleum stocks after the various fuels are removed. The refining process removes impurities, such as dirt, asphalt, and wax, and also certain portions of the oil itself that are undesirable, because they are unstable and tend to oxidize or deteriorate otherwise under the severe conditions encountered in the engine. The purified oils are graded according to characteristics that determine the applications for which they are suitable. Physical and chemical tests are made in the laboratory with test apparatus designed and operated in accordance with ASTM test codes. These tests form the basis for specifications that limit certain properties and characteristics of the oil to narrow ranges that have been found proper for the application covered by the specification. Viscosity, which is one of the most important of the items included, expresses the resistance of the oil to flow in terms of the number of seconds required for 60 cc of the oil to flow through the orifice of the Saybolt universal viscosimeter at temperatures of 100 F and 210 F. When plotted on American Petroleum Institute viscosity-temperature coordinate charts, viscosities lie in a straight line which is extrapolated to find the value at 0 F. Oil tends to decrease in viscosity or thin out at higher temperatures.

The rate of change in the viscosity of an oil with temperature is expressed by its viscosity index. Viscosity index is important because excessive flow resistance at low temperature seriously hinders starting and early lubrication; conversely, excessive thinning at high temperature impairs lubrication and increases oil consumption. Straight mineral oils refined from paraffin-base crude oils have been assigned a viscosity index of 100, and naphthenic oils, which lose viscosity at increased temperature to a much greater extent, have an index of 0. A comparison with the characteristics of these two oils establishes the viscosity index of any oil. Viscosity-index improvers are added to some commercial lubricants, and the oils refined by solvent processes have high viscosity indexes. Also related to the viscosity is the pour point of an oil. The lowest temperature at which an oil flows determines its pour point. Crankcase oil must have an appropriate pour point if an engine is to be started readily and without damage after exposure to extremely low temperature. Dewaxing operations in the refining process and the addition of pour-point depressants are the methods used to procure oils with low pour points.

Detergent-dispersant additives are desirable in oils used in Diesel engines or in spark-ignition engines operated under unusually severe conditions. They impart cleansing properties to the oil and cause it to pick up and retain in suspension particles of carbon, dirt, and oxidized hydrocarbons, which would otherwise accumulate in the piston-ring grooves, resulting in stuck rings when solidified by the high temperature, or would be deposited on metal surfaces in the form of lacquer. Corrosion inhibitors, anti-foam agents, and anti-oxidants are among the other chemical additives used in heavy-duty oils to meet the needs of some modern engines.

Five classifications of motor oils are designated by the API, according to the severity of the service for which the oils are suitable. Each type is identified by a two-letter code marking on factory-sealed containers. The first letter is M for automobile lubricants or D for Diesel lubricants. For the M oils, the second letter of the code designation is L for light and favorable conditions, M for moderate conditions, or S for severe conditions. Continuous high speeds, start-and-stop operation, heavy loads, or sensitive design features, such as hydraulic valve lifters, are some of the causes of severe conditions. For the D lubricants, the second letter is G for ordinary Diesel-engine operating conditions or S for more severe conditions or for use with fuels that tend to increase wear and deposits.

The ML, MM, and MS oils correspond to those that were formerly classified as regular, premium, and heavy duty. Essentially all MS oils are also DG oils of heavy-duty type, containing detergent-dispersant additives, anti-oxidants, and corrosion inhibitors. DS oils have much greater additive content than DG oils.

Lubricating oils are graded by SAE viscosity numbers from 5W to 50, each number including oils whose viscosities lie within a certain range of Saybolt seconds at 0 F for W oils and at 210 F for others. The W oils are low-viscosity grades, suitable for cold-weather lubrication of automobile engines. The viscosity at 0 F of a 5W oil must be less than 4000 sec, that of a 10W oil between 6000 and 12,000 sec, and that of a 20W oil between 12,000 and 48,000 sec. Oils suitable for engines requiring higher-viscosity oils and for higher-temperature operation are graded at 210 F. The viscosity ranges specified for these oils are as follows: No. 20—45 to 58 sec, No. 30—58 to 70 sec, No. 40—70 to 85 sec, and No. 50—85 to 110 sec at 210 F.

Multigrade oils have high viscosity indexes, attained by adding viscosity-index improvers, and meet specifications at both 0 F and 210 F. A viscosity index of 140 enables the viscosity of a 5W–20 oil to meet the 5W limitation at 0 F and also to lie within the 20 range at 210 F. Similarly, a 10W oil may also be a 30 oil. A 10W–30 oil is thinner at 0 F and thicker at 210 F than a 20W oil, thus providing easier starting at low temperature and less oil consumption at high temperature.

Lubrication systems differ considerably between extreme types of engines. Large slow-speed engines commonly use mechanical lubricators consisting of a large number of metering pumps which deliver an adjustable number of drops of oil to each separate lubricated surface. Automotive-type engines use circulating systems in which a quantity of oil is carried in a sump formed in the lower extremity of the crankcase. A pump, usually of the gear type, takes oil from the sump and delivers it under pressure to a system of tubes and passages leading to the bearings of the engine. In most instances the crankshaft is drilled so that the oil which is delivered to its main bearings flows through drilled passages in the cheeks of the crank to the crankpin bearings. Frequently, the connecting rods have oil passages drilled through them to convey oil up to the piston pins. In other instances, the oil that issues from the crankshaft bearings and is sprayed throughout the engine enclosure reaches the cylinder walls in sufficient quantities to lubricate the pistons.

Figure 5.43 shows a typical system applied to a four-cylinder gasoline engine. Various parts of the engine, such as the valve mechanism and front-end gears or chains, are supplied with oil by suitable connections. After passing through the engine, the oil drains back to the sump to be recirculated. The filter, shown in the diagram, continuously by-passes a portion of the oil pumped and passes it through a filtering medium which removes the more injurious particles of foreign matter picked up by the oil, thus maintaining it in good condition. The pressure under which the oil is circulated is regulated by a spring-loaded relief valve to prevent excessive pressures at high speed.

The lubricating oil in an engine tends to oxidize and form solid compounds that are called sludge. When the sludge concentration becomes too great, lubrication is interfered with. There is also a

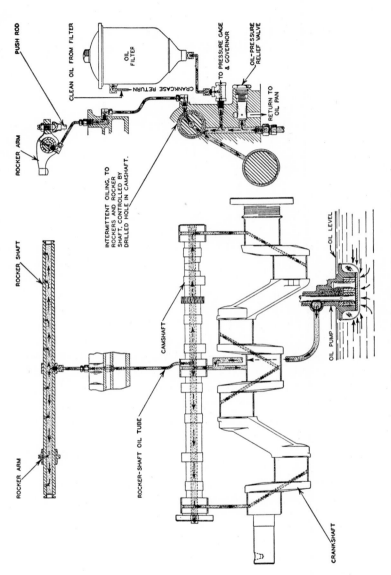

**Fig. 5.43.** Pressure lubrication system of Le Roi engine.

possibility that decomposition will yield corrosive substances. The quality of the oil also suffers from contamination by products of combustion and heavy ends of unburned fuel that pass the piston rings. Because of these accumulations, the oil must be purified and renewed at regular intervals, or it will fail to protect the working parts against damage.

### 5.43  Regulation

The fuel input to an engine must be regulated in accordance with the load applied and the desired speed. The speed will increase with decreases in load unless the amount of fuel supplied is decreased proportionately. The fuel admitted to the cylinder of the internal-combustion engine may be adapted to the load and speed requirements by two methods known as quantity and quality governing.

The gasoline engine is regulated by a throttle or butterfly valve at the entrance to the intake manifold. This throttle valve is incorporated in the carburetor and in the control unit of the gasoline-injection device. The throttling method of regulation is called quantity governing, because the fuel and air supplied to the engine are both varied in substantially constant proportion.

Compression-ignition engines do not use throttle valves in the air-induction system, but are controlled by varying only the amount of fuel supplied. This type of regulation changes the air–fuel proportions and is called quality governing, since the mixture becomes richer in fuel at increased loads. The air–fuel ratio of the gasoline engine varies with throttle position only by the amount necessary to meet the changes in mixture requirements and remains so within a narrow range above and below the chemically correct ratio. The Diesel engine always operates with excess air, and the air–fuel ratio varies from about 20 to 25 at maximum load to several hundred at light load because of quality governing.

Throttling causes a drop in inlet pressure below the outside condition, and the cylinder is filled at the end of the intake stroke at a reduced pressure. The charge volume is the same at all loads, but the pressure is progressively lower as the throttle is closed. The decreased pressure lowers the specific weight of the charge, and the cylinder is filled with an expanded charge of smaller weight. A corresponding reduction in the weight of fuel is brought about by the metering action of the carburetion device.

Throttling is not an efficient means of regulating an engine. Compression begins at a much lower cylinder pressure at light load than at full load because of throttling action, and the pressure at ignition is correspondingly lower. The entire cycle is thus executed at lower temperatures and pressures which result in lower efficiencies and increased specific fuel consumption. The fuel economy of the Diesel engine is not decreased so much at light load as that of the gasoline engine because of its more efficient method of regulation.

The controls which regulate an engine may be operated manually by a throttle lever or by the accelerator pedal of a vehicle. Constant speed may be automatically maintained, if desired, by a mechanical or hydraulic governor. Stationary and marine engines operate at governed speeds that are maintained with the degree of accuracy demanded by the installation. Diesel engines of all types, stationary, marine, or automotive, are equipped with governors which may be of the constant-speed type, or they may permit the operator to select any speed within the useful range of a variable-speed engine. The accelerator pedal of a Diesel-powered vehicle acts on the governor which, in turn, acts on the fuel-injection pumps, but that of a gasoline-powered vehicle acts directly on the throttle valve of the engine. Speed governors are incorporated in the fuel-injection units used by several Diesel manufacturers, and others provide separate governing units.

### 5.44 Performance

The performance of an engine is expressed by the relationship of its power, speed, and fuel consumption. Several other relationships based upon the power, size, and weight of the engine are of interest, because they affect the usefulness of the engine for exacting applications. Among these are the ratio of the engine weight in pounds to the horsepower delivered and the horsepower developed per cubic inch of piston displacement. The brake mean effective pressure is also used as a standard for measuring the effectiveness of the engine.

At any speed within the operating range of an engine, which may be a very narrow range in the stationary power-plant engine or may cover a wide range in the automobile-type engine, it is capable of developing a certain definite amount of power at full throttle. Since the speed at which cycles are executed in the cylinders is a factor in determining the power developed, more power is produced at increased speeds, provided other factors remain unchanged. When the speed becomes extremely high, however, the ability of an unsupercharged engine to draw in as full a charge as at lower speeds causes the mean effective pressure in the cylinders to decrease and offset, to some extent, the effect of increased speed. A point is eventually reached above which the mean effective pressure decreases at a rate higher than that at which the speed increases, and the peak horsepower of the engine is reached. At speeds above that at which the maximum horsepower is developed, the effect of reduced mean effective pressure will predominate, and the full-throttle power output will be less with further increases in speed.

The relationship between the full-throttle brake horsepower of an automobile engine and its speed in revolutions per minute is shown by the brake-horsepower curve in Fig. 5.44. This curve is typical of those plotted from data taken in testing an automobile engine loaded by a dynamometer. It shows that for any speed within the range

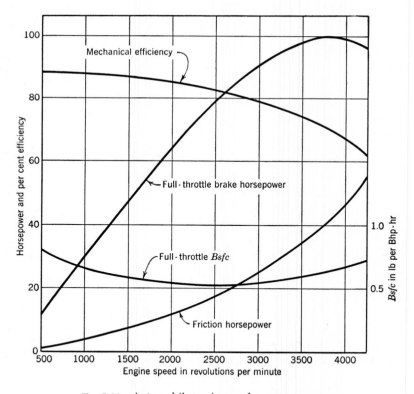

**Fig. 5.44.** Automobile-engine performance curves.

tested there is a corresponding horsepower which is the maximum the engine can develop. Any amount of power less than the maximum can be developed by partially closing the throttle, but more than that cannot be produced at that speed. Below the speed at which the horsepower curve peaks, more power can be had by increasing the speed, but, when that peak is reached, the full-power capacity of the engine is being developed. The curve shows that the engine for which it was plotted attained its maximum output of 100 bhp at 3800 rpm. Those values of horsepower and speed would ordinarily be stated as the rating of the engine.

Heavy-duty engines intended for continuous operation at full load are commonly rated at less than the maximum horsepower they are able to produce. This is especially true of the Diesel engine, because it can be supplied larger amounts of fuel than it can burn efficiently and can develop more power than is compatible with long continuous and trouble-free service. Such engines usually are given an intermittent horsepower rating which exceeds that at which the engine is most economical in fuel consumption and in maintenance expense. A Diesel engine should be operated at its intermittent rating only to carry peak loads of short duration. A second and lower continuous rating is given the engine which represents the maximum power output at which the engine can be expected to give satisfactory operation at steady loads. Above the continuous rating the Diesel exhaust will show smoke, and the cylinders will accumulate carbon deposits. The fuel rate will also be appreciably higher than at lighter load.

The fuel consumed by an engine can be weighed at the same time that the horsepower of the engine is determined. The specific fuel rate can then be determined. The brake-specific-fuel-rate curve of Fig. 5.44 shows the pounds of fuel per brake-horsepower-hour corresponding to operation at full throttle. The fuel rate decreases to a minimum as the speed is increased to about 2200 rpm and then increases again at higher speeds. It is especially important that both the horsepower curve and the fuel-rate curve shown in this diagram be recognized as those of full-throttle conditions only. Partially closing the throttle at any speed will decrease the horsepower below the values plotted and will also cause important changes in the fuel rate.

Figure 5.45 shows the effect of load upon the fuel rate at any constant speed. The typical engine is most economical at about 75 to 85 per cent load. The fuel rate usually increases slightly as the load is increased to 100 per cent, and it increases at a progressively faster

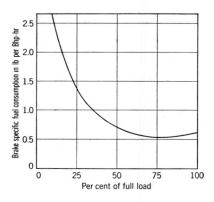

**Fig. 5.45.** Variation in fuel rate of automobile engine with changing load at constant speed.

rate as the load is decreased, becoming much higher at very small loads than the minimum rate. This curve indicates that it is very uneconomical to operate an engine at small percentages of its full-throttle output and that it should operate at slightly less than maximum capacity if its fuel consumption is to be low.

Since the fuel economy is a function of both load and speed, it is evident that there is some intermediate speed and per cent load at which the combined effects of speed and load will produce the best fuel economy of which the engine is capable. This speed and load combination is not likely to be the one at which an engine would be selected to operate, however, because a larger and more costly engine would be required for a given power output than if it were operated at a higher speed. The speed ratio by which an engine is connected to its load must be carefully chosen in order that the engine may operate at favorable load and speed combinations as much as possible. Because of the increased fuel rates at light load, it is frequently desirable to install several small engines instead of one large one for fluctuating loads. The minimum number of engines necessary to carry the load can then be operated with each carrying an economical load.

The minimum fuel rate attained depends upon the size, type, and speed of the engine. In general, the large slow-speed Diesel engines have better fuel economy than other engines. The average engine of this type will operate at a fuel rate of 0.39 lb per bhp-hr or less at its best operating range. High-speed Diesels do not ordinarily operate at fuel rates lower than about 0.45 lb per bhp-hr, and many engines of this type consume more than that amount of fuel. A range of 0.47

to 0.60 lb per bhp-hr may be taken as the range of minimum fuel rates for gasoline engines of the ordinary types, but high-output aviation engines are known to show superior economy on high-octane fuel under optimum conditions.

In comparing the fuel consumption of Diesel and gasoline engines, the fact that fuel rates are calculated on a weight basis whereas fuels are marketed on a volume basis should be considered. Diesel fuels are higher in specific weight than gasoline by a ratio of about 7 to 6, their approximate respective weights per gallon. The somewhat lower cost per gallon of Diesel fuel, the greater weight per gallon, and the lower specific fuel consumption all combine to make the cost of fuel per power unit substantially less for the Diesel than for the gasoline engine. This saving is partially offset in competing fields by the higher initial investment in the Diesel, resulting from its more expensive construction throughout. The net saving of the Diesel increases with increased output in horsepower-hours over a period of time.

Weight–horsepower ratios range from 0.85 lb per bhp for modern aircraft engines to about 200 lb per bhp for stationary Diesels where no attempt is made to reduce weight. Horsepower–displacement ratios range from about 0.02 bhp per cu in. for slow-speed engines to about 2.0 bhp per cu in. for high-output automotive engines. Brake mean effective pressures at rated load are between 75 and 125 psi for a wide range of engines. Aircraft-engine brake mean effective pressures of 250 psi are not unusual, and values much higher than that are obtained experimentally.

The lowest fuel cost per brake horsepower-hour of any type engine is attained by Diesel engines operating on gaseous fuel, usually natural gas or by-product gases. In geographical regions near gas wells and along pipe lines, natural gas is quite low in price and is the most economical fuel to burn. All hydrocarbon gases have relatively high self-ignition temperatures, resulting from their small compact molecular structure. Consequently, their use in compression-ignition engines caused ignition difficulties, and for many years gas-burning internal-combustion engines were limited to the spark-ignition type. Gases are very suitable for spark-ignition engines because of their high-octane ratings.

The ignition problem, which prevented satisfactory use of gases in Diesel engines, has now been solved by the injection of a pilot charge of ordinary Diesel fuel oil, which is easily ignited, into the gaseous mixture. In operation, the gas Diesel inducts a charge of air and gas

**Fig. 5.46.** Sixteen-cylinder V-type Cooper-Bessemer Diesel engine.

into the cylinders during the intake stroke. The mixture is very lean, partly because the Diesel is inherently a lean-mixture engine, and also because the gaseous charge is only a fraction of the total fuel to be burned, the remainder being supplied later in the pilot charge of fuel oil. The fuel-injection pumps are timed to deliver the pilot charge at the proper point in the compression stroke to cause combustion at the optimum time. The inducted charge of gas and air does not ignite at the temperature of compression to which it is exposed before the injection of the pilot charge because of the low fuel–air ratio and the high self-ignition temperature of the gas.

Most of the large stationary power-plant Diesels now manufactured are equipped to burn gas because of the large savings in power costs effected. Figure 5.46 illustrates a large stationary engine of the gas- or oil-burning type. The unit is a model LSV 16 Cooper-Bessemer turbosupercharged four-stroke-cycle gas-Diesel engine with 16 cylinders arranged in two banks of 8 forming an angle of 36 degrees. The cylinders are of 15.5-in. bore and 22-in. stroke. Two exhaust-gas-driven turbochargers are provided, one at each end of the engine serving separate banks of cylinders. The engine is equipped with a complete Diesel-fuel system and also with gas-admission equipment, permitting operation entirely on Diesel fuel or on proportions of fuel oil and gas up to 96 per cent gas and 4 per cent pilot fuel oil. At any time while the engine is running, the fuel mixture may be changed to the proportions desired. Gas is taken into the cylinders with the intake air, where it is ignited by the pilot oil. A governor controls the quantity of the two fuels needed to meet engine demands. Because of its lower cost, the gaseous portion of the charge is normally made as large as possible, but at times the available gas may be inadequate to supply the power demanded of the engine and it is necessary to increase the proportion of fuel oil in order to carry the load.

The heating value of the charge that can be burned in a gas Diesel is less than that of the fuel-oil charge of the conventional Diesel, and the horsepower rating is therefore lower. The rating of the engine illustrated, when supercharged, is 3700 hp when burning fuel oil at 327 rpm and 3300 hp when burning 96 per cent gas at the same speed. Unsupercharged, the rating is 2300 hp. These ratings correspond to brake mean effective pressures of 135, 120, and 85 psi, respectively. The engine is 37 ft 7 in. long and at maximum rating develops over 12 hp per sq ft of floor space occupied.

Figure 5.47 shows the brake specific fuel consumption of the Cooper-Bessemer LSV engine. When it is operating on gas, the fuel

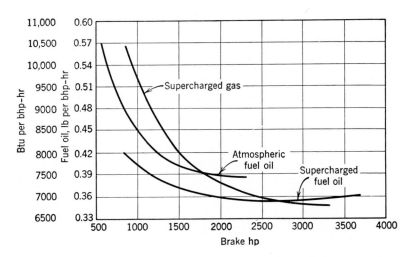

**Fig. 5.47.** Fuel-rate curves for 16-cylinder model LSV Cooper-Bessemer Diesel engine.

rate is expressed in Btu per brake horsepower-hour and is calculated
on the basis of the lower heating value of the gas plus the higher heating value of the pilot fuel oil.   The scale of pounds of fuel oil per
brake horsepower-hour applies to the curves for operation on fuel oil
and is approximately equivalent to the Btu scale on the higher-
heating-value basis.   Compared to the automobile-engine performance of Fig. 5.44, these curves show that, at its most economical load,
the Diesel operating turbocharged on fuel oil uses less than three
fourths the weight of fuel per brake horsepower-hour that is used by
the automobile engine at its most economical load.   The economy of
the engine when burning gas is especially high at heavy loads, but
the fuel rate increases more at light loads than it does on fuel oil.
Since the price of natural gas in many parts of the United States is
somewhat lower per Btu than that of fuel oil, appreciable savings in
the cost of power generation are effected by engines of this type.
The turbosupercharger is also shown to reduce substantially the fuel
consumption of this engine.

The performance of an engine is best analyzed by means of an
energy balance.   A summation of energy quantities for the Cooper-
Bessemer LSV engine running on gas and minimum pilot fuel can be
made from routine power-plant data as follows:

|                                           | Btu per Bhp-Hr | Per Cent |
|-------------------------------------------|:--------------:|:--------:|
| Energy utilized in bhp                    | 2545           | 37.5     |
| Energy dissipated by jacket water         | 1200           | 17.7     |
| Energy dissipated by oil coolers          | 285            | 4.2      |
| Energy dissipated by aftercoolers         | 150            | 2.2      |
| Energy to exhaust, radiation, etc.        | 2620           | 38.4     |
| Energy supplied                           | 6800           | 100      |

The brake thermal efficiency of 37.5 per cent indicated is based
upon the lower heating value of the gas burned plus the higher heating value of the fuel-oil pilot charge, and would be decreased if the
higher heating values of both were used.   It is customary in the
United States to charge engines with the higher heating value of the
fuel, assuming that it is a fault of the engine that it cannot utilize
the energy equivalent to the latent heat of the water vapor formed by
burning hydrogen.

A more detailed energy balance may be calculated from test data
in the same manner as is discussed in Article 4.23.   The results of a

complete test of an automobile engine are stated in the following data and the energy balance determined by the accompanying calculations.

**Example 11.** An automobile engine developing 100 bhp at 3600 rpm at full throttle burns 8.9 gal of gasoline per hr. Cooling water enters the jacket at a rate of 66 gpm at a temperature of 70 F and leaves at 180 F. Fuel analysis by mass: 15.8% hydrogen and 84.3% carbon. Specific gravity of fuel: 0.731. Higher heating value of fuel: 20,750 Btu per lb. Exhaust gas analysis by volume: 11.5% $CO_2$, 4.1% CO, 0.4% $O_2$, 2.3% $H_2$, 0.2% $CH_4$, and 81.5% $N_2$. Exhaust temperature, 1100 F. Ambient temperature, 70 F.

*Solution:* Contents of 100 moles of exhaust gas:

|  | Dry Gases | | Carbon | | Hydrogen |
|---|---|---|---|---|---|
| $CO_2$ | $11.5 \times 44 =$ | 505 | $11.5 \times 12 =$ | 138 | |
| CO | $4.1 \times 28 =$ | 115 | $4.1 \times 12 =$ | 49.2 | |
| $O_2$ | $0.4 \times 32 =$ | 12.8 | | | |
| $H_2$ | $2.3 \times 2 =$ | 4.6 | | | $2.3 \times 2 = 4.6$ |
| $CH_4$ | $0.2 \times 16 =$ | 3.2 | $0.2 \times 12 =$ | 2.4 | $0.2 \times 4 = 0.8$ |
| $N_2$ | $81.5 \times 28 =$ | 2280 | | | |
| | 100.0 moles $=$ | 2920.6 lb | | 189.6 lb | 5.4 lb |

$$\text{Lb of dry gas per lb fuel} = \frac{\text{lb of 100 moles of dry gas}}{\text{lb of C in 100 moles of gas}} \times \text{lb of C per lb fuel}$$

$$= \frac{2920.6}{189.6} \times 0.842 = 13 \text{ lb dry gas per lb fuel}$$

$$\text{Lb of fuel per 100 moles of dry gas} = \frac{\text{lb of C in 100 moles of dry gas}}{\text{lb of C per lb fuel}}$$

$$= \frac{189.6}{0.842} = 225.2 \text{ lb fuel}$$

Lb of H in fuel per 100 moles of dry gas $=$ lb of fuel per 100 moles of dry gas $\times$ lb of $H_2$ per lb fuel $= 225.2 \times 0.158 = 35.6$ lb $H_2$ in fuel

Lb of $H_2$ in water vapor formed by combustion $=$ lb of $H_2$ in fuel $-$ lb of $H_2$ in dry gas $= 35.6 - 5.4 = 30.2$ lb $H_2$ in water vapor
9 lb water vapor are formed per lb $H_2$

$9 \times 30.2 = 271.8$ lb water vapor in exhaust per 100 moles of dry gas

$$\text{Lb of water vapor per lb fuel} = \frac{271.8}{225.2} = 1.2 \text{ lb } H_2O \text{ per lb fuel}$$

Lb dry gas $+$ lb $H_2O$ $-$ 1 lb fuel $=$ lb air per lb fuel

$13 + 1.2 - 1 = 13.2$ lb air per lb fuel

$$\text{Bsfc} = \frac{\text{lb fuel per hr}}{\text{bhp}} = \frac{8.9 \times 0.731 \times 8.33}{100} = 0.548 \text{ lb per bhp-hr}$$

Fuel burned per min $= \dfrac{0.548 \times 100}{60} = 0.9133$ lb

Energy utilized, Btu per lb fuel $= \dfrac{\text{Btu per hp-hr}}{\text{bsfc}} = \dfrac{2545}{0.548} = 4644$ Btu

% of energy supplied utilized in Btu $= \dfrac{4644}{Q_H} = \dfrac{4644}{20,750} = 22.38\%$

Energy absorbed by coolant $= 66 \times 8.33 \times 10 = 5498$ Btu per min

Energy to coolant per lb fuel $= \dfrac{5498}{0.9133} = 6022$ Btu

% of energy supplied absorbed by coolant $= \dfrac{6022}{20,750} = 29.02\%$

Dry-gas loss = (exhaust-gas temperature − air temperature) × specific heat of gas × lb gas per lb fuel = $(1100 - 70)0.254 \times 13.0 = 3400$ Btu

% of energy supplied lost in sensible heat $= \dfrac{3400}{20,750} = 16.4\%$

Heating value of combustibles in exhaust $= Q_H$ for each combustible constituent × lb of each constituent per lb fuel

$Q_H = 4330$ Btu per lb CO, 62,000 Btu per lb $H_2$, and 23,700 Btu per lb $CH_4$

Heating value of CO in exhaust $= 4330 \times \dfrac{115}{225.2} = 2210$ Btu per lb fuel

Heating value of $H_2$ in exhaust $= 62,000 \times \dfrac{4.6}{225.2} = 1265$ Btu per lb fuel

Heating value of $CH_4$ in exhaust $= 23,700 \times \dfrac{3.2}{225.2} = 327$ Btu per lb fuel

Energy in combustibles in exhaust $= 2210 + 1265 + 327 = 3812$ Btu

% of energy supplied in combustibles in exhaust $= \dfrac{3812}{20,750} = 18.37\%$

Energy in $H_2O$ formed by combustion of $H_2$ in fuel = lb of $H_2O$ (1066 + 0.5 × exhaust temperature − air temperature) = 1.2 (1066 + 0.5 × 1100 − 70) = $1.2 \times 1546 = 1855$ Btu per lb fuel

% of energy supplied in $H_2O$ formed by combustion $= \dfrac{1855}{20,750} = 8.94\%$

Radiation and unaccounted-for losses, by difference = 1017 Btu

% of energy supplied unaccounted for $= \dfrac{1017}{20,750} = 4.89\%$

Summary

|  | Btu per Lb Fuel | Per Cent of Energy Supplied |
|---|---|---|
| I. Energy utilized, Btu | 4,644 | 22.38 |
| II. Losses |  |  |
| 1. Loss due to sensible heat in dry gaseous products | 3,400 | 16.40 |
| 2. Loss due to combustibles in dry gaseous products | 3,812 | 18.37 |
| 3. Loss due to heat transfer to jacket cooling water | 6,022 | 29.02 |
| 4. Loss due to water vapor formed from the hydrogen in the fuel | 1,855 | 8.94 |
| 5. Loss due to radiation and unaccounted-for losses | 1,017 | 4.89 |
| Totals | 20,750 | 100.00 |

## PROBLEMS

**1.** An engine of 3.75-in. bore and 3.5-in. stroke has 5.0 cu in. clearance volume. What is the thermal efficiency of an Otto cycle as applied to this engine?

**2.** A Diesel-engine indicator diagram drawn with a 180-lb spring and a one-half normal size piston has an area of 1.23 sq in., and its length is 2.6 in. What is the mep of the cylinder?

**3.** A four-cylinder four-stroke-cycle engine has a bore of 4 in. and a stroke of 3.25 in. What ihp is produced when it is running at 2300 rpm with 95 psi mep?

**4.** An engine loaded by a prony brake with a 54-in. brake arm runs at 500 rpm, and the net scale reading is 130 lb. What are the torque in lb-ft and the bhp delivered?

**5.** A dynamometer with a 21-in. torque arm indicates a force of 150 lb when absorbing the output of an engine running at 2700 rpm. What is its bhp?

**6.** Find the thermal efficiency of a theoretical Diesel cycle with a compression ratio of 15 and a cut-off ratio of 4.

**7.** Find the thermal efficiency of a Diesel cycle when applied to an engine of 5-in. bore and 6-in. stroke which has 8.5 cu in. clearance volume with cut-off occurring when the piston has descended 1 in. on its stroke.

**8.** An engine operating on the Otto cycle has a compression ratio of 14 and a bsfc of 0.48 when burning cracked fuel oil of 36 degrees API gravity. Find its brake thermal efficiency and ideal cycle efficiency.

**9.** If the engine of Problem 5 is motored by the dynamometer, and a reactive force of 25 lb is produced on the scales at 2700 rpm, what are the fhp and the ihp when it is delivering 75 bhp?

**10.** A six-cylinder four-stroke engine of 3.25-in. bore and 3-in. stroke delivers 56 bhp at 2500 rpm. Find the bmep.

**11.** An engine burns 0.5 gal of gasoline in a 4-min test while delivering 85 bhp. What is its bsfc if gasoline weighs 6.1 lb per gal?

**12.** Find the brake thermal efficiency of an engine that burns 1.6 gal of gasoline in 12 min while delivering 100 bhp. Assume 20,000 Btu per lb heating value and 6.1 lb per gal.

**13.** Write the combining-weight equation for combustion of heptane, $C_7H_{16}$, in air. Find the weight of air required to burn 1 lb of heptane in chemically correct proportions.

**14.** Heptane vapor has a higher heating value of 5475 Btu per cu ft. Find the heating value of 1 cu ft of a chemically correct heptane and air mixture.

**15.** The Ford engine has eight cylinders of $3\frac{9}{16}$-in. bore and $3\frac{3}{4}$-in. stroke. When it was running at 2300 rpm, it was found to consume 140 cu ft of air per min, which was metered at 120 F and 14 psia. If the room temperature was 70 F and the barometric pressure 29.2 in. Hg, what volumetric efficiency was indicated?

**16.** If the engine of Fig. 5.44 develops 60 hp at 2500 rpm, and the curve of Fig. 5.45 is assumed to be plotted for that speed, find the gal of gasoline it will consume per hr at 6.1 lb per gal.

**17.** When a supercharger was applied to a spark-ignition engine, its compression ratio was decreased from 8 to 7 to avoid knock. The volumetric efficiency was increased from 77% to 98% by supercharging, and the mechanical efficiency remained at 85%. (*a*) By what per cent should the bhp have increased? (*b*) What per cent change in fuel economy should have resulted?

**18.** An engine with the performance characteristics of Fig. 5.44 is installed in an automobile that requires 31 bhp to propel at 60 mph. The direct-drive gear ratio allows the engine to turn 3000 revolutions per mile. Assume that the bsfc changes with load at the same rate as in Fig. 5.45 and that gasoline weighs 6.1 lb per gal. (*a*) Find the miles per gallon of fuel at 60 mph. (*b*) Find the miles per gallon at 60 mph if an overdrive transmission reduces the engine speed by a factor of 0.7.

**19.** Determine the possible firing orders for the cylinders of a six-cylinder automobile engine with the crank throws arranged as in Fig. 5.28.

**20.** Determine the length of dynamometer arm for which the constant $C$ in the horsepower equation is equal to $\frac{1}{5000}$.

**21.** The General Motors six-cylinder two-stroke-cycle Diesel engine has a bore of 4.25 in. and a stroke of 5 in. When it is delivering 100 bhp at 1000 rpm, its fuel rate is 0.46 lb per bhp-hr. The blower delivers a volume of air, measured at the outside temperature of 70 F and pressure of 14.7 psia, 35 per cent in excess of the piston displacement volume. $Q_H$ for the fuel is 19,600 Btu per lb, and specific gravity is 0.85.

(*a*) Find the brake thermal efficiency.

(*b*) Find the air–fuel ratio supplied to the engine. Note that this is not the actual air–fuel ratio of the mixture burned because not all the air remains in the cylinder.

(c) Find the volume of fuel in cu mm supplied to each cylinder per cycle.

22. A six-cylinder automobile engine of 3.25-in. bore and 3.5-in. stroke is tested on a dynamometer which has a 15.75-in. arm. The scale reading of the dynamometer is 110 lb when loading the engine at 3800 rpm and 30 lb when motoring the engine at the same speed; 9 gal of gasoline are burned per hr. $Q_H = 20,000$ Btu per lb. Find (a) the bhp delivered, (b) the bmep, (c) the torque, (d) the ihp, and (e) the mechanical efficiency. The license rating used in many states as the basis for fees assumes that the bhp per cylinder equals 40 per cent of the square of the cylinder bore in in. (f) Find the license rating of this engine, and compare with the actual bhp.

23. An eight-cylinder automobile engine of 3.375-in. bore and 3.25-in. stroke with a compression ratio of 7 is tested at 4000 rpm on a dynamometer which has a 21-in. arm. During a 10-min test at a dynamometer scale beam reading of 90 lb, 10 lb of gasoline for which $Q_H$ is 20,000 Btu per lb are burned, and air at 70 F and 14.7 psia is supplied to the carburetor at the rate of 12 lb per min. Find (a) the bhp delivered, (b) the bmep, (c) the bsfc, (d) the brake specific air consumption, (e) the brake thermal efficiency, (f) the volumetric efficiency, and (g) the air–fuel ratio.

24. A supercharged six-cylinder four-stroke-cycle Diesel engine of 4.125-in. bore and 5-in. stroke has a compression ratio of 15. When it is tested on a dynamometer with a 21-in. arm at 2500 rpm, the scale beam reads 180 lb, 6.3 lb of 19,700 Btu per lb $Q_H$ are burned during a 6-min test, and air is metered to the cylinders at the rate of 24 lb per min. Find (a) the bhp developed, (b) the bmep, (c) the bsfc, (d) the air–fuel ratio, (e) the brake thermal efficiency, and (f) the ideal-cycle efficiency based on the theoretical Otto cycle.

25. A 200-hp gasoline truck engine operates at an average 65 per cent load and bsfc of 0.72 lb per bhp-hr. Find the saving in fuel cost that would result during an 8-hr run if a Diesel engine of the same rating were substituted with a bsfc of 0.52 under the same conditions. Assume gasoline of 0.73 specific gravity at 30 cents per gal and fuel oil of 0.85 specific gravity at 20 cents per gal.

26. A Diesel engine burns 9.0 gal of fuel oil per hr while delivering 150 bhp at 2000 rpm. A total of 70 gal of cooling water per min enters the jackets at 154 F and leaves at 165 F. The fuel contains 15.2% hydrogen and 84.8% carbon by weight, has a specific gravity of 0.85, and $Q_H$ of 19,650 Btu per lb. The exhaust-gas analysis by volume is 9.1% $CO_2$, 8.0% $O_2$, 0.04% CO, 0.05% $CH_4$, and 82.81% $N_2$. The exhaust-gas temperature is 900 F, and the ambient temperature is 70 F. Calculate an energy balance for the engine.

27. An engine loaded by a prony brake with a 4.5-ft arm runs at 300 rpm, and a platform-scale beam indicates a net torque reaction of 135 lb. The acceleration of gravity at the location is 30 ft per sec². Find the bhp.

# CHAPTER 6

# Fossil-Fuel-Burning Equipment; Nuclear Reactors

### 6.1   Introduction

The combustion of fossil fuel in the cylinder of an internal-combustion engine and the conversion into work of part of the energy thus released have been discussed in Chapter 5. This present chapter is concerned primarily with the principles involved and the equipment used for the economical combustion of fossil fuels for steam generation and the release, control, and removal of energy from nuclear reactors.

The functions of the furnace, the construction and methods of cooling furnace walls, the behavior of fossil fuels on grates and in furnaces, and the equipment used to burn solid, liquid, and gaseous fuels are considered. Although the discussion is concerned primarily with the combustion of fossil fuel under steam boilers, either for power generation or for steam heating, the same principles apply, and the same equipment with modifications is used in industrial furnaces such as drying or baking ovens; ceramic, cement, or lime kilns; soaking pits; open hearth furnaces; annealing furnaces; forge furnaces; and even domestic heating plants.

From the standpoint of ease of regulation, cleanliness, and control of combustion, gas is the ideal fossil fuel. It is used in many industrial furnaces even though it is often necessary to manufacture it from coal. Gas can be burned efficiently if a *burner* is provided to proportion and mix the gas with the air required for its complete combustion and to introduce this mixture into a hot *furnace* or compartment of sufficient size and shape to permit the combustion reactions to proceed to completion. Liquid fuels are atomized in burners which are designed to produce a fine spray of oil intimately mixed with air. This mixture is introduced into a hot furnace and burned in suspension like a gaseous fuel. Coal may be pulverized or reduced to a fine dust, mixed with air, and burned in suspension in a suitable furnace in a manner similar to the burning of oil or gas. Lump coal is placed, either by hand or mechanically, on a *grate* through which air may pass to react with the combustible constituents of the coal. Although part of the energy is released by combustion in the fuel bed, the gases which leave the fuel bed may contain as much as 40 to 60 per cent of the heating value of the fuel, and a furnace or combustion space of proper design is necessary if the fuel is to be burned completely.

U-233, U-235, and Pu-239 are nuclear fuels that are burned in a reactor under controlled conditions. The nuclear reactor is therefore a particular type of fuel-burning equipment in which energy is released as a result of the nuclear-fission reaction. The major types of reactors for power generation and their construction, control, cooling, and shielding will be discussed.

## 6.2 Furnaces

A furnace is a fairly gas-tight and well-insulated space in which gas, oil, pulverized coal, or the combustible gases from solid-fuel beds may be burned with a minimum amount of excess air and with reasonably complete combustion. Near the exit from the furnace, at which place most of the fuel has been burned, the furnace gases will consist of inert gases such as $CO_2$, $N_2$, and $H_2O$ vapor, together with some $O_2$ and some combustible gases such as CO, $H_2$, hydrocarbons, and particles of free carbon (soot). If combustion is to be complete, the combustible gases must be brought into intimate contact with the residual oxygen in a furnace atmosphere composed principally of inert gases. Also, the oxygen must be kept to a minimum if the loss due to heating the excess air from room temperature to chimney-gas temperature is to be low. Consequently, *the major function of the furnace is to provide space in which the fuel may be burned with a minimum amount of excess air and with a minimum loss due to the escape of unburned fuel.*

The design of a satisfactory furnace is based upon the "three T's of combustion": *temperature, turbulence,* and *time.*

For each particular fossil fuel, there is a minimum *temperature,* known as the *ignition temperature,* below which the combustion of that fuel in the correct amount of air will not take place. The igni-

tion temperature of a fuel in air as reported by various investigators
depends somewhat upon the methods used to determine it and, for
some common gases, is as follows:

| | |
|---|---|
| Hydrogen ($H_2$) | 1075–1095 F |
| Carbon monoxide (CO) | 1190–1215 F |
| Methane ($CH_4$) | 1200–1380 F |
| Ethane ($C_2H_6$) | 970–1165 F |

If the combustible gases are cooled below the ignition temperature, they will not burn, regardless of the amount of oxygen present.
A furnace must therefore be large enough and be maintained at a high
enough temperature to permit the combustible gases to burn before
they are cooled below the ignition temperature. In other words, the
relatively cool heat-transfer surfaces must be so located that they do
not cool the furnace gases below the ignition temperature until after
combustion is reasonably complete.

*Turbulence* is essential if combustion is to be complete in a furnace
of economical size. Violent mixing of oxygen with the combustible
gases in a furnace increases the rate of combustion, shortens the flame,
reduces the required furnace volume, and decreases the chance that
combustible gases will escape from the furnace without coming into
contact with the oxygen necessary for their combustion. The amount
of excess oxygen or air required for combustion is decreased by effective mixing. Turbulence is obtained, in the case of oil, gas, and
powdered coal, by using burners which introduce the fuel–air mixture
into the furnace with a violent whirling action. High-velocity steam
or air jets and mixing arches may be used to increase the turbulence
in furnaces fired with coal on stokers.

Since combustion is not instantaneous, *time* must be provided for
the oxygen to find and react with the combustible gases in the furnace. In burning fuels such as gas, oil, or pulverized coal, the incoming fuel–air mixture must be heated above the ignition temperature
by radiation from the flame or hot walls of the furnace. Since gaseous fuels are composed of molecules, they burn very rapidly when
thoroughly mixed with oxygen at a temperature above the ignition
temperature. However, the individual particles of pulverized coal or
atomized oil are very large in comparison with the size of molecules,
and many molecules of oxygen are necessary to burn one particle of
coal or droplet of oil. Time is required for the oxygen molecules to
diffuse through the blanket of inert products of combustion which
surround a partially burned particle of fuel and to react with the un-

burned fuel. Consequently, oil and pulverized coal burn with a longer flame than gaseous fuels.

The required furnace volume is dependent, therefore, upon the kind of fuel burned, the method of burning the fuel, the quantity of excess air in the furnace, and the effectiveness of furnace turbulence. The shape of the furnace depends upon the kind of fuel burned, the equipment employed to burn the fuel, and the type of boiler used to absorb the energy if the fuel is burned for steam generation.

Industrial furnaces in which the objective is to create and maintain a region at a high temperature and the furnaces of small steam boilers are constructed of fire brick, a brick that has been developed to withstand high temperatures without softening, to resist the erosive effects of furnace atmospheres and particles of ash, and to resist spalling when subjected to fluctuating temperatures. Low vertical walls may be constructed of fire brick in the conventional manner. High walls which are subject to considerable expansion may be tied to and sectionally supported by an external steel frame.

Where it is necessary to provide a brick roof over part of a furnace, or where brick mixing or reflecting arches are required, the suspended arch as shown in Fig. 6.1 may be used. Slotted refractory tile are hung from T-headed supporting castings which are attached to a structural-steel framework. Some industrial furnaces employ this type of suspended-arch construction for flat or inclined roofs up to 20 ft or more in width and over 80 ft in length.

When a boiler furnace is operated at high capacity, the temperature may be high enough to melt or fuse the ash which is carried in suspension by the furnace gases. Molten ash will chemically attack and erode the fire brick with which it comes into contact. Also, if

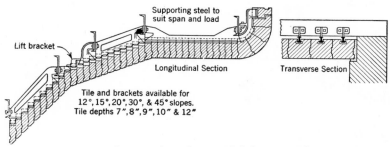

**Fig. 6.1.** Construction of suspended furnace arches.

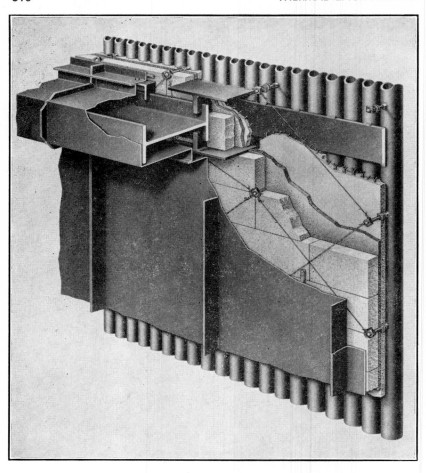

**Fig. 6.2.** Typical furnace-wall construction, showing boiler tubes, steel supports, insulation, and gas-tight casing.

the ash particles are not cooled below the temperature at which they are plastic or sticky before they are carried into the convection tube banks of the boiler, they will adhere to these surfaces, obstruct the gas passages, and force a shutdown of the unit. Moreover, the function of a boiler is to generate steam, and the most effective heat-transfer surface is that which can "see" the high-temperature flame and absorb radiant energy. The rate of heat absorption expressed in Btu per hour per square foot of projected wall area may be from 1000 to 10,000 times as great as the heat-transfer rate in the boiler surface with which the products of combustion are in contact last before being discharged up the chimney. Consequently, the walls of furnaces for large steam boilers are constructed of boiler tubes. Figure 6.2 shows a typical wall construction built around vertical boiler tubes in which steam is being generated. The tubes are welded to steel supports to maintain alignment. A backing of refractory and insulating materials and a gas-tight steel casing are anchored to them from studs

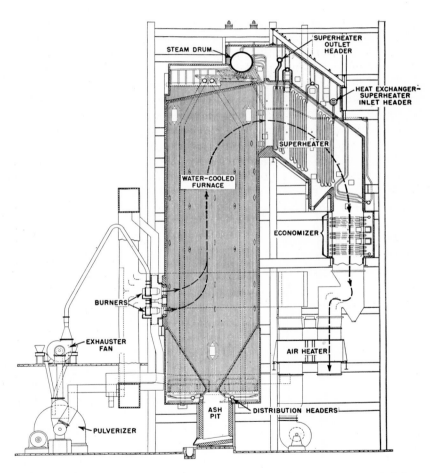

**Fig. 6.3.** Single-drum steam-generating unit with completely water-cooled furnace. 800,000 lb per hr at 1350 psig and 955 F.

that are welded to the back side of the tubes.  This illustration shows the tangent wall construction in which the tubes touch each other and form a continuous water-cooled surface.  Under some conditions, tubes may be spaced on vertical centers equal to two to four times their diameters as shown in Fig. 8.7.  Such construction is used where it is desirable to reduce the heat absorption below that which would exist with a fully cooled wall.

Figure 6.3 illustrates a large modern steam-generating unit fired by pulverized coal in which the hopper bottom and the four sides of the furnace are constructed entirely of boiler tubes.  Furnaces of this kind may be as high as a ten-story office building.

The furnace energy-release rate is expressed in Btu per hour per cubic foot of furnace volume.  It is computed by multiplying the quantity of fuel burned per hour by the heating value per pound of fuel and dividing this product by the furnace volume in cubic feet. Furnaces used for steam generation are usually proportioned for an energy-release rate of 20,000 to 50,000 Btu per hr per cu ft.  Excessive maintenance of furnace brick work, troubles due to clinker formation on stokers, and plugging of gas passages between boiler tubes by particles of plastic ash can be avoided by the use of liberal furnace volumes and adequate water cooling of furnace walls.  Large pulverized-coal-fired furnaces intended for long periods of trouble-free operation are usually designed for a furnace energy-release rate of about 20,000 Btu per hr per cu ft, especially if the ash has a low fusing temperature.  Higher values of furnace energy-release rates may be used in small units, stoker-fired installations, cyclone burners, and furnaces burning oil or gas which is relatively free of ash.

### 6.3   Gas Burners

Gas is burned in many industrial furnaces because of its cleanliness, ease of control of furnace atmosphere, ability to produce a long slow-burning flame with uniform and gradual energy liberation, and ease of temperature regulation.  Natural gas is used for steam generation in gas-producing areas and in areas served by natural-gas transmission lines where coal is not available at a competitive price.  It is also burned extensively in coal- or oil-fired units during the summer months in districts served by natural-gas pipe lines, at which time the absence of the domestic heating load creates a temporary surplus of natural gas.  By-product gas such as blast-furnace gas may be available at the steel mills for steam generation.  Because of the variable or seasonal supply of gaseous fuels, combination burners have been developed to permit the simultaneous burning of the available gas together with pulverized coal or oil in an amount sufficient to produce the required steam.

When a molecule of combustible gas is mixed with the oxygen necessary for its combustion at a temperature above the ignition temperature, combustion is practically instantaneous. For steam generation, where a short flame is desired in order to reduce the required furnace volume, the burner should provide for rapid and thorough mixing of the fuel and air in the correct proportions for good combustion. For such applications, a good burner is primarily a proportioner and mixing device. In industrial furnaces where long "lazy" flames are desired, slow and gradual mixing of the air and fuel in the furnace is necessary. Figure 6.4 shows the construction of a number of typical gas burners. In the burner illustrated in Fig. 6.4A, the gas, under pressure in the supply line, enters the furnace through a burner port and induces a flow of air through the port. Mixing is poor, and a fairly long flame results. The flame can be shortened by use of the ring burner (Fig. 6.4B), in which the gas flows through an annular ring and induces air flow both around and within the annulus of gas. Where both air and gas are under pressure, an arrangement as shown in Fig. 6.4C may be used. In each of these burners, the gases must flow through the port into the furnace at a velocity high enough to prevent the flame from burning back into the

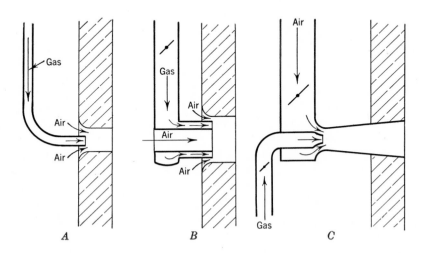

**Fig. 6.4.** Typical gas burners.

burner. The arrangement of burners and the shape of the furnace should be such that the flame does not impinge on the furnace walls or the heat-transfer surfaces.

### 6.4   Oil Burners

Most of the fuel oil that is used for steam generation is the residuum that remains after crude petroleum has been refined to produce gasoline, kerosene, Diesel-engine fuels, and lubricants. Because of the ease of storage and handling, smaller storage space required compared with coal, absence of ash, high capacity obtainable from small furnaces, and small amount of labor required, oil is the only fossil fuel used for steam generation in naval vessels and many merchant-marine ships. It is burned in steam power plants in regions where the price is competitive with the price of coal. Since the residual fuel oil that is used for steam generation contains the ash and most of the sulphur originally present in the crude petroleum before refining, serious difficulties with corrosion and slag formation on furnace walls and boiler tubes have been encountered with some grades of fuel oil.

To burn fuel oil successfully, it must first be atomized or broken up into very small droplets of oil. In a hot furnace, these small drops of oil will be partially vaporized to form gases which will burn very rapidly if mixed with the correct amount of air. The complex hydrocarbon molecules of which the oil is composed are unstable at high temperatures and may decompose upon heating to form gaseous compounds which are readily burned and carbon particles which burn slowly with a long flame. Decomposition of molecules to form carbon can be kept to a minimum by thorough mixing of the fuel with an adequate supply of air. A successful oil-burning installation is one that produces a spray of finely atomized oil which is thoroughly mixed with air in a furnace hot enough to vaporize the oil quickly and large enough to permit combustion to be completed without flame impingement on the boiler-tube bank or on the furnace walls.

A mechanical atomizing oil burner and air register is illustrated in Fig. 6.5. All the air required for combustion is supplied to the burner through the air register. The air, which is normally supplied under slight pressure by a fan, flows through adjustable air doors in the register in such a manner as to impart a whirl to the air. The position of the air doors is shown in the left cutaway section of the front view of the register. Oil at a pressure up to 300 psig is admitted to the mechanical atomizer which is essentially a tube with a removable

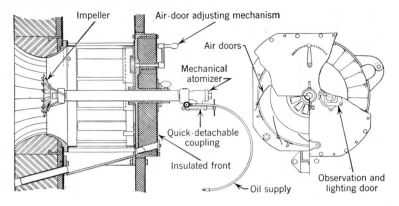

**Fig. 6.5.** Mechanical atomizing oil burner.

sprayer plate in the tip through which the oil is delivered in a finely atomized spray having the shape of a hollow cone. An impeller is mounted on a hollow distance piece within which the mechanical atomizer is placed. It may be moved in or out of the burner throat to control the amount of air passing through and around the impeller.

A minimum oil pressure at the burner tip of about 100 psig is necessary for proper atomization of the heavy grades of fuel oil. With a maximum oil pressure of 300 psig, the range of capacity of the mechanical atomizing burner is about 1.0 to 1.7; that is, the maximum capacity is 170 per cent of the minimum capacity. Greater capacity ranges are obtainable by changing sprayer plates or by cutting burners in or out of service. Since all the air is admitted through the registers, several rows of burners can be installed across a boiler front and removing a burner from service is easy.

Number 6 fuel oil (see Table 4.5) is most commonly used for steam generation because of its low cost. It must be heated in a steam-oil heat exchanger to reduce the viscosity sufficiently so that it can be atomized properly.

A steam atomizing burner is similar to the mechanical atomizing burner except that the high-pressure oil pump is replaced by a low-pressure pump, the heat exchanger may be omitted in some cases, and the mechanical atomizing tube as shown in Fig. 6.5 is replaced by a steam atomizing tube having a similar external appearance. A high-velocity steam jet is used to atomize the oil instead of a high-pressure mechanical sprayer. Since from 1 to 2 per cent of the output of the boiler may be required to atomize the oil, the mechanical atomizer is preferred where the loss of this steam would be objectionable, as in ocean-going vessels.

Wide-range oil burners having a capacity range of 1 to 7 often employ mechanical atomization for the high range of operation and steam atomization for the range in which the oil pressure is inadequate to produce a good spray.

### 6.5   Combustion of Coal on a Hand-Fired Grate

Except for domestic heating plants, coal is seldom fired by hand at the present time.  However, a knowledge of the behavior of coal on a hand-fired grate is important to an understanding of the reasons for the particular construction of stokers.

Since combustion is primarily a surface effect, a lump of coal must be supported on a grate for a considerable period of time while oxygen (air) flows across its surface under such conditions as to burn the combustible constituents and leave a residue of ash.  Let Fig. 6.6 represent conditions that exist in a level fuel bed in which lump bituminous coal is fired by hand.  The fuel bed is supported on a grate through which primary air is admitted.  Above the ash zone is an

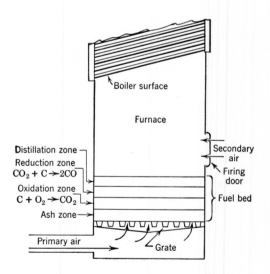

**Fig. 6.6.** Zones in a hand-fired fuel bed.

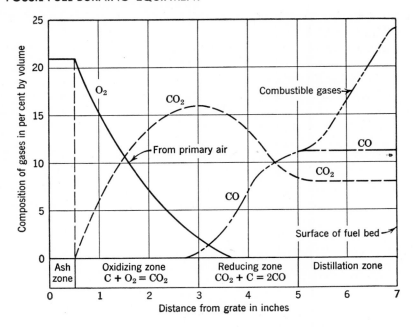

**Fig. 6.7.** Composition of gases in a hand-fired fuel bed.

oxidizing zone in which the fixed carbon in the fuel combines with the oxygen in the air to form $CO_2$. In a level fuel bed, all the oxygen will combine with carbon in a vertical distance of about 4 in. Above the oxidation zone is a reducing zone in which, in the absence of oxygen, the carbon combines with $CO_2$ to form CO. The freshly fired coal, which is placed on top of the hot fuel bed, forms a distillation zone in which the volatile matter is distilled from the coal. The combustible gases rising from the fuel bed consist of CO and volatile matter and may represent 40 to 60 per cent of the total heating value of the fuel. If the furnace is of adequate size and is maintained at a high enough temperature, and proper mixing occurs, these gases are burned in the furnace by secondary air supplied over the fuel bed and by leakage of air through holes in the fuel bed.

Figure 6.7, based on experiments of the U. S. Bureau of Mines, shows the composition of gases in a level hand-fired fuel bed of coal. Practically all the oxygen will disappear within 2½ to 4 in. of the grate, regardless of the rate of burning or the kind of coal. The $CO_2$

content reaches a maximum at the point where most of the oxygen has combined with the carbon, after which the percentage of $CO_2$ decreases, owing to the reaction $CO_2 + C = 2CO$. The combustible gases at the surface of the fuel bed consist of the CO formed in the reducing zone plus the volatile matter driven out of the coal in the distillation zone. The line labeled "combustible gases" in Fig. 6.7 represents the condition in the fuel bed shortly after firing. Three or four minutes later, most of the volatile matter will have been distilled from the coal, leaving fixed carbon, and the line labeled "CO" more nearly represents the combustible gases leaving the fuel bed. It is apparent that only part of the carbon is burned in the fuel bed, that much of the carbon is gasified to CO, and that this CO plus the volatile matter must be burned in the furnace by secondary air supplied over the fuel bed.

It is difficult to burn high-volatile bituminous coal on a hand-fired grate without producing objectionable *smoke*. Black smoke is caused by particles of carbon floating in the colorless gases such as $CO_2$, CO, $O_2$, and $N_2$. *These carbon particles are caused by the cracking or thermal decomposition of the unstable hydrocarbon compounds which are distilled from the coal as volatile matter. If these compounds can find the oxygen and temperature necessary for their combustion as soon as they are driven from the coal, they will burn to $CO_2$ and $H_2O$. If this oxygen is not present, they will decompose to form soot which is almost impossible to burn once it is formed.* It should be noted that in a hand-fired installation the green coal is thrown on top of the hot fuel bed where it is *heated rapidly in the absence of oxygen,* all the primary oxygen having combined with carbon in the oxidizing zone of the fuel bed. Conditions are therefore favorable for the decomposition of this volatile matter and the formation of free carbon or soot. If the coal is fired intermittently in heavy charges, there will be a period of 3 or 4 min following firing during which large quantities of volatile matter are produced and smoke will be heavy. Then, the volatile matter having been driven off, the carbon which remains will burn to CO and $CO_2$ with little or no smoke. The extremes of this condition are shown in Fig. 6.7 in the curves labeled "combustible gases" and "CO." The intermittent evolution of volatile matter makes it difficult to adjust the secondary air to the fluctuating quantity of combustible gas in the furnace, resulting in alternate periods of deficient and of excess air. This condition can be helped by more frequent firing of smaller amounts of coal, but this involves more labor.

### 6.6 Spreader Stokers

The spreader stoker is designed to throw coal *continuously* onto a stationary or moving grate. Figure 6.8 shows a spreader stoker equipped with a moving grate which travels toward the feeder mechanism and discharges the refuse continuously. Coal is fed from the hopper by means of a reciprocating feeder plate having a variable-speed drive which for best performance should be regulated automatically to feed coal in accordance with the demand for energy. The coal is delivered by the feeder to a rapidly revolving drum or rotor on which are fastened specially shaped blades which throw the fuel into the furnace and distribute it uniformly over the grate. Coal can be distributed thus for a total distance of about 22 ft. The feeder mechanism is built in standardized widths, and several units may be installed across the front of the larger furnaces.

Air is supplied by means of a blower to the space under the moving grate through an adjustable damper. The active fuel bed is normally not over 1½ in. deep so that an adequate supply of air can penetrate the fuel bed and enter the furnace. Active fuel beds much thicker than 1½ in. will produce excessive amounts of smoke. Much of the volatile matter is distilled from the coal before it strikes the fuel bed, and the caking properties of the fuel are thus destroyed, thereby making it possible to burn even the strongly caking bituminous coals. Since the fuel bed is thin and undisturbed and the ash is cooled by the flow of air through it, trouble with clinkering or fusing

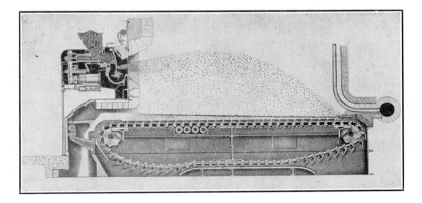

**Fig. 6.8.** Spreader stoker with continuous ash discharge.

of the ash is uncommon, and this stoker can burn almost any kind of bituminous coal. Since the finer sizes of coal are burned in suspension, large furnaces are required, and objectionable quantities of dust may be discharged from the installation if it is not designed correctly and if dust collectors are not installed to clean the gases leaving the steam-generating unit. Also, it is standard practice to install high-velocity steam jets in the furnace to promote turbulence, improve combustion, and reduce smoke.

Large units provided with continuous ash-discharge grates as shown in Figs 6.8 and 8.9 are capable of burning 12 to 15 tons of coal per hr. Small units may have stationary grates with clean-out doors through which the ashes may be removed manually with a hoe, or they may have dump grates operated by a power cylinder in which grate sections may be tilted periodically to dump the ashes as illustrated in Fig. 8.7.

The spreader stoker is simple in construction and reliable in operation. It can burn a wider variety of coal successfully than any other type of stoker. Maximum continuous combustion rates of 45 to 60 psf of grate area per hr are normally used. When provided with automatic regulation of fuel and air in accordance with the demand for energy, this stoker is very responsive to rapidly fluctuating loads. However, it is not so adaptable to light-load operation as other types of stokers because of the difficulty of maintaining ignition and combustion in the very thin fuel bed with a cold furnace. It is because of the thin fuel bed and the continuous, uniform firing of coal that the spreader stoker overcomes the smoke-producing problem associated with the thick intermittently hand-fired fuel bed.

### 6.7   Chain- and Traveling-Grate Stokers

A chain-grate stoker, as illustrated in Fig. 6.9, has a moving grate in the form of a continuous chain. The upper and lower runs of the chain are supported on a structural steel frame. The chain is driven from the stoker front by means of sprockets mounted on a rotating shaft which is actuated by a ratchet mechanism and hydraulic cylinder. The grate bars are made of heat-resistant cast iron, are cooled by the air supplied for combustion, and form a flat undisturbed surface for the fuel bed.

Coal from the stoker hopper is placed on the moving grate in a uniform layer, the depth of which is controlled by the vertical movement of an adjustable fuel gate. The depth of the fuel bed is usually

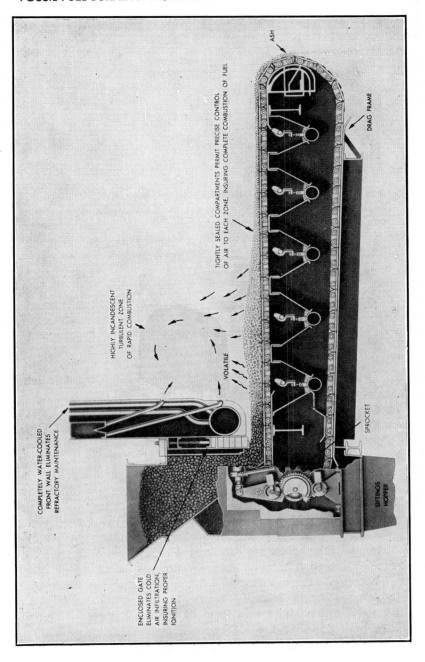

**Fig. 6.9.** Chain-grate stoker.

between 3 and 8 in., depending upon the kind of fuel being burned. The speed of the grate may be adjusted, usually between the limits of 4 and 20 in. per min, so that the combustible material is burned before the ash is discharged from the rear end into the ashpit. The shearing action of adjacent grate bars as they pass around the curved supporting member at the rear of the stoker provides a self-cleaning action for the grate bars. Air is supplied under adjustable pressure to several compartments under the grate. Five such compartments are shown in Fig. 6.9. Thus the supply of air to various sections of the fuel bed may be adjusted to suit the combustion requirements.

When bituminous and other high-volatile coals are burned, high-velocity air jets are installed in the front furnace wall as illustrated in Fig. 6.9. The volatile matter that is released from the incoming green coal is mixed with the swirling turbulent air that is introduced above the distillation zone. Two important results are thereby accomplished: (1) the volatile matter is burned smokelessly, and (2) a high-temperature zone is formed which provides for stable ignition of the incoming coal. The existence of this highly incandescent zone of turbulent combustion over the front end of the stoker makes mixing arches in the furnace unnecessary, and an open furnace with vertical walls similar to the spreader-stoker furnace of Fig. 8.9 may be used.

The small sizes of anthracite which cannot be sold for a domestic fuel and the small sizes of coke which are too small to charge into the blast furnace, called coke breeze, are important stoker fuels in certain localities. These fuels contain practically no volatile matter. Because of the fine size and large total surface of the incandescent carbon in the fuel bed, all the oxygen combines with carbon a short distance above the grate as illustrated in Fig. 6.7 unless fuel-bed air velocities are so high as to almost lift the fuel from the grate. Under these conditions, large amounts of fine particles of carbon are blown upward into the furnace. With an open furnace such as shown in Fig. 8.9, excessive amounts of carbon would be blown into the ashpit. Also, it is necessary to maintain a hot zone above the entering fuel to ignite the fuel on the grate. Accordingly, furnaces for burning anthracite and coke breeze are constructed with a long rear arch and over-fire air injection through the rear arch as illustrated in Fig. 6.10. The net effect is to maintain a hot zone over the incoming fuel and to blow the fine particles of carbon onto the front of the stoker so as to assist ignition and retain them in the combustion zone until they are burned. Over-fire air injection and a high furnace are necessary to burn the CO that is formed in the fuel bed.

The *traveling-grate* stoker is similar in general appearance and operation to the chain-grate stoker except that individual grate bars or keys are mounted on carrier bars which extend across the width of the stoker and are attached to and driven by several parallel chains. Since adjacent grate bars have no relative motion with respect to each

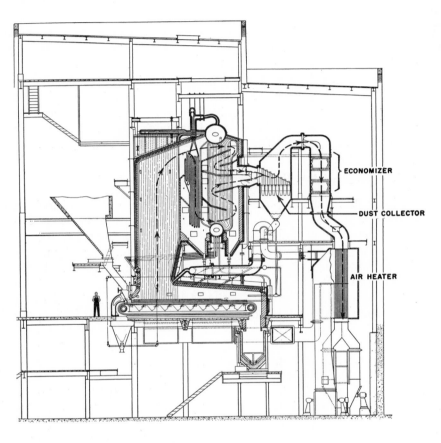

**C-E STEAM GENERATING UNIT**

MAX. CONT. CAP. — 175,000 LB PER HR AT 875 PSI AND 910 F TOTAL TEMP.

**HICKLING STATION**

**New York State Electric & Gas Corporation, E. Corning, N. Y.**

DESIGNED AND MANUFACTURED BY COMBUSTION ENGINEERING COMPANY, INC.

GILBERT ASSOCIATES, INC., Consulting Engineers

**Fig. 6.10.** Steam-generating unit fired by anthracite-burning traveling-grate stoker.

other, this stoker is particularly applicable to the burning of the fine sizes of anthracite and coke breeze in which all the fuel may pass through a screen having $\frac{3}{16}$-in. round openings.

The chain- and traveling-grate stokers are particularly adapted to the burning of the free-burning or noncaking bituminous coals of the Midwest, subbituminous coal, lignite, and the steam sizes of anthracite and coke breeze. The development of the method of over-fire air injection illustrated in Fig. 6.9 makes it possible to burn caking coals also. The fuel should preferably contain at least 8 per cent of ash in order to provide a protective covering from furnace radiation for the rear of the grate. Combustion rates of 30 to 45 lb of coal per sq ft of grate per hr are common with peak burning rates of 60 lb. This stoker is not as responsive to rapid load fluctuations as the spreader stoker but can be operated at lighter loads. It has been built in sizes large enough to generate 250,000 lb of steam per hr.

### 6.8   Single-Retort Underfeed Stokers

Figure 6.11 shows a photograph taken from the rear and looking down on a single-retort underfeed stoker before installation in a furnace. Figures 6.12 and 6.13 are longitudinal and cross sections, respectively, of this type of stoker. Coal is fed from a hopper by means of a reciprocating plunger into a horizontal retort or trough. Auxiliary reciprocating pusher blocks in the bottom of the retort have an adjustable stroke and are designed to assist in pushing the coal upward out of the retort and in distributing it uniformly along the length of the retort, as indicated in Fig. 6.12.

The behavior of the fuel on the single-retort underfeed stoker may be understood by reference to Fig. 6.13. The raw coal in the retort is pushed upward by the plunger and auxiliary pusher blocks and, as it rises above the level of the top of the retort, is heated by the incandescent fuel at the top of the fuel bed. Air is supplied to the air chamber beneath the retort under pressure by means of a fan. Air flows through openings in the tuyères or grate bars to the distillation zone. Thus the volatile matter is distilled from the coal *uniformly in the presence of air,* and the combustible mixture which results from this process is heated to the ignition temperature and burned as it travels upward through the high-temperature fuel bed above the distillation zone. The coke which remains after distillation of the volatile matter from the coal is forced upward and to the sides where it receives air through the grates and burns. The ash and some un-

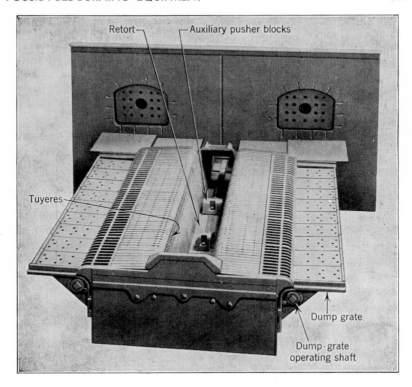

Retort — ⌐ Auxiliary pusher blocks

Tuyeres —

Dump grate

Dump-grate operating shaft

**Fig. 6.11.** Single-retort underfeed stoker as seen from above and at the rear of the stoker.

burned carbon accumulate on the dumping grates. Air can be supplied to the dumping grates periodically to burn out the carbon before the grates are rotated about a shaft to drop the refuse into the ashpit.

A constant-speed motor drives the blower or fan and the crankshaft at uniform speed. Dampers on the fan inlet and lost-motion adjustments on the plunger operating rod permit variable feed of air and coal as needed. Smaller stokers are often provided with a rotary-feed screw instead of a reciprocating plunger and distributing blocks. Larger stokers often have some laterial reciprocating motion of the side grates to assist in the distribution of the burning coke over a wider grate.

Single-retort stokers have been built in sizes from domestic stokers to units capable of burning over 3000 lb of coal per hr. They require

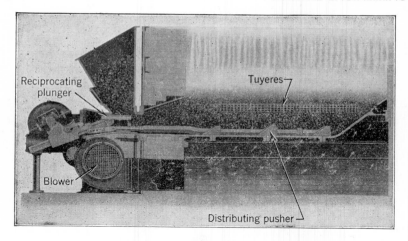

**Fig. 6.12.** Longitudinal section through the retort of a single-retort underfeed stoker.

little more head room than good hand-fired grates and have largely replaced the hand-fired grates for small and medium-sized industrial power boilers and for apartment house and industrial heating boilers because of their ability to burn bituminous coal efficiently and smokelessly with a minimum of attention. Maximum continuous combustion rates of 20 to 30 psf of grate area per hr are recommended, the

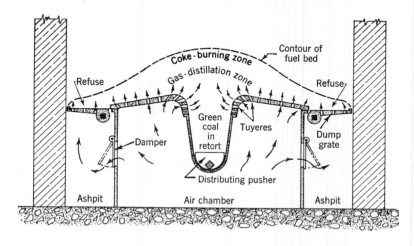

**Fig. 6.13.** Cross section through the retort of a single-retort underfeed stoker.

higher rates being used on the wider stokers and with the better coals.

Many multiple-retort underfeed stokers are in operation but few are being sold. Essentially, they consist of a number of parallel retorts combined into one machine with the retorts set at an angle of about 20 degrees with the horizontal. Fuel is supplied to the upper end of each retort and the refuse is discharged from the lower end. They are particularly adapted to the burning of coking bituminous coal in large installations. Their former place is currently being taken by spreader stokers and pulverized-coal units.

### 6.9  Pulverized Coal

Coal may be pulverized or ground to a dust, mixed with air, blown into a hot furnace, and burned in suspension. In the furnace, the combustion of a particle of coal takes place in four steps as follows: (1) evaporation of any moisture, (2) gasification of the volatile matter, (3) combustion of the gaseous volatile matter, and (4) combustion of the residue of coke dust. These four steps are not distinct but tend to overlap. Even with fine grinding of the coal, the size of the individual particles is great compared with the size of a molecule of gas, and each particle of coal dust will require for its complete combustion a volume of air equal to about *14,000* times its own volume. As combustion progresses, the particle of fuel is surrounded by a layer of products of combustion, consisting mainly of $CO_2$ and $N_2$, through which the oxygen in the furnace atmosphere must penetrate by diffusion or turbulence in order to react with the carbon on the surface of the particle of coke dust. Considering the relative volumes of air and coke dust involved in the combustion process, it is apparent that combustion of powdered coal does not occur so rapidly as the combustion of gas or oil and that large furnaces are required. A high degree of turbulence is necessary in a pulverized-coal furnace to accelerate the process of bringing the oxygen into contact with the suspended particles of coke dust if a furnace of reasonable size is to be used.

Figure 6.14 illustrates a typical system for preparing and firing pulverized coal. Raw coal which has been crushed to pass through a screen having ¾-in. openings is supplied by gravity from a coal bunker through a feeder to the pulverizer or mill. The feeder automatically regulates the rate of flow of coal in proportion to demand. Practically all pulverized-coal installations include air heaters in which the air for combustion is heated to between 350 and 600 F by

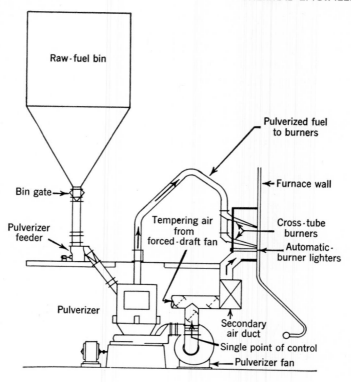

**Fig. 6.14.**  Pulverized-coal firing system.

the hot products of combustion before they are discharged up a chimney. Hot primary air is supplied to the pulverizer by means of a fan as shown in Fig. 6.14. In passing through the mill, the hot air dries the coal and also picks up the fine particles and carries them in suspension to the burners. The use of preheated air in the pulverizer is essential to satisfactory operation, since the power consumption is reduced and the capacity of the mill is increased when dry coal is pulverized. Tempering or cold air is supplied to the fan under automatic control so that the air leaving the pulverizer will be between 130 and 200 F, depending on the moisture content and kind of coal. The rate at which coal is delivered to the burners is determined by the amount of air that the fan delivers to the pulverizers, and this is under automatic control. The primary air and fuel from the pulverizer are mixed with secondary air at the burners which are designed to deliver the fuel and all the air required for combustion to the furnace with a high degree of turbulence.

A typical pulverizer or mill is illustrated in Fig. 6.15. This mill operates on the principle of the ballbearing. A set of balls rolls between a stationary top grinding ring and a rotating bottom grinding ring which is driven by a vertical shaft geared to a motor. Adjustable springs load the stationary grinding ring from above and control the pressure on the balls. The raw coal is admitted through a feeder to maintain automatically the desired coal supply in the mill. Preheated air is admitted at the bottom of the mill, passes upward

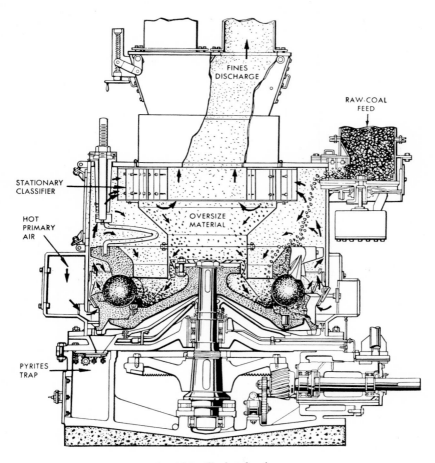

**Fig. 6.15.** Coal pulverizer.

around the balls where it dries the coal and entrains the pulverized-coal particles, flows upward through a rotating classifier where the oversize particles of coal are rejected to the mill for further grinding, and is discharged from the top of the mill to the burners.

The power consumption of pulverizers and the blowers associated with them varies from 9 to 16 kw-hr of electric energy per ton of coal pulverized, depending upon the type of mill, the moisture content of the coal, the required fineness of pulverization, and the "grindability" of the coal.

The fineness to which the coal is pulverized is expressed as the per cent by weight of the pulverized coal that will pass a 200 mesh sieve, that is, a screen having 200 openings per linear inch or 40,000 openings per square inch. In general, the required fineness increases with the decreasing volatile content of coal because of the greater ease of igniting high-volatile coal. The approximate range for optimum performance is indicated in Table 6.1 with the lower figures applying to the larger installations.

Figure 6.16 shows the effect of fineness of pulverization upon the power consumption per ton pulverized and the mill maintenance. It may be noted that, on the basis of unity for 70 per cent fineness, a coal that must be ground to 80 per cent fineness will require 40 per cent more power and maintenance.

The grindability of a particular coal is expressed as an index in which the relative ease of grinding the coal is compared to a standard

## TABLE 6.1

### Required Pulverized Coal Fineness at Maximum Rating

(For stationary water-cooled boilers)

| Rank of Fuel * | Per Cent through 200-Mesh Sieve |
|---|---|
| Anthracite | 80–85 |
| Low-volatile bituminous | 75–80 |
| Medium-volatile bituminous | 70–75 |
| High-volatile $A$ and $B$ bituminous | 70–75 |
| High-volatile $C$ bituminous | 65–70 |
| Subbituminous | 60–65 |

* See Table 4.3 for definitions of these ranks.

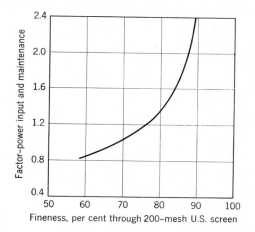

**Fig. 6.16.** Effect of fineness of pulverization on power consumption and mainte-nance.

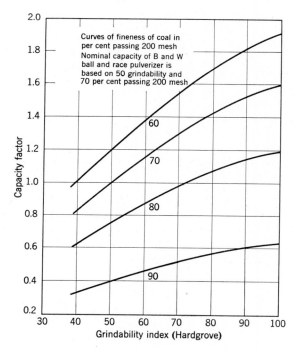

**Fig. 6.17.** Effect of grindability and fineness on capacity. Based on a pulverizer rating of unity for 70 per cent through a 200-mesh sieve and a grindability index of 50.

coal having a grindability index of 100. The lower grindability index number, the easier it is to grind a coal, the greater will be the capacity of a given mill, and the lower will be the cost to grind a ton of coal. Figure 6.17 shows how the capacity of a given pulverizer is affected by the grindability index of coals have indices between 40 and 100 and by the required fineness of grind. The curves are based on rated capacity for a coal having a grindability index of 50 and 70 per cent through a 200-mesh sieve. It will be noted that actual capacity may readily vary from 50 to 150 per cent of rated capacity, depending on the grindability index of the coal and the required fineness.

Although there is a wide variety of pulverized-coal burners, most of them operate on the principle that the stream of hot primary air and coal from the mill is mixed thoroughly with the preheated secondary air at the burner and the mixture is introduced into the hot furnace with a maximum degree of turbulence. A typical high-capacity burner of the cross-tube type is illustrated in Fig. 6.18. The fuel and primary air are delivered to the furnace through a thin, wide, horizontal fan-shaped port. The furnace end of the fuel port is

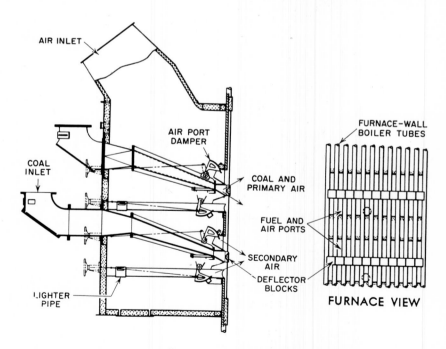

**Fig. 6.18.** Cross-tube pulverized-coal burner.

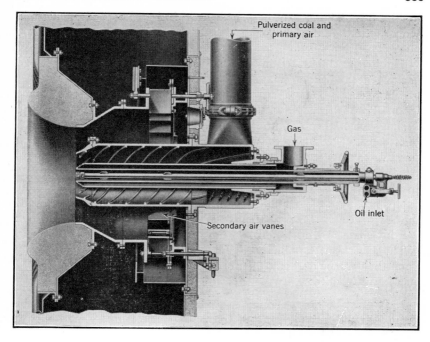

**Fig. 6.19.** Circular multiple-fuel burner.

shielded from the hot furnace by the vertical boiler tubes. Suitable deflectors, mounted in the plane of the fuel-discharge port, protect the tubes from fuel impingement and split the air-fuel stream into two streams, one of which is deflected upward and the other downward. The secondary air enters the burner through adjustable dampers that are located above and below the fuel port. Thus the streams of primary air-fuel and secondary air are made to impinge on each other in such a manner as to promote thorough mixing and turbulence. The direction of the flame can also be controlled by the adjustment of the secondary air dampers so as to distribute the secondary air unequally between the two streams.

A typical burner of the circular type is illustrated in Fig. 6.19. The primary air and fuel are delivered to the furnace horizontally through an annular tube in which spiral ribs impart a whirling motion to the stream. The secondary air is supplied through adjustable vanes that likewise impart a whirling motion. The arrangement for supplying secondary air is similar to that used in the oil burner shown in Fig. 6.5. The stream of primary air-fuel and secondary air mix violently in the burner throat and produce an intense flame in the furnace.

The burner illustrated in Fig. 6.19 is also arranged to burn oil and gas. A mechanical atomizing oil-burner tube is mounted in the center of the burner. This is surrounded by an annular space terminating in small ports through which gas may be supplied. It is possible to design such multifuel burners so that the entire load or any fraction of it can be carried by coal, oil, or gas. This arrangement makes it possible to shift from one fuel or combination of fuels to another, depending on market conditions. Also, there are many industrial plants in which varying amounts of liquid and gaseous fuels are available as by-products of the manufacturing process. Such fuels can be burned up to the limit of their availability and the rest of the load can be carried by coal, all under automatic control.

The success of a pulverized-coal installation depends very largely upon the furnace design. The following conditions must be met by a successful furnace:

1. It must be large enough so that the particles of carbon can be burned in the furnace with 15 to 20 per cent of excess air and less than 2 per cent loss of heating value in the unburned carbon.

2. It must be of such shape and size that the walls are not subjected to intensive flame impingement.

3. It must be sufficiently hot to maintain stable ignition and smokeless combustion at the minimum desired rating.

4. It must have sufficient water-cooled surface so that the ash particles leaving the furnace are cooled below a temperature at which they might adhere to convection tube banks.

5. The bottom of the furnace and the burners must be arranged so that the ash which drops out of the gas may be removed either as a dry dust or as a liquid.

The furnace shown in Fig. 6.3 is typical of many large installations in which circular burners are mounted in the front wall of a high, completely water-cooled furnace. The front and rear walls at the bottom of the furnace are shaped to form an ash hopper with a steep slope so that any ash particles that settle out in the furnace will drop into the ashpit as a dry dust. In a furnace of this type, about 85 per cent of the ash will be carried out of the furnace with the products of combustion. Ample space and cooling are provided to solidify the ash particles before they enter the boiler-tube banks. Also, mechanical or electrostatic dust collectors or both must be installed ahead of the chimney to prevent the discharge of excessive quantities of fly-ash into the atmosphere.

For burning coals having low-fusion-temperature ash, a wet-bottom or slag-tap furnace is often used. The burners are set close to a flat completely water-cooled floor in which the floor tubes are armored with cast-iron blocks to protect them from molten ash and to form a liquid-tight floor. The use of turbulent burners set close to the furnace floor results in a hot zone which melts the ash and permits its removal from a slag-tapping hole in a continuous molten stream.

As compared to stokers, the pulverized-coal installation is more expensive for the same capacity and requires more power for its operation. On the other hand, it is more efficient because of its flexibility, ease of control, and ability to achieve complete combustion with less excess air. In general, for units having a steaming capacity of 100,000 lb per hr or less, the stoker is preferred. For units of 250,000 lb or more, pulverized coal is generally preferred. Pulverized-coal units have been built to generate over 3,000,000 lb of steam per hr which require burning more than 2 tons of coal per min. It is physically impossible to build stokers for even 20 per cent of such capacity. In the range between 100,000 and 250,000 lb per hr, careful analysis of all factors may be necessary before the more economical installation can be determined.

### 6.10    The Cyclone Furnace

The cyclone furnace has been developed to burn coals having a high content of low-fusing-temperature ash in which the liquid ash has a low viscosity. Over 50 per cent of the bituminous coal and lignite mined in the United States is of this type.

Figure 6.20 is a schematic diagram showing the principles of operation of a cyclone furnace. In effect, the furnace is a horizontally cylindrical, completely water-cooled furnace in which the coal is burned at extremely high rates of combustion. The coal is first crushed in an ordinary coal crusher to pass a $\frac{1}{4}$-in. screen. The coal is fed as needed into the entrance section of the furnace. The primary air, amounting to about 15 per cent of the total air, is supplied tangentially at very high velocity to the entrance section of the burner where it imparts a whirling motion to the coal. The secondary air is also supplied tangentially at high velocity. Furnace tempera-

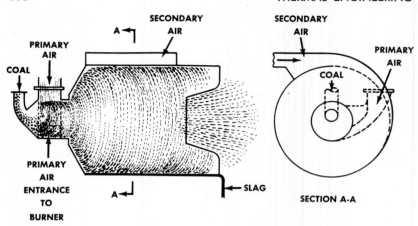

**Fig. 6.20.** Schematic diagram of cyclone furnace.

tures in excess of 3000 F are developed by the high energy-release rate, which exceeds 500,000 Btu per cu ft per hr. As a result, the water-cooled cylindrical wall of the furnace is covered with a layer of liquid slag (molten ash). The incoming fuel particles that are too heavy to burn in suspension are thrown outward by centrifugal force and float on the surface of the molten slag. Here they are scrubbed by the intense whirling action of the air stream and are burned rapidly. The gaseous products of combustion are discharged through the re-entrant water-cooled throat while the slag drains out of a tap hole in a continuous stream. About 85 per cent of the ash in the coal is removed from the cyclone furnace in the liquid state, whereas in the dry-bottom-furnace pulverized-coal installation about 85 per cent of the ash is carried out of the furnace as a fine dust with the products of combustion and must be removed by an ash precipitator.

Figure 6.21 illustrates the actual arrangement of the boiler tubes which make up the cylindrical furnace and its re-entrant outlet. The cyclone furnace is built in diameters of 5 to 9 ft and has a steam-generating capacity of 100,000 to 200,000 lb per hr. Several furnaces are installed under the larger boilers. Figure 8.20 shows a unit having a capacity of 2,100,000 lb of steam per hr which is fired from both front and rear by cyclone furnaces that face each other. The furnaces discharge into a common water-cooled boiler furnace in which combustion is completed and any ash that is carried in suspension in the products of combustion is cooled and frozen before the gases enter the boiler-tube banks. The slag is drained continuously from an opening in the center of the flat floor of the boiler furnace into a slag tank partially filled with water, where the molten slag solidifies in small pieces similar to sand.

The power consumption required for coal preparation is much less for the cyclone furnace than for the pulverized-coal unit, but the fan power required to move the combustion air at the high velocities utilized in the cyclone furnace is so high that the total power consumption for the two systems is about the same. The cyclone furnace is particularly attractive when burning the poorer grades of high-volatile bituminous coal of high ash content and low fusion temperature which may give trouble in pulverized-coal installations because of the "sticky" character of the ash.

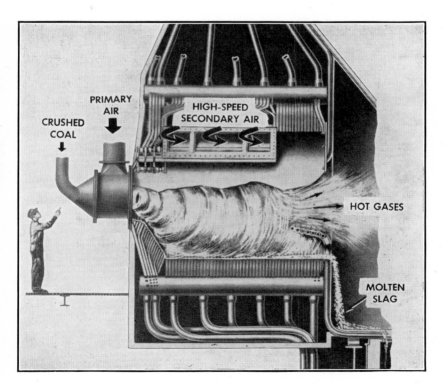

**Fig. 6.21.** Cyclone furnace.

## NUCLEAR REACTORS

### 6.11   Introduction

In keeping with the objectives of this book, the discussion of nuclear reactors will be limited to those types of reactors which at the
present time seem to have the most promise for use in power generation.

A nuclear reactor may be considered as a fuel-burning device in
which a fuel in the form of fissionable material is subjected to neutron bombardment and is reduced to a waste material with a decrease
in mass and the release of large quantities of energy.  A coolant must
be used to remove this energy as rapidly as it is released; otherwise
the temperature of the system will increase until it is destroyed.  The
coolant may be a compressible fluid which is capable of expansion in
a prime mover to perform useful work, or it may be a fluid which is
used in a heat exchanger to produce a second fluid that is capable of
expansion in a prime mover.

### 6.12   Classification of Power Reactors

There are many ways in which power reactors may be classified.
One of the most useful classifications is by the type of coolant as
indicated in Table 6.2.  The coolant may be an inert gas such as
helium or carbon dioxide which may be circulated through the reactor
under a pressure of perhaps 100 psig.  The hot gas may be used to

### TABLE 6.2

#### Classification of Reactors by Coolant

| State of Coolant | Fluid |
| --- | --- |
| Gaseous | He or $CO_2$ under pressure |
| Nonboiling liquid | Pressurized water<br>Liquid sodium<br>Organic liquids |
| Boiling liquid | Water undergoing evaporation |

produce steam in a boiler external to the reactor, or it may be expanded directly in a gas turbine. Unfortunately, gases have poor heat-transfer characteristics as compared to liquids.

If the reactor is liquid cooled, the coolant may be water under a pressure in excess of the saturation pressure at the maximum water temperature in the system. Such a reactor is called a *pressurized-water reactor* (PWR). It is customary to use water at about 2000 psig at a maximum temperature of about 540 F, which is about 100 F below the saturation temperature at 2000 psig. The pressurized water may be circulated through an external heat exchanger or boiler in which the energy that it gives up is used to produce steam at a pressure of 400 to 600 psig. In order to avoid the high pressures that are required to prevent water from boiling in a reactor, liquid sodium may be used as a coolant. In addition to having excellent heat-transfer properties, liquid sodium boils at 1620 F at atmospheric pressure. Unfortunately, liquid sodium becomes highly radioactive in the reactor, and it is necessary to cool it in a closed circuit in an intermediate heat exchanger through which another fluid, usually a sodium-potassium compound, is circulated. The hot NaK solution may be used to generate steam in a steam boiler. Also, sodium will react violently with water and thus creates a serious problem in safety.

Recent developments have shown that it is possible to operate a stable reactor with the coolant in the form of boiling water. The reactor is supplied with high-purity water that is evaporated in the reactor itself, which then becomes a steam boiler. This system of cooling eliminates external heat exchangers, pumps, and piping. Such a reactor is known as a *boiling-water reactor* (BWR).

Reactors may also be classified as to the form of the nuclear fuel into *heterogeneous* or *homogeneous* reactors. In the homogeneous reactor, the fuel is in the form of a liquid such as uranyl sulphate ($UO_2SO_4$) dissolved in water. The solution of nuclear fuel plus coolant is pumped through the reactor and an external heat exchanger in which steam may be generated. In the heterogeneous reactor, the fuel is in the solid form and consists of pellets of $UO_2$ or metallic uranium alloyed with such metals as molybdenum. The fuel is encased in a cladding which confines the fuel and the fission products. The cladding must be constructed from corrosion-resistant materials having low neutron absorption, such as stainless steel or alloys of zirconium. The clad fuel in the form of tubes or bars is assembled into bundles called *fuel assemblies* which are securely mounted in the reactor and are cooled by a suitable coolant that flows through passages in them.

Reactors may also be classified on the basis of operation with *fast* or *thermal* neutrons. If thermal neutrons are used, they may be classified further with regard to the moderator employed. Ordinary water is a good moderator so that, in the pressurized and boiling-water reactors, the coolant is also the moderator. Sodium- and gas-cooled reactors generally use graphite as the moderator.

The *breeder reactor* is a special type of reactor in which the objective is not only to produce energy for power generation but to produce more nuclear fuel than is burned.

### 6.13    The Pressurized-Water Reactor

The pressurized-water reactor (PWR) will be described in some detail because it is currently the most extensively used type of power reactor in the United States, and its construction illustrates the basic principles employed in all the common types of power reactors except the homogeneous reactor. Water is circulated through the reactor under a pressure of about 2000 psig and leaves the reactor at about 540 F, or approximately 100 F below the saturation temperature, so as to insure that serious local boiling does not occur at any place in the reactor core. In order to withstand this pressure and contain the material in the reactor in case of an accident, the reactor parts are housed in a heavy *reactor vessel* as illustrated in Fig. 6.22. The reactor vessel is in the form of a cylinder closed at each end by a hemispherical head. The upper head is bolted to the barrel and can be removed to gain access to the interior of the vessel. The reactor vessel may be about 10 ft in internal diameter, 35 ft high, and from 7 to 9 in. thick, is lined with stainless steel, and weighs about 250 tons. The nuclear fuel is contained in fuel assemblies that are mounted in a *cage* in the center of the reactor vessel. The cage consists of a *core barrel* of cylindrical shape closed at the top and bottom by horizontal *core support plates*.

A typical fuel assembly is illustrated in Fig. 6.23. It has an overall length of 6 to 8 ft, is mounted rigidly in a vertical position between the core support plates, and can be removed vertically from the reactor vessel. The fuel may consist of pellets of $UO_2$ or an alloy of uranium which may be enriched by more than the normal 0.7 per cent of U-235 that is found in natural uranium. The nuclear fuel is encased in stainless steel tubes that are seal-welded at the ends to

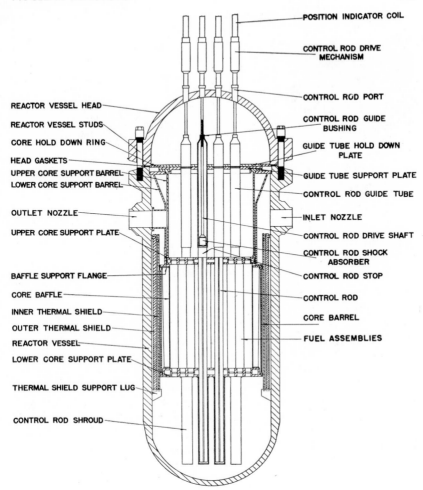

**Fig. 6.22.** Vertical section through a pressurized-water reactor.

produce a corrosion-resistant gas-tight cladding. These tubes are kept in accurate parallel alignment by means of baffle plates mounted in a housing of rectangular cross section that is constructed of some corrosion-resistant alloy such as Zircalloy 2. Since the energy is released in the nuclear fuel within the small tubes, they are so spaced that the cooling water flows around them in quantities adequate to keep them cool and prevent local boiling of the water.

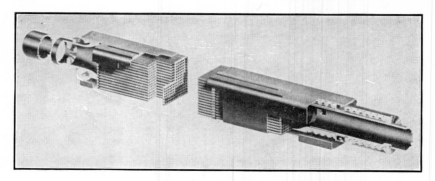

**Fig. 6.23.** Tube type of fuel assembly.

A somewhat different type of fuel element consists of flat bars of a uranium alloy which are clad in a stainless steel, aluminum, or zirconium sheath and are mounted in parallel positions in a rectangular housing in such a manner that water can flow across all surfaces of the bars or plates.

*Thermal shields* are constructed from steel plates in the form of thin-walled cylinders and are supported from the reactor vessel by lugs in such a manner as to be parallel to the reactor vessel wall and to extend for some distance below and above the core support plates. The function of these thermal shields is to absorb gamma rays and neutrons that are not reflected back into the core, thus protecting the walls of the pressure vessel from excessive radioactivity. The thermal shields must be cooled to prevent them from overheating as a result of the energy that is absorbed from the neutrons and gamma rays.

Figure 6.24 is a horizontal section through a typical reactor core. This illustration shows the position of the thermal shields with respect to the wall of the reactor vessel. The core illustrated in Fig. 6.24 consists of 120 fuel assemblies. Each of the fuel assemblies is of the type illustrated in Fig. 6.23 except that on two adjacent sides part of the outer row of tubes has been omitted. This permits forming of the fuel-assembly housing to provide open spaces between fuel assemblies into which *control rods* of cruciform shape may be inserted and withdrawn.

The control rods are made of a material such as boron or cadmium that has a high neutron absorption. Their function was discussed in

Article 4.34.  In general, there are two classes of control rods, known as *regulating* or *shim* rods, and *scram* rods.  The shim rods are moved in and out of the reactor core under close automatic control in such a manner as to control the number of neutrons available to produce fission and thus regulate the rate of energy release.  The safety rods can be inserted fully in a very short period of time in case the shim rods fail to control the rate of reaction.  A sudden insertion of the safety rods is called a scram.

Figure 6.22 illustrates the arrangement of the control rods.  The drive mechanism which automatically controls the position of the rods in the reactor core is located above the reactor vessel.  Control-rod drive shafts extend through control-rod ports in the head of the

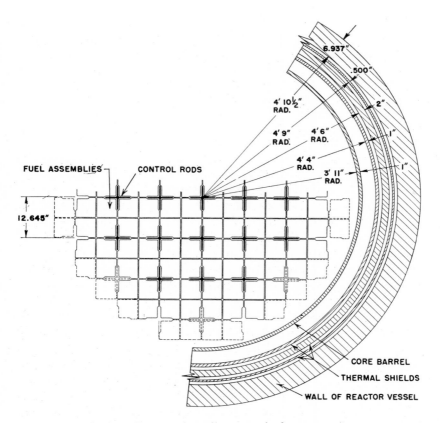

**Fig. 6.24.**  Cross section of a pressurized-water reactor.

reactor vessel and through internal guide tubes and are attached to the control rods by shock absorbers. Below the lower core support plate are control-rod shrouds into which the control rods extend in the "out" position. The control-rod mechanism raises or lowers the shim rods in accordance with the demand for energy and quickly raises the safety rods if a scram becomes necessary.

The coolant, pressurized water, enters the reactor vessel illustrated in Fig. 6.22 through an inlet nozzle located below the flanged head. The water flows downward along the walls of the reactor vessel and through the thermal shields to the bottom of the vessel. It then flows upward through the fuel elements in the reactor core and out the discharge nozzle which is located opposite the inlet nozzle. Baffles prevent the water from flowing directly from the inlet to the outlet nozzle. The heated water that leaves the outlet nozzle is circulated through from one to as many as four external heat-exchanger systems in parallel and back to the reactor inlet nozzle as illustrated schematically in Fig. 8.27.

The nuclear fuel is enclosed in cladding which is designed to retain the radioactive waste products of the nuclear reactions with the exception of radiation. Alpha and beta particles have low penetrating power as discussed in Article 4.28 and do not present a problem. Most of the neutrons will be reflected by the moderator or stopped by the thermal shield. A biological shield is necessary to absorb the gamma radiation and remaining neutron flux. This shield generally takes the form of a deep pit made of thick walls of dense concrete containing iron aggregate in which the reactor vessel is placed. The top of the pit extends well above the top of the reactor. When the head of the reactor is removed for repairs or changing of fuel elements, the pit may be filled with borated water which will absorb the dangerous radiation. Elaborate provisions are required for the removal of spent fuel elements from the reactor vessel and their safe storage and handling, and the insertion of new fuel elements into the reactor vessel.

Since the coolant in the pressurized-water reactor is radioactive, the reactor vessel and all associated piping, pumps, boilers, etc., should be housed in a containment vessel of sufficient size and strength to hold all the material that might be released as a result of any accident.

The fuel that is used in the pressurized-water reactor consists at least partly of enriched U-235. The enriched fuel may be concentrated in fuel assemblies, called *seed assemblies,* which are placed in the form of a hollow square in the reactor core. Assemblies of normal uranium, called *blanket assemblies,* may be placed in the core around the space occupied by the seed assemblies. Part of the neutrons from the seed assemblies will be absorbed by U-238 in the blanket assemblies. The Pu-239 thus produced is fissioned and contributes to the total fuel being burned in the reactor. The fuel may also be thorium enriched with U-235. Some of the neutrons will convert thorium to U-233, which in turn fissions and releases energy. In any case, sufficient neutrons must always be present to maintain the chain reaction for a period of perhaps one year of full-load operation before the burn-up of fuel requires that the fuel elements be replaced.

The pressurized-water reactor is at the present time the most highly developed type of power reactor. The success of the submarines of the U. S. Navy in long periods of continuous underwater operation under the Polar ice cap, and in operation over 100,000 miles without refueling, attests to the reliability of this type of power reactor.

### 6.14 The Boiling-Water Reactor

In internal construction, the general arrangement of the fuel core, thermal shields, and control rods is similar in the pressurized-water and in the boiling-water reactor. A definite water level must be maintained in the boiling-water reactor. Provisions must be made in the upper part of the reactor vessel for the installation of suitable separating devices so that the steam may be separated from the water and delivered to the turbine as relatively dry steam. Internal baffles must be installed so as to provide for rapid and positive circulation of the steam-water mixture through the reactor core. For delivery of steam at the same pressure as in the pressurized-water reactor, the reactor vessel may be designed for an operating pressure of 400 to 600 psig instead of 2000 psig. The external circulating pumps, valves, steam boilers, and piping that are associated with the pressurized-water reactor are eliminated. Thus the boiling-water reactor results in a cheaper and simpler reactor installation. Unfortunately, the steam from the boiling-water reactor is radioactive. This presents a very serious problem in the shielding and maintenance of all the equipment through which the steam flows, including turbines, condensers, extraction heaters, pumps, and piping.

### 6.15    The Sodium-Cooled Graphite-Moderated Reactor

Liquid sodium is an excellent heat-transfer medium, in addition to which its boiling point is so high (1620 F) that the only pressure required is that which is necessary to overcome the frictional resistance in the circuit through which it is being pumped. Consequently, the reactor vessel, heat exchangers, pumps, and piping may be designed for low pressure. Since the coolant is not a moderator, it is necessary to use a separate moderator, usually graphite, in the reactor core. The coolant temperature leaving the reactor is in the range of 800 to 950 F. This permits the generation of steam with some superheat at a pressure higher than that which can be realized with a pressurized-water reactor. The corrosion problems that always exist with high-temperature water are much less serious with liquid sodium. Unfortunately liquid sodium becomes highly radioactive in the reactor. Also, it reacts violently with water. Consequently, a double heat-exchanger system is used as illustrated schematically in Fig. 6.25. The liquid sodium is circulated through the reactor and an intermediate heat exchanger. A second fluid, usually an alloy of sodium and potassium (NaK), is heated in the intermediate heat exchanger and generates steam at about 800 psig and 825 F in the steam boiler.

When compared to the pressurized-water reactor, the sodium-graphite reactor has the advantages of much lower pressures except

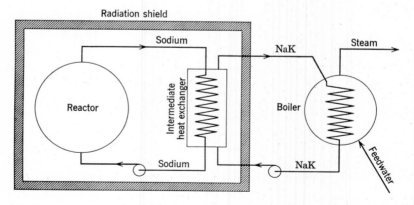

**Fig. 6.25.**  Schematic arrangement of coolant-steam generator circuits of a sodium-cooled reactor.

in the steam boiler, higher temperatures, and steam generation at higher pressures. Its disadvantages are a more complex reactor core because of the graphite moderator, the requirement of an intermediate heat exchanger because of the intense radioactivity of sodium, and the violent chemical reaction between sodium and water.

### 6.16 The Fast-Breeder Reactor

The fast-breeder reactor may have many of the structural characteristics of the pressurized-water reactor except that the coolant is liquid sodium, which is not a moderator, pressures are lower, and the construction of the reactor core is more complicated. The reactor core may be in the form of a right circular cylinder enclosed within a core barrel and flat upper and lower core support plates as in the pressurized-water reactor. In the center of the reactor core is a group of fuel assemblies which occupy the volume of a right circular cylinder about 2.5 ft in diameter and 2.5 ft high. The fuel in this region consists of U-238 highly enriched with U-235. Surrounding this region of highly enriched fuel on all sides, including top and bottom, are fuel assemblies of natural U-238 known as blanket assemblies.

In the central region of highly enriched fuel, the density of the flux of fast neutrons is so great that U-235 undergoes fission rapidly enough to support the chain reaction. In addition to the chain reaction, part of the neutrons will convert U-238 to Pu-239, some of which may in turn undergo fission.

Many of the neutrons that escape from the central core convert U-238 in the blanket assemblies into Pu-239. Some of the neutrons that escape from the blanket assemblies are reflected back into the assemblies by the thermal shield. Successful operation of the breeder reaction depends upon the maintenance of the chain reaction in the central region of highly enriched fuel and the conversion of fertile material in the surrounding blanket assemblies into more fissionable material than is burned in the reactor. When the reactor is shut down for replacement of fuel assemblies, the blanket fuel assemblies are processed to separate and recover the Pu-239, which may be used as a reactor fuel or as material for a nuclear bomb. Th-232 may also be used in the blanket assemblies, in which case fissionable U-233 is produced.

Since, through the breeding process, the relatively abundant fertile materials, Th-232 and U-238, may be converted into more fissionable material than is burned, the breeder reactor is believed to have greater long-range possibilities for development than any other type of power reactor.

### 6.17   The Aqueous Homogeneous Reactor

In the aqueous homogeneous reactor, the fuel may consist of a chemical compound such as uranyl sulphate ($UO_2SO_4$) dissolved in water. Also, enriched uranium in a solid form that has been finely ground may be carried in suspension in water as a slurry. To prevent boiling, the operating pressure is about the same as that used in the pressurized-water reactor. The reactor core is essentially a sphere of sufficient size to contain a mass in excess of the critical mass of the solution. At all other points in the system, the mass must be distributed in a manner to keep it below the critical mass. The fuel solution is circulated through a closed circuit consisting of the reactor core, steam boiler, pump, and piping. The solution is highly radioactive and corrosive. The fuel elements of the heterogeneous reactors and the attendant problems of fabrication, corrosion of fuel elements, heat transfer, and replacement are eliminated. The continuous chemical processing of the irradiated fuel outside the reactor is possible.

The core of the homogeneous reactor may be surrounded by a blanket of Th-232 in which part of the neutrons that escape from the core breed fissionable U-233 in the blanket.

The homogeneous reactor is inherently stable and control rods are unnecessary. The entire system is much simpler than that of the heterogeneous reactors. The homogeneous reactor offers many advantages provided that the difficulties associated with the corrosiveness of the solution can be solved.

### 6.18   Gas-Cooled, Graphite-Moderated Reactor

The British reactor program is being developed around the gas-cooled, graphite-moderated reactor, utilizing natural uranium as the fuel. The fuel is metallic uranium clad in a corrosion-resistant alloy of magnesium in the form of elements 40 in. long and 1⅛ in. in diameter. The Calder Hall reactor core is 35 ft in diameter and 25 ft high and is mounted in a reactor vessel 36 ft in diameter and 65 ft high with a wall thickness of 2 in. The coolant is carbon dioxide ($CO_2$). The large core size is required because of the volume necessary to contain a critical mass of natural uranium in a graphite moderator with gas as a coolant.

The $CO_2$ is circulated at a pressure of about 100 psia. It enters the reactor at 284 F and leaves at 637 F. Because of the low coolant temperature, it is used to generate steam at 210 psia and 63 psia. These low pressures result in a steam plant of relatively poor efficiency. Higher steam pressures would require higher gas temperatures. Unfortunately, the heat-transfer characteristics of a gas such as $CO_2$ are poor is comparison with those of water and liquid sodium. Operation with natural uranium necessitates maximum possible neutron conservation, which prevents the use of fuel cladding material, such as stainless steel, that will stand higher temperatures. Undoubtedly progress will be made in overcoming some of the temperature limitations and reducing the size of the reactor.

### 6.19 United States Power-Reactor Program

Because of the abundance and relatively low cost of fossil fuel in the United States, the various power reactors in operation and under construction cannot produce power as cheaply as plants burning fossil fuels. The Atomic Energy Commission has authorized several groups of companies to design, build, and operate large units of most of the types of reactors that have been discussed in this chapter, often with a government subsidy to partially defray the development costs. It is expected that operation of these plants will show what types of reactors are best suited for economical power generation and will result in important improvements in technology. It is anticipated that the cost of fossil fuels will continue to increase whereas the cost of nuclear fuels and nuclear plants will decrease as a result of technological developments. Perhaps within a period of 5 to 15 years, power can be generated more cheaply from nuclear than fossil fuels, at least in certain parts of the United States.

### PROBLEMS

1. A company wishes to install a steam-generating unit having a maximum continuous output of 150,000 lb of steam per hr. A total of 1100 Btu is absorbed in converting 1 lb of feedwater into steam. The unit is to be designed for an efficiency of 80 per cent at maximum load (efficiency = energy absorbed ÷ heating value of fuel burned). After consideration

of the characteristics of the available coals and the nature of the load, it has been decided that a continuous-ash-discharge spreader stoker should be installed. Experience with this type of installation when fired with the coals available to this plant indicates that the maximum continuous combustion rate on the stoker should be limited to an energy-release rate of 700,000 Btu per sq ft of stoker area per hr (lb of fuel burned per sq ft per hr × heating value of fuel in Btu per lb). Also, experience indicates that, with the fuels available, the maximum furnace energy-release rate should be limited to 26,000 Btu per cu ft per hr. (a) Calculate (1) the grate area, and (2) the furnace volume, required to satisfy these conditions. (b) Calculate the mean vertical distance from the grate to the boiler tubes, assuming a furnace with vertical walls closed at the top by a bank of inclined straight boiler tubes.

**2.** An oil-fired furnace is to be designed for a steam boiler having a maximum continuous output of 75,000 lb of steam per hr. A total of 1150 Btu is absorbed in converting 1 lb of feedwater into steam. Experience indicates that, for the fuels available and the type of furnace construction and service required, the furnace energy-release rate should be limited to 45,000 Btu per cu ft per hr. The unit is to be designed to operate at an efficiency of 78 per cent when burning fuel oil having a gravity of 18 degrees API. The heating value of the fuel is 148,000 Btu per gal. Calculate (a) the gal of oil required per hr, and (b) the furnace volume required in cu ft.

**3.** A pulverized-coal-fired furnace is being designed for a steam-generating unit capable of producing a maximum continuous output of 900,000 lb of steam per hr with an efficiency of 85 per cent. A total of 1150 Btu is absorbed in converting 1 lb of feedwater into steam. The furnace is of completely water-cooled construction, 26 ft long by 22 ft wide with vertical walls. Experience indicates that, with the coals available, trouble-free operation for long periods of time requires that the furnace energy-release rate should not exceed 21,000 Btu per cu ft per hr. The furnace is to be designed to burn coal having a heating value of 11,000 Btu per lb. It has been found by experience that an average of 13 kw-hr of electric energy will be required to pulverize 1 ton of this coal. Calculate (a) the required furnace volume, (b) the required furnace height, and (c) the power input to the coal pulverizer.

**4.** A boiler having a maximum continuous output of 70,000 lb of steam per hr is to be fired by a chain-grate stoker. A total of 1130 Btu is absorbed in converting 1 lb of feedwater into steam. Experience in similar plants burning the same fuels as are available to this plant under comparable conditions has shown that, for trouble-free reliable service, the maximum combustion rate should be limited to 55 lb of coal per sq ft of grate area per hr, and the maximum furnace energy-release rate for the type of furnace under consideration should be limited to 28,000 Btu per cu ft per hr when burning the available coals, which have an average heating value of 10,500 Btu per lb. The unit is to be designed for an efficiency of 78 per cent. Calculate the minimum required grate area and furnace volume.

**5.** A total of 10,000,000 Btu of energy is released per min in a pressurized-water reactor as a result of the fission reactions. Water is supplied to the reactor at 500 F and leaves at 540 F. Assume that the specific heat of the water in this temperature range, $c$, is 1.22 Btu $lb^{-1} F^{-1}$. How many gal of water must be circulated per min through the reactor to maintain a constant temperature in the system?

**6.** A total of 8,000,000 Btu of energy is released per min in a liquid-sodium-cooled reactor. The liquid sodium is supplied to the reactor at 550 F and leaves at 800 F. The mean specific heat of liquid sodium, $c$, is 0.30 Btu $lb^{-1} F^{-1}$. How many lb of sodium must be pumped through the reactor per min?

# CHAPTER 7

# Heat Transfer

## 7.1 Introduction

Boilers, superheaters, economizers, condensers, evaporators, coolers, and heaters are types of equipment that are used to transfer energy from one fluid to another through a metal surface that prevents the fluids from mixing. Since most of this equipment operates at temperatures that are considerably different from room temperature, the equipment and interconnecting piping are insulated to prevent transfer of energy to or from the atmosphere. The design of the amount of heat-transfer surface and its arrangement and the selection of the insulation to be applied to the equipment are based on the laws of heat transfer and economics.

A working knowledge of the science and art of heat transfer can be obtained only by a thorough study of several textbooks, reference works, an extensive periodical literature, and considerable experience in applying the principles to actual practice. The purpose of this brief treatment of the subject is to acquaint the student with some of the basic principles as applied to a limited number of simple cases in order that he may have some appreciation of the major factors that control the design of heat-transfer equipment.

## 7.2 Modes of Heat Transfer

Heat has been defined as energy that is being transferred across the boundaries of a system because of a temperature difference. This transfer may occur through the mechanism of conduction, convection, or radiation, either separately or in combination.

Heat is transferred by *conduction* through a solid, partly as a result of molecular collisions but primarily as a result of a flow of electrons which is induced by a temperature difference. Metals that are good conductors of electricity are also good conductors of heat. Poor conductors (good insulators) are solids that have low density because of the presence of large numbers of small pores or pockets containing air which reduce to a minimum the cross-sectional area of the solid material through which the electrons may flow. Conduction also occurs in liquids and gases at rest, that is, where there is no motion other than the random motion of the molecules. Since the energy is transferred as a result of random molecular collisions, the conductivity of liquids and gases is low as compared to the conductivity of solids.

*Convection* occurs when, either because of a difference in density or because of the operation of a fan or pump, a fluid flows across a hot or cold surface and exchanges energy with that surface. The heated or cooled fluid may then flow to some other region. Since convective heat transmission always involves a flowing fluid, the laws governing heat transfer by convection are closely related to the laws of fluid dynamics.

*Radiation* involves the transfer of energy through space in the form of electromagnetic waves that are different from light waves only in their length (frequency). Since radiant energy travels in straight lines with the velocity of light and may be absorbed, reflected, or transmitted by the receiving surface in a manner similar to the action of light, the laws of optics are important in the study of radiant-energy transfer.

In general, a heat exchanger consists of a metal wall through which heat flows from one fluid to another. Heat transfer through the wall follows the laws of conduction. Heat transfer between the moving fluid and the wall involves convection, in addition to which radiation may be important at high temperatures.

Heat transfer under steady-state conditions only will be considered in this book. Steady-state heat transfer occurs when the temperature at all points in the system remains constant.

### 7.3 Conduction through a Plane Homogeneous Wall

Let Fig. 7.1 represent a section of a plane wall of sufficient area so that the flow of heat is perpendicular to the surface. The basic equation for heat transfer by conduction may be written as follows:

$$q = -kA \frac{dt}{dx} \tag{7.1}$$

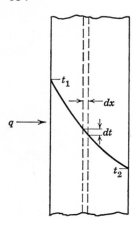

**Fig. 7.1.** Heat transfer by conduction.

where $q$ = heat transferred, Btu per hr (Btu hr$^{-1}$)

$A$ = area through which heat is being transferred, sq ft

$dt$ = differential temperature difference, F

$dx$ = differential thickness measured in the direction of heat flow, ft

$k$ = thermal conductivity of the material of which the wall is composed, Btu hr$^{-1}$ ft$^{-1}$ F$^{-1}$

The negative sign indicates that the flow of heat is in the direction of decreasing temperature.

If it is assumed that the material is homogeneous and the value of $k$ is constant, then the temperature gradient through the wall will be

**Fig. 7.2.** Conduction through a plane wall. $k$ is constant.

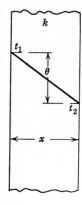

represented by a straight line as shown in Fig. 7.2, and Equation 7.1 reduces to the following form:

$$q = Ak\frac{t_1 - t_2}{x} = Ak\frac{\theta}{x} \qquad (7.2)$$

where $t_1$ and $t_2$ are the surface temperatures of the wall, F

$\qquad \theta = t_1 - t_2$, the temperature difference, F

$\qquad x = $ wall thickness, ft

$\qquad \theta/x = $ temperature gradient, deg F per ft of thickness measured in the direction of heat flow

In general, the coefficient of thermal conductivity, $k$, varies with temperature, and a mean value must be used in Equation 7.2. Typical values of $k$ are given in Table 7.1 for some representative materials. Considerable variation in values will be found in the literature. In metals, small amounts of impurities or alloying elements may produce large differences in conductivity. For insulating materials, density (or porosity) will affect the conductivity markedly.

Serious errors can result from confusion in units when using published values of thermal conductivity. Equation 7.2 may be written as follows:

$$k = \frac{q}{A\dfrac{\theta}{x}}$$

In this book, the foot is used as the basic unit of length. Therefore, $k$ is expressed in the following units:

$$k = \frac{q}{A\dfrac{\theta}{x}} = \frac{\dfrac{\text{Btu}}{\text{hr}}}{\text{ft}^2\,\dfrac{\text{F}}{\text{ft}}} = \frac{\text{Btu}}{\text{hr ft F}} = \text{Btu hr}^{-1}\,\text{ft}^{-1}\,\text{F}^{-1}$$

In many published tables, the values of $k$ are given for a temperature gradient, $\theta/x$, of 1 deg F per in. of thickness instead of 1 deg F per ft of thickness. Such values must be divided by 12 before using Equation 7.2, or the thickness, $x$, in Equation 7.2 must be expressed in

## TABLE 7.1

### Thermal Conductivity of Some Solids, Liquids, and Gases

| Material | Temperature, F | $k$, Btu hr$^{-1}$ ft$^-$ F$^{-1}$ |
|---|---|---|
| Metals | | |
| Aluminum | 212 | 119 |
| Brass (70% Cu, 30% Zn) | 212 | 60 |
| Copper | 212 | 218 |
| Nickel | 212 | 34 |
| Silver | 212 | 238 |
| Steel, mild | 212 | 26 |
| Steel, austenitic | 212 | 8–12 |
| Building materials | | |
| Asbestos sheet | 124 | 0.096 |
| Brick work | 68 | 0.4 |
| Diatomaceous earth (natural) | 400 | 0.05–0.08 |
| Cork, ground | 86 | 0.025 |
| Felt, wool | 86 | 0.03 |
| Mineral wool | 86 | 0.023 |
| 85% magnesia | 200 | 0.036 |
| Liquids | | |
| Water | 100 | 0.363 |
| Water | 300 | 0.395 |
| Petroleum oils | 212 | 0.075–0.08 |
| Mercury | 212 | 5.9 |
| Sodium | 392 | 46.3 |
| Gases | | |
| Air | 212 | 0.0184 |
| Ammonia | 212 | 0.0192 |
| Carbon dioxide | 212 | 0.0128 |
| Hydrogen | 212 | 0.124 |
| Oxygen | 212 | 0.0188 |

Data taken from *Heat Transmission*, 3rd Ed., by William H. McAdams, 1955, by permission of the author.   McGraw-Hill Book Company.

inches.   Confusion and errors will be avoided by consistently using the foot as the unit of length.

**Example 1.**   What will be the rate of heat transfer in Btu per sq ft of area per hr through a flat steel plate ¼ in. thick for a temperature difference of 15 F between the surfaces of the plate?   Assume that, for steel, $k = 26$ Btu hr$^{-1}$ ft$^{-1}$ F$^{-1}$.

*Solution:* From Equation 7.2,

$$\frac{q}{A} = k\frac{\theta}{x}$$

$$\frac{q}{A} = (26 \text{ Btu hr}^{-1} \text{ ft}^{-1} \text{ F}^{-1})\,\frac{15 \text{ F}}{\dfrac{1}{4 \times 12}\, \text{ft}} = 18,700 \text{ Btu hr}^{-1} \text{ ft}^{-2}$$

### 7.4   Conduction through a Plane Composite Wall

Let Fig. 7.3 represent a plane wall composed of three different materials, *a*, *b*, and *c*.   The following assumptions will be made:

1. Each of the materials is homogeneous.
2. The thermal conductivity of each material is constant with respect to temperature.
3. The surfaces of the different materials are in intimate contact

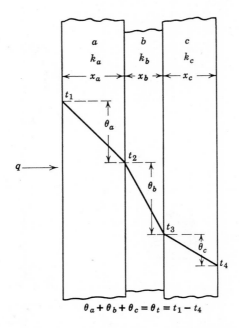

$$\theta_a + \theta_b + \theta_c = \theta_t = t_1 - t_4$$

**Fig. 7.3.**   Conduction through a composite plane wall.

so that no resistance to heat transfer exists at the contact surfaces.

4. The flow of heat is steady and perpendicular to the surface.

For material $a$, Equation 7.2 may be written as follows:

$$q = Ak_a \frac{\theta_a}{x_a} \quad \text{or} \quad \theta_a = \left(\frac{q}{A}\right) \frac{x_a}{k_a} \tag{7.3}$$

For material $b$,

$$q = Ak_b \frac{\theta_b}{x_b} \quad \text{or} \quad \theta_b = \left(\frac{q}{A}\right) \frac{x_b}{k_b} \tag{7.4}$$

For material $c$,

$$q = Ak_c \frac{\theta_c}{x_c} \quad \text{or} \quad \theta_c = \left(\frac{q}{A}\right) \frac{x_c}{k_c} \tag{7.5}$$

In accordance with the assumption that the flow of heat is perpendicular to the plane of each material, the term $q/A$ has the same numerical value in Equations 7.3, 7.4, and 7.5. Also, $\theta_a + \theta_b + \theta_c = t_1 - t_4 = \theta_t$, where $\theta_t$ is the temperature difference across the entire wall. Therefore, adding Equations 7.3, 7.4, and 7.5 results in the following equation:

$$\theta_t = \frac{q}{A}\left(\frac{x_a}{k_a} + \frac{x_b}{k_b} + \frac{x_c}{k_c}\right)$$

and

$$q = \frac{A\theta_t}{\dfrac{x_a}{k_a} + \dfrac{x_b}{k_b} + \dfrac{x_c}{k_c}} \tag{7.6}$$

An example will illustrate the use of Equation 7.6 to calculate the heat transferred through a plane composite wall and the use of Equation 7.2 to determine the temperatures at the interfaces between the various materials in the wall.

**Example 2.** A plane composite wall is composed of three homogeneous materials, $a$, $b$, and $c$, having thicknesses of 8 in., 2 in., and 4 in., respectively, and thermal conductivities, $k$, of 0.40, 0.02, and 0.15 Btu hr$^{-1}$ ft$^{-1}$ F$^{-1}$, respectively. The hot surface is at a temperature of 1000 F and the cold surface at a temperature of 150 F. Calculate the heat-transfer rate in Btu hr$^{-1}$ ft$^{-2}$ and the temperatures at the interfaces between materials $a$ and $b$ and materials $b$ and $c$.

*Solution:* From Equation 7.6,

$$\frac{q}{A} = \frac{\theta_t}{\dfrac{x_a}{k_a} + \dfrac{x_b}{k_b} + \dfrac{x_c}{k_c}}$$

$$= \frac{(1000 - 150)\ \text{F}}{\left(\dfrac{\frac{8}{12}\ \text{ft}}{0.4\ \text{Btu hr}^{-1}\ \text{ft}^{-1}\ \text{F}^{-1}}\right) + \left(\dfrac{\frac{2}{12}\ \text{ft}}{0.02\ \text{Btu hr}^{-1}\ \text{ft}^{-1}\ \text{F}^{-1}}\right)}$$

$$+ \left(\dfrac{\frac{4}{12}\ \text{ft}}{0.15\ \text{Btu hr}^{-1}\ \text{ft}^{-1}\ \text{F}^{-1}}\right)$$

$$= \frac{850\ \text{F}}{1.67\ \text{Btu}^{-1}\ \text{hr ft}^2\ \text{F} + 8.33\ \text{Btu}^{-1}\ \text{hr ft}^2\ \text{F} + 2.22\ \text{Btu}^{-1}\ \text{hr ft}^2\ \text{F}}$$

$$= \frac{850\ \text{F}}{12.22\ \text{Btu}^{-1}\ \text{hr ft}^2\ \text{F}} = 69.5\ \text{Btu hr}^{-1}\ \text{ft}^{-2}$$

To calculate the temperature at the interface between materials $a$ and $b$, apply Equation 7.2 to material $a$ to solve for the temperature difference, $\theta_a$, across this material. Thus

$$\theta_a = \left(\frac{q}{A}\right)\left(\frac{x_a}{k_a}\right) = (69.5\ \text{Btu hr}^{-1}\ \text{ft}^{-2})\left(\frac{\frac{8}{12}\ \text{ft}}{0.4\ \text{Btu hr}^{-1}\ \text{ft}^{-1}\ \text{F}^{-1}}\right) = 116\ \text{F}$$

Then $t_2 = 1000\ \text{F} - 116\ \text{F} = 884\ \text{F}$.

To calculate the temperature at the interface between materials $b$ and $c$, apply Equation 7.2 to material $b$ as follows:

$$\theta_b = \left(\frac{q}{A}\right)\left(\frac{x_b}{k_b}\right) = (69.5\ \text{Btu hr}^{-1}\ \text{ft}^{-2})\left(\frac{\frac{2}{12}\ \text{ft}}{0.02\ \text{Btu hr}^{-1}\ \text{ft}^{-1}\ \text{F}^{-1}}\right) = 578\ \text{F}$$

Then $t_3 = t_2 - \theta_b = 884\ \text{F} - 578\ \text{F} = 306\ \text{F}$.

The solution can be checked by calculating the temperature drop through material $c$ and subtracting this value from $t_3$ to see if the result agrees with the specified cold-surface temperature of 150 F.

## 7.5   Conduction through a Cylindrical Wall

Let Fig. 7.4 represent a cylindrical wall such as a thick-walled tube or a section of insulation on a pipe. Assume that the material is homogeneous with a constant value of $k$ and that the length of the section, $l$, is such that the flow of heat is radial throughout the length under consideration.

Equation 7.1 may be expressed as follows:

$$q = -kA\,\frac{dt}{dx} = -k(2\pi r l)\,\frac{dt}{dr}$$

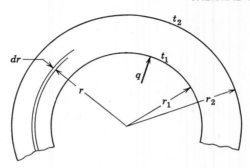

**Fig. 7.4.** Conduction through a cylindrical wall.

Then
$$\frac{q}{2\pi l}\frac{dr}{r} = -k\,dt$$

Integrating and substituting the boundary conditions that at $r_1$ the temperature is $t_1$ and at $r_2$ the temperature is $t_2$ gives the following equation:

$$\frac{q}{2\pi l}(\ln r_2 - \ln r_1) = k(t_1 - t_2)$$

$$q = \frac{k2\pi l(t_1 - t_2)}{\ln \dfrac{r_2}{r_1}} \tag{7.7}$$

### 7.6 Laminar and Turbulent Flow

Heat transfer by convection occurs between a surface and a fluid that is moving over the surface. If the flow is created by the difference in the density of the fluid that results from the temperature change, heat transfer is said to take place by *natural convection*. If a fan or pump is used to create the flow, then the heat transfer is said to take place by *forced convection*. Heat transfer to the atmosphere from the surface of an insulated pipe or heat exchanger involves natural convection. Heat transfer from one fluid to another through a separating wall in a heat exchanger normally involves forced convection.

*Laminar* or *stream-line* flow occurs when a fluid flows across a stationary surface at such low velocities that the flow pattern consists of a large number of thin sheets or laminae of fluid that are sliding

over each other without mixing and with a velocity profile that varies
from zero at the surface to a maximum value at considerable distance
from the surface. Where a fluid flows across a cooler surface, the temperature profile is as illustrated in Fig. 7.5A. The transfer of heat is
by conduction between successive nonmixing layers of fluid.

At higher velocities, there will be an adherent film of relatively
stationary fluid at the metal surface, adjacent to which will be a
layer of fluid moving under laminar conditions. This layer is known
as the laminar layer. Beyond the laminar layer will be a buffer or
transition layer of fluid in which incipient eddies tend to form and the
laminar condition tends to break up. The laminar and buffer layers
are known as the boundary layer. Beyond the buffer zone is the zone
of fully developed turbulent flow which is full of whirls and eddies
that involve velocities perpendicular to the stationary surface. As
shown in Fig. 7.5B, the temperature difference from the bulk of the
moving fluid to the wall is made up of a relatively large temperature
drop through the laminar layer where heat transfer occurs by conduction, followed by another temperature drop through the buffer zone

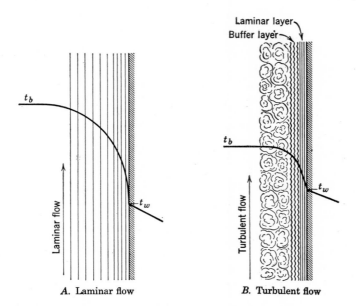

A. Laminar flow          B. Turbulent flow

**Fig. 7.5.** Temperature gradient through surface films for laminar and turbulent
flow. $t_b$ = bulk stream temperatures; $t_w$ = wall temperature.

to the turbulent region where mixing maintains a fairly uniform temperature. In the turbulent region, energy flows in the direction of the wall by the movement of the fluid in whirls and eddies. High velocities reduce the thickness of the boundary layer and the resistance to heat transfer.

It will be noted from Table 7.1 that the thermal conductivity of gases and liquids is low as compared to the thermal conductivity of metals and, in fact, approaches the values for insulating materials. Thus, for most fluids, the boundary layer creates a film of high resistance to the flow of heat.

### 7.7   Heat Transfer by Convection

The basic equation for heat transfer by convection from a moving fluid to a stationary wall may be expressed as follows:

$$q = Ah(t_b - t_w) \tag{7.8}$$

where $t_b$ = bulk or mean temperature of the main stream of fluid
   $t_w$ = wall temperature
   $h$ = coefficient of convective heat transfer, Btu $hr^{-1}$ $ft^{-2}$ $F^{-1}$, often called the film coefficient

The equation which is used to correlate the many variables that influence heat transfer by *forced convection for turbulent flow* is usually written as follows:

$$\left(\frac{hD}{k}\right) = a\left(\frac{DG}{\mu}\right)^m\left(\frac{c_p\mu}{k}\right)^n \tag{7.9}$$

where $h$ = convective heat transfer or film coefficient, Btu $hr^{-1}$ $ft^{-2}$ $F^{-1}$
   $D$ = some characteristic dimension such as the diameter of the tube, ft
   $k$ = thermal conductivity of the fluid, Btu $hr^{-1}$ $ft^{-1}$ $F^{-1}$
   $G$ = mass velocity, $lb_m$ $hr^{-1}$ $ft^{-2}$
   $\mu$ = dynamic viscosity of the fluid, $lb_m$ $hr^{-1}$ $ft^{-1}$
   $c_p$ = specific heat of the fluid, Btu $lb_m^{-1}$ $F^{-1}$
   $a$, $m$, and $n$ are experimentally determined constants

If the dimensions for each of the terms in Equation 7.9 are substituted in the equation, it will be found that each of the three terms appearing in parentheses in Equation 7.9 is dimensionless. This is the reason for grouping the variables in the form of Equation 7.9. An examination of any standard textbook on heat transfer will show that much of the book is devoted to the particular values to use for the constants $a$, $m$, and $n$ in this equation and the temperature at which the physical properties of the fluid should be evaluated. Specific values for $a$, $m$, and $n$ are recommended for individual cases such as flow inside tubes, flow in annuli, flow parallel to tubes, flow across tubes, and flow across flat plates.

In general, the coefficient of convective heat transfer, $h$, is affected by the more important physical variables as follows:

1. The thermal conductivity of the fluid, $k$, determines the temperature difference through the laminar layer. High values of $k$ results in higher rates of heat transfer.

2. The mass velocity, $G$, or the rate of flow in $lb_m$ per sq ft of cross-sectional flow area per hr is determined by the velocity with which the fluid is flowing across the surface and the density of the fluid. Higher mass-flow rates result in thinner films on the metal surface and higher heat-transfer rates.

3. The viscosity of the fluid, $\mu$, is a measure of its resistance to flow and determines the thickness of the boundary layer for a given mass rate of flow. Viscous fluids will have thicker films and lower heat-transfer rates than less viscous fluids if other physical properties are the same.

4. The specific heat determines the amount of energy delivered by a unit mass of fluid per degree of temperature change. Consequently, high specific heats are conducive to somewhat higher rates of heat transfer.

McAdams (1) recommends the following simplified dimensional equations for calculating film coefficients for *forced convection* for the specific cases indicated:

Turbulent flow of diatomic gases inside clean smooth tubes:

$$h = \frac{0.024 c_p G^{0.8}}{D_i^{0.2}} \tag{7.10}$$

Turbulent flow of water inside clean tubes:

$$h = \frac{160(1 + 0.012 t_f) V^{0.8}}{D_i^{0.2}} \tag{7.11}$$

Gas flow normal to a bank of staggered tubes:

$$h = \frac{0.36 c_p G_{max}^{0.6}}{D_o^{0.4}} \tag{7.12}$$

where $G$ = mass velocity, $lb_m\ hr^{-1}\ ft^{-2}$, of cross section of fluid channel

$G_{max}$ = mass velocity through minimum free area normal to fluid stream, $lb_m\ hr^{-1}\ ft^{-2}$

$V$ = average velocity, $ft\ sec^{-1}$

$D_i$ = inside diameter of tube, ft

$D_o$ = outside diameter of tube, ft

$t_f$ = arithmetic mean of wall temperature and bulk fluid temperature, F

When heat is transferred by *natural convection*, the change in density of the fluid with temperature and the geometry of the system are important considerations in addition to the physical properties of the fluid. McAdams (1) recommends the following equations for calculating film coefficients for *natural-convection* heat transfer from a surface to air at normal temperature and pressure for the specific cases indicated:

Horizontal pipes:

$$h = 0.5 \left(\frac{\Delta t}{D'}\right)^{0.25} \tag{7.13}$$

Vertical plates more than 1 ft high:

$$h = 0.27(\Delta t)^{0.25} \tag{7.14}$$

where $\Delta t$ = temperature difference between the surface and surroundings, F

$D'$ = diameter, in.

The literature contains many equations that are applicable to specific situations. The determination of the correct value of the film coefficient, $h$, for a particular application is beyond the scope of this book and, in fact, is a major function of the research laboratory of the equipment manufacturer. The designer, through his arrangement of the spacing of heat-transfer surface, has control over the mass velocity. Other factors being constant, the film coefficient is essentially a function of the 0.8 power of the mass velocity. Doubling the mass velocity will increase the film coefficient by about 75 per cent and reduce the required surface proportionately where the controlling resistance to heat transfer is in the fluid film. On the other hand, doubling the mass velocity will increase by a factor of about four the power required to produce flow. Figure 7.6 illustrates the effect of increased velocity on the cost of power, annual fixed charges, and total annual cost of a heat exchanger to do a specified job. There

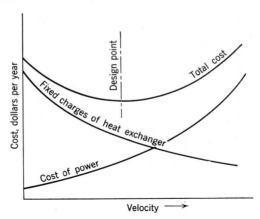

**Fig. 7.6.** Effect of fluid velocity on annual cost of a heat exchanger.

will be some mass velocity, determined by the arrangement of heat-transfer surface, which will result in a minimum total cost, and it is the function of the designer to achieve this economical balance between power cost and cost of surface.

## 7.8  Heat Transfer in Boiling and Condensation

When boiling or condensation occurs on a metal surface, the latent heat of vaporization is involved. Film coefficients are therefore much higher than for the heating or cooling of a fluid without change of phase.

Condensation may be *film type* or *dropwise*. In most commercial heat exchangers, the surface is rough and is covered with a film of condensate that drains from the tube by gravity as the vapor condenses. The tube is therefore partially insulated by a film of liquid. In dropwise condensation, the liquid collects in drops which fall from the tube by gravity and much of the surface is comparatively bare, resulting in higher film coefficients. The range of film coefficients for film-type and dropwise condensation is shown in Table 7.2. Since dropwise condensation occurs mainly on polished surfaces, it is not important in most commercial heat exchangers.

Boiling occurs as *nucleate* or *film boiling* which is analogous to dropwise or film-type condensation. When boiling takes place at a normally rough surface, the vapor bubbles grow in size from points

## TABLE 7.2

### Approximate Range of Film Coefficients

| Fluid | $h$, Btu hr$^{-1}$ ft$^{-2}$ F$^{-1}$ |
|---|---|
| Air, heating or cooling | 0.2–10 |
| Steam, superheating | 5–20 |
| Oils, heating or cooling | 10–300 |
| Water, heating | 50–3000 |
| Organic vapor, condensing | 200–400 |
| Water, boiling | 300–9000 |
| Steam, film-type condensation | 1000–3000 |
| Steam, dropwise condensation | 5000–20,000 |

Data taken from *Heat Transmission*, 3rd Ed., by William H. McAdams. 1955, by permission of the author. McGraw-Hill Book Company.

on the surface, break away, and rise through the liquid. Liquid is actually in contact with most of the surface. If the rate of heat transfer is increased until the surface is entirely surrounded by vapor, then film-type boiling takes place. The vapor film is a fairly good insulator as may be inferred from the $k$ value for vapors in Table 7.1. The effect of this insulating blanket is to cause a rapid increase in the metal temperature since it is no longer in contact with liquid, and a tube burnout may result. The avoidance of film boiling is one of the major considerations in the design of nuclear reactors of the liquid-cooled or boiling-water type.

In general, film coefficients for boiling water and condensing steam are in the range of 1000 to 3000 Btu hr$^{-1}$ ft$^{-2}$ F$^{-1}$ and are many times the film coefficients for gases in heating or cooling.

### 7.9   Effect of Dirt and Scale

The thermal conductivity of scale that is deposited on metal surfaces by ordinary water is about 1.3 Btu hr$^{-1}$ ft$^{-1}$ F$^{-1}$, which compares with an average value of 26 for mild steel and 220 for copper. Dirt from the air, soot and ashes in the products of combustion of fossil fuels, and algae growth in water also produce layers of insulating material on heat-transfer surfaces. Allowances for fouling of surfaces must be made in the design of heat exchangers, and the correct factors to be used are based on experience and judgment.

Transfer of heat through a layer of scale or dirt is by conduction. Since the thickness is not known, it is customary to calculate the heat transferred by means of the following equation:

$$q = Ah_s\theta_s \qquad (7.15)$$

where $h_s$ = coefficient of heat transfer through the layer of scale or dirt, Btu hr$^{-1}$ ft$^{-2}$ F$^{-1}$

$\theta_s$ = temperature drop across the layer of scale or dirt

$h_s$ varies from 200 to 2000 or more, depending upon the thickness and composition of the scale.

## 7.10 Heat Transfer by Convection and Conduction through a Plane Homogeneous Wall

Let Fig. 7.7 represent a plane metal wall through which heat is being transferred from a hot gas in turbulent flow to boiling water.

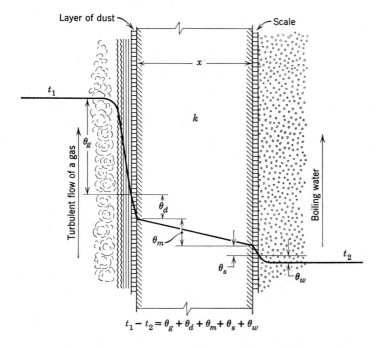

**Fig. 7.7.** Heat transfer from a gas to boiling water by convection and conduction.

Assume that there is a thin layer of scale adhering to the water side of the wall and a layer of dirt on the gas side. The following symbols apply to Fig. 7.7:

$t_1$ = bulk temperature of hot gas, F

$t_2$ = bulk temperature of boiling water, F

$\theta_t$ = total temperature drop, $t_1 - t_2$

$\theta_g$ = temperature drop through the gas film from the bulk temperature of the gas, $t_1$, to the surface temperature of the film of dirt, F

$\theta_d$ = temperature drop through the layer of dirt, F

$\theta_m$ = temperature drop through the metal wall, F

$\theta_s$ = temperature drop through the layer of scale on the water side of the metal wall, F

$\theta_w$ = temperature drop through the water film to the bulk temperature of the boiling water

$h_g$ = film coefficient for heat transfer by convection through the gas film, Btu $hr^{-1}$ $ft^{-2}$ $F^{-1}$

$h_d$ = coefficient for heat transfer through the layer of dirt, Btu $hr^{-1}$ $ft^{-2}$ $F^{-1}$

$k$ = thermal conductivity of the metal in the wall, Btu $hr^{-1}$ $ft^{-1}$ $F^{-1}$

$h_s$ = coefficient of heat transfer through the layer of scale, Btu $hr^{-1}$ $ft^{-2}$ $F^{-1}$

$h_w$ = film coefficient for heat transfer by convection through the water film, Btu $hr^{-1}$ $ft^{-2}$ $F^{-1}$

The same reasoning that was used in developing Equation 7.6 for the flow of heat through a composite plane wall by conduction results in the following equations:

For the gas film:

$$q = Ah_g\theta_g \quad \text{or} \quad \theta_g = \left(\frac{q}{A}\right)\frac{1}{h_g}$$

For the layer of dirt:

$$q = Ah_d\theta_d \quad \text{or} \quad \theta_d = \left(\frac{q}{A}\right)\frac{1}{h_d}$$

For the metal wall:

$$q = A\frac{k}{x}\theta_m \quad \text{or} \quad \theta_m = \left(\frac{q}{A}\right)\frac{x}{k}$$

For the layer of scale:

$$q = Ah_s\theta_s \quad \text{or} \quad \theta_s = \left(\frac{q}{A}\right)\frac{1}{h_s}$$

For the water film:

$$q = Ah_w\theta_w \quad \text{or} \quad \theta_w = \left(\frac{q}{A}\right)\frac{1}{h_w}$$

If it is assumed that the area of the wall is such that the flow of heat takes place normal to the surface, then $q/A$ is the same in each of the above equations, and

$$\theta_g + \theta_d + \theta_m + \theta_s + \theta_w = \theta_t = t_1 - t_2 = \frac{q}{A}\left(\frac{1}{h_g} + \frac{1}{h_d} + \frac{x}{k} + \frac{1}{h_s} + \frac{1}{h_w}\right)$$

Therefore

$$q = \frac{A(t_1 - t_2)}{\dfrac{1}{h_g} + \dfrac{1}{h_d} + \dfrac{x}{k} + \dfrac{1}{h_s} + \dfrac{1}{h_w}} \tag{7.16}$$

Frequently, an overall coefficient of heat transfer is used in which the combined effects of the various films and the metal wall are included in a single term so that Equation 7.16 reduces to the following form:

$$q = AU(t_1 - t_2) = AU\theta \tag{7.17}$$

where $\theta$ = total temperature difference, fluid to fluid

$U$ = an overall coefficient of heat transfer, expressed in Btu $hr^{-1}$ $ft^{-2}$ $F^{-1}$, and

$$U = \frac{1}{\dfrac{1}{h_g} + \dfrac{1}{h_d} + \dfrac{x}{k} + \dfrac{1}{h_s} + \dfrac{1}{h_w}} \tag{7.18}$$

where the various film coefficients, $h_g$, etc., have meanings as defined in developing Equation 7.16.

The calculation of the rate of heat transfer by conduction and convection through a dirty metal wall may be illustrated by Example 3.

**Example 3.** Calculate (a) the rate of heat transfer in Btu $ft^{-2}$ $hr^{-1}$, (b) the temperature profile, and (c) the overall heat-transfer coefficient for heat transfer from a hot gas to boiling water through a dirty metal wall as illustrated in Fig. 7.7 for the following conditions: the plane wall is of steel $\frac{1}{4}$ in. thick which has a thermal conductivity of $k = 26$ Btu $hr^{-1}$ $ft^{-1}$ $F^{-1}$; $t_1 = 800$ F; $t_2 = 400$ F; $h_g = 10$ Btu $hr^{-1}$ $ft^{-2}$ $F^{-1}$; $h_d = 400$ Btu $hr^{-1}$ $ft^{-2}$ $F^{-1}$; $h_s = 1000$ Btu $hr^{-1}$ $ft^{-2}$ $F^{-1}$; and $h_w = 2000$ Btu $hr^{-1}$ $ft^{-2}$ $F^{-1}$.

*Solution: (a)* From Equation 7.16,

$$\frac{q}{A} = \frac{t_1 - t_2}{\dfrac{1}{h_g} + \dfrac{1}{h_d} + \dfrac{x}{k} + \dfrac{1}{h_s} + \dfrac{1}{h_w}}$$

$$= \frac{800 - 400}{\dfrac{1}{10} + \dfrac{1}{400} + \dfrac{\frac{4 \times 12}{26}}{1} + \dfrac{1}{1000} + \dfrac{1}{2000}}$$

$$= \frac{400}{0.100 + 0.0025 + 0.0008 + 0.001 + 0.0005}$$

$$= \frac{400}{0.1048} = 3817 \text{ Btu hr}^{-1} \text{ ft}^{-2}$$

(*b*) To calculate the temperature profile, consider first the gas film.

$$q = A h_g \theta_g \quad \text{or} \quad \theta_g = \left(\frac{q}{A}\right)\frac{1}{h_g}$$

$$\theta_g = (3817)\tfrac{1}{10} = 381.7 \text{ F}$$

For the layer of dirt,

$$q = A h_d \theta_d \quad \text{or} \quad \theta_d = \left(\frac{q}{A}\right)\frac{1}{h_d}$$

$$\theta_d = (3817)\tfrac{1}{400} = 9.5 \text{ F}$$

For the metal wall,

$$q = A\frac{k}{x}\theta_w \quad \text{or} \quad \theta_m = \left(\frac{q}{A}\right)\frac{x}{k}$$

$$\theta_m = (3817)\frac{\frac{1}{4 \times 12}}{26} = 3.1 \text{ F}$$

For the layer of scale,

$$q = A\frac{1}{h_s}\theta_s \quad \text{or} \quad \theta_s = \left(\frac{q}{A}\right)\frac{1}{h_s}$$

$$\theta_s = (3817)\tfrac{1}{1000} = 3.8 \text{ F}$$

For the water film,

$$q = A\frac{1}{h_w}\theta_w \quad \text{or} \quad \theta_w = \left(\frac{q}{A}\right)\frac{1}{h_w}$$

$$\theta_w = (3817)\tfrac{1}{2000} = 1.9 \text{ F}$$

(*c*) From Equation 7.17,

$$U = \frac{q}{A(t_1 - t_2)} = \frac{q}{A}\frac{1}{t_1 - t_2} = \frac{3817}{400} = 9.53 \text{ Btu hr}^{-1} \text{ ft}^{-2} \text{ F}^{-1}$$

The temperature profile in Example 3 was found to be as follows:

Temperature drop through the gas film      = 381.7 F
Temperature drop through the layer of dirt    =   9.5 F
Temperature drop through the steel wall    =   3.1 F
Temperature drop through the layer of scale =   3.8 F
Temperature drop through the water film    =   1.9 F
                      ─────────

Total temperature drop         = 400.0 F

It will be noted that the *controlling factor* in determining the rate of heat transfer is the *boundary layer* of the dry gas which acts as an effective insulator.

Consider next the heat transfer from condensing steam to water as illustrated by Example 4.

**Example 4.** Heat is being transferred through a plane metal wall from condensing steam to water. The wall is $\frac{1}{8}$ in. thick and is made of brass, for which $k = 60$ Btu hr$^{-1}$ ft$^{-1}$ F$^{-1}$. The wall is clean on the steam side. The film coefficient for heat transfer on the steam or vapor side, $h_v$, is 2500 Btu hr$^{-1}$ ft$^{-2}$ F$^{-1}$. The film coefficient for heat transfer on the water side, $h_w$, is 500 Btu hr$^{-1}$ ft$^{-2}$ F$^{-1}$. There is a layer of scale on the water side having a coefficient of heat transfer, $h_s$, = 1000 Btu hr$^{-1}$ ft$^{-2}$ F$^{-1}$. The steam temperature is 100 F and the bulk water temperature is 70 F. Calculate (*a*) the rate of heat transfer in Btu hr$^{-1}$ ft$^{-2}$, (*b*) the temperature profile, and (*c*) the overall coefficient of heat transfer.

*Solution:* (*a*) To calculate the rate of heat transfer, Equation 7.16 may be written for this problem as follows:

$$\frac{q}{A} = \frac{t_1 - t_2}{\dfrac{1}{h_w} + \dfrac{1}{h_s} + \dfrac{x}{k} + \dfrac{1}{h_v}}$$

$$= \frac{100 - 70}{\dfrac{1}{500} + \dfrac{1}{1000} + \dfrac{1}{\dfrac{8 \times 12}{60}} + \dfrac{1}{2500}}$$

$$= \frac{30}{0.0020 + 0.0010 + 0.00017 + 0.0004} = \frac{30}{0.00357} = 8400 \text{ Btu hr}^{-1} \text{ ft}^{-2}$$

(*b*) The temperature drop through the steam film may be calculated by using the equation

$$q = A \frac{\theta_s}{h_v} \quad \text{or} \quad \theta_s = \left(\frac{q}{A}\right)\frac{1}{h_v}$$

$$\theta_s = (8400)\frac{1}{2500} = 3.3 \text{ F}$$

For the metal wall,

$$\theta_m = \left(\frac{q}{A}\right)\frac{x}{k} = 8400\frac{\dfrac{1}{8 \times 12}}{60} = 1.5\ \text{F}$$

For the scale,

$$\theta_s = \left(\frac{q}{A}\right)\frac{1}{h_s} = (8400)\frac{1}{1000} = 8.4\ \text{F}$$

For the water film,

$$\theta_w = \left(\frac{q}{A}\right)\frac{1}{h_w} = (8400)\frac{1}{500} = 16.8\ \text{F}$$

(c) The overall coefficient of heat transfer may be computed from Equation 7.17 as follows:

$$U = \frac{q}{A(t_1 - t_2)} = \frac{q}{A}\frac{1}{t_1 - t_2} = \frac{8400}{30} = 280\ \text{Btu hr}^{-1}\ \text{ft}^{-2}\ \text{F}^{-1}$$

The calculations for the temperature profile through the brass tube from condensing steam to cooling water as determined in Example 4 may be summarized as follows:

Temperature drop across the film on the steam side = 3.3 F
Temperature drop across the metal wall = 1.5 F
Temperature drop across the layer of scale = 8.4 F
Temperature drop across the water film = 16.8 F

Total temperature drop                        30.0 F

The following conclusions may be drawn from an examination of the results of Examples 3 and 4 and from general practice:

1. The controlling factor in determining the heat-transfer rate is the resistance to the flow of heat through the boundary layer of the gas or water.

2. The boundary layer of a dry gas presents much greater resistance to the flow of heat than does the boundary layer of a liquid such as water. This may be verified by noting the range of values for the film coefficients for convective heat transfer of water and gas in Table 7.2.

3. The films created by condensing steam and boiling water offer comparatively little resistance to heat transfer.

4. The resistance of the metal wall is a minor factor in determining the heat-transfer rate. In selecting a metal for use in a heat exchanger, resistance to corrosion or erosion, strength, weight, ease of

### TABLE 7.3

**Approximate Range of Values of the Overall Coefficient of Transfer from One Fluid to Another through a Metal Wall**

| Fluid | $U$, Btu hr$^{-1}$ ft$^{-2}$ F$^{-1}$ |
|---|---|
| Dry gas to dry gas | 1–10 |
| Dry gas to water without phase change | 2–20 |
| Dry gas to boiling water or condensing steam | 2–30 |
| Water to water | 250–500 |
| Water to boiling water or condensing steam | 250–1000 |
| Boiling water or condensing steam to boiling water or condensing steam | 500–1500 |

fabrication, and cost may be more important than thermal conductivity.

5. The presence of layers of scale or dirt may or may not be important and depends upon ability to maintain clean surfaces.

Approximate ranges for the value of the overall coefficient of heat transfer from one fluid to another through a metal wall are listed in Table 7.3 in order of increasing magnitude of the value of $U$. In general, the controlling factor is the film coefficient of the dry gas or water where such exists.

### 7.11 Temperature Difference

The overall equation for the flow of heat through a wall from one fluid to another is Equation 7.17:

$$q = AU\theta \tag{7.17}$$

where $\theta$ is the temperature difference from fluid to fluid.

Let Fig. 7.8 represent a double-pipe heat exchanger with one fluid flowing through the inner pipe and another fluid flowing through the annular space between the two pipes. In the general case, one fluid is being heated and the other is being cooled or one of the fluids is undergoing evaporation or condensation. Perhaps a mean value of the overall coefficient of heat transfer, $U$, may be determined. What

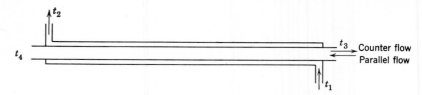

**Fig. 7.8.** Double-pipe heat exchanger.

is the temperature difference that should be used in Equation 7.17? Four special cases are illustrated in Fig. 7.9.

In Fig. 7.9, Case 1 illustrates the temperatures of the fluids in a heat exchanger in which evaporation of a liquid is occuring at constant pressure by heat transfer from a hot gas or liquid that is flowing across the heat-transfer surface. A common illustration would be the generation of steam in a boiler tube as the hot products of combustion flows across the tube.

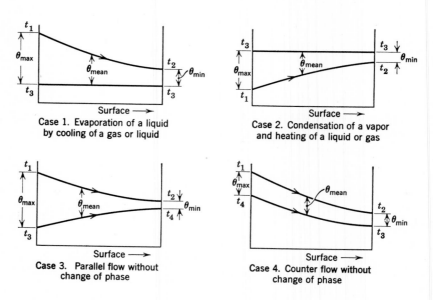

**Fig. 7.9.** Mean temperature difference for various types of heat exchangers.

$$\theta_{mean} = \frac{\theta_{max} - \theta_{min}}{\ln \dfrac{\theta_{max}}{\theta_{min}}}$$

Case 2 represents the temperature distribution in a heat exchanger in which heating of a liquid or gas occurs as it flows across a surface on the other side of which a vapor is condensing at constant pressure. The steam condenser and feed-water heater are common illustrations of this situation.

Case 3 represents the temperature distribution in a heat exchanger in which transfer of heat from one fluid to another occurs without change of phase when the flow takes place in a heat exchanger as illustrated in Fig. 7.8 with both fluids flowing in the *same* direction. This is known as a *parallel-flow* heat exchanger. With an infinite amount of surface, it would be possible to discharge both fluids at the same final temperature.

Case 4 represents the temperature distribution in a heat exchanger in which the fluids are flowing in opposite directions through a heat exchanger as illustrated in Fig. 7.8 without change of phase. This is known as a *counter-flow* heat exchanger.

Any standard textbook on heat transfer will show the limitations and development of the following equation for determining the mean temperature difference to be used in Equation 7.17 for any of the cases illustrated in Fig. 7.9:

$$\theta_m = \frac{\theta_{max} - \theta_{min}}{\ln \dfrac{\theta_{max}}{\theta_{min}}} \tag{7.19}$$

where $\theta_m$ = mean temperature difference

$\theta_{max}$ = temperature difference between the two fluids at that end of the heat exchanger at which the temperature difference is a maximum

$\theta_{min}$ = temperature difference between the two fluids at that end of the heat exchanger at which the temperature difference is a minimum

This temperature difference as calculated from Equation 7.19 is known as the "log mean temperature difference." For complex heat exchangers having mixed or cross flow, this equation must be modified by methods that are discussed in textbooks on heat transfer but are beyond the scope of this chapter.

### 7.12   Heat Transfer by Radiation

Every surface, whether it be the surface of the sun, the incandescent filament in an electric light bulb, a sheet of warm metal, or a billiard ball, emits radiant energy in the form of electromagnetic waves that travel through space in straight lines at 186,000 miles per sec. At low temperatures, the wave lengths of the radiated energy are in the infra-red region; that is, they are too long to affect the human eye. At high temperatures, a portion of the radiated energy is in the form of wave lengths to which the human eye is sensitive, and this radiation is called *light*.

When radiant energy is incident upon a surface, much of it may be transmitted as light is transmitted through a clear pane of window glass, or much of it may be reflected as light is reflected from the surface of a mirror, or much of it may be absorbed as light is absorbed by a dull black surface. The distribution of the total incident radiation is given by the following equation:

$$\alpha + \rho + \tau = 1.0 \qquad\qquad (7.20)$$

where $\alpha$ = the absorptivity, or the fraction of the total incident radiation that is absorbed by the surface

$\rho$ = reflectivity, or the fraction of the total incident radiation that is reflected by the surface

$\tau$ = transmissivity, or the fraction of the total incident radiation that is transmitted through the surface

For some gases like oxygen and nitrogen, $\tau = 1.0$. For most opaque materials such as the walls of heat exchangers, $\tau = 0$; part of the radiant energy is absorbed and part is reflected, or $\alpha + \rho = 1.0$.

When all the incident radiation of all wave lengths is absorbed by a surface, that surface is said to be a *black body*. In other words, for a black body, $\alpha = 1.0$. Consider a well-insulated cavity of uniform temperature with one small opening through which radiant energy is being transmitted to the interior of the cavity. Part of the radiant energy which enters the cavity will be absorbed by the wall opposite the opening, and part will be reflected to other surfaces of the cavity where much of the reflected energy will be absorbed but some will be reflected again, etc. Through the process of successive absorption of much of the reflected energy, all the incident energy will be absorbed. This hollow cavity is therefore a black body. The term "black body" refers to the complete absorption of all incident radiation and not to the appearance of the surface which might actually seem to be bright if at high temperature.

Now assume that the opening in the cavity is closed by an opaque plug. The inner surface of the plug will receive radiant energy from the rest of the surface of the cavity and will absorb part of this energy and reflect the rest. Ultimately it will reach temperature equilibrium with the surface of the cavity. When this equilibrium or steady-state condition has been attained, the inner surface of the plug must emit radiant energy at the same rate that it absorbs radiant energy or the equilibrium conditions will not be maintained.

The *emissive power* of a unit area of surface is the total amount of radiant energy that this surface can emit per unit of time. The *emissivity*, $\epsilon$, of the surface is the ratio of the emissive power of a surface to the emissive power of a black body at the same temperature. It has been pointed out that, for equilibrium conditions in a cavity, the radiant energy absorbed by a unit area of surface must equal the radiant energy emitted by this same area in the same interval of time. Therefore, for *equilibrium conditions, the emissivity of a surface is equal to its absorptivity.* This is known as Kirchhoff's law.

The emissive power of a black body may be calculated from the following equation:

$$I_b = \sigma T^4 \tag{7.21}$$

where $I_b$ = total radiant energy emitted per unit of surface per unit of time

$\sigma$ = the Stephan-Boltzman constant

$T$ = the absolute temperature of the surface

Equation 7.21 may be expressed in English engineering units as follows:

$$q_b = A\sigma T^4 = A(0.174 \times 10^{-8})T^4$$

$$= 0.174A \left(\frac{T}{100}\right)^4 \tag{7.22}$$

where $q_b$ = radiant energy emitted by a black body, Btu hr$^{-1}$

$A$ = area, sq ft

$\sigma$ = $0.0174 \times 10^{-8}$ Btu hr$^{-1}$ ft$^{-2}$ R$^{-4}$, the Stephan-Boltzman constant

$T$ = temperature of the surface, degrees Rankine

### 7.13   Heat Transfer as Radiant Energy between Equal Areas Located in Large Plane Parallel Black-Body Surfaces

Let Fig. 7.10 represent two large plane parallel black-body surfaces, 1 and 2, at absolute temperatures $T_1$ and $T_2$, which are rela-

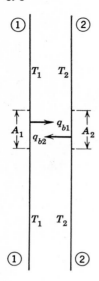

**Fig. 7.10.** Heat transfer by radiation between two equal areas in the center of two large, closely spaced, parallel, flat plates.

tively close together and in which two equal and opposite areas, $A_1$ and $A_2$, are so located that none of the radiant energy emitted by either surface escapes into outer space around the edges of the surfaces. The black-body radiation from area $A_1$ at $T_1$ may be calculated from Equation 7.22 as follows:

$$q_{b1} \rightarrow \; = 0.174A_1 \left(\frac{T_1}{100}\right)^4 \tag{7.23}$$

Similarly, the black-body radiation from area $A_2$ at $T_2$ is as follows:

$$q_{b2} \rightarrow \; = 0.174A_2 \left(\frac{T_2}{100}\right)^4 \tag{7.24}$$

Since $A_1 = A_2$, then the net amount of heat transferred per hour as radiant energy from the hotter surface to the cooler surface is the difference between Equations 7.23 and 7.24, or

$$q_{net} = 0.174A_2 \left[\left(\frac{T_1}{100}\right)^4 - \left(\frac{T_2}{100}\right)^4\right] \tag{7.25}$$

The solid curve in Fig. 7.11 shows the heat transferred in Btu per hour per square foot between two black-body surfaces, one at a constant temperature of 400 F and the other at higher temperatures, as calculated from Equation 7.25. Since the net exchange of radiant energy is a function of the difference of the fourth powers of the absolute temperatures, it will be noted that very high rates of heat transfer occur at surface temperatures above 2000 F. Radiant heat transfer becomes very important, therefore, in all heat-exchanger surfaces that are so located that they can "see" the flame which is produced by burning fuel or, in other words, are located in the walls of furnace cavities.

### 7.14   Effect of Surface Geometry on Radiant-Heat Transfer

Equation 7.25 is valid for black-body radiation under such conditions that all the radiant energy emitted by surface $A_1$ is absorbed by

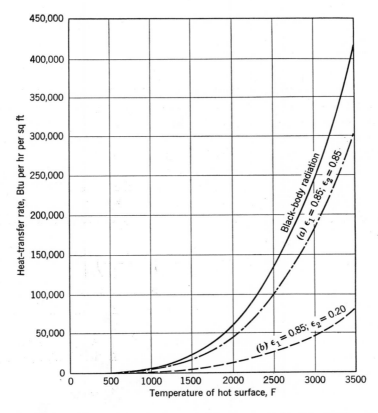

**Fig. 7.11.** Heat transfer by radiation between two large, parallel, closely spaced, flat planes for a constant cold-surface temperature of 400 F.

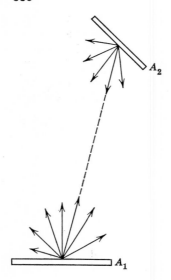

**Fig. 7.12.** Radiant-energy exchange between two surfaces in space.

surface $A_2$ and all the radiant energy emitted by surface $A_2$ is absorbed by surface $A_1$. Now consider two surfaces that are located as shown in Fig. 7.12. Only a part of the radiant energy emitted by surface $A_1$ is absorbed by surface $A_2$, and only a part of the radiant energy emitted by surface $A_2$ is absorbed by surface $A_1$. The net radiant-energy exchange between the two surfaces is dependent, among other factors, upon the geometry of the system. For black-body radiation, Equation 7.25 must be modified by a configuration factor, $F_c$, which is a function of the geometry of the system, as follows:

$$q_{net} = 0.174 A_2 F_c \left[ \left( \frac{T_1}{100} \right)^4 - \left( \frac{T_2}{100} \right)^4 \right] \qquad (7.26)$$

Values for the configuration factor, $F_c$, have been published for many of the simpler geometrical arrangements such as parallel squares, circles, and rectangles, rectangles normal to each other, and rectangular surfaces in furnace cavities composed partly of refractory surfaces. For the parallel planes considered in Article 7.13, $F_c = 1.0$.

### 7.15   Effect of Emissivity on Radiant-Heat Transfer

The surfaces of commercial materials absorb a part of the incident radiation and reflect the remainder, or $\alpha + \rho = 1.0$. The oxidized surface of steel absorbs most of the incident radiation, whereas the highly polished surface of silver reflects most of the incident radiation. The emissivity, $\epsilon$, of a surface is the ratio of the energy radiated by a surface at a given temperature to the energy radiated by a black body at the same temperature. The emissivity of a surface varies with its temperature and roughness and, for metals, the degree of oxidation. Table 7.4 shows the range of emissivity values for representative materials as affected by surface conditions and temperature. The emissivity of most surfaces increases with temperature, but for some surfaces emissivity decreases with increasing temperature.

According to Kirchhoff's law, the emissivity of a surface at a given temperature is equal to the absorptivity of that surface to

### TABLE 7.4

#### Emissivity of Surfaces

| Surface | Temperature Range, F | Emissivity |
|---|---|---|
| A. Metals and their oxides | | |
| Aluminum | | |
| Highly polished | 440–1070 | 0.039–0.057 |
| Oxidized at 1110 F | 390–1110 | 0.11 –0.19 |
| Brass | | |
| Highly polished | 494–710 | 0.033–0.037 |
| Oxidized at 1100 F | 390–1110 | 0.61 –0.59 |
| Steel | | |
| Highly polished pure Fe | 300–1800 | 0.05 –0.37 |
| Well oxidized, smooth | 70–2000 | 0.80 –0.90 |
| Silver, pure, polished | 440–1160 | 0.20 –0.032 |
| Tin, bright | 76 | 0.043–0.064 |
| Zinc-galvanized iron | 80 | 0.23 –0.28 |
| B. Building materials, refractories, paints | | |
| Aluminum paint | 212 | 0.27 –0.67 |
| Asbestos | 100–700 | 0.93 –0.95 |
| White enamel paint, gypsum, lime plaster | 70 | 0.87 –0.91 |
| White paper, glazed porcelain, polished marble, rough red brick, smooth glass, water | 70 | 0.92 –0.96 |

Data taken from *Heat Transmission*, 3rd Ed., by William H. McAdams, 1955, by permission of the author. McGraw-Hill Book Company.

radiation from a source at the same temperature. As the temperature of a source of radiation increases, not only does the amount of radiated energy increase but also the distribution of the radiant energy with respect to wave length changes, with an increasing proportion of the total radiation occurring in the shorter wave lengths. For many engineering applications, the effect of the temperature of the surface upon the distribution of the radiant energy by wave length is neglected. Under such conditions, the surface is said to be a gray body. The absorptivity of a gray body is equal to its emissivity although the temperature of the radiating source is above the temperature of the absorbing surface.

In order to account for the influence of the emissivity of the surfaces upon radiant heat transfer, it is necessary to assume gray-body radiation and to introduce a factor known as the emissivity factor, $F_\epsilon$, which is some function of the emissivities of the surfaces. For the case of the parallel planes discussed in Article 7.13, but modified to consider gray-body surfaces having emissivities less than 1.0, the emissivity factor may be calculated as follows:

$$F_\epsilon = \frac{1}{\dfrac{1}{\epsilon_1} + \dfrac{1}{\epsilon_2} - 1} \tag{7.27}$$

where $\epsilon_1$ and $\epsilon_2$ are the emissivities of the two gray-body surfaces.

Equation 7.25 may then be written as follows for this special case:

$$q_{net} = 0.174 A_2 \left( \frac{1}{\dfrac{1}{\epsilon_1} + \dfrac{1}{\epsilon_2} - 1} \right) \left[ \left( \frac{T_1}{100} \right)^4 - \left( \frac{T_2}{100} \right)^4 \right] \tag{7.28}$$

The dotted lines in Fig. 7.11 show the heat transferred in Btu per hour per square foot between equal gray-body surfaces in the center of large parallel plane surfaces when (a) the emissivity of each surface is 0.85 and (b) the emissivity of one surface is 0.85 and the emissivity of the other surface is 0.20. It is apparent that the emissivities of the surfaces greatly affect the amount of heat transferred as radiant energy.

### 7.16   Heat Transfer by Radiation and Convection

In many applications involving the flow of a high-temperature gas across a heat-transfer surface, energy is transferred to the surface by radiation and also by forced convection. Heat is transferred from the surface of well-insulated equipment to the atmosphere by natural convection and also by radiation. In such cases, it may be necessary to evaluate the energy transferred by radiation and by convection separately and to combine them to obtain the overall result.

## REFERENCES

1. W. H. McAdams, *Heat Transmission,* McGraw-Hill Book Company.
2. Lionel S. Marks, *Mechanical Engineers' Handbook,* McGraw-Hill Book Company.
3. Max Jakob and G. A. Hawkins, *Elements of Heat Transfer,* John Wiley & Sons.
4. D. Q. Kern, *Process Heat Transfer,* McGraw-Hill Book Company.
5. Max Jakob, *Heat Transfer,* John Wiley & Sons.

## PROBLEMS

1. A vertical plane brick wall is 8 in. thick. The temperature difference between the hot and cold surfaces of the brick is 1000 F. What is the heat transmission in Btu $hr^{-1}$ through 40 sq ft of wall surface?

2. What will be the difference in temperature between the hot and the cold surfaces of a flat copper plate $\frac{1}{4}$ in. thick if heat is being transferred through the plate at the rate of 100,000 Btu $ft^{-2}$ $hr^{-1}$?

3. A flat surface is insulated with a 2-in. layer of 85 per cent magnesia. The temperature difference between the hot and cold surfaces of the insulation is 500 F. What is the heat transmission in Btu $hr^{-1}$ through 20 sq ft of insulation?

4. A plane composite wall is composed of 9 in. of fire brick, $4\frac{1}{2}$ in. of insulating brick, and 9 in. of common brick. The thermal conductivities of these types of brick are 0.8, 0.1, and 0.4 Btu $hr^{-1}$ $ft^{-1}$ $F^{-1}$, respectively. The hot surface of the wall is at 2000 F and the cold surface is at 120 F. Determine (a) the heat transmission in Btu $ft^{-2}$ $hr^{-1}$ and (b) the temperatures at the interfaces between the fire brick and insulating brick and the insulating brick and common brick.

5. A plane composite wall consists of a flat mild-steel plate $\frac{1}{4}$ in. thick insulated with a layer of 85 per cent magnesia 3 in. thick. The hot surface of the steel plate is at 600 F and the cold surface of the insulation is at 110 F. Determine (a) the heat transmission in Btu $ft^{-2}$ $hr^{-1}$, and (b) the temperature at the interface between the steel and the insulation.

6. A steel pipe having an outside diameter of 10.75 in. is covered with 3 in. of 85 per cent magnesia. The temperature difference between the hot and cold surfaces of the insulation is 400 F. How much heat is trans-

ferred per hr from the contents of the pipe to the surrounding atmosphere per 100 linear feet of pipe?

**7.** A straight run of steel pipe is 200 ft long and 6.625 in. in diameter. It is covered with 2 in. of 85 per cent magnesia. The temperature difference between the hot and cold surfaces of the insulation is 300 F. How much heat is transferred to the surroundings in Btu $hr^{-1}$?

**8.** A long cylinder having an outside radius $r_1$ is covered by two tightly fitting concentric layers of insulation. The outside radius of the inner layer is $r_2$, and the outside radius of the outer layer is $r_3$. The thermal conductivities of the two layers are $k_a$ and $k_b$ Btu $hr^{-1}$ $ft^{-1}$ $F^{-1}$, respectively. The temperature difference between the inner surface of the inner layer and the outer surface of the outer layer of insulation is $\theta_t$. Prove that the heat transferred per linear foot of the cylinder is given by the following equation:

$$q' = \frac{2\pi\theta_t}{\dfrac{\ln \dfrac{r^2}{r_1}}{k_a} + \dfrac{\ln \dfrac{r^3}{r_2}}{k_b}}$$

**9.** Air is flowing across a hot metal surface which is at 400 F. The bulk temperature of the air is 130 F. The coefficient of heat transfer by convection is 3.8 Btu $hr^{-1}$ $ft^{-2}$ $F^{-1}$. How much heat is transferred in Btu $ft^{-2}$ $hr^{-1}$?

**10.** Air is flowing at constant pressure through a clean steel tube having an internal diameter of 2 in. at a mass velocity of 1500 lb $ft^{-2}$ $hr^{-1}$. Calculate the coefficient of heat transfer by convection in Btu $hr^{-1}$ $ft^{-2}$ $F^{-1}$.

**11.** Water at 100 F flows through a clean tube having an inside diameter of 1 in. at a velocity of 6 ft per $sec^2$. What is the coefficient of heat transfer by convection in Btu $hr^{-1}$ $ft^{-2}$ $F^{-1}$?

**12.** Air is flowing normal to a bank of staggered heating tubes at a mass velocity through the minimum free area normal to the stream of 3000 lb $ft^{-2}$ $hr^{-1}$. The outside diameter of the tube is 2 in. What is the coefficient of heat transfer by convection in Btu $hr^{-1}$ $ft^{-2}$ $F^{-1}$?

**13.** A horizontal uninsulated steam pipe has an outside diameter of 3.5 in. The outside temperature of the pipe is 210 F. The ambient air temperature is 80 F. Calculate the coefficient of heat transfer by natural convection from the pipe in Btu $hr^{-1}$ $ft^{-2}$ $F^{-1}$.

**14.** The outside surface of a vertical wall is at 140 F. The ambient temperature of the air is 70 F. Calculate the coefficient of heat transfer by natural convection in Btu $hr^{-1}$ $ft^{-2}$ $F^{-1}$.

**15.** A clean flat steel plate ¼ in. thick has condensing steam at 300 F on the hot side and water at 100 F on the cold side. The coefficients of heat transfer by convection for the steam and water films are 2000 and 1500 Btu $hr^{-1}$ $ft^{-2}$ $F^{-1}$, respectively. Calculate (a) the rate of heat transfer in Btu $ft^{-2}$ $hr^{-1}$, (b) the temperature profile, and (c) the overall coefficient of heat transfer.

**16.** Heat is being transferred through a flat steel plate ⅜ in. thick from a hot gas at 600 F to water at 250 F. The coefficients of heat transfer by convection for the gas and water films are 8 and 500 Btu hr$^{-1}$ ft$^{-2}$ F$^{-1}$, respectively. The thermal conductivity of the dirt on the gas side of the metal plate is 400 Btu hr$^{-1}$ ft$^{-2}$ F$^{-1}$. The thermal conductivity of the scale on the water side of the tube is 800 Btu hr$^{-1}$ ft$^{-2}$ F$^{-1}$. Calculate (a) the rate of heat transfer in Btu hr$^{-1}$ ft$^{-2}$ F$^{-1}$, (b) the temperature profile, (c) the overall coefficient of heat transfer.

**17.** If the water side of the surface in Problem 15 is cleaned and the layer of scale is removed, what will be (a) the heat-transfer rate, (b) the temperature profile, and (c) the overall coefficient of heat transfer after cleaning of the water-side surface?

**18.** If through an increase in velocity of the gas flowing across the hot surface in Problem 15, the coefficient of heat transfer by convection on the gas side can be increased from 8 to 12 Btu hr$^{-1}$ ft$^{-2}$ F$^{-1}$, what will be the new rate of heat transfer in Btu hr$^{-1}$ ft$^{-2}$ F$^{-1}$?

**19.** If a copper plate is substituted for the steel plate in Problem 15 and all other conditions remain the same, what will be the rate of heat transfer in Btu hr$^{-1}$ ft$^{-2}$ F$^{-1}$? On a percentage basis, how much is the heat-transfer rate increased by substituting copper for steel?

**20.** Water is boiling at a constant temperature of 400 F on one side of the metal surface of a heat exchanger. At 1500 F, 100,000 lb hr$^{-1}$ of gas ($c_p = 0.25$) enter the heat-exchanger surface. The overall coefficient of heat transfer is 12 Btu hr$^{-1}$ ft$^{-2}$ F$^{-1}$. (a) How many sq ft of surface are required to cool the gas to 600 F? (b) How much additional surface is required to cool the gas to 500 F?

**21.** Steam is being condensed at a constant temperature of 100 F in a steam condenser. Cooling water is supplied to the condenser at 70 F and leaves at 85 F, and 1000 Btu are transferred in condensing each lb of steam. The overall coefficient of heat transfer is 400 Btu hr$^{-1}$ ft$^{-2}$ F$^{-1}$. Per hr, 100,000 lb of steam are condensed. (a) How much heat-transfer surface is required in sq ft? (b) How many gal of water are required per min, assuming that the specific heat of the cooling water is 1.0 Btu lb$^{-1}$ F$^{-1}$?

**22.** Water is heated from 200 to 300 F in a counter-flow heat exchanger by means of the hot products of combustion which leave a steam boiler and enter the heat exchanger at 700 F. Assume that the specific heats of the water and gas are 1.0 and 0.25 Btu lb$^{-1}$ F$^{-1}$, respectively. Per hr, 100,000 lb of water are being heated and 120,000 lb of gas are being cooled. The overall coefficient of heat transfer is 10 Btu hr$^{-1}$ ft$^{-2}$ F$^{-1}$. How much heating surface is required in sq ft?

**23.** If a parallel-flow heat exchanger is substituted for the counter-flow heat exchanger in Problem 21 and all other data remain the same, how much heating surface will be required?

**24.** In a counter-flow heat exchanger 150,000 lb of air at 70 F are heated to 300 F by means of the hot products of combustion which leave a boiler at 600 F. The flow of the gas is 175,000 lb per hr. The specific heats of the air and gas are 0.24 and 0.26 Btu lb$^{-1}$ F$^{-1}$, respectively. The overall coefficient of heat transfer is 4.1 Btu hr$^{-1}$ ft$^{-2}$ F$^{-1}$. Calculate the surface required in the heat exchanger in sq ft.

**25.** How much surface will be required if a parallel-flow heat exchanger is substituted for the counter-flow heat exchanger in Problem 23, all other data remaining the same?

**26.** A double-tube heat exchanger is constructed by placing one tube concentrically within a large tube. One fluid flows through the inner tube and the other fluid flows in the opposite direction through the annular space between the tubes. Let $r_i$ and $r_o$ represent the inside and outside radii of the inner tube, $\theta_t$ the bulk temperature difference between the fluids in the heat exchanger, $h_i$ and $h_o$ the coefficients of heat transfer by convection on the inside and outside surfaces of the tube, and $k$ the thermal conductivity of the material of which the inner tube is made. Assume clean surfaces. Show that the heat transfer per linear foot of tube is given by the following equation:

$$q' = \frac{\theta_t}{\dfrac{1}{h_i 2\pi r_i} + \dfrac{\ln \dfrac{r_o}{r_i}}{k 2\pi} + \dfrac{1}{h_o 2\pi r_o}}$$

**27.** Two large plane parallel black-body surfaces are at 2000 and 100 F, respectively. Calculate the net radiant-energy transfer in Btu ft$^{-2}$ hr$^{-1}$.

**28.** Same as Problem 26 except that the hot surface is at 1000 F.

**29.** Same as Problem 26 except that the cold surface is at 400 F.

**30.** For two parallel square surfaces not connected by walls and having a ratio of length of side to distance between planes of 2.0, the configuration factor is 0.408. If the hot surface is at 2000 F and the cold surface is at 500 F, what is the net radiant-energy exchange between surfaces in Btu ft$^{-2}$ hr$^{-1}$, assuming black-body radiation?

**31.** For two parallel square surfaces connected by nonconducting but reradiating walls and having a ratio of length of side to distance between planes of 2.0, the configuration factor is 0.692. If the hot surface is at 2000 F and the cold surface is at 500 F, what is the net radiant-energy exchange between surfaces in Btu ft$^{-2}$ hr$^{-1}$, assuming black-body radiation?

**32.** If the data in Problem 26 are unchanged except that the emissivity of the hot surface is 0.9 and the emissivity of the cold surface is 0.4, calculate the net radiant-energy exchange in Btu ft$^{-2}$ hr$^{-1}$.

# CHAPTER 8

# Steam Generation

## 8.1 Introduction

Steam is used for space heating, in manufacturing processes, and for power generation. Except for hydroelectric power plants, practically all the central-station generating capacity in the United States is in the form of steam turbines. Because of the magnitude of the load and the economies that are effected through the use of the smallest possible number of largest machines, most central-station turbines now being built are in the size range of 100,000 to 600,000 kw. It is standard practice to install one steam-generating unit per turbine. Consequently, these turbines require steam-generating units in the capacity range of 750,000 to over 3,000,000 lb of steam per hr.

The steam boiler is a pressure vessel in which feedwater can be converted into *saturated steam* of high quality at some desired pressure. When other heat-transfer surfaces such as superheater, air heater, or economizer surfaces are combined with boiler surface into a unified installation, the name *steam-generating unit* is applied to the complete unit.

Since the *function of the boiler is to convert water into relatively dry saturated steam,* it must be so constructed that the steam bubbles can be separated from the water effectively. This is normally accomplished in a cylindrical horizontal boiler *drum* in which a definite water level is maintained, either manually or automatically. The drum must have sufficient steam-disengaging surface or separator capacity to allow the steam to escape with the entrainment of very little water in the form of a mist or fog. The boiler must be so designed as to have as much heat-absorbing surface as possible for the amount of steel in the boiler. This surface is usually in the form of tubes which are 2 to 4 in. in outside diameter (OD) with a wall thickness that depends upon the steam pressure at which the unit is designed to operate. The heating surfaces must be so arranged as to provide adequate circulation of the boiler water in order to prevent overheating of the metal at high rates of heat transfer.

Boilers in which the water is inside the tubes are called *water-tube* boilers, whereas boilers that have the hot products of combustion in the tubes and the water outside the tubes are called *fire-tube* boilers. Boiler heating surface is defined as that surface which receives heat from the flame or hot gases and is in contact with water. The area is based on the surface receiving the heat, that is, the outside area of water tubes and the inside area of fire tubes.

Because of the large energy storage in boiler water under pressure and the disastrous results of a sudden release of the pressure in an explosion, the American Society of Mechanical Engineers has developed a boiler construction code which specifies the design rules, methods of construction, and materials to be used. Most states have laws that require all boilers installed within their borders to be constructed in accordance with this code. Within the limits of the code, many arrangements of drums and heating surfaces are possible. Some of the more important types of boilers are discussed in subsequent articles of this chapter. These discussions are limited to power boilers, that is, boilers designed for pressures above 15 psig.

### 8.2   The Horizontal-Return Tubular Boiler

The horizontal-return tubular (HRT) fire-tube boiler is illustrated in Fig. 8.1. The boiler shell is a horizontal cylinder closed at each end by a flat tube sheet or head. The fire tubes, which are usually 3 to 4 in. in diameter, extend through the boiler from one tube sheet to the other and are rolled or expanded into the tube sheets at each end, thus serving not only as flues through which the hot combustion products flow but also as tie rods to hold the flat tube sheets in place against the steam pressure in the boiler. The flat surfaces of the heads above the tubes are braced to the boiler shell by diagonal braces.

The boiler illustrated in Fig. 8.1 is supported by a steel frame, is provided with a brick setting which encloses the furnace, and is fired by a single-retort underfeed stoker. The gaseous products of combustion from the stoker pass over a bridge wall at the rear of the stoker which is intended to promote turbulence, then through the brick furnace under the boiler shell to the rear of the boiler. They then flow through the boiler tubes to the front of the boiler after which they pass a damper and are discharged to a chimney.

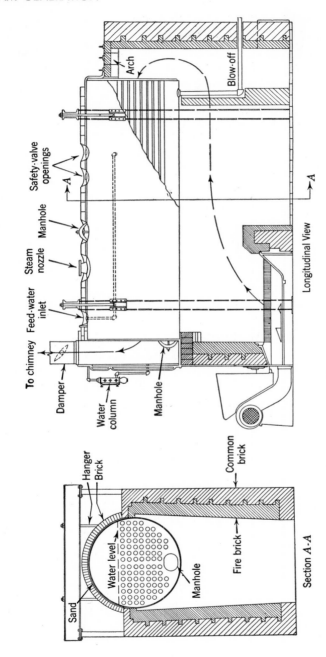

**Fig. 8.1.** Horizontal-return tubular boiler set over a single-retort underfeed stoker.

A water level is maintained a short distance above the top tubes so as to provide adequate surface for the separation of the steam from the water and, at the same time, to keep water in contact with all surfaces across which hot gases are flowing. The water level in the boiler is indicated by a water column which is connected to the boiler by two pipes, one above and one below the water level. The water in the water column is thus maintained at the same level as in the boiler, and this level is indicated by a glass tube attached to the water column.

A blow-off line is connected to the bottom of the drum at the rear. Valves in this line are opened periodically and some of the boiler water is blown down to a sewer, thus carrying out of the system the impurities that are coming into the boiler in the feedwater. It is common practice in these small boilers to add chemicals to the feedwater. These chemicals are intended to prevent the scale-forming impurities in the feedwater from precipitating on the heating surfaces as an adherent scale. If the boiler produces dry steam, all these impurities remain in the boiler. They must be removed by periodic blowdown in order to maintain the concentration in the boiler water below a level that will cause scale formation.

The boiler shell is provided with suitable openings for the attachment of spring loaded safety valves, feed-water inlet, a steam outlet nozzle, and manholes or cleanouts.

Since this boiler is provided with a brick furnace which is external to the boiler itself, it is known as an *externally fired boiler*.

### 8.3   Package Type of Internally Fired Fire-Tube Boiler

The fire-tube boiler illustrated in Fig. 8.2 is typical of the many package-type units that are shipped completely assembled and are ready to operate upon placement on a floor and connection of feedwater and steam lines, fuel lines, and electric wiring. These units are fired by oil or gas and are provided with completely automatic control, including automatic shutdown in the case of low water, failure of the fuel to ignite in the furnace, etc. The justification for this design lies in the reduced labor costs that are possible through automatic controls.

The boiler consists of a horizontal cylindrical shell closed by flat heads and surrounded by a well-insulated metal-clad jacket. The distinctive feature of the boiler is an internal cylindrical furnace which is secured at its ends to the two boiler heads and is corrugated like a road culvert to prevent collapse under the externally applied boiler pressure. An oil or gas burner delivers fuel and air to the front end of this water-cooled furnace. The gases pass through the furnace to a rear insulated compartment from which they flow through fire tubes to the front of the boiler and then to the chimney. A water level is maintained slightly above the top row of fire tubes. The most effective heat-transfer surface is that of the furnace. The boiler is therefore internally fired and very compact.

## 8.4   The Water-Tube Boiler

For steam-generating capacities in excess of about 5000 lb per hr, the water-tube boiler is used. The heat-transfer surface is composed of tubes in which steam is generated. The tubes are normally 2, 3, or $3\frac{1}{4}$ in. OD. They are connected to suitable drums or headers to provide for adequate circulation and steam liberation.

Figure 8.3 illustrates schematically the arrangement of a water-tube boiler. A *steam drum* is provided for the separation of the steam from the water. One or more headers or a drum is placed at the low points in the system. One or more *downcomers* provide for the flow of water from the steam drum to the bottom headers or drum. A large number of boiler tubes, of which only one is illustrated in Fig. 8.3, connect the lower headers or drum with the steam drum and are so arranged that they are located in the furnace walls or in tube banks across which the hot products of combustion flow. Steam is generated in the heated *risers* and is separated from the water in the drum. Circulation occurs because of the difference in the density of the relatively steam-free water in the downcomers and the steam-water mixture in the heated risers.

Let $z$ = vertical distance between the water level in the steam drum and the center line of the lower drum, ft

$\gamma_d$ = mean specific weight of fluid in the downcomers, $lb_f$ per cu ft

$\gamma_r$ = mean specific weight of fluid in the risers, $lb_f$ per cu ft

$\Delta p$ = differential pressure available for creating circulation in the circuit, $lb_f$ per sq ft

Then
$$\Delta p = z(\gamma_d - \gamma_r)$$

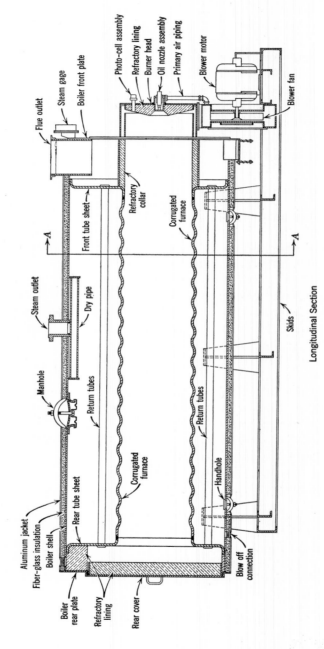

**Fig. 8.2.** Package type of internally fired fire-tube boiler.

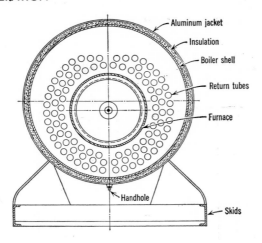

Section $A-A$

**Fig. 8.2 (Continued).** Sectional view.

Since physical properties such as steam-table data are tabulated on a unit mass basis, let

$\rho_d$ = mean density of fluid in the downcomers, $lb_m$ per cu ft
$\rho_r$ = mean density of fluid in the risers, $lb_m$ per cu ft

From Equation 1.10$b$,
$$\gamma = \frac{g}{g_c}\rho$$

Therefore,
$$\Delta p = \frac{g}{g_c}z(\rho_d - \rho_r) \tag{8.1}$$

The vertical distance, $z$, may be from 15 to 150 ft, the higher values being used for the higher pressures and capacities. The downcomers should be supplied with water as free as possible from entrained steam in order that the density may be a maximum. The steam-water mixture being delivered to the drum by the risers will be within the limits of 50 and 80 per cent steam by volume at maximum capacity with the higher values being associated with the lower pressures.

The circulation head is used to overcome fluid friction in the circuits. This fluid friction is a function of the square of the velocity of the fluid in the circuit. There may be from two to several dozen downcomers and several hundred heated risers in a large boiler. The parallel circuits must be designed for approximately equal resistance,

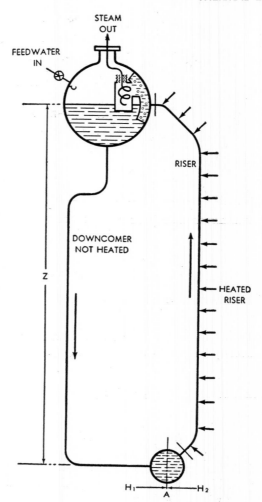

**Fig. 8.3.** Schematic arrangement of drums, downcomers, and heat-transfer surface in a water-tube boiler.

and the total resistance of the circuit must be such that the differential pressure will maintain adequate and stable circulation in all risers at maximum load.

It was pointed out in Chapter 7 that in nucleate boiling the bubbles of vapor break away from the surface as rapidly as they are formed and the surface is continuously wetted by the liquid.  At high differences between the temperature of the metal surface and the liquid, film boiling occurs in which the liquid is continuously separated from the metal surface by a film of vapor.  It was further pointed out that the film conductance of a liquid is many times the film conductance of a vapor or gas.  Conservative boiler design requires that, for proper cooling of boiler tubes, the circuits be so designed as to operate in the range of nucleate boiling in order to maintain a film of water in contact with the metal surface.  It is easier to accomplish this in a vertical tube than in an inclined tube in which the steam tends to separate from the water and flow through the upper part of the tube cross section.  Thus, for vertical tubes, entering velocities of 1 to 5 ft per sec are adequate to maintain nucleate boiling, whereas, for tubes that are nearly horizontal, entering velocities of 5 to 10 ft per sec are necessary.  In general, from 5 to 20 per cent of the mass of water entering a circuit may be evaporated in the circuit, the higher figure applying to the higher pressures where the specific volume of the steam is lower.

Figure 8.4 shows the variation of the density of saturated water and dry saturated steam with pressure.  It will be noted that the difference in the densities decreases with pressure and becomes zero

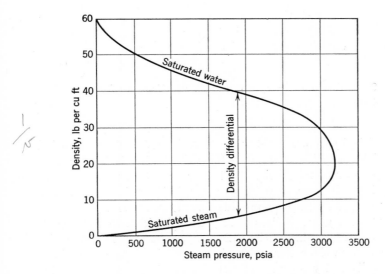

**Fig. 8.4.**  Variation in density of saturated water and saturated steam with pressure.

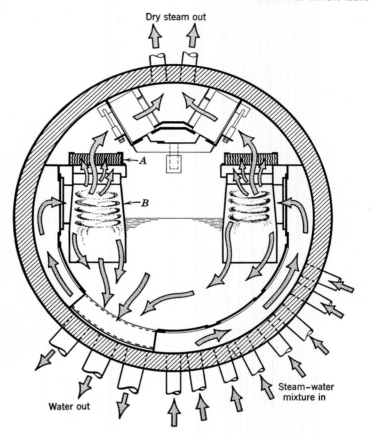

**Fig. 8.5.** Section through boiler drum showing centrifugal separators $B$ and scrubber plates $A$.

at the critical pressure, 3206 psia.  Experience has shown that boilers can be designed with adequate natural circulation for operating pressures up to about 2650 psig.  At higher pressures, the differential in density is inadequate to maintain natural circulation and it becomes necessary to use pumps to provide forced circulation.

Until recent years, boilers were provided with two or three steam drums so as to obtain sufficient disengagement area for separation of the steam from the water.  Effective separation is necessary in order to (1) deliver dry steam and (2) obtain maximum density in the downcomers.  The high cost of boiler drums led to the development of steam separators, and this in turn has made it possible to design even the largest boilers with only one steam drum.

Figure 8.5 is a cross-sectional view of a boiler drum that is provided with centrifiugal separators and scrubber plates. The steam-water mixture from all the heated risers is directed to one of two rows of centrifugal separators by means of baffle plates. The steam-water mixture enters the centrifugal separators horizontally with a high tangential velocity. The water droplets, because of their mass, are thrown outward by centrifugal force and flow downward along the walls of the separator while the steam leaves at the top of the separator. The steam then passes through a battery of corrugated scrubber plates to which the remaining droplets of water adhere. Thus the steam drum delivers dry steam and the downcomers are provided with water of maximum density.

Figure 8.6 is a shop view of a completed boiler drum. The cylindrical part of the drum is usually made by bending two flat plates into the form of half-cylinders, machining the edges, and making longitudinal welds in automatic multiple-pass arc-welding machines to join them into a cylinder. The hemispherical heads are formed hot between dies in a large press. The ends of the drum and the edges of the heads are machined and welded by automatic welders. Forged-steel nozzles are welded to the drums for all piping connections. After all welding operations have been completed, the drum is stress relieved in an annealing oven and every inch of weld is X-rayed for possible defects. A manhole is formed in one or both of the heads during the forming operation. It is closed by a door which opens inward and is held on its seat by yokes and bolts. Boiler drums are

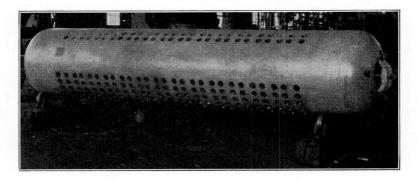

**Fig. 8.6.** Shop view of a drum for a bent-tube boiler.

built in diameters up to 72 in., lengths up to 60 ft, thicknesses up to 10 in., and weights up to 150 tons or more.

Radial tube holes are drilled in the drum for the installation of the boiler tubes, and circumferential grooves are machined in the tube seats. The boiler tubes are bent at the factory to enter the drum radially and are rolled during erection to a pressure-tight seat by an expanding internal roller operated from within the drum.

In the following articles in this chapter, some of the major typical types of water-tube boilers currently being manufactured will be discussed. Because of the large number of arrangements of drums and tubes that are possible, no effort will be made to cover all types. In particular, no mention will be made of the many multi-steam-drum boilers that are still in operation but are not being sold in quantities since the development of the steam separator and the single-steam-drum boiler.

### 8.5   The Two-Drum Water-Tube Boiler

Figure 8.7 illustrates a typical small two-drum water-tube boiler, fired by a spreader stoker equipped with a dump grate. By means of baffles, the gases are forced to follow a path from the furnace to the boiler exit as indicated by the dotted lines. This arrangement of gas flow is known as a "three-pass" design. A water level is maintained slightly below the mid-point in the steam drum. Water circulates from the steam drum to the lower or mud drum through the six rows of tubes in the rear of the boiler-tube bank where the comparatively low gas temperature results in a low heat-transfer rate. Circulation is from the mud drum to the steam drum through the front boiler tubes and the side-wall furnace tubes. The side-wall furnace tubes are supplied with water from the mud drum by means of circulators connected to rectangular water boxes located in the side walls at the level of the grate. Water for the front-wall tubes is supplied to a round front-wall header by downcomer tubes connected to the steam drum and insulated from the furnaces by a row of insulating brick. Most of the steam is generated in the furnace-wall tubes and in the first and second rows of boiler tubes which can "see" the flame in the furnace and absorb energy by radiation.

Boilers of this type have been standardized in a range of sizes capable of generating 8000 to 50,000 lb of steam per hr. The position of the drums and the shape of the tubes result in a compact unit having a well-shaped and economically constructed furnace. By

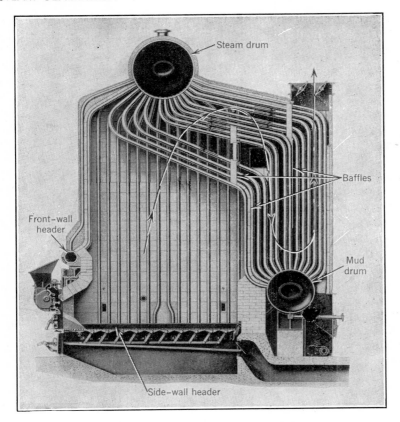

**Fig. 8.7.** Two-drum bent-tube boiler with spreader stoker and water-cooled furnace.

simple changes in the arrangement of furnace-wall tubes, the design can be adopted to almost any kind of firing equipment and fuel.

### 8.6 Superheaters

Superheated steam is produced by causing saturated steam from a boiler to flow through a heated tube or superheater as illustrated in Fig. 3.7, thereby increasing the temperature, enthalpy, the specific volume of the steam. It should be noted that in an actual superheater there will be a decrease in steam pressure due to fluid friction in the superheater tubing.

In Articles 2.12 and 3.15 it was shown that maximum work is obtained when a fluid expands at constant entropy, that is, without friction and without heat transfer to the surroundings. By calculations similar to those illustrated in Example 6 of Chapter 3, it will be found that the constant-entropy expansion of 1 lb of dry saturated steam at 1000 psia to a final pressure of 1.0 psia will result in the conversion into work of 417 Btu, whereas the expansion of superheated steam at the same initial pressure, 1000 psia but *at 1000 F*, to the same final pressure of 1.0 psia will result in the conversion into work of 581 Btu, an increase of 39.3 per cent.

In addition to the theoretical gain in output due to the increased temperature of superheated steam as compared to saturated steam, there are additional advantages to the use of superheated steam in turbines. The first law of thermodynamics states that all the work done by the turbine comes from the energy in the steam flowing through the turbine. Thus, if steam enters the turbine with an enthalpy of 1300 Btu per lb and the work done in the turbine is equivalent to 300 Btu per lb of steam, the enthalpy of the exhaust steam will be $1300 - 300 = 1000$ Btu per lb, neglecting heat transfer to the surroundings. If sufficient energy is converted into work to reduce the quality of the steam below about 88 per cent, serious blade erosion results because of the sand-blasting effect of the droplets of water on the turbine blades. Also, each 1 per cent of moisture in the steam reduces the efficiency of that part of the turbine in which the wet steam is expanding by 1 to $1\frac{1}{2}$ per cent. It is necessary, therefore, that high-efficiency steam turbines be supplied with superheated steam. The minimum recommended steam temperature at the turbine throttle of condensing turbines for various initial steam pressures is as follows:

| Throttle Steam Pressure, psig | Minimum Steam Temperature, F |
|:---:|:---:|
| 400 | 725 |
| 600 | 825 |
| 850 | 900 |
| 1250 | 950 |
| 1450 | 1000 |
| 1800 | 1050 |

Large power plants currently being built in regions of high fuel cost are designed for operation at pressures of more than 1500 psig. At these high pressures, a reduction in the annual fuel cost of 4 to 5 per cent can be made by expanding the steam in the turbine from the initial pressure and 1000 to 1100 F to an intermediate pressure of about 30 per cent of the initial pressure, returning the steam to the steam-generating unit, and passing it through a second superheater, known as a reheater, where it is superheated to 1000 to 1100 F, and then completing the expansion of the steam in the turbine. For initial steam pressures above the critical pressure (3206 psia), a second stage of reheating is employed.

The decreased strength of steel at high temperature makes it necessary to use alloy steels for superheater tubing where steam temperatures exceed 800 F. Alloy steels containing 0.5 per cent of molybdenum and 1 to 5 per cent of chromium are used for the hot end of high-temperature superheaters at steam temperatures up to 1050 F, and austenitic steels such as those containing 18 per cent chromium and 8 per cent nickel are used for higher temperatures.

Superheaters may be classified as convection or radiant superheaters. *Convection* superheaters are those that receive heat by direct contact with the hot products of combustion which flow around the tubes. *Radiant* superheaters are located in furnace walls where they "see" the flame and absorb heat by radiation with a minimum of contact with the hot gases.

A typical superheater of the *convection* type is shown in Fig. 8.8. Saturated steam from the boiler is supplied to the upper or inlet header of the superheater by a single pipe or by a group of circulator tubes. Steam flows at high velocity from the inlet to the outlet header through a large number of parallel tubes or elements of small diameter. Nipples are welded to the headers at the factory, and the tube elements are welded to the nipples in the field, thus protecting the headers from temperature stresses due to uneven heating during

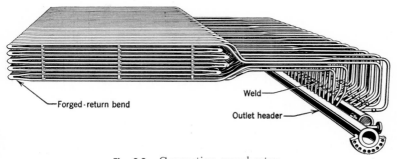

**Fig. 8.8.** Convection superheater.

final welding. In the superheater shown in Fig. 8.8, the individual elements are fabricated from tubes whose ends are connected by forged return bends. When space permits, the forged return bends may be replaced by 180-degree U bends.

The amount of surface required in the superheater depends upon the final temperature to which the steam is to be superheated, the amount of steam to be superheated, the quantity of hot gas flowing around the superheater, and the temperature of the gas. In order to keep the surface to a minimum and thus reduce the cost of the superheater, it should be located where high-temperature gases will flow around the tubes. On the other hand, the products of combus-

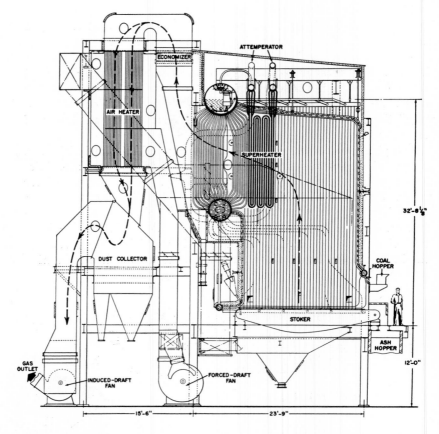

**Fig. 8.9.** Two-drum boiler with superheater, economizer, and air heater, fired by a continuous-ash-discharge spreader stoker.

tion must be cooled sufficiently before they enter the superheater tubes so that any ash that may be present has been cooled to a temperature at which it is no longer sticky or plastic and will not adhere to the superheater tubes. Figure 8.9 illustrates the location of the superheater in a modern two-drum steam-generating unit fired by a continuous-ash-discharge spreader stoker. In this installation, the superheater is located ahead of the boiler convection surface and at the gas exit from the furnace. In installations burning coal having a high content of low-fusing-temperature ash, it may be necessary to place a few boiler tubes ahead of the superheater.

The location of superheaters and reheaters in modern high-pressure units will be discussed further in subsequent articles in this chapter.

### 8.7 Economizers and Air Heaters

It has been pointed out in Article 4.23 that the largest loss that occurs when fuel is burned for steam generation is the so-called "sensible heat" carried away in the hot flue gas. The efficiency of a steam-generating unit provided with good fuel-burning equipment is a function of the flue-gas temperature.

Theoretically, the minimum temperature to which the products of combustion may be cooled is the temperature of the heat-transfer surface with which they are last in contact. In the conventional boiler such as is illustrated in Fig. 8.7, the theoretical minimum flue-gas temperature would be the saturation temperature of the water in the boiler tubes. Figure 8.10 shows the relative amount of boiler heat-transfer surface required to cool the products of combustion from 1500 F to lower temperatures, based on saturated water in the boiler tubes at 1000 psia (544.61 F). It will be noted that, as the temperature difference decreases, each increment of added surface becomes less effective and that the amount of surface required to cool the gases from 700 to 600 F is about 60 per cent of that required to cool the gases from 1500 to 700 F.

In general, it is not economical to install sufficient boiler surface to cool the gases to within less than 150 F of the saturation temperature of the water in the tubes, because sufficient heat cannot be transmitted to the tubes at such low temperature differences to pay for the cost of the boiler surface. The curve labeled "minimum gas temperature from boiler" in Fig. 8.11 is 150 F above the saturation temperature. The minimum gas temperature increases with boiler pressure and is from 600 to 850 F in the range of boiler pressures used in effi-

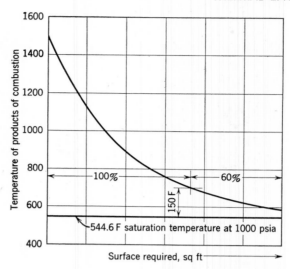

**Fig. 8.10.** Relative surface required to cool products of combustion from 1500 F. Based on 544.6 F water in tubes, constant overall coefficient of heat transfer, and constant specific heat of gases.

cient plants, that is, 400 to 4000 psi. To allow the gases to escape at such high temperatures would be very wasteful.

The minimum allowable flue-gas temperature is 275 to 350 F, depending upon the sulphur content of the fuel, and is the temperature below which corrosion of heat-transfer surfaces will take place. In Fig. 8.11 the desirable flue-gas temperature has been assumed to be 325 F. The gases must be cooled from the boiler exit-gas temperature shown in Fig. 8.11 to the flue-gas temperature required for high efficiency (325 F) by means of heat exchangers supplied with fluids at temperatures less than the saturation temperature at the boiler pressure. This can be done in an *air heater* supplied with the air required for combustion at room temperature or in an *economizer* supplied with boiler feedwater at a temperature considerably below the saturation temperature, or both. In many installations, it is economical to install a small boiler and a large economizer and air heater and to deliver the gases to the economizer at temperatures as high as 900 F rather than to cool the gases to lower temperatures by a larger boiler.

A typical economizer is shown in Fig. 8.12. Feedwater is supplied to the inlet header from which it flows through a number of parallel circuits of 2-in. OD tubes of considerable length to the discharge

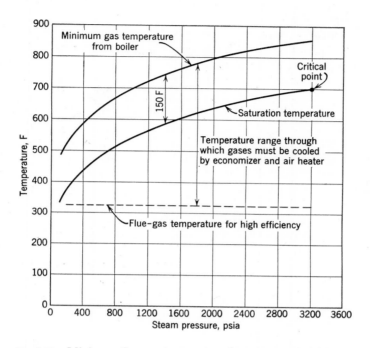

**Fig. 8.11.** Minimum flue-gas temperature from conventional boiler.

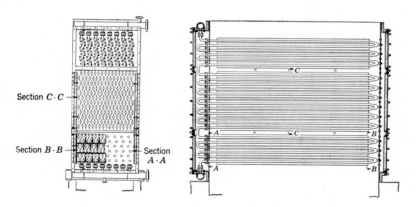

**Fig. 8.12.** Front and longitudinal sections through Elesco fin-tube economizer.

header. If the inlet header is at the bottom so that the water rises as it flows from tube to tube, the hot gas normally enters at the top and flows downward. Thus the coldest gas will be in contact with the coldest tubes, and it is possible to cool the gas to within 125 to 150 F of the temperature of the *inlet water* if sufficient surface is installed. This counter-flow arrangement was discussed in Article 7.11 and illustrated in Fig. 7.9. Since the economizer has water in the tube and a dry gas around the tube, the major resistance to heat transfer is on the gas side. In order to increase the surface exposed to the gas per linear foot of tube and thus increase the effectiveness of the tubular surface, the economizer shown in Fig. 8.12 has fins welded to the top and bottom of each tube. This increases the surface available for heat transfer from the gas without substantially increasing the pressure drop of the gas as it flows across the surface. The gas flows at right angles to the tubes, and the 2-in. finned tubes are staggered to promote effective scrubbing of the outside surface by the gas so as to improve the overall heat-transfer coefficient.

Where scale-free feedwater is available or acid cleaning of heat-transfer surfaces is used to remove scale, the flanged return bends shown on the left side of the longitudinal section in Fig. 8.12 may be eliminated. The flow circuits then consist of continuous welded tubing between inlet and outlet headers.

The tubular air heater, as illustrated in Fig. 8.13, is constructed by expanding vertical tubes into parallel tube sheets which form the top and bottom surfaces, respectively, of the gas inlet and outlet boxes. The tube bank is enclosed in an insulated casing so constructed that the inlet air at room temperature can be admitted to the heating surfaces at the upper end from a fan or blower. The air passes downward around the tubes in a direction opposite to the flow of the hot gases and leaves the air heater at the lower end of the tube bank. Deflecting baffles are installed to guide the air and reduce frictional resistance at the turns. A by-pass damper and baffle permit by-passing the air around the upper half of the tube surface on light loads when there is danger of corrosion due to low flue-gas temperatures. Long tubes closely spaced to maintain high air and gas velocities and countercurrent flow of gases and air make it possible in many installations to cool the gases to a temperature 100 to 200 F below the temperature at which the hot air is discharged.

Another type of air heater which operates on the *regenerative* principle is shown in Fig. 8.14. A drum filled with corrugated-sheet-steel plates is rotated about a vertical shaft at about 3 rpm by means

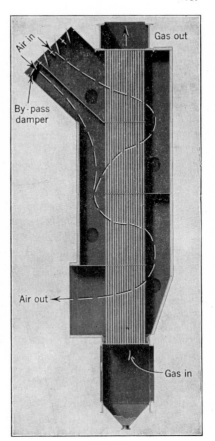

**Fig. 8.13.** Tubular air heater with by-pass damper for light-load operation.

of a small motor. Hot flue gas passes downward through the right side of the rotor from a duct connected to the economizer or boiler. An *induced-draft fan* (not shown in Fig. 8.14) may be connected by a duct to the lower side of the air-heater casing. This fan induces a flow of the gases through the boiler, economizer, and air-heater surfaces, and discharges them to waste up the chimney. The cold air from a *forced-draft* fan flows upward through the left side of the rotor in Fig. 8.14 where the air is heated, after which it is delivered through suitable duct work to the stoker or burner in the furnace. Any point on the corrugated sheet-metal surface of the rotor is rotated alternately into the hot descending gas stream and the cold ascending air stream, thus transferring energy from the hot gas to

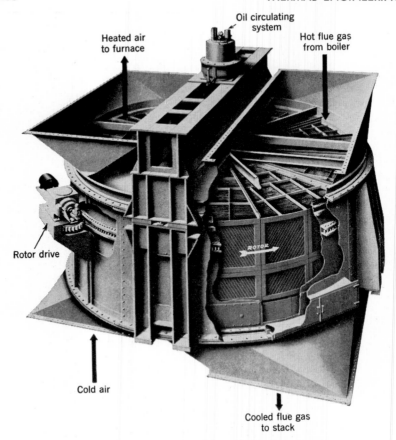

**Fig. 8.14.** Regenerative air heater.

the cold air. Radial seals with rubbing surfaces on them are mounted on the rotor and make contact with a flat section of the casing between the hot-gas and cold-air ducts, thus minimizing leakage between the two streams of fluid. The depth of the rotor is normally between 3 and 4 ft. The unit is also made for operation about a horizontal shaft with horizontal flow of gas and air where building space makes such an arrangement desirable.

The maximum air temperature that can be used in stoker-fired installations without increasing grate maintenance is about 300 F, since the grate surface which supports the hot fuel bed must be cooled by the air to a temperature below which the iron grates will not be damaged. Air temperatures of 600 F are often used with pulverized coal. Since the stoker limits the heat-recovery possibilities of the air heater, both economizers and air heaters are usually installed in stoker-fired high-pressure steam-generating units. Where oil, gas, or pulverized coal is burned, an air heater is often installed without an economizer, although in many high-pressure units it may be more economical to reduce the boiler surface and use an economizer. The air heater is necessary in modern pulverized-coal plants since the coal is dried in the pulverizer by hot air to reduce power consumption and increase the capacity of the mill.

### 8.8   The Steam-Generating Unit

For operation at pressures below the critical pressure, a steam-generating unit consists of a boiler, superheater, air heater, and/or economizer. The furnace walls are either partially or fully covered with boiler tubes. In general, most of the steam is generated in the furnace-wall tubes since they can absorb radiant energy from the high-temperature flame.

Figure 8.9 shows a typical stoker-fired steam-generating unit in the smaller size range, this unit having a capacity of 72,500 lb of steam per hr. The gases as they leave the completely water-cooled furnace pass across the superheater surface, then the convection tubes of the boiler, then upward through a small economizer, downward through a tubular air heater, dust collector, and fan, to the chimney. The boiler is of the two-drum type without gas baffles; that is, it is a single-pass boiler. The internal baffles in the steam drum are so arranged that the last four rows of boiler tubes in which the heat-transfer rate is quite low (see Fig. 8.9) are downcomers. Since a major item in the cost of a boiler is the drums, as many boiler tubes as possible have been placed between the drums. A large amount of surface is required to cool the gases from the temperature at which they leave the superheater to the final temperature.

Figure 8.15 shows the effect of pressure on the distribution of energy absorption between economizer, boiler, and superheater, for a constant final superheat temperature of 1000 F. It is further assumed that, depending upon the steam pressure, the feedwater has been heated in regenerative feed-water heaters to 275 F to over 600 F, depending on pressure, before being admitted to the economizer. Essentially, the economizer raises the feed-water temperature almost

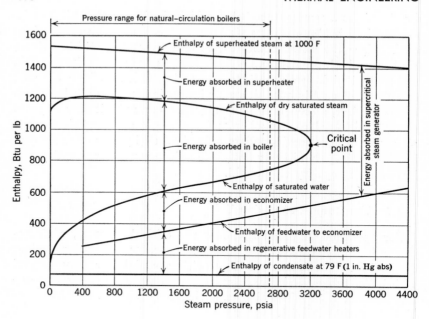

**Fig. 8.15.** Effect of pressure on energy available for absorption in economizer, boiler, and superheater.

to the saturation temperature, the boiler supplies the latent heat, and the superheater supplies the superheat. It will be noted that, as the pressure increases, a decreasing portion of the total energy absorption occurs in the boiler and that, for pressures above the critical, there is no boiler. Supercritical-pressure steam generators essentially are economizers connected to superheaters. There is no steam drum since there is no boiling and no steam to be separated from water at a constant temperature. Figure 8.15 does not take into consideration the use of steam temperatures above 1000 F at the higher pressures or the energy absorbed by reheaters (intermediate-pressure superheaters) which are generally used at pressures above 1500 psia. It may be concluded that, at the higher pressures at which natural-circulation boilers may be used, the boiler becomes a smaller part of the installation and the superheater and reheater become a larger portion of the total heat-transfer surface.

Figure 8.16 illustrates the temperature distribution of the fluids that are flowing in a steam-generating unit. In the water-cooled furnace, combustion and heat transfer occur simultaneously, thus accounting for the shape of the curve. For the superheater, the conditions illustrated would occur in installations such as are shown in Figs. 6.3 and 8.9 where there is no boiler convection surface between the furnace and the superheater. In general, the feed-water temperatures to the economizer are too high to permit cooling the products of combustion to the desired temperature, 275 to 350 F, and an air heater must be used. Since air heaters have a gas film on each side of the metal surface, overall coefficients of heat transfer are much lower than for economizer or boiler surface. However, air-heater surface is not exposed to boiler pressure and therefore is relatively cheap. The actual distribution of surface between the air heater, economizer, and boiler convection surface beyond the superheater depends, therefore, upon many factors and is determined for each installation on the basis of the operating conditions and the smallest total investment necessary to achieve the desired efficiency.

Modern high-capacity steam-generating units have been developed to the point that they can be depended upon to carry heavy loads continuously for months at a time. Their reliability is approximately equal to that of modern steam turbines. Consequently, most new central-station power plants are built on the unit system: that is, with each turbine generator supplied with steam from its own steam-generating unit. Thus, turbine-generator units in capacities up to

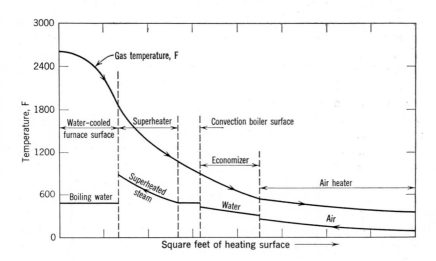

**Fig. 8.16.** Temperature distribution in a steam-generating unit.

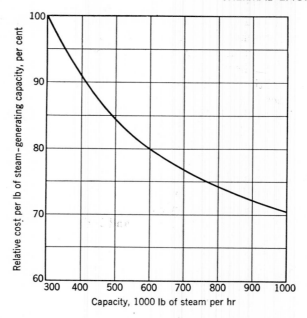

**Fig. 8.17.** Effect of capacity of steam-generating units upon the cost per unit of capacity.

500,000 kw are being supplied with steam from a single steam-generating unit. One of the major reasons for this arrangement is the decreased cost per unit of capacity which results from increased size. Figure 8.17 shows how the cost per pound of steam-generating capacity decreases with increased size for a line of modern steam-generating units which are alike except for size, and which operate at the same steam pressure and temperature, feed-water temperature, and efficiency. The cost per unit of capacity of turbine-generator units likewise decreases with increased size. Moreover, doubling the size of a unit requires only a moderate increase in building space, building and foundation cost, piping cost, erection expenses, and operating labor. The installation of large units has made it possible to build new central stations at an investment not much greater per kilowatt of capacity than was required before World War II and is one of the major reasons why electric energy is still being sold at approximately prewar rates in spite of the greatly increased cost of fuel and labor.

## 8.9   Superheat Control

In modern high-capacity, high-efficiency steam-generating units, superheater outlet steam temperatures of 1050 F are common. At these temperatures, the hot end of the superheater must be made of alloy tubing containing up to 5 per cent chromium. Steam temperatures much in excess of this require the use of austenitic steels of the 18-8 stainless steel type, which are much more expensive than the ferritic alloy tubing. The strength of steel drops substantially with a 50-degree increase in temperature at these temperatures, and the material is subject to creep or plastic deformation. Maximum efficiency requires that the superheater deliver steam at the maximum design temperature over a wide range of load. Long life and trouble-free operation require that the temperature of the steam does not exceed the design temperature, and in this temperature range 25 F above design temperature may lead to serious trouble. Consequently, high-temperature steam-generating units must be provided with means for controlling the final temperature of the superheated steam automatically over a wide range of loads.

The following methods are in extensive use, either singly or in combination, for controlling superheat steam temperature:

1. The final steam temperature from a convection superheater increases with increased load. Radiant superheaters, that is, those located in the furnace walls where they receive energy by direct radiation from the flame, have the characteristic that final superheat temperature decreases with increased load. Advantage can be taken of the opposite characteristics of these types of superheaters by combining them in series. Figure 8.18 shows the manner in which these superheaters may be combined to produce a flat temperature curve.

2. The superheater may be designed with enough surface to deliver steam at the desired temperature at the minimum load, and a heat exchanger can be built into the system to provide cooling to neutralize the excess capacity at heavy loads. Figure 8.19 shows diagrammatically a superheater that is built in two sections with an intermediate header in which coils are installed. Boiler feedwater is pumped through the coils to cool the steam to such a temperature that it leaves the secondary superheater at the desired temperature. Control may be obtained by using a steam by-pass to vary the amount of steam flowing through the heat exchanger. A variation of this method of control is to spray boiler feedwater directly into the

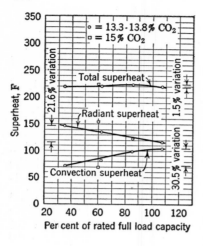

**Fig. 8.18.** Characteristic performance of combination radiant and convection superheater.

intermediate header.   Evaporation of this water will provide the desired cooling effect.

3. When burning pulverized coal, oil, or gas, the burners may be designed so that they can be tilted upward during light loads to confine the flame to the top of the furnace and thereby reduce the cooling effect of the furnace cavity.   At maximum load, the burners are tilted downward to use the entire furnace volume.

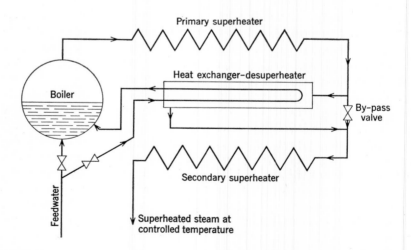

**Fig. 8.19.**  Schematic arrangement of superheaters and desuperheater for control of final steam temperature.

4. Differentially fired divided furnaces may be used in which the furnace is divided by a row of boiler tubes into two furnaces that are fired independently. By unequal arrangement of the superheater surface in the path of the gases from the two furnaces, or by the installation of radiant superheaters in the walls of one furnace, the rate of firing can be controlled to regulate the final steam temperature.

5. By-pass dampers may be installed in such a manner that at light loads all the products of combustion pass across the superheater. As the load increases and the final steam temperature tends to rise, a damper opens and part of the gases by-pass the superheater. Such an installation is shown in Fig. 6.10.

6. Gas recirculation may be employed to increase the superheat temperature at light loads. The superheater is designed to produce the desired temperature at maximum loads. At loads below the maximum, products of combustion from a location ahead of the air heater are recirculated to the furnace by means of a fan. The increased gas flow across the superheater raises the steam temperature.

### 8.10   High-Capacity, High-Efficiency Steam-Generating Units

Such units are currently being designed for capacities from 750,000 to 3,000,000 or more lb of steam per hr at pressures of 1200 to 5000 psia and temperatures of 950 to 1200 F. Because of the quantity of fuel burned, they are designed for efficiencies of 87 to 90 per cent and always include a large air heater. They are fired by pulverized coal or cyclone furnaces, or, where the economics of the situation permit, by gas or oil or a combination of these fuels. Since it is standard practice to install one steam generator per turbine, they are very carefully designed to insure reliable and continuous operation for long periods of time. Depending on boiler-insurance requirements and state laws, they may be operated for two to three years without a major shutdown for cleaning and overhaul.

Figure 6.3 shows a single-drum unit having a capacity of 800,000 lb per hr at 1350 psig and 955 F superheat temperatures. Two large downcomers deliver water from the steam drum to the four headers that supply the furnace-wall tubes in the front, rear, and side walls of the furnace. These furnace tubes deliver their steam-water mixture to the boiler drum. Practically all the steam is generated in the furnace walls. The steam flows from the boiler drum to a heat exchanger that is used for superheat control and then through the counter-flow superheater. It should be noted that the hot end of the superheater is next to the furnace. There are four rows of boiler tubes between the superheater and the economizer. Final cooling of the gases occurs in a regenerative air heater.

Figure 8.20 illustrates a single-drum natural-circulation steam generator rated at 2,100,000 lb of steam per hr, 2700 psig design pressure,

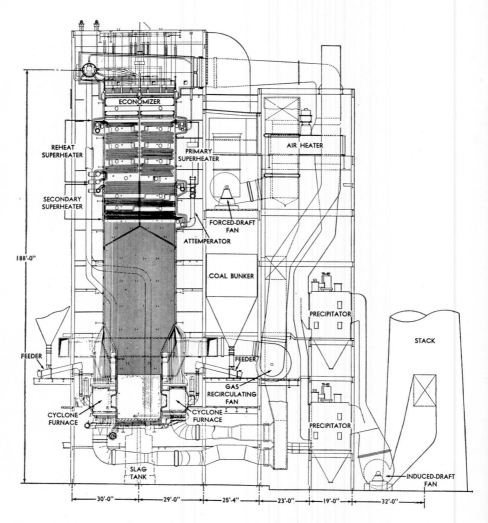

**Fig. 8.20.** Single-drum steam-generating unit fired by cyclone furnaces. 2,100,000 lb per hr, 2700 psig design pressure, 1050 F steam temperature, 1000 F reheat temperature.

1050 F total steam temperature, and 1000 F reheat steam temperature. The single boiler drum is set 188 ft above the ground floor, a distance equal to the height of a fifteen-story office building. Water from the drum flows downward through two large unheated downcomers to distribution headers located below the furnace. The unit is fired by cyclone furnaces which discharge from opposite sides into a common secondary water-cooled furnace. All the steam is generated in the cyclone and secondary furnace-wall tubes.

Some of the front- and rear-wall furnace tubes are brought together to provide a vertical division wall which separates the superheating zone into two parallel passes. Steam from the drum flows downward through the primary superheater in the right pass, then through an attemperator or heat exchanger which provides for superheat temperature control, and finally through the secondary superheater which extends across the top of the entire furnace. The reheat superheater is located in the upper part of the left pass. The large amount of space occupied by these superheaters should be noted. A small economizer is placed above the superheaters. The gases from the economizer then pass downward through a regenerative air heater, precipitators that remove entrained fly ash, and an induced-draft fan, to the base of the chimney. In addition to the heat exchanger between the primary and secondary superheaters, superheat temperature control is provided by a gas-recirculating fan which takes hot gas from ahead of the air heater and introduces it into the boiler furnace above the cyclones.

A considerable number of large steam generators of the *forced-circulation* type have been installed for operation at pressures from 1800 to 2700 psig. A forced-circulation unit is shown diagrammatically in Fig. 8.21. Feedwater is fed through a conventional counter-flow economizer $G$ to a boiler drum. Also, steam from the boiler drum flows through a conventional superheater $H$. Water from the boiler drum flows by gravity to a circulating pump which discharges into a distributing header $D$. Water from the distributing header flows through long small-diameter boiler tubes located in the walls and roof of the furnace to the drum, where the steam is separated and the water returns to the pump. Orfices at the inlet to each circuit at the distributing header correctly proportion the water among the many parallel circuits so that each one receives its proper share. The circulating pump raises the water pressure to about 40 psi above the drum pressure, this being sufficient to overcome the resistance of the flow-controlling orifices and the long circuits of small-

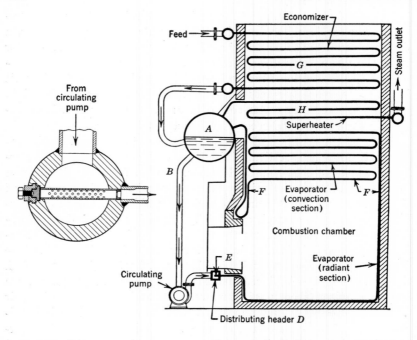

**Fig. 8.21.** Diagrammatic arrangement of forced-circulation steam generator.

diameter tubing. These tubes may be constructed of thinner walls than would be required by the larger tubes that are used in natural-circulation boilers and may be arranged so that the flow is upward, horizontal, downward, or any combination thereof.

In the conventional forced-circulation boiler, the amount of water circulated is four to five times the amount of steam generated and an effective steam-separating drum is as essential as in the natural-circulation boiler. Recent developments in feed-water treatment have resulted in feedwater of high purity. This has made it possible to build steam-generating units in which the large horizontal steam drum has been eliminated. Figure 8.22 shows a so-called "monotube" unit which is designed for a capacity of 1,150,000 lb of steam per hr at 2610 psi, 1060 F superheat temperature, and 1060 F reheat temperature. The conventional steam drum has been replaced by a vertical water separator. Water is pumped through small-diameter tubes in the furnace walls and is converted into steam of high quality which is discharged into the water separator. Here the small amount of un-

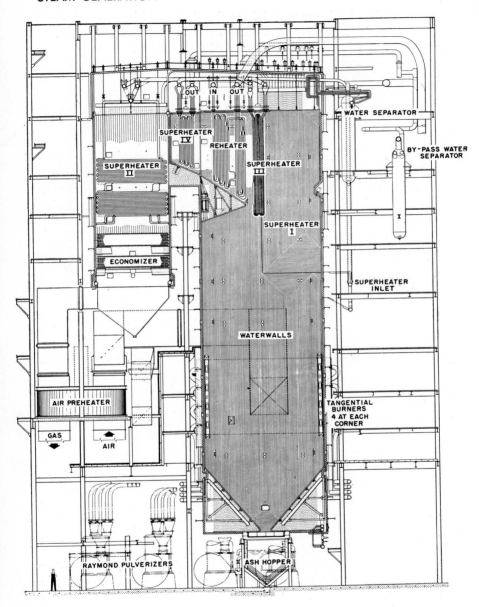

**Fig. 8.22.** "Monotube" steam-generating unit fired by pulverized coal. 1,150,000 lb per hr at 2610 psi and 1060 F. Reheat to 1060 F.

evaporated water is separated from the steam and is blown down to a lower pressure, carrying out with it any impurities that have been concentrated in the water as a result of evaporation. The dry steam from the separator then passes through four sections of superheater tubing, designated as superheater I, II, III, and IV, to the turbine. The steam is resuperheated in the reheater to the initial temperature at a pressure of about 30 per cent of the initial pressure. An economizer and air heater are provided to cool the products of combustion to the low temperature necessary for high efficiency.

For operation at pressures above the critical pressure, 3206 psia, water does not boil (see Fig. 8.15). No boiler drum is required, and the steam generator becomes essentially a continuous circuit of seamless steel tubing with intermediate headers of small diameter. Such a unit is known as a "once-through" steam generator. It is illustrated in Fig. 8.23, which shows a unit having a capacity of 2,900,000 lb per hr, a design pressure of 4500 psi, a superheater outlet pressure of 3625 psi, a high-pressure superheated steam temperature of 1050 F, and two stages of reheat, each to 1050 F.

The unit is fired by eight 10-ft-diameter cyclone furnaces arranged to discharge from opposite sides into a common secondary furnace. The gases flow upward through the secondary furnace and through three parallel vertical passes to three air heaters. The walls of the three parallel vertical passes are constructed of closely spaced steam-generating tubes. The primary and secondary superheater, the first and second reheaters, and the economizer are located in these passes.

The feedwater flows from the economizer section to the cyclones through outside downcomers. From the cyclones, the fluid flows upward through the secondary furnace-wall tubes and the convection section baffle-wall tubes which form the walls of the three parallel gas passes above the furnace. From the baffle walls, the fluid flows through the primary superheater, a heat exchanger or attemperator that is used for superheat control, and then through the secondary superheater to the superheater outlet and turbine. The transition from water to steam occurs in the upper part of the furnace enclosure.

After expansion in the turbine to an intermediate pressure, the steam is reheated in the first-stage reheater to 1050 F. After further expansion in the turbine, it is reheated in the second-stage reheater to 1050 F. It should be noted that the superheaters and reheaters occupy a major part of the total volume of the installation.

Final superheat and reheat temperatures are controlled by (a) a heat exchanger between the primary and secondary superheater, (b)

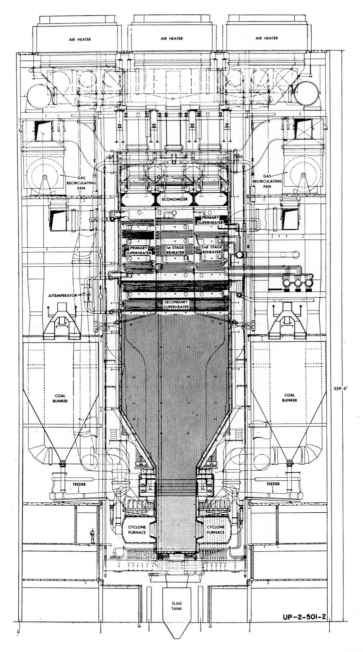

**Fig. 8.23.** Once-through steam-generating unit fired by cyclone furnaces. 2,900,000 lb per hr at 3625 psig and 1050 F with two stages of reheat to 1050 F.

421

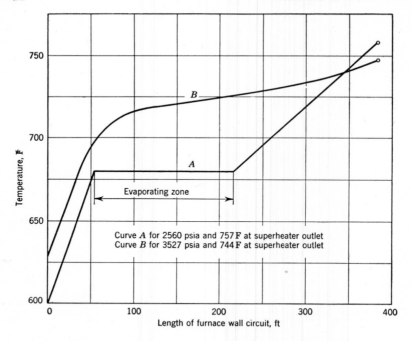

**Fig. 8.24.** Water and steam temperatures in the furnace-wall circuits of the Purdue once-through steam-generating unit.

dampers above the economizers in the three parallel vertical gas passes, and (c) recirculation of gas from a location beyond the economizer to the secondary furnace above the cyclones.

Figure 8.24 shows the change in temperature of the fluid as it flows through the circuits in the once-through steam-generating unit in the laboratory at Purdue University. Curves *A* and *B* illustrate the performance when the discharge pressure is below and above the critical pressure, respectively. A definite evaporating zone exists at a discharge pressure of 2560 psia as indicated by the horizontal portion of curve *A*. At a discharge pressure of 3527 psia, the latent heat is zero, no evaporation occurs, and the transition from water to steam is accompanied by a slowly rising temperature. This particular unit operates equally well at any pressure between 1500 and 3500 psi.

The commercial use of the once-through steam-generating unit had to wait until feed-water treatment had advanced to the stage where noncorrosive feedwater free of scale-forming impurities could be maintained in the system.

### 8.11  Steam Generators for Nuclear Reactors

The boiling-water reactor is a steam generator. It produces saturated steam that is delivered directly to the steam turbine. If the reactor is cooled by pressurized (nonboiling) water, liquid metal, an organic liquid, or an inert gas, the hot coolant is circulated through a heat exchanger in which steam is generated and the temperature of the coolant is reduced.

Figure 8.25 illustrates a steam generator for a pressurized-water reactor. A U-shaped heat exchanger is mounted in a horizontal position. A cut-away view shows the location of one of the tube sheets. The other tube sheet occupies a similar position at the other end of the U-shaped heat-exchanger shell. The ends of the U-shaped heat-

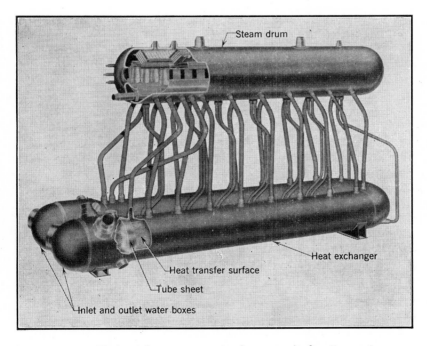

**Fig. 8.25.**  Horizontal steam generator for pressurized-water reactor.

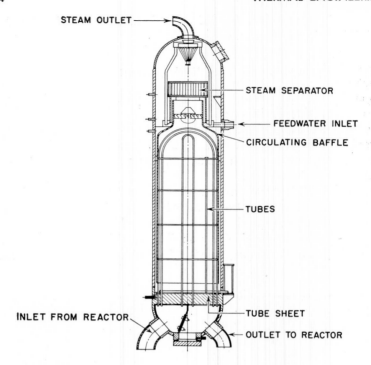

STEAM OUTLET

STEAM SEPARATOR

FEEDWATER INLET

CIRCULATING BAFFLE

TUBES

INLET FROM REACTOR

TUBE SHEET

OUTLET TO REACTOR

**Fig. 8.26.** Vertical steam generator for pressurized-water reactor.

exchanger tubes are fastened in the tube sheets so that the reactor coolant circulates from the inlet header through the tubes to the outlet header. The U-shape of the tubes and heat-exchanger body allows for free expansion of the shell and tube bundle. The inlet and outlet headers are connected to the reactor by suitable piping that is welded to the headers and contains a circulating pump. A conventional boiler drum with internal separators is located above the heat exchanger and connected to it by risers and downcomers. A water level is maintained slightly below the mid-point of the steam drum so that the space around the heat-exchanger tubes is filled with boiling water. The boiler produces saturated steam for delivery to the turbine.

Figure 8.26 illustrates a vertical type of steam generator for use with a pressurized-water reactor. The inlet and outlet connections for the reactor coolant are located at the bottom of the heat exchanger. Vertical U-shaped tubes are fastened in a heavy horizontal tube sheet. The coolant flows through these tubes from the inlet to the outlet water box. A steam separator occupies much of the space above the water level. Circulation is controlled by a baffle which is so located that the circulated water plus the feedwater flows downward between the shell of the vessel and the baffle and then upward around the tubes which constitute the heat-transfer surface. This design has the advantage of occupying less floor area than is required for the unit illustrated in Fig. 8.25 but has a smaller space for steam separation.

In a typical pressurized-water reactor system, the cooling water enters the heat exchanger at about 2000 psi and 540 F and leaves at about 505 F. Saturated steam is generated at about 600 psi. Figure 8.27 illustrates schematically the manner in which a large nuclear reactor is connected to the steam boilers. Four steam generators are shown in this sketch. Full output is obtained from three steam generators; the fourth one is held in reserve. Suitable piping and valves are installed so that any one or more of the steam generators may be taken out of service. Each circuit contains its own circulating pump. Because of the danger of discharging radioactive vapor in case of a

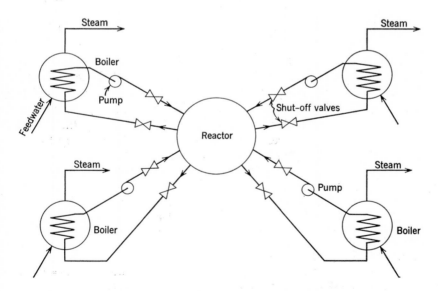

**Fig. 8.27.** Schematic arrangement of coolant-steam generation circuits of a large pressurized-water reactor.

failure of any pressure parts in the reactor coolant circuit, it is customary to house the reactor, the steam generators, the pumps, and associated piping in a containment vessel of sufficient strength to contain all material in the reactor and coolant circuits in the event of a failure of any pressure parts. In some nuclear power plants, the containment vessel is a steel sphere, generally between 125 and 175 ft in diameter and about 2 in. thick.

### 8.12   Capacity and Efficiency of Steam-Generating Units

The energy absorbed per hour in converting feedwater into steam may be computed as follows:

$$\text{Energy absorbed, Btu per hr} = m_s(h_2 - h_1) \qquad (8.2)$$

where $m_s$ = steam generated, lb per hr
  $h_2$ = enthalpy of steam leaving the unit, Btu per lb
  $h_1$ = enthalpy of feedwater entering the unit, Btu per lb

The enthalpy of the feedwater is the enthalpy of *compressed* water at the feed-water temperature and pressure and is determined in accordance with the procedures outlined in Article 3.13. For pressures below 400 psia, the effect of compression may be neglected in performing slide-rule computations and the enthalpy of *saturated* water at the feed-water temperature may be used.

If the unit includes a superheater, the enthalpy of the steam leaving the superheater is obtained from the superheated steam table, Table A.3 of the Appendix, at the pressure and temperature at which the steam is delivered by the unit. If the steam generator does not include a superheater, then saturated steam is delivered by the boiler. Some moisture will be present in the steam, and it is necessary to use a throttling calorimeter as discussed in Article 3.12 to determine the enthalpy (and quality, if desired) of the steam leaving the unit.

**Example 1.** A total of 10,000 lb of steam is generated per hr in a boiler operating at a pressure of 200 psia. A throttling calorimeter connected to the boiler outlet indicates a steam temperature of 260 F for a calorimeter pressure of 20 psia. Feedwater is supplied at 220 F. Compute the energy absorbed in Btu per hr.

*Solution:* From the calorimeter pressure and temperature given, $h_2 = 1172$ (see Fig. 3.9). At 220 F, $h_f = 188$ (from Table A.1). Then,

Energy absorbed = 10,000 (1172 − 188) = 9,840,000 Btu per hr

**Example 2.** A total of 150,000 lb of steam is generated per hr at a pressure of 1000 psia and a final steam temperature of 900 F from feedwater at 200 F. Compute the energy absorbed in Btu per hr.

*Solution:* For 1000 psia and 900 F, $h_2 = 1448.2$ from Table A.3.

For 200 F feedwater at 1000 psia, $h_c = h_f$ at 200 F + correction from Fig. 3.11 = 167.99 + 2.2 = 170.2. Then,

Energy absorbed = 150,000 (1448.2 − 170.2) = 191,700,000 Btu per hr

The *rated capacity* of steam-generating units was originally expressed in terms of boiler horsepower, 1 boiler hp being equal to 10 sq ft of heating surface. Thus, a boiler having 5000 sq ft of heating surface was *rated* as a 500-hp boiler. This arbitrary rating has been abandoned, and steam-generating units are now rated in terms of the steam they can deliver continuously in pounds per hour. It is now standard practice in large central-station power plants to install one steam-generating unit per turbine and to operate the group as a unit. Where this is done, the capacity of the steam-generating unit is sometimes expressed in kilowatts equal to the kilowatt rating of the turbine generator to which it supplies the steam.

The *output* of steam-generating units was originally expressed in terms of *boiler horsepower*, where 1 boiler hp was defined as the absorption of 33,475 Btu per hr. Since no work is done in a boiler, the work being done by the expansion of the steam in an engine or turbine, the term "boiler horsepower" is a misnomer and has been abandoned. The output of steam-generating units is now expressed as the energy absorbed per hour in units of 1000 Btu, this unit being given the symbol $kB$.

The *output* of a steam-generating unit may be computed as follows:

$$\text{Output} = \frac{\text{energy absorbed per hr in generating steam}}{1000} kB \quad (8.3)$$

**Example 3.** Express the output of the steam-generating unit of Example 2 in $kB$.

*Solution:* From Example 2,

Energy absorbed = 191,700,000 Btu per hr

Then
$$\text{Output} = \frac{191,700,000}{1000} = 191,700 \, kB$$

The *efficiency* of a steam-generating unit,

$$\eta = \frac{\text{energy absorbed, Btu per hr}}{\text{heating value of fuel burned per hr}}$$

or
$$\eta = \frac{m_s(h_2 - h_c)}{m_f Q_H} \times 100 \text{ in } \%$$
(8.4)

where $m_f$ = fuel burned, lb per hr

$Q_H$ = higher heating value of the fuel, Btu per lb

**Example 4.** Compute the efficiency of the steam-generating unit of Example 2 if 21,000 lb of coal having a heating value of 12,000 Btu per lb are burned per hr.

*Solution:*

$$\eta = \frac{\text{energy absorbed, Btu per hr}}{\text{heating value of fuel burned per hr}} = \frac{191,700,000}{21,000 \times 12,000} \times 100$$

$$= 76.3\%$$

The efficiency obtainable from a steam-generating unit depends upon the kind of fuel burned, the method of firing the fuel, the characteristics of the furnace, the arrangement and extent of heat-transfer surfaces, the load on the unit, and the care exercised in operating it. The maximum efficiency is determined by the design of the unit. The extent to which this capability is realized depends on the care exercised in operating it. The amount of the various losses that prevent attainment of 100 per cent efficiency may be determined from an energy balance. The methods used in calculating the energy balance were discussed in Article 4.23. These calculations, together with the computations for determination of the capacity and efficiency of a steam-generating unit, are illustrated in Example 5 as applied to a typical set of test data.

**Example 5.** The following data were collected during the test of a steam-generating unit consisting of a stoker-fired boiler, superheater, and air heater.

Ultimate analysis of coal, %: C = 66.0; H = 4.4; O = 7.9; N = 1.5; S = 1.1; ash = 7.6; moisture = 11.5; total 100%

Heating value of coal = 11,780 Btu per lb

Orsat analysis of dry gaseous products of combustion, %: $CO_2$ = 14.5; CO = 0.2; $O_2$ = 4.4; $N_2$ = 80.9; total = 100%

Coal burned = 6480 lb per hr

Dry refuse collected = 622 lb per hr; heating value of refuse = 3070 Btu per lb

Steam generated = 55,000 lb per hr

Feed-water temperature = 300 F; feed-water pressure = 800 psig

Boiler pressure = 605 psig

Calorimeter pressure and temperature at boiler outlet = 5 psig and 310 F, respectively

Superheater outlet pressure = 585 psig; superheater outlet temperature = 800 F

Barometric pressure = 29.9 in. of Hg = 14.7 psia

Air temperature = 70 F; exit flue-gas temperature = 400 F

Compute (a) Energy absorbed in boiler per lb of steam generated.
(b) Energy absorbed in superheater per lb of steam generated.
(c) Energy absorbed in steam-generating unit per lb of steam generated.
(d) Output of steam-generating unit in $kB$.
(e) Efficiency of steam-generating unit.
(f) Energy balance.

*Solution:* (a) Enthalpy of compressed feedwater at 300 F and 800 psig = $h_f$ at 300 F plus correction factor for 815 psia from Fig. 3.11 = 269.6 + 1.5 = 271.1 Btu per lb of feedwater

Enthalpy of wet steam at the boiler outlet can be determined from the pressure and temperature in the throttling calorimeter. At 310 F and 19.7 psia, $h_2$ = 1196.5 Btu per lb = $h_1$

Then the energy absorbed in the boiler = 1196.5 − 271.1 = 925.4 Btu per lb of steam generated

(b) Enthalpy of the steam leaving the superheater at 599.7 psia (585 psig + 14.7 psi barometric pressure) = 1407.7 Btu per lb

Then the energy absorbed in the superheater = enthalpy of steam leaving superheater − enthalpy of steam leaving boiler = 1407.7 − 1196.5 = 211.2 Btu per lb of steam

(c) Energy absorbed in steam-generating unit per lb of steam generated = enthalpy of steam leaving superheater − enthalpy of feedwater = 1407.7 − 271.1 = 1136.6 Btu per lb of steam generated

Check: energy absorbed in boiler + energy absorbed in superheater = 925.4 + 211.2 = 1136.6 Btu per lb of steam generated

(d) Output of steam-generating unit in $kB$

$$= \frac{\left\{ \begin{array}{c} 55,000 \text{ lb of steam generated per hr} \times 1136.6 \text{ Btu} \\ \text{absorbed per lb of steam generated} \end{array} \right\}}{1000} = 62,513 \; kB$$

(e) Efficiency of steam-generating unit

$$= \frac{\text{Energy absorbed, Btu per hr}}{\text{Heating value of fuel supplied, Btu per hr}}$$

$$= \frac{\left\{ \begin{array}{l} 55{,}000 \text{ lb of steam generated per hr} \times 1136.6 \text{ Btu} \\ \text{absorbed per lb of steam} \end{array} \right\}}{6480 \text{ lb of coal burned per hr} \times 11{,}780 \text{ Btu per lb of coal}} \times 100$$

$$= 81.8\%$$

(f) Energy balance; computed for 1 lb of fuel:

Carbon input = 6480 lb of coal burned per hr × 66% C in coal = 4275 lb of C per hr

Carbon in refuse = 622 lb of dry refuse per hr × 3070 Btu per lb of refuse ÷ 14,600 Btu per lb of C = 131 lb of C per hr

Carbon burned to CO and $CO_2$ per lb of coal, $C_b$

$$= \frac{4275 \text{ lb of C in fuel} - 131 \text{ lb of C in refuse}}{6480 \text{ lb of fuel}}$$

$$= 0.64 \text{ lb of C per lb of fuel}$$

| Dry Flue-Gas Analysis, % | Lb per 100 Moles | Lb of C in 100 Moles |
|---|---|---|
| $CO_2$ = 14.5 | 14.5 × 44 = 638 | 14.5 × 12 = 174 |
| CO = 0.2 | 0.2 × 28 = 6 | 0.2 × 12 = 2.4 |
| $O_2$ = 4.4 | 4.4 × 32 = 141 | |
| $N_2$ = 80.9 | 80.9 × 28 = 2265 | |
| Total 100.0 | 3050 | 176.4 |

Dry products of combustion per lb of fuel

$$= \frac{3050 \text{ lb of dry gas per 100 moles}}{176.4 \text{ lb of C per 100 moles}} \times 0.64 \text{ lb of C per lb of fuel}$$

$$= 11.05 \text{ lb of dry gaseous products of combustion per lb of fuel}$$

(1) Loss due to sensible heat in dry gaseous products of combustion = lb of dry flue gas per lb of fuel × $c_p$ $(t_g - t_a)$ = 11.05 lb of dry gas per lb of fuel × 0.24 (400 − 70) = 876 Btu per lb of fuel

(2) Loss due to CO in dry products of combustion

$$= \frac{2.4 \text{ lb of C in CO per 100 moles of dry gas}}{176.4 \text{ lb of C in CO and CO}_2 \text{ per 100 moles of dry gas}}$$

× 0.64 lb of C per lb of fuel

× 10,160 Btu per lb of C in the form of CO

$$= 89 \text{ Btu per lb of fuel}$$

(3) Loss due to C in the refuse

$$= \frac{622 \text{ lb of refuse per hr} \times 3070 \text{ Btu per lb of refuse}}{6480 \text{ lb of coal burned per hr}}$$

$$= 292 \text{ Btu per lb of fuel}$$

(4) Loss due to evaporating moisture in fuel

$$= 0.115 \text{ lb of moisture in fuel} (1089 - t_a + 0.46t_g)$$

$$= 0.115 (1089 - 70 + 0.46 \times 400) = 138 \text{ Btu per lb of fuel}$$

(5) Loss due to water vapor formed from H $= 0.0444$ lb of H per lb of fuel $\times 9$ lb of $H_2O$ per lb of H $\times (1089 - t_a + 0.46t_g) = 0.0444 \times 9 \times 1203$

$$= 482 \text{ Btu per lb of fuel}$$

(6) Energy absorbed in generating steam

$$= \frac{55,000 \text{ lb of steam generated per hr} \times 1136.6 \text{ Btu per lb of steam}}{6480 \text{ lb of coal burned}}$$

$$= 9640 \text{ Btu per lb of fuel}$$

The energy balance may be summarized as follows in Btu per lb of fuel and in per cent of the heating value of the fuel:

|  | Btu per Lb | Per Cent |
|---|---|---|
| I. Energy absorbed | 9,640 | 81.8 |
| II. Losses: |  |  |
| 1. Sensible heat in dry flue gas | 876 | 7.4 |
| 2. CO in dry flue gas | 89 | 0.8 |
| 3. Unburned C in refuse | 292 | 2.5 |
| 4. Moisture in coal | 138 | 1.2 |
| 5. Water vapor formed from hydrogen | 482 | 4.1 |
| 6. Radiation and unaccounted for (by difference) | 263 | 2.2 |
| Total | 11,780 | 100.0 |

## PROBLEMS

**1.** The vertical distance between the water level in the boiler drum and the center line of the lower drum of a two-drum boiler is 30 ft. The downcomers contain water at an average pressure and temperature of 450 psia and 456 F, respectively. The risers contain a steam-water mixture at an average pressure of 450 psia and an average quality of 3 per cent. Calculate the differential pressure in the circuit in lb per sq in. which is available for creating circulation in the boiler.

**2.** The vertical distance between the water level in a boiler drum and the center line of the furnace-wall distribution headers in a single-drum boiler is 90 ft. The downcomers contain saturated water at 567 F.

The risers contain a steam-water mixture at an average pressure and quality of 1200 psia and 5 per cent, respectively. Calculate the differential pressure in the circuit in lb per sq in. which is available for creating circulation in the boiler.

**3.** Steam at 400 psia and 99.5 per cent quality is expanded isentropically to 2 psia. If a superheater is installed in the system to deliver the steam at 380 psia and 700 F, what is the increase in energy available for conversion into work upon isentropic expansion to 2 psia?

**4.** How much energy must be supplied to convert 1 lb of water at 200 F and 100 psia into steam at 100 psia and 99 per cent quality?

**5.** How much energy must be supplied to convert 1 lb of water at 200 F and 2000 psia into steam at 2000 psia and 1000 F?

**6.** How much energy must be supplied to convert 1 lb of steam at 400 psia and 98 per cent quality into steam at 380 psia and 700 F?

**7.** What are the enthalpy and quality of steam at 300 psia if this steam upon being throttled to 20 psia is at a temperature of 270 F?

**8.** A total of 120,000 lb of gas is to be cooled by means of convection boiler surface. Boiler pressure is 200 psia. Assume a constant specific heat for the gas, $c_p = 0.27$ Btu $lb^{-1}$ $F^{-1}$, and a constant overall coefficient of heat transfer, $U = 13$ Btu $hr^{-1}$ $ft^{-2}$ $F^{-1}$. How many sq ft of boiler surface are required to cool the gases to (a) 700 F, (b) 600 F, (c) 500 F, (d) 400 F? (e) Plot the results with heating surface as abscissa and gas temperature as ordinate.

**9.** A total of 250,000 lb of products of combustion per hr is to be cooled in the convection tube bank of a boiler from 1200 to 700 F. The boiler pressure is 600 psia. The mean specific heat of the gas is 0.27 Btu $lb^{-1}$ $F^{-1}$. The overall coefficient of heat transfer is 14 Btu $hr^{-1}$ $ft^{-2}$ $F^{-1}$. Calculate the area of the boiler surface required in sq ft.

**10.** (a) How much additional surface would be required to cool the gas in Problem 9 from 700 to 600 F? (b) To cool the gas from 600 to 500 F?

**11.** A steam-generating unit delivers 200,000 lb of steam per hr at 1200 psia and 1000 F. Neglect pressure drop in the superheater, assume that the steam enters the superheater as dry saturated steam, and that the flow is counter-current. Per hr, 245,000 lb of products of combustion enter the superheater at a temperature of 1800 F. The mean specific heat of the gas, $c_p$, is 0.26 Btu $lb^{-1}$ $F^{-1}$. The overall coefficient of heat transfer is 12 Btu $hr^{-1}$ $ft^{-2}$ $F^{-1}$. (a) What is the temperature of the gas leaving the superheater? (b) How many sq ft of superheater surface are required?

**12.** Change the initial temperature of the gas entering the superheater in Problem 11 from 1800 to 1600 F. If all other conditions remain the same, how many sq ft of superheater surface are required?

**13.** A total of 150,000 lb per hr of feedwater at 300 F and 1000 psia is supplied to a counter-flow economizer. In the economizer 190,000 lb per hr of products of combustion are cooled from 700 to 450 F. The mean specific heat of the gas, $c_p$, is 0.25 Btu $lb^{-1}$ $F^{-1}$. The overall coefficient of heat transfer is 9 Btu $hr^{-1}$ $ft^{-2}$ $F^{-1}$. Calculate (a) the final feed-water temperature; (b) the sq ft of economizer surface required.

**14.** How much additional surface would be required for the economizer in Problem 13 if the products of combustion were cooled to 375 F?

**15.** In an air heater 300,000 lb of products of combustion are cooled from 700 to 325 F; 275,000 lb of air are supplied to the air heater at 80 F. The mean specific heats of the gas and air are 0.26 and 0.25 Btu lb$^{-1}$ F$^{-1}$, respectively. The overall coefficient of heat transfer is 4.1 Btu hr$^{-1}$ ft$^{-2}$ F$^{-1}$. Determine (a) the discharge air temperature and (b) the sq ft of air-heater surface required.

**16.** A steam boiler operates at a pressure of 200 psig and delivers 50,000 lb of steam per hr. A throttling calorimeter connected to the steam line from the boiler indicates a temperature of 250 F at a pressure of 5 psig. Feed-water temperature and pressure are 220 F and 240 psig. Barometric pressure is 29.8 in. of Hg. Calculate (a) the energy absorbed in Btu per hr, (b) the output of the unit in kB.

**17.** Steam at 1200 psia and 900 F flows through a pipe having an internal diameter of 12 in. at the rate of 400,000 lb per hr. Calculate the average steam velocity in the pipe.

**18.** Compute (a) the energy absorbed in Btu per hr, (b) the output in kB, and (c) the efficiency of a boiler operating under the following conditions: feed-water temperature, 200 F; boiler pressure, 150 psig; barometric pressure, 14.7 psia; a throttling calorimeter connected to the boiler outlet line shows 240 F at a calorimeter pressure of 5 psig; 20,000 lb of steam generated per hr; 2400 lb of coal burned per hr; heating value of the coal, 13,000 Btu per lb.

**19.** The following data were obtained during a test of a steam-generating unit: feed-water temperature, 280 F; feed-water pressure, 1500 psig; boiler pressure, 1250 psig; calorimeter pressure at boiler outlet, 5 psig; calorimeter steam temperature, 260 F; superheater outlet pressure, 1185 psig; superheater outlet temperature, 900 F; 400,000 lb of steam generated per hr; 42,800 lb of coal burned per hr; heating value of coal, 13,400 Btu per lb. Compute (a) energy absorbed in the boiler in Btu per hr, (b) energy absorbed in the superheater in Btu per hr, (c) energy absorbed in the steam-generating unit in Btu per hr, (d) output of the unit in kB, and (e) efficiency of the steam-generating unit.

**20.** The following data were obtained from tests of coal-fired steam-generating units:

| Item | | Unit 1 | Unit 2 | Unit 3 | Unit 4 |
|---|---|---|---|---|---|
| Steam generated, lb per hr | | 80,000 | 100,000 | 120,000 | 324,500 |
| Coal fired, lb per hr | | 9,850 | 11,300 | 12,800 | 30,690 |
| Ash-pit refuse, lb per hr | | 970 | 1,150 | 700 | 2,510 |
| Heating value of refuse, Btu per lb | | 2,920 | 2,190 | 1,460 | 2,920 |
| Ultimate analysis of coal as fired, %: | C | 67 | 68 | 76 | 80.6 |
| | H | 5 | 4 | 5 | 4.2 |
| | O | 6 | 7 | 5 | 2.9 |
| | N | 2 | 2 | 2 | 1.3 |
| | S | 2 | 2 | 1 | 1.2 |
| | Ash | 8 | 9 | 5 | 6.6 |
| | Moisture | 10 | 8 | 6 | 3.2 |

| Item | | Unit 1 | Unit 2 | Unit 3 | Unit 4 |
|---|---|---|---|---|---|
| Heating value of fuel, Btu per lb | | 12,500 | 11,900 | 12,700 | 14,120 |
| Orsat analysis of dry flue gas, % by volume: | $CO_2$ | 13.4 | 14.0 | 14.5 | 14.7 |
| | $O_2$ | 6.2 | 5.8 | 5.2 | 3.8 |
| | CO | 0.2 | 0.1 | 0.1 | 0.02 |
| Air tempeature, F | | 80 | 70 | 90 | 54 |
| Flue-gas temperature, F | | 440 | 380 | 360 | 360 |
| Feed-water temperature, F | | 190 | 350 | 280 | 250 |
| Feed-water pressure, psig | | 500 | 1,200 | 750 | 600 |
| Superheater-outlet pressure, psig | | 385 | 890 | 585 | 411 |
| Superheater-outlet temperature, F | | 750 | 900 | 800 | 720 |

For each of the sets of test data, compute (a) the output of the unit in kB, (b) the efficiency, and (c) a complete energy balance.

21. The following data were obtained from a test on the Purdue once-through boiler: Volumetric analysis of dry flue gas in per cent: $CO_2 = 12.0$; $O_2 = 5.8$; $CO = 0.1$; $N_2 = 82.1$; barometric pressure = 29.3 in. of Hg; feed-water pressure = 2790 psia; feed-water temperature = 245 F; steam-outlet pressure = 2560 psia; final steam temperature = 751 F; water evaporated per hr = 2225 lb; 97 F air temperature; 50 per cent relative humidity; 165 F oil temperature; 650 F flue-gas temperature; 164 lb of oil burned per hr; ultimate analysis of oil in per cent: $C = 88.54$; $H = 10.75$; $N = 0.14$; $S = 0.57$; heating value of the oil = 18,650 Btu per lb. Compute (a) the output in kB, (b) the efficiency of the unit, and (c) a complete energy balance.

22. The following data were obtained from a steam-generating unit fired by natural gas: Volumetric analysis of fuel in per cent: $H_2 = 2$; $CO = 1$; $CH_4 = 93$; $O_2 = 2$; $N_2 = 2$; heating value = 900 Btu per cu ft at 14.7 psia and 58 F; volumetric analysis of dry products of combustion in per cent: $CO_2 = 10.5$; $O_2 = 2.0$; $N_2 = 87.3$; $CO = 0.2$; temperature of air = 90 F; temperature of fuel gas = 70 F; temperature of flue gas = 350 F; feed-water temperature = 300 F; feed-water pressure = 1500 psia; superheater-outlet temperature = 900 F; superheater-outlet pressure = 1200 psia; steam generated = 300,000 lb per hr; fuel burned = 481,000 cu ft per hr measured at 70 F and 16.3 psia. Compute (a) the output of the unit in kB, (b) the efficiency, and (c) a complete energy balance.

23. A total of 2,940,000 lb of water at 2000 psig and 540 F is supplied per hr to a steam boiler from a nuclear reactor. The water leaves the boiler at 500 F and 1900 psig. Feedwater is supplied to the boiler at 700 psig and 400 F. Steam is generated at 600 psig and 99.8 per cent quality. How many lb of steam are generated per hr?

24. A total of 50,000 lb of liquid sodium-potassium (NaK) is circulated per min through a steam boiler from the intermediate heat exchanger attached to a sodium-graphite nuclear reactor. NaK enters at 750 F and leaves at 500 F. Feedwater is supplied to the boiler at 400 F and 600 psia and steam leaves at 600 psia and 700 F. The mean specific heat of NaK is 0.23 Btu $lb^{-1}$ $F^{-1}$. How many lb of steam are generated per hr?

# CHAPTER 9

# Heat Exchangers

## 9.1 Introduction

As stated in Chapter 7, all power and refrigeration plants contain equipment which has as its major function the transfer of heat from one fluid to another. This equipment includes boilers, superheaters, economizers, heaters, coolers, condensers, and evaporators and are called heat exchangers. The same laws of heat transfer, fluid flow, and economics apply to all heat exchangers. Heat exchangers differ in design characteristics only because of the different functions which they perform and conditions under which they operate.

In the preceding chapter the boiler, superheater, economizer, and air preheater were discussed in some detail. These units are normally associated with the power-production industry. However, coolers, heaters, condensers, and evaporators are four types of heat exchangers which have extensive applications in most industries. For example, in the chemical and petroleum industries many processes require the evaporation of liquids, the condensation of vapors, or the heating and cooling of liquids and gases. Oil coolers are required for the lubricating oils used in engines, turbines, gears and machine tools, and for oils used in transformers or hydraulic presses. Evaporators are necessary equipment in such industries as dyestuffs, dairies, food processing, petroleum, and pharmaceuticals. Condensers find application in the same industries.

In discussing heat exchangers emphasis will be placed on the design of those units which find common use in power plants. The principles of heat transfer and fluid flow can be applied equally well to all heat exchangers.

Two heat exchangers commonly found in stationary power plants are the steam condenser and feed-water heater. They are distinct and separate pieces of equipment, and they differ in their relative positions and primary functions in the cycle. The purpose of the feed-water heater is to increase the overall efficiency of the cycle. This is accomplished by heating the boiler water before it enters the boiler with either waste steam or steam extracted from the turbine. With the feedwater entering the boiler at high temperatures the boiler is relieved of a part of its load and the temperature stresses within the boiler are reduced. Feed-water heaters are designed as direct-contact heaters or surface heaters.

### 9.2    Direct-Contact Feed-Water Heaters

The direct-contact heater is often called an open heater, although it may operate at pressures above atmospheric pressure. A typical direct-contact heater is illustrated in Fig. 9.1. It consists mainly of an outer shell in which are placed trays or pans. Water enters at the top of the shell. It feeds by gravity over rows of staggered trays which break up the solid stream of water. Steam entering near the center of the shell intimately mingles with the water and condenses.

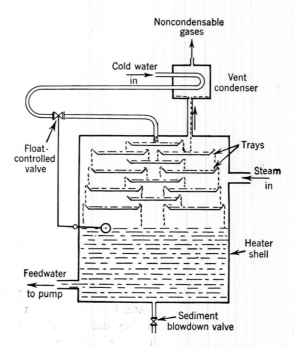

**Fig. 9.1.**  Direct-contact heater.

In condensing, the steam gives up heat to the water. The heated water and condensate mixture is collected at the bottom of the shell and is removed by a boiler feed pump. A float control operating the inlet water valve maintains a constant level in the feed-water tank. A vent at the top removes the excess steam and the noncondensable gases. In the larger heaters where the vented steam is appreciable, a vent condenser may be employed. Water, before it enters the tray section of the feed-water heater, is passed through coils in the vent condenser. Heat is transferred from the vented steam to the water as the steam is condensed. The condensate from the vent condenser is returned to the heater. Noncondensable gases are expelled to the atmosphere.

Because of the stress limitations of the heater shell, the steam pressure is limited to a few pounds per square inch above atmospheric pressure, although pressures to 70 psia have been used. Consequently, the feedwater is rarely heated above 220 F. If direct-contact heaters are used in series, a feed-water pump must be installed ahead of each heater. The advantages of the direct-contact feed-water heater are:

1. Complete conversion of the steam to water is accomplished.
2. Noncondensable corrosive gases are removed from the feedwater.
3. The removal of impurities in the water is possible.
4. The water is brought to the temperature of the steam.
5. The heater acts as a small reservoir.

Even with a lack of knowledge of the events which take place within a direct-contact feed-water heater, one can analyze its performance by writing an energy balance for it. The design details, such as its size, shape, and tray configuration, are beyond the scope of this book.

Assume that an open heater represented by Fig. 9.2 is to be used to heat subcooled water entering from various sources to a saturated liquid condition at a predetermined pressure. Steam is introduced at this pressure, is condensed on coming into contact with the water, and leaves as condensate with the heated water. If a vent condenser is a part of the heater, all the steam introduced will be condensed.

If no heat is gained or lost through the walls of the heater and because the change in mechanical potential and kinetic energy of the water and steam is negligible, the energy balance of the unit can be written as:

$$\text{Energy in} = \text{energy out}$$

or
$$m_5 h_5 = m_1 h_1 + m_2 h_2 + m_3 h_3 + m_4 h_4 \qquad (9.1)$$

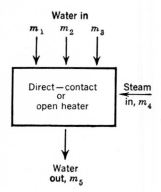

**Fig. 9.2.** Flow diagram for a direct-contact heater.

where $m_5$ = quantity of water out, lb, = $m_1 + m_2 + m_3 + m_4$
$h_5$ = enthalpy of the water out, Btu per lb
$m_1, m_2, m_3$ = quantities of water in at the points 1, 2, 3, respectively, lb
$h_1, h_2, h_3$ = enthalpies of the water in at the points 1, 2, 3, respectively, Btu per lb
$m_4$ = steam in, lb
$h_4$ = enthalpy of steam in, Btu per lb

If the pressure effect is neglected, the enthalpy of the water entering the heater is equal to the enthalpy of the saturated water, $h_f$, at the inlet temperature. It is customary to assume, unless otherwise known, that the water leaving the heater is saturated water at the heater pressure.

**Example 1.** Saturated steam at a pressure of 20 psia enters the feed-water heater shown in Fig. 9.2. Water enters from three sources at temperatures, $t_1 = 70$ F, $t_2 = 100$ F, and $t_3 = 150$ F, and in amounts, $m_1 = 1000$ lb per hr, $m_2 = 5000$ lb per hr, and $m_3 = 3000$ lb per hr. If the water leaves at the saturation temperature corresponding to 20 psia, find the pounds of steam entering the heater.

*Solution:* Since

$$m_5 h_5 = m_1 h_1 + m_2 h_2 + m_3 h_3 + m_4 h_4$$

where $m_1 = 1000$, $m_2 = 5000$, $m_3 = 3000$
$m_5 = m_1 + m_2 + m_3 + m_4 = 9000 + m_4$

Also, $h_5 = h_f$ at 20 psia = 196.16 Btu per lb

and $h_1 = 38.04$, $h_2 = 67.97$, $h_3 = 117.89$

and $h_4 = h_g$ at 20 psia = 1156.3 Btu per lb

Then,

$(9000 + m_4)196.16$

$$= 1000 \times 38.04 + 5000 \times 67.97 + 3000 \times 117.89 + m_4 \times 1156.3$$

and $\qquad\qquad\qquad 960m_4 = 1,034,000$

$$m_4 = 1080 \text{ lb per hr}$$

It should be recognized that the maximum temperature to which the water can be heated in a direct-contact heater is the saturation temperature corresponding to the pressure of the steam if sufficient steam is available. In Example 1 it was assumed that the water left at the maximum temperature.

### 9.3   Closed Feed-Water Heaters

Closed heaters or surface-type feed-water heaters are of the shell and tube design. Generally, the water is introduced to the heater through tubes around which the steam circulates. Closed heaters may be classified as single or multipass and straight tube or bent tube. In a single-pass heater the water flows in only one direction. In a multipass heater the water reverses direction as many times as there are passes. A two-pass straight-tube type of closed feed-water heater is illustrated in Fig. 9.3. Water enters at the bottom of one end of the heater and flows through the lower bank of tubes to the opposite end where its direction is reversed. The water returns through the upper bank of tubes to the outlet at the top. Steam enters the shell at the top and flows toward each end, and condensate leaves the shell at the bottom.

A floating head is provided to permit the tubes to expand. Vents at the top are provided to remove gases trapped in the shell. The heater shown in Fig. 9.3 is designed for a water pressure of 1100 psi. Closed heaters placed in series require only one feed-water pump unless the pressure drop through the heaters is high. If bent tubes are used in place of the straight tubes, as shown in Fig. 9.4, no floating head is necessary. However, the bent tubes may be difficult to clean.

In closed heaters the feedwater can never be heated to the temperature of the steam, but generally the terminal temperature difference at the outlet is not greater than 15 F. To maintain a high overall heat transfer for the heater the water velocity should be high, but pumping costs will limit the velocity. A balance between pumping costs and the amount of heat transferred will result in water veloci-

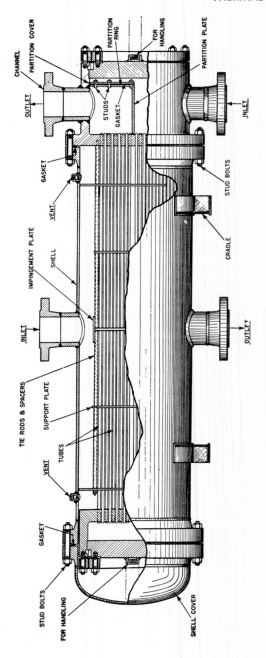

**Fig. 9.3.** A two-pass closed feed-water heater

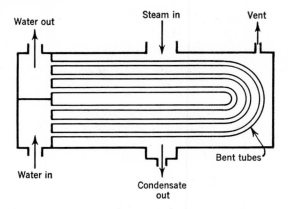

**Fig. 9.4.** A bent-tube closed feed-water heater.

ties of 3 to 8 fps. Generally, the heaters are rated in terms of the square feet of heat-transfer surface and of the quantity of heat transferred.

**Example 2.** It is desired to provide a surface, feed-water heater with sufficient surface to raise the temperature of 100,000 lb of water per hr from 100 F to within 12 F of the steam temperature. If steam enters the heater at 14.7 psia and saturated and leaves at a saturated liquid condition, calculate the heat-transfer area, sq ft, assuming average values for the steam and water-film coefficients. Neglect the resistance due to fouling and the tube wall.

*Solution:* Referring to the sketch of the surface heater, Fig. 9.5, steam enters with a temperature of 212 F (corresponding to the pressure of 14.7 psia) and

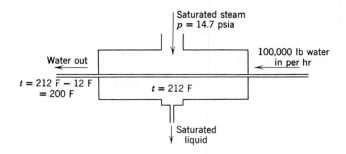

**Fig. 9.5.** Flow diagram for Example 2.

leaves at 212 F and saturated liquid. The water enters at 100 F and leaves at a temperature of 12 F less than the steam temperature, 212 F, or 200 F. Calculating the total heat transferred to the water gives

$$q = m_w c_p \, \Delta t_w$$

$$= 100,000 \times 1.0(200 - 100)$$

$$= 10,000,000 \text{ Btu per hr}$$

Applying the general equation for heat transfer,

$$q = UA\theta$$

in which $q = 10,000,000$ Btu per hr, and

$$\theta = \frac{\theta_g - \theta_l}{\ln \dfrac{\theta_g}{\theta_l}}$$

where

$$\theta = \frac{112 - 12}{\ln \frac{112}{12}} = \frac{100}{\ln 9.3}$$

$$= 45 \text{ F}$$

Also,

$$U = \frac{1}{\dfrac{1}{h_s} + \dfrac{1}{h_w}}$$

where                          $h_s = 1500$ Btu hr$^{-1}$ ft$^{-2}$ F$^{-1}$

and                            $h_w = \phantom{0}500$ Btu hr$^{-1}$ ft$^{-2}$ F$^{-1}$

The values of the film coefficients, $h_s$ and $h_w$, are arbitrarily assumed to be correct and are taken from Table 7.2 as low average values.

Thus,                          $U = \dfrac{1}{\frac{1}{1500} + \frac{1}{500}}$

$$= 380 \text{ Btu hr}^{-1} \text{ ft}^{-2} \text{ F}^{-1}$$

Substituting the values of $U$ and $\theta$ in the heat-transfer equation gives

$$q = 10,000,000 = 380 \times 45 \times A$$

$$\therefore \; A = 590 \text{ sq ft (approximately)}$$

NOTE: The selection of $h_s$ and $h_w$ requires heat-transfer experience. The value for $h_w$ depends on the velocity of the water. It will be high for high water velocities, but high water velocity will increase the resistance to flow and, thus, the pumping costs. The design of a heater is far more complicated than indicated in this simple example and is beyond the scope of this book.

### 9.4 Condensers

The primary function of a condenser is to reduce the exhaust pressure of the prime mover. A reduction in the exhaust pressure will increase the pressure and temperature drop through the prime mover and will result in a corresponding increase in efficiency and output. Secondary functions of the condenser are:

1. To reduce the amount of make-up boiler feedwater by condensing the steam in order that it can be returned to the boiler.

2. To remove air or other noncondensable gases which are corrosive.

Like feed-water heaters, condensers are classed as direct-contact or surface types.

The direct-contact condenser is a jet condenser consisting of water nozzles, a steam-and-water-mixing chamber, and a Venturi section or a tailpipe. The jet condenser may be used where it is not necessary to reclaim the condensate. Although it requires more cooling water than a surface condenser, the jet condenser has the following advantages:

1. Construction and operation are simple.

2. No vacuum pump is required to remove noncondensable gases from the steam.

The jet condenser is used mainly for small prime-mover installations in industry.

The conventional surface condenser is of a shell and tube construction. Cooling water passes through the tubes, and steam circulates around the tubes and is condensed and removed. At no time do the steam and condensate come into contact with the cooling water. Condensers, like feed-water heaters, are classified as single or multipass and straight or bent tube. Generally, condensers used with prime movers are the straight-tube single- or two-pass type.

A cut-away view of a single-pass surface condenser is shown in Fig. 9.6. Water enters from the bottom left, passes through the tubes, and leaves at the upper right. Steam enters the condenser shell from above, circulates around the nest of tubes, and then flows toward the center or core which is the zone of lowest pressure. Air and other noncondensable gases are removed from one end of the core at the vents. The condensed steam or condensate flows by grav-

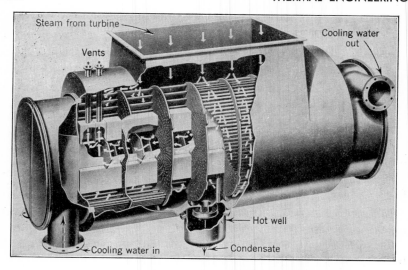

**Fig. 9.6.** A cut-away view of a single-pass surface condenser.

ity to the condensate well or hot well. The condensate is then removed from the well by a pump. An end view and a side view of the condenser are shown in Fig. 9.7.

Because cooling water is usually corrosive in nature, condenser tubes are often made of special alloys of copper or aluminum. Among these are Admiralty metal, Muntz metal, arsenical copper, and aluminum brass, to name a few.

The tubes may be rolled into each end plate. In this case, expansion is taken care of by bowing the tubes. The tubes of some condensers are rolled into and keyed to one end plate and are free to move in the other end plate. Leakage between the tube and end plate is prevented by packing, as illustrated in Fig. 9.8. Expansion and contraction of the condenser shell may be taken care of by providing an expansion joint in the shell wall at one end, as shown in Fig. 9.9.

Owing to the expansion and contraction of the exhaust line or nozzle leading from the turbine to the condenser, all condensers are either rigidly suspended from the turbine or connected to the turbine by an expansion joint. In the former case, the condenser may be placed on spring supports. The spring supports permit the condenser to rise or fall without overloading the turbine exhaust line. In the

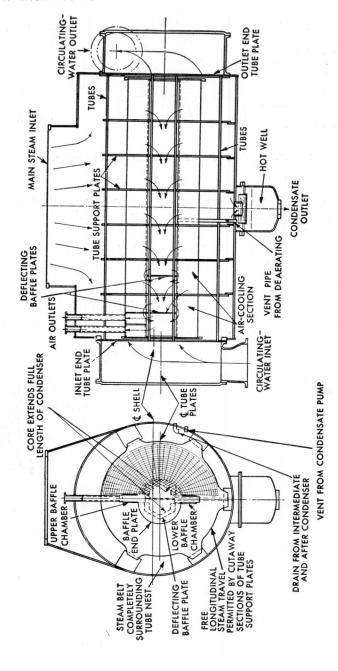

CIRCULATING-WATER OUTLET

OUTLET END TUBE PLATE

TUBES

MAIN STEAM INLET

HOT WELL

TUBES

CONDENSATE OUTLET

DEFLECTING BAFFLE PLATES

TUBE SUPPORT PLATES

AIR-COOLING SECTION

VENT PIPE FROM DE-AERATING

AIR OUTLETS

INLET END TUBE PLATE

CIRCULATING-WATER INLET

CORE EXTENDS FULL LENGTH OF CONDENSER

₵ SHELL

₵ TUBE PLATES

VENT FROM CONDENSATE PUMP

UPPER BAFFLE CHAMBER

BAFFLE END PLATE

LOWER BAFFLE CHAMBER

DRAIN FROM INTERMEDIATE AND AFTER CONDENSER

STEAM BELT COMPLETELY SURROUNDING TUBE NEST

DEFLECTING BAFFLE PLATE

FREE LONGITUDINAL STEAM TRAVEL PERMITTED BY CUTAWAY SECTIONS OF TUBE SUPPORT PLATES

**Fig. 9.7.** End and side views of the radial-flow condenser shown in Fig. 9.6.

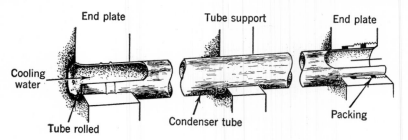

**Fig. 9.8.** Method of installing tubes in a condenser.

latter case, the condenser will be rigidly anchored to the floor. All expansion or contraction in the turbine exhaust line will be taken up in the expansion joint.

There are a number of condenser auxiliaries that are essential to the proper functioning of the condenser. They are illustrated in Fig. 9.10 and are as follows:

1. A condensate hot well for collecting the condensate.

2. A condensate pump to return the condensate to a surge tank where it can be reused as boiler feedwater.

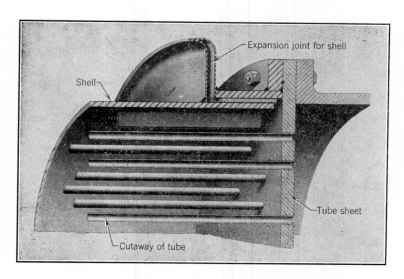

**Fig. 9.9.** Expansion joint for condenser shell.

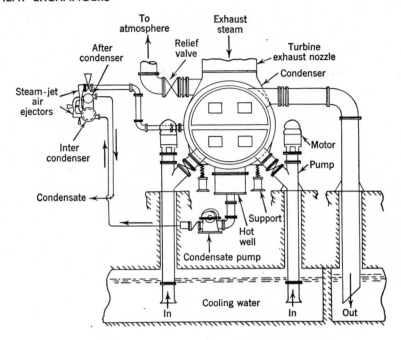

**Fig. 9.10.** An end view of a condenser and its auxiliaries.

3. A circulating pump for circulating the cooling water.

4. An atmosphere relief valve for relieving the pressure in the condenser in case the condenser or auxiliaries do not function properly.

5. An air ejector or a vacuum pump for removing the noncondensable gases from the condenser.

The condensate pump and circulating-water pump are generally of the centrifugal type. If the source of the cooling water is a lake or river, there is no need for water conservation. However, in many localities, the water supply may be low. In such a case, the cooling water, after passing through the condenser, is pumped to a cooling pond or cooling tower where it is cooled by contact with air and then is recirculated through the condenser.

If noncondensable gases are permitted to collect in the condenser, the vacuum in the condenser will decrease. A decrease in the vacuum will result in a decrease in the pressure drop through the turbine and will affect adversely the turbine efficiency. Also, the noncondensable gases are highly corrosive. Thus, their removal in the condenser is essential. They may be removed by a vacuum pump or by a steam-jet air ejector as described in Chapter 11 (see Fig. 11.30). A diagrammatic sketch of a two-stage air ejector is shown in Fig. 9.11.

Steam enters the first and second stages through nozzles where it acquires a high velocity. The air and some vapor from the main condenser are entrained by the high-velocity steam and are compressed in the first stage, forcing tube. The forcing tube is the Venturi-shaped section shown in Fig. 9.11. The steam and vapor are condensed in the intercondenser and drained to the hot well of the main condenser.

Air in the intercondenser is then entrained by high-velocity steam leaving the second-stage nozzles and is compressed further in the second stage, forcing tube. Steam is condensed in the aftercondenser and is drained to the main condenser. The air is vented to the atmosphere. Normally, condensate from the turbine condenser is used as cooling water to condense the steam in the ejector. Both the condensate and cooling water will then be returned to a surge tank.

Characteristic curves for typical ejectors are shown in Fig. 9.12. For a given capacity the multistage air ejector produces lower pressures than the single-stage ejector.

The design and performance of a condenser depend upon many factors involving a knowledge of the fundamentals of thermody-

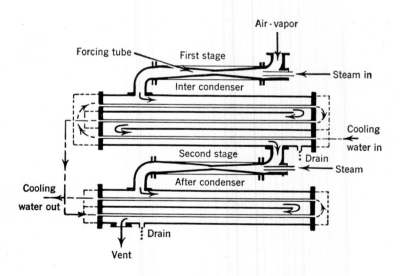

**Fig. 9.11.** A two-stage steam-jet air ejector.

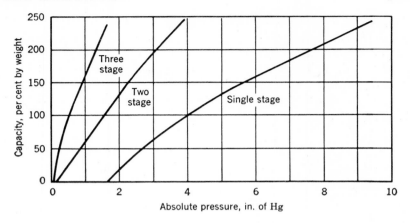

**Fig. 9.12.** Characteristic curves for typical ejectors.

namics and of heat transfer. The efforts of the engineer are directed toward increasing the amount of heat that can be transferred from the exhaust steam to the cooling water and toward increasing the heat-transfer rate to reduce the size of the condenser. The amount of heat that can be transferred for a given heat-transfer area is affected by (1) the flow paths of the water and steam, (2) the velocities of the steam and water, and (3) the various resistances to heat flow produced by layers of gas, steam vapor, dirt, and scale. Normally, for each square foot of condensing surface and for each degree of temperature difference between the steam and water, the heat-transfer rate will vary from 500 to 700 Btu per hr.

### 9.5 Surface-Condenser Calculations

As in the case of feed-water heaters, the first law of thermodynamics can be applied to a condenser without a knowledge of events occurring within the condenser. It is customary to assume that the changes in kinetic and potential energies are negligible and that the heat lost to the surroundings is negligible. Because no work is done by or on a condenser the energy balance becomes a heat balance. Thus, the heat transferred by the steam in condensing must equal the sensible heat acquired by the cooling water or

$$m_s \, \Delta h_s \; = \; m_w c_p \, \Delta t_w \tag{9.2}$$

In most preliminary condenser designs $\Delta h_s$ is assumed to be 950 Btu per lb of steam. Although this is not the actual change in enthalpy of the steam for any particular condenser, it represents an excellent average value. The error in using such a value would not be greater than 2 per cent.

The heat-transfer relations for a condenser are taken from Chapter 7. They are:

$$q = AU\theta \qquad\qquad (9.3)$$

where $\qquad \theta = $ mean temperature difference, and

$$\theta = \frac{\theta_g - \theta_l}{\ln \dfrac{\theta_g}{\theta_l}} \qquad\qquad (9.4)$$

It has been explained in Chapter 7 that $U$ is the overall heat-transfer coefficient and may be calculated from the film coefficients, the conductivity of the tube wall, and the conductivity of scale or dirt. Such a method for obtaining the value of $U$ is seldom used in steam-condenser design. Rather, $U$ is obtained directly from experimental data assembled by the Heat Exchange Institute. The data are pre-

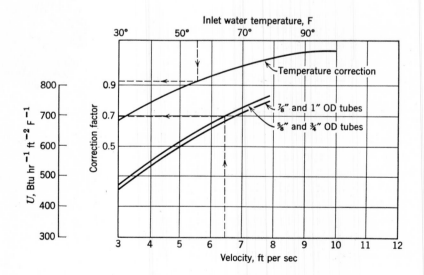

**Fig. 9.13.** Experimental determination of $U$.

sented in Fig. 9.13 as curves for each common condenser tube size and a correction curve for different inlet temperatures.

**Example 3.** Find the overall heat-transfer coefficient, $U$, for a condenser having an inlet water temperature of 55 F, ¾-in. OD tubes, and a water velocity of 6.5 fps.

*Solution:* Starting at 6.5 fps, trace a line vertically until it intersects the curve for ¾-in. OD tubes. Read this point on the left-hand scale. Note that $U = 700$ Btu per hr per sq ft per deg LMTD. Multiply the $U = 700$ by the temperature correction obtained at 55 F. Thus, the corrected $U = 700 \times 0.92 = 644$.

A further correction is made by the type of tube metal used and the wall thickness of the tube. This correction value is shown for two types of condenser-tube alloy and tube-wall gage or thickness.

|  | 18 BWG | 16 BWG |
|---|---|---|
| Admiralty | 1.0 | 0.96 |
| Aluminum brass | 0.96 | 0.91 |

Thus, for the example just discussed, if 16-gage BWG aluminum brass tubes were used, the correction factor would be 0.91 and the corrected $U$ would be $644 \times 0.91 = 585$ Btu hr$^{-1}$ ft$^{-2}$ F$^{-1}$.

The overall heat-transfer coefficient obtained in the example is for clean, bright, oxide-free tubes and needs still further correction for the condenser which has been in operation for more than a few days. Common practice in condenser design is to reduce the coefficient by multiplying it by a cleanliness factor of 0.85. Thus, the coefficient for the example would be:

$$U = 585 \times 0.85$$
$$= 496 \text{ Btu hr}^{-1} \text{ ft}^{-2} \text{ F}^{-1}$$

NOTE: The overall heat-transfer coefficient, $U$, obtained by using the curves in Fig. 9.13 applies *only* to steam-turbine condenser design.

**Example 4.** Steam exhausts from a turbine at a pressure of 1 in. Hg abs and at a rate of 100,000 lb per hr. Inlet cooling-water temperature is 55 F. The terminal temperature difference of the condenser is to be 12 F. The condenser has a cleanliness factor of 0.85 and is constructed of aluminum brass tubes, ¾ in.-OD, 16 BWG gage. Calculate: (a) the pounds per hour of cooling water required to condense the steam; (b) the heat-transfer area required.

*Solution:* (a) To calculate the water required apply the heat balance, Equation 9.2, to the condenser. Thus,

$$m_s \, \Delta h_s = m_w c_p \, \Delta t_w$$

in which $m_s = 100,000$ lb per hr

$$\Delta h_s = 950 \text{ Btu per lb (assumed)}$$
$$c_p = 1.0 \text{ Btu per lb per F}$$
$$\Delta t_w = (79 \text{ F} - 12 \text{ F}) - 55 \text{ F}$$
$$= 12 \text{ F}$$

The 79 F corresponds to the saturated steam temperature at 1 in. Hg abs. The 12 F is the temperature difference between the cooling-water outlet temperature and the steam temperature. Substituting the values in the heat-balance equation gives

$$100,000 \times 950 = m_w 12$$

$$\therefore m_w = 7,900,000 \text{ lb per hr}$$

(b) The area can now be calculated by applying the general heat-transfer equation, Equation 9.3, and assuming a water velocity of 6.5 fps.

$$q = UA\theta$$

in which $q = 100,000 \times 950$ Btu per hr

$U = 496$ Btu hr$^{-1}$ ft$^{-2}$ F$^{-1}$ (as obtained in previous example)

$$\theta = \frac{\theta_g - \theta_l}{\ln \dfrac{\theta_g}{\theta_l}} \quad \text{(Equation 9.4)}$$

$$\theta_g = 79 - 55 = 24 \text{ F}$$

$$\theta_l = 79 - 67 = 12 \text{ F}$$

$$\therefore \theta = \frac{24 - 12}{\ln \frac{24}{12}} = 17.3$$

and

$$A = \frac{q}{U\theta} = \frac{100,000 \times 950}{496 \times 17.3}$$

$$= 11,000 \text{ sq ft (approximately)}$$

This area is that required of the total outside surface of all the tubes used in the condenser. It is equal to the product of the number of tubes times their circumference times their length or

$$A = 11,000 = nl\pi d \tag{9.5}$$

The choice of the number of tubes to provide an area of 11,000 sq ft depends on the selection of the length, $l$, and the diameter, $d$, of the tubes. The selection of length and diameter can be made only by applying practical, economic, engineering judgment. For example, an absurd selection would be to design the condenser having only one tube, 11,000 ft long and a diameter of $1/\pi$. This design would provide the required 11,000 sq ft of heat-transfer surface, but a tube 2 miles in length certainly would be impractical. Equally absurd would be to select 11,000 tubes of 1-ft length each and a diameter of $1/\pi$. Obviously, there is some selection intermediate between the two cases which would be practical.

Standard tube lengths are from 6 ft for condensers of 100 sq ft area to 30 ft for condensers of 100,000 sq ft. Tube diameters are standard at $\frac{5}{8}$ in., $\frac{3}{4}$ in., $\frac{7}{8}$ in., and 1 in. Other things being equal, the tubes with the smaller diameter will produce a greater resistance to flow of the water passing through the tubes. Also, the resistance to flow increases with an increase in tube length. Condenser manufacturers recommend the most economical tube sizes to satisfy a given set of operating conditions.

In the case discussed in Example 4, if the condenser is to have a single pass, the velocity is fixed at 6.5 fps, and the tube diameter is $\frac{3}{4}$ in. OD, the number of tubes, $n$, becomes fixed and may be calculated by use of the law of continuity of flow. Since

$$m = AV\rho \qquad (9.6)$$

in which, for Example 4,

$$m = \frac{7,900,000}{3600} \text{ lb of water per sec}$$

$A$ = cross-sectional area for flow of all the tubes

$$= \frac{n\pi d^2}{4} \; (d = \text{inside-tube diameter})$$

$V = 6.5$ fps

$\rho = 62.4$ lb per cu ft

Thus, for this case,

$$A = \frac{m}{V\rho} = \frac{7,900,000}{6.5 \times 62.4 \times 3600} = 5.42 \text{ sq ft}$$

NOTE: This is the inside area perpendicular to the axis of the tubes and is the sum of the areas of all the tubes used, $n$.

But the number of tubes, $n$, equals the total cross-sectional area, in this case, 5.42 sq ft, divided by the area of each tube, $\dfrac{\pi d^2}{4 \times 144}$, in which

$d$ is the inside diameter, in feet, of the tube. If one neglects the tube-wall thickness, then the inside diameter will be equal to the outside diameter or $\frac{3}{4}$ in. The cross-sectional area is, thus,

$$\frac{\pi d^2}{4 \times 144} \text{ sq ft} = \frac{\pi(\frac{3}{4})^2}{4 \times 144}$$

or

$$\frac{\pi 9}{16 \times 4 \times 144}$$

Therefore,

$$n = 5.42 \text{ sq ft} \div \frac{\pi 9}{16 \times 4 \times 144}$$

$$= \frac{5.42 \times 16 \times 4 \times 144}{9\pi}$$

$$= 1770 \text{ tubes}$$

Now, referring to Equation 9.5 and solving for the length of tube,

$$l = \frac{A}{\pi n d}$$

in which $A = 11,000$ sq ft

$n = 1770$ tubes

$d = \frac{3}{4}$ in. or $\frac{3}{4} \times \frac{1}{12}$ ft

Thus,

$$l = \frac{11,000 \times 4 \times 12}{\pi \times 3 \times 1770}$$

$$= 31.6$$

If a two-pass condenser were used with no change in the specifications established for Example 4, the number of tubes would be $2 \times 1770$ or 3540 tubes, and their length would be cut in half, or 16 ft. The actual economical design would probably be between these two extremes by controlling the flow of water and the temperature differences.

## PROBLEMS

1. It is desired to heat 50,000 lb of water per hr from 80 F to 212 F in a direct-contact heater. If steam supplied to the heater is saturated, what must be its minimum pressure? Calculate the minimum amount of steam required per hr.

2. A direct-contact heater operates at a pressure of 30 psia. Water enters from two sources at temperatures of 180 F and 90 F and at flow rates of 5000 lb per hr and 10,000 lb per hr, respectively. Steam is supplied at 30 psia and 280 F. Calculate the maximum temperature of the water leaving the heater and the amount of steam required.

3. A surface or closed heater is supplied with saturated steam at 100 psia. The condensed steam leaves as saturated liquid. Water enters the heater at 275 F and leaves at a temperature 8 F less than the steam temperature. Calculate the lb of steam supplied per lb of water.

4. If it is desired to heat 10,000 lb of water per hr in the heater of Problem 3, estimate the heat-transfer surface of the heater in sq ft.

5. By use of Fig. 9.13 find the value of $U$ for a condenser for which the following data are available: inlet water temperature, 53 F; water velocity, 7 fps; tubes, 18 BWG and of admiralty metal; cleanliness factor, 85 per cent or 0.85.

6. Compute the approximate heat-transfer area required to condense 70,000 lb of steam per hr under the conditions given in Problem 5.

7. Compute the cross-sectional area for flow of the cooling water of the condenser of Problems 5 and 6 if the terminal temperature difference is to be 10 F and the cooling water leaves at a temperature of 71 F.

8. Compute the number of tubes necessary for the condenser of Problems 5, 6, and 7 for a tube length of 20 ft. For a single-pass condenser will this provide a water velocity through the tubes equal to the assumed value of 7 fps?

9. List the conditions or factors in designing a condenser which may be controlled by the design engineer.

# CHAPTER 10

# Turbines

## 10.1 Introduction

The steam turbine is a prime mover in which a part of that form of energy of the steam evidenced by a high pressure and temperature is converted into kinetic energy of the steam and then into shaft work.

The concept of the steam turbine is not new, for about 120 B.C. Hero of Alexandria described a sphere which revolved owing to escaping steam, but its utilization was limited to a form of toy. An old sketch of Hero's turbine is shown in Fig. 10.1. Steam generated in a boiler passed through hollow trunnions into a hollow sphere. Mounted on the sphere were two outlets which were directed tangentially to the sphere and from which the steam issued in the form of a jet. The reactive force of the steam leaving the outlets turned the sphere about its axis.

In 1791, John Barber obtained a patent in England on a gas turbine which had most of the essential elements of a modern gas turbine.

In 1882, a Swedish mechanical engineer, Dr. Carl de Laval, con-

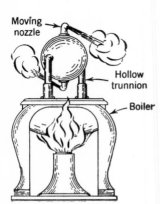

**Fig. 10.1.** Hero's reaction turbine principle.

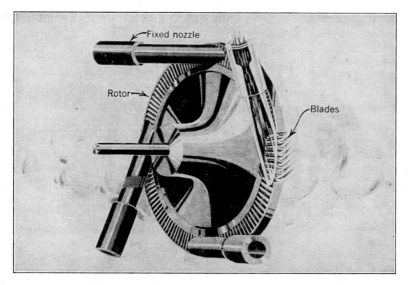

**Fig. 10.2.** de Laval's impulse turbine.

structed a steam turbine based upon the Hero reaction principle that was used for driving a cream separator. Later he developed the turbine shown in Fig. 10.2. Steam generated in a boiler passed through the stationary nozzle or nozzles where it acquired a high velocity. The high-velocity jet was then directed against a row of buckets or blades mounted on a wheel. The impulse force of the jet on the blades turned the wheel.

In 1884 an English inventor and shipbuilder, Sir Charles Parsons, developed a steam turbine based on Hero's reaction principle which was later applied to practical maritime use. Many of the features of the modern turbine were introduced by de Laval and Parsons.

Early in the twentieth century, serious attention was given to the development of the gas turbine as a prime mover. Dr. A. J. Buchi of Sulzer Bros. in Switzerland and Dr. S. A. Moss of the General Electric Company, respectively, pioneered the development of the exhaust-gas-turbine-driven supercharger for Diesel and aircraft engines. World War II greatly accelerated the development of the exhaust-gas-turbine-driven supercharger for the aircraft piston engine and also produced the gas turbine for use in high-speed jet-propelled fighter aircraft.

The development of the gas turbine had been retarded by the low efficiency of compressors and the inability of available metals to operate at the required high temperatures without failure. During the past two decades, highly efficient air compressors have been built. Advances in high-temperature metallurgy are making alloys available which can be operated for reasonable periods of time at temperatures of 1500 to 1600 F. As a result of these advances the gas turbine has become a competitor in the field of mobile power plants and small- to medium-size stationary power plants.

The simple gas-turbine power plant has the advantages of high rotative speed, light weight, small space requirements, few auxiliaries, simplicity, and ability to operate without cooling water. The efficiency can be increased substantially by the addition of heat exchangers at the sacrifice of simplicity, space, weight, and cost.

### 10.2  Fundamentals of the Turbine

The basic advantage of the turbine over other forms of prime movers is the absence of any reciprocating parts. With only rotating motion involved, high speeds are attainable. Since power is directly proportional to torque times speed, an increase in the rotative speed materially decreases the value of the torque required for a given power output. A decrease in the required torque permits a reduction in the size of the prime mover by reducing the length of the torque arm or the force acting on the torque arm. Also, with the absence of any reciprocating parts, vibration is greatly minimized. Owing to the high rotative speeds available with relatively little vibration, the size and cost of the driven machinery, of the building space, and of the foundations are greatly reduced. These advantages are most apparent in large prime movers and permit the steam turbine to be built in sizes of over 350,000 hp in single units, and 760,000 hp in compound units.

Fundamentally, two types of turbines exist, based on the method by which the kinetic energy of the fluid (gas or vapor) is converted into shaft work. They are the impulse turbine and the reaction turbine. The impulse turbine consists of two basic elements: a fixed *nozzle* to convert the energy of the fluid into kinetic energy, and a *rotor*, consisting of blades mounted on the periphery of a disk or wheel, to absorb the kinetic energy of the fluid jet and to convert it into rotary motion. The reaction turbine has the same two basic parts but differs from the impulse turbine in that the blades of the

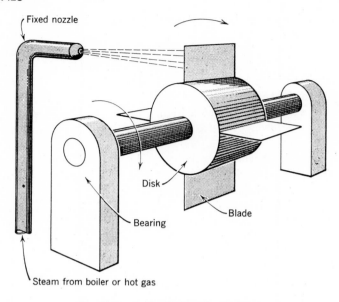

**Fig. 10.3.** A simple impulse turbine.

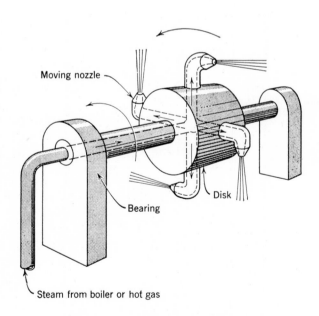

**Fig. 10.4.** A simple reaction turbine.

impulse turbine are replaced by nozzles, mounted on the disk and rotating with it.

The principles underlying the two types of turbines are illustrated in Figs. 10.3 and 10.4 and apply equally well to all gases or vapors. Only the essential parts are shown. Steam is produced in a boiler and piped to the turbine. Before reaching the nozzles, the steam has a very low velocity but is at a high pressure and temperature. On expanding to a lower pressure and temperature, the steam increases in velocity.

In the impulse turbine steam expands in fixed nozzles (see Fig. 10.3). The high-velocity steam jet is directed against the paddle wheel or rotor. An impulse or force is produced by the steam against the blades which, if sufficient to overcome friction, will cause the rotor to turn. Thus work is done on the rotor.

In the reaction turbine (Fig. 10.4) steam expands in moving nozzles. Steam enters the hollow shaft and rotor and flows to the nozzles. When it expands through the nozzles from a high pressure to a low pressure, a reactive force is produced on nozzles, disk, and shaft. The reactive force will produce rotation opposite to the direction of the

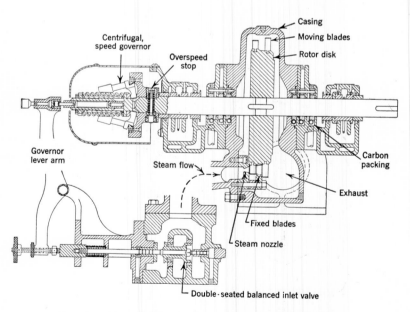

**Fig. 10.5.** Cross-sectional view of a small impulse turbine.

steam jet, and work will be done. An everyday application of this principle is the lawn sprinkler which rotates because of the reactive force of the water leaving the sprinkler nozzles.

In the actual turbine the nozzles do not resemble those shown in Fig. 10.3 or 10.4, but the principle of operation is the same. Also, a *casing* is added to confine and direct the flow of steam. The disk, blades, nozzle, and casing are shown in the small turbine in Fig. 10.5. Steam enters from the left, passes through a valve controlled by a *governor* system, and enters the turbine proper. The governor automatically regulates the flow of steam to the nozzle to maintain constant speed. A *lubricating* system supplies oil to the bearings on which the rotor shaft is mounted. Shaft packing is used to prevent steam leakage.

### 10.3   Entropy and the Mollier Diagram

In Article 2.12 and again in Article 3.15 it was stated that in an adiabatic, frictionless compression or expansion process the property, entropy, remains constant. Also, in Chapter 2 the application of the general-energy equation to certain flow processes showed that a change in enthalpy equaled work and for other processes a change in enthalpy equaled the change in kinetic energy. Turbine design requires analysis of both types of processes. To speed the calculations a chart has been developed, having as its ordinates enthalpy and entropy, called a Mollier chart. Such a chart is shown in skeleton form in Fig. 10.6. A working diagram for the solution of simple problems is shown in Appendix Fig. A.1. A much larger diagram will be found in Keenan and Keyes' *Thermodynamic Properties of Steam.*

Lines of constant pressure extend diagonally across the chart from the lower left-hand corner to the upper right-hand corner as shown in Fig. 10.6. Across the center of the diagram is a curved line known as the "saturation line" which is a plot of $h_g$ versus $s_g$ from the tabular values in Table A.2. The enthalpy and entropy of 1 lb of *dry saturated steam* at any pressure can be read directly from the chart by locating the point of intersection of the saturation line and the appropriate constant-pressure line.

The saturation line divides the chart into a region below this curve which is the wet-steam region and a region above the curve which is the superheat region. A series of lines of constant-moisture content are located below and roughly parallel to the saturation line. The line of 5 per cent moisture is a line of 95 per cent quality. A series of

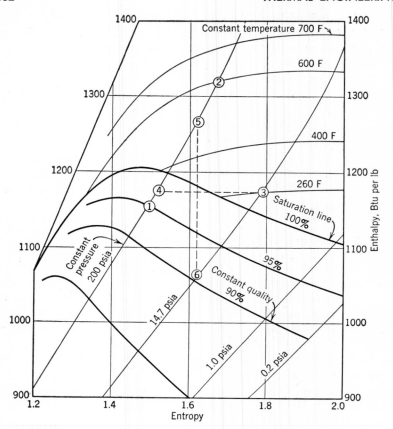

**Fig. 10.6.** Mollier diagram for steam.

constant-temperature lines is plotted in the superheat region of the chart.

The use of this chart may be illustrated by a few examples.

**Example 1.** Determine the enthalpy and entropy of 1 lb of wet steam at 95 per cent quality and 200 psia.

*Solution:* The point is illustrated as point 1 on Fig. 10.6 and is determined by locating the intersection of the 200-psia constant-pressure line with the line of 95 per cent quality. From the Mollier chart in the Appendix (Fig. A.1) the values can be read as follows: $h_x = 1156$; $s_x = 1.495$.

**Example 2.** Determine the enthalpy and entropy of 1 lb of superheated steam at 200 psia and 600 F.

*Solution:* The point is illustrated as point 2 on Fig. 10.6 and is determined by locating the intersection of the 200-psia constant-pressure line with the line of 600 F total temperature. From the Mollier chart in the Appendix the values can be read as follows: $h_s = 1322$; $s_s = 1.676$.

The reader should note that this problem could have been solved more quickly and accurately by the use of the superheated-steam tables and that the steam tables must be used to obtain any data on specific volume.

**Example 3.** Steam at 200 psia is throttled in a throttling calorimeter to atmospheric pressure where the temperature is found to be 260 F. Determine the enthalpy and quality of the steam at 200 psia.

*Solution:* For the throttling process, $h_1 = h_2$. Point 3 on Fig. 10.6 represents the condition of the superheated steam in the calorimeter at 14.7 psia and 260 F. The enthalpy is 1174 Btu per lb. Since the enthalpy at 200 psia is also 1174, point 4 is located on Fig. 10.6 by drawing a horizontal or constant-enthalpy line from point 3 to the 200-psia line. The moisture content is found on the Mollier chart in the Appendix to be 2.8 per cent, or the quality is 97.2 per cent.

**Example 4.** Steam at 200 psia and 500 F is expanded at constant entropy to 14.7 psia. Determine the final quality and the change in enthalpy.

*Solution:* Point 5 is located on Fig. 10.6 at 200 psia and 500 F and a vertical line (constant entropy) is drawn to the final pressure line, 14.7 psia where point 6 is located. The initial and final enthalpies are found from the Mollier chart in the Appendix to be 1269 and 1063, respectively, and the final quality is 91 per cent. The change in enthalpy = $1269 - 1063 = 206$ Btu per lb.

### 10.4 The Turbine Nozzle

The turbine nozzle performs two functions:

1. It transforms a portion of the energy of the fluid, acquired in the heat exchanger and evidenced by a high pressure and temperature, into kinetic energy.

2a. In the impulse turbine it directs the high-velocity fluid jet against blades which are free to move in order to convert the kinetic energy into shaft work.

2b. In the reaction turbine the nozzles, which are free to move, discharge high-velocity fluid. The reactive force of the fluid against the nozzle produces motion, and work is done.

For the first function to be performed efficiently, the nozzle walls must be smooth, streamlined, and so proportioned as to satisfy the changing conditions of the steam or gas flowing through the nozzle.

For the second function the nozzle should discharge the fluid at the

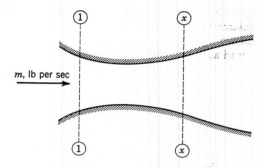

**Fig. 10.7.** An arbitrary flow passage.

correct angle with the direction of blade motion to allow a maximum conversion of kinetic energy into work.

The main consideration in nozzle design is to provide a nozzle of proper wall contour. The contour of the walls depends upon the conditions of the fluid required by the turbine and upon certain properties of the fluid which are influenced by these established conditions. For nozzle design the engineer has at his disposal four fundamental *tools* or relations. They are:

1. The first law of thermodynamics.
2. The equation of continuity of flow.
3. The characteristic equation of state of the fluid.
4. The equation of the process.

The application of the first law of thermodynamics to nozzle flow has been discussed in Article 2.15. For a flow passage such as that shown in Fig. 10.7 the energy equation per pound of fluid (Equation 2.19) can be written as

$$\frac{V_1{}^2}{2gJ} + h_1 = \frac{V_x{}^2}{2gJ} + h_x \qquad (10.1)$$

where $V_1$ = initial velocity, fps
     $h_1$ = initial enthalpy, Btu per lb
     $V_x$ = velocity at any arbitrary point, $x$, downstream from the initial section, fps
     $h_x$ = downstream enthalpy, Btu per lb

From Article 2.15 it will be recalled that Equation 10.1 was developed on the assumption that no work is done between sections 1–1 and $x$–$x$ and that the heat transferred and the change in potential energy are negligible. Furthermore, if it is assumed that the initial velocity, $V_1$, is small, then Equation 10.1 reduces to

$$\frac{V_x^2}{2gJ} = h_1 - h_x \tag{10.2}$$

and
$$V_x = 223.8\sqrt{h_1 - h_x} \tag{10.3}$$

The equation of continuity of flow states that the mass rate of flow of the fluid is constant for all cross sections in the flow passage and is equal to the volume rate of flow divided by the specific volume of the fluid if the flow passage is completely filled with the fluid. Thus, for the passage shown in Fig. 10.7 the continuity equation can be written as

$$m = \frac{A_1 V_1}{v_1} = \frac{A_x V_x}{v_x} \tag{10.4}$$

where $m$ = mass rate of flow, lb per sec
$A$ = area at sections designated, sq ft
$V$ = velocity at sections designated, fps
$v$ = specific volume at sections designated, cu ft per lb

The characteristic equation of state is the equation that relates the properties of the fluid. For a perfect gas it is $pv = RT$. For steam the equation is not so simple, and the steam tables or Mollier chart are used in place of an equation.

The equation of the process depends upon the flow conditions which the fluid undergoes. In actual flow processes the conditions may be quite complex. For this reason the simplifying assumption is often made that there is no friction. Thus, if there is no friction and no heat flow, the process becomes an isentropic process or one in which the entropy remains constant.

The application of these four basic tools to a nozzle can be illustrated best by an example.

**Example 5.** Steam enters a nozzle at 200 psia and 500 F with a rate of flow of 1 lb per sec. It expands without friction or heat flow to a pressure of 1 psia. If it is assumed that the initial velocity of the steam is negligible, find the proper areas of the nozzle at (a) 140 psia, and (b) 1 psia.

*Solution:* (a) The proper area of the nozzle is that area which will satisfy the continuity-flow equation at the point in question or $A_x = mv_x/V_x$. In this equation, the point $x$ is reached by steam expanding from $p = 200$ psia and $t = 500$ F to $p_x = 140$ psia at constant entropy.

The process may be represented on the Mollier diagram as shown in Fig. 10.8. The point $x$ for $p = 140$ psia is identified as point 4. Now, both the

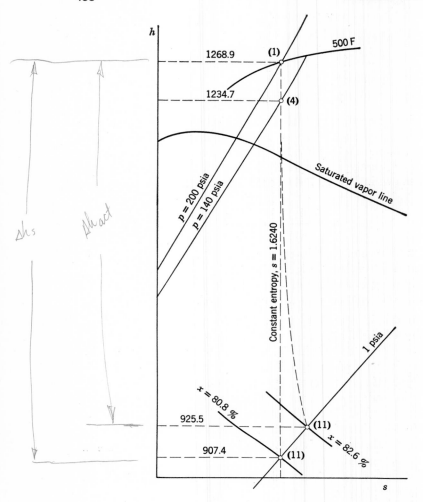

**Fig. 10.8.**  Expansion process for steam in the nozzle of Examples 5 and 6.

initial and final enthalpies may be obtained and are: $h_1 = 1268.9$ Btu per lb and $h_4 = 1234.7$ Btu per lb.  From the general-energy equation, the velocity, $V_4$, can be calculated.  Thus,

$$V_4 = 223.8\sqrt{h_1 - h_4}$$

$$= 223.8\sqrt{1268.9 - 1234.7}$$

$$= 1320 \text{ fps}$$

Also, the value of the specific volume of the steam, $v_4$, may be obtained from the steam tables at the point where $p = 140$ psia and $s = 1.6240$. From Table A.3, $v_4 = 3.584$ cu ft per lb.

Substituting these values in the continuity equation gives

$$A_4 = \frac{mv_4}{V_4}$$

$$= \frac{1 \text{ lb per sec} \times 3.584 \text{ cu ft per lb}}{1320 \text{ fps}}$$

$$= 0.00271 \text{ sq ft}$$

(b) In a similar manner the area at the point $x$ where $p = 1$ psia can be determined. If the point in the flow path where $p = 1$ psia and the entropy, $s, = 1.6240$, is identified as point 11 (see Fig. 10.8), then the final enthalpy, $h_{11}, = 907.4$ Btu per lb. Also, the quality of the steam, $x$, at point 11 is approximately 80.8 per cent, as read on the Mollier chart.

From the general-energy equation, the velocity, $V_{11}$, can now be calculated. Thus,

$$V_{11} = 223.8\sqrt{h_1 - h_{11}}$$

$$= 223.8\sqrt{1268.9 - 907.4}$$

$$= 4240 \text{ fps}$$

At a pressure of 1 psia and a quality $x$ of 80.8 per cent, the specific volume can be calculated by

$$v_{11} = v_f + xv_{fg}$$

where $v_f = 0.01614$ (Table A.2 at $p = 1$ psia)

$$v_{fg} = v_g - v_f = 333.6 - 0.01614 = v_g \text{ approximately} = 333.6$$

Thus, $\qquad v_{11} = 0.808 \times 333.6$ approximately

or $\qquad v_{11} = 270$ cu ft per lb

Substituting these values in the continuity equation

$$A_{11} = \frac{mv_{11}}{V_{11}}$$

gives $\qquad A_{11} = \dfrac{1 \text{ lb per sec} \times 270 \text{ cu ft per lb}}{4240 \text{ fps}}$

$$= 0.0637 \text{ sq ft}$$

In a similar manner the required area of the nozzle at any other point in the flow path can be determined. The results of the calculations of areas for some intermediate points are shown in Table 10.1.

## TABLE 10.1

### Nozzle and Steam Values

for

$m = 1$ lb per sec; $p_1 = 200$ psia; $t_1 = 500$ F

| (1) Point of Flow | (2) Pressure $p$, psia | (3) Enthalpy $h$, btu per lb | (4) Velocity $V$, fps | (5) Specific Volume $v$, cu ft per lb | (6) Area $A$, sq ft | (7) Diameter $d$, in. | (8) Remarks |
|---|---|---|---|---|---|---|---|
| 1 | 200 | 1268.9 | 0 | 2.726 | ... | ... | Entrance |
| 2 | 190 | 1263.4 | 546 | 2.830 | 0.00518 | 0.973 | |
| 3 | 170 | 1252.4 | 920 | 3.083 | 0.00335 | 0.793 | |
| 4 | 140 | 1234.7 | 1320 | 3.584 | 0.00271 | 0.703 | |
| 5 | 110 | 1213.0 | 1680 | 4.307 | 0.00256 | 0.683 | Throat |
| 6 | 80 | 1186.2 | 2040 | 5.484 | 0.00268 | 0.699 | |
| 7 | 50 | 1148.4 | 2450 | 8.26 | 0.00337 | 0.782 | |
| 8 | 20 | 1082.4 | 3050 | 18.6 | 0.00608 | 1.05 | |
| 9 | 15 | 1061.4 | 3220 | 24.3 | 0.00755 | 1.17 | |
| 10 | 7 | 1014.0 | 3560 | 47.0 | 0.0132 | 1.55 | |
| 11 | 1 | 907.4 | 4240 | 270 | 0.0637 | 3.42 | Exit |
| 12 | 0.49 | 872.5 | 4462.6 | 506. | 0.1134 | 4.56 | |

In Fig. 10.9*A* the values of specific volume and velocity of the steam and diameter of the tube section, as shown in Table 10.1, are plotted against pressure. The curves show that the specific volume and velocity increase with a decrease in pressure but that they do not increase at the same rate. The velocity increases more rapidly than the volume at the high pressures and less rapidly at the low pressures. Thus, the required diameter *d* will at first decrease, reach a minimum value, and then increase. By using the values of the diameters calculated, a nozzle can now be constructed (see Fig. 10.9*B*) having a contour which will give a constant pressure drop per unit of nozzle length, but which will not produce a well-defined jet.

A nozzle constructed to give a more concentrated jet is shown in Fig. 10.9*D*. The pressure, velocity, and diameter changes are shown in Fig. 10.9*C*. The point at which the diameter is a minimum is called the "throat." The section leading up to the "throat" is called the converging section, and the exit section is called the diverging section. A nozzle of this type is a converging–diverging nozzle. The nozzles shown in Figs. 10.9*B* and 10.9*D* would expand 1 lb of steam per sec from 200 psia and 500 F to 1 psia. As seen in Table 10.1 they would produce an exit velocity of 4240 fps if no friction were encountered. If a nozzle is desired which will produce an exit velocity of 3050 fps under the same entrance conditions, it should be constructed as the ones previously designed but with only those sections up to section 8 included. Thus, the exit diameter of the new nozzle would be only 1.05 in. Then the exit pressure should be maintained at 20 psia. A nozzle constructed as the originals but extending only to section 5, the throat section, would produce an exit velocity of 1680 fps under the same inlet conditions. The exit pressure should be maintained at 110 psia. A purely converging nozzle results. Thus, if a further reduction in velocity is desired, a nozzle of only converging shape is necessary.

In Table 10.1 the ratio of the pressure at the throat section to the pressure at inlet is equal to 110 ÷ 200 or 0.55. This ratio is called the "critical ratio," and the pressure at the throat is called the critical pressure. The "critical ratio" is constant for any particular gas, 0.53 for air, 0.55 for superheated steam, and 0.58 for wet steam. Thus, the throat pressure can be calculated readily if the initial pressure is known, and it can be shown experimentally that the flow of steam always reaches its maximum value when the exhaust pressure of a nozzle is equal to 0.55 of the initial pressure. Reduction in the ex-

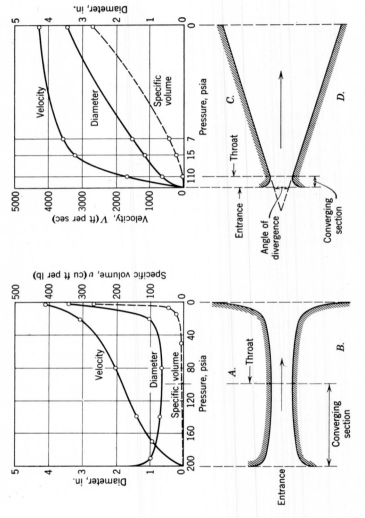

**Fig. 10.9.** Two types of nozzles constructed from steam values shown in Table 10.1. (Figure 10.9 and Table 10.1 are similar to those appearing in E. F. Church's *Steam Turbines*.)

haust pressure below 0.55 of the initial pressure will not increase the mass rate of flow of the steam.

The kinetic energy of the steam leaving the nozzles shown in Figs. 10.9B and 10.9D would be

$$\frac{V_1{}^2}{2 \times 32.2} = \frac{1}{64.4} (4240)^2 = 279,000 \text{ ft-lb per lb of steam}$$

If friction were encountered, as in the actual case, the exit velocity of the steam would be less than 4240 fps, and correspondingly the kinetic energy would be less. The efficiency of the actual nozzle would be the ratio of its kinetic energy to that of a nozzle operating under the same conditions but having no losses. The nozzle is designed to keep the losses, particularly the friction loss, small by the use of smooth streamlined walls and by proper adjustment of the length of the nozzle.

The lengths of the nozzles shown in Figs. 10.9B and 10.9D have been arbitrarily selected. Neither nozzle would perform efficiently under actual steam-flow conditions. Experimentally, it has been found that the converging section should be short and well rounded, as in Fig. 10.9D. The walls of the nozzle from the throat to exit are straight to simplify manufacture, and they are reamed or cast so that the angle of divergence between them (the total included angle) is about 10 degrees. The angle of divergence of the nozzle is selected through experience, and the length of the nozzle is thus fixed for a given throat and exit diameter or area.

The actual turbine nozzle is constructed to direct the jet at an angle with the direction of motion of the turbine blades as shown in Fig. 10.10. The angle subtended by the axis of the nozzle and the direction of motion of the blade is called the nozzle angle and is designated by $\alpha$. The nozzle angle is made as small as satisfactory flow conditions permit in order that maximum thrust on the blade be obtained.

If more than one nozzle is necessary to handle a large steam or gas volume, nozzles are generally manufactured in blocks. Two such sets of nozzle blocks are shown in Figs. 10.11A and 10.11B. The nozzles in Fig. 10.11A are converging–diverging nozzles, which operate at an exhaust pressure below the critical pressure. The nozzles in Fig. 10.11B are converging nozzles which exhaust to a pressure above the critical pressure.

The engineer, working with the theoretical calculations shown in Example 5 and with certain experimental data on actual nozzles, can

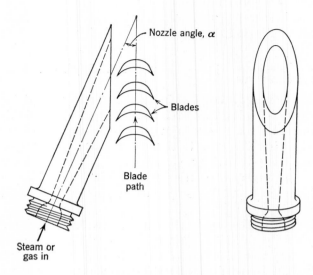

**Fig. 10.10.** An actual turbine nozzle and its position relative to the moving blades.

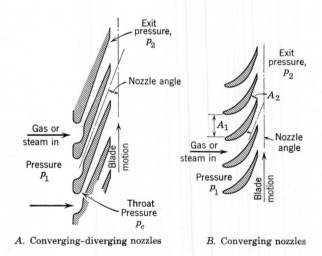

*A.* Converging-diverging nozzles   *B.* Converging nozzles

**Fig. 10.11.** Nozzle blocks.

design an actual nozzle. The performance of an actual nozzle is expressed as a ratio in one of two forms:

$$(1) \text{ Nozzle efficiency, } \eta = \frac{\Delta h_{\text{actual}}}{\Delta h_{s=c}} \qquad (10.5)$$

$$(2) \text{ Velocity coefficient, } K = \frac{V_{\text{actual}}}{V_{\text{ideal}}} \qquad (10.6)$$

From the first relation it follows that

$$\Delta h_{\text{actual}} = \eta \, \Delta h_{s=c}$$

where $\Delta h_{\text{actual}}$ = actual enthalpy drop across the nozzle
$\Delta h_{s=c}$ = enthalpy drop across the nozzle for a constant-entropy process

or for which no friction occurs and heat is transferred.

**Example 6.** Assume that experimental data indicate that the nozzle efficiency, $\eta$, is 1.0 for all nozzle sections up to and including the throat section and that $\eta$ is 0.95 from inlet to exhaust for the entire converging–diverging nozzle. Applying this added information to Example 5, calculate the actual nozzle for the same operating conditions.

*Solution:* The velocity and the nozzle area at the throat would be the same as for the point 5 shown in Table 10.1 or 1680 fps and 0.00256 sq ft, respectively.
The actual enthalpy drop to the exit, point 11, would be

$$\Delta h_{\text{actual}} = 0.95(1268.9 - 907.4)$$
$$= 361.4 \times 0.95$$
$$= 343.4 \text{ Btu per lb}$$
$$V_{11(\text{actual})} = 223.8\sqrt{343.4}$$
$$= 4140 \text{ fps}$$

Point 11 in Fig. 10.8 for the actual nozzle would fall on the 1-psia pressure line and at an enthalpy of $1268.9 - 343.4 = 925.5$. This enthalpy line cuts the 1-psia pressure line at $x = 82.6$ per cent. Thus, the condition for the actual nozzle at exit is $V = 4140$, $h = 925.5$, $x = 82.6$ per cent. The specific volume is $0.826 \times 333.6 = 274$ cu ft per lb. Compare these values with the values shown in Example 5. Now, the exit area becomes

$$A_{11} = \frac{mv}{V_{11}}$$
$$= \frac{1 \text{ lb per sec} \times 274 \text{ cu ft per lb}}{4140 \text{ fps}}$$
$$= 0.066 \text{ sq ft}$$

In the design of a nozzle for a gas turbine the same general principles may be applied as were used in the design of a steam nozzle. However, the working medium is a mixture of air plus the products of combustion resulting from the burning of a fuel. Because the properties of the mixture are approximately the same as for air, it is common practice in preliminary design to assume that the working medium is *air* and that the air acts *as a perfect gas*.

The fundamental relations which apply to the gas-turbine nozzle are:

(1) The energy balance

$$V_x = 223.8\sqrt{h_1 - h_x} \qquad (10.7)$$

(2) The continuity equation

$$m = \frac{A_1 V_1}{v_1} = \frac{A_x V_x}{v_x} \qquad (10.8)$$

(3) The equation of state of a perfect gas

$$pv = RT \qquad (10.9)$$

(4) The change in enthalpy for a perfect gas

$$h_1 - h_x = c_p(T_1 - T_x) \qquad (10.10)$$

(5) The equation of an adiabatic, frictionless process for a perfect gas

$$\frac{T_x}{T_1} = \left(\frac{p_x}{p_1}\right)^{\frac{k-1}{k}} = \left(\frac{v_1}{v_x}\right)^{k-1} \qquad (10.11)$$

Equations 10.8 and 10.9 have been discussed in Article 3.6, Case IV. The basic calculations for a nozzle design can be illustrated best by an example.

**Example 7.** Air enters a nozzle at 60 psia and 1500 F with a flow rate of 1 lb per sec. It expands without friction or heat transfer to a pressure of 15 psia (approximately atmospheric). Assume that air is a perfect gas and that $k = 1.4$; thus $c_p = 0.24$. (This assumption is not quite true because of the high temperature.) Calculate the area which the nozzle should have at the throat section.

*Solution:* The critical pressure ratio for air is 0.53; that is, the pressure at the throat will be 0.53 times the initial pressure. Thus,

$$p_x = 0.53 p_1$$

or
$$\frac{p_x}{p_1} = 0.53$$

The temperature ratio can be found by Equation 10.11. Thus,

$$\frac{T_x}{T_1} = \left(\frac{p_x}{p_1}\right)^{\frac{k-1}{k}}$$

$$= (0.53)^{\frac{1.4-1}{1.4}}$$

$$= 0.835$$

$$T_x = 0.835T_1, \text{ where } T_1 = 1500 + 460 = 1960 \text{ R}$$

$$= 0.835(1500 + 460)$$

$$= 1640 \text{ R} \quad \text{or} \quad 1180 \text{ F}$$

Now, the velocity at the throat can be calculated by combining Equations 10.7 and 10.10. Thus,

$$V_x = 223.8\sqrt{c_p(T_1 - T_x)}$$

Substituting $c_p = 0.24$, $T_1 - T_x = 1500 - 1180 = 320$ F,

$$V_x = 223.8\sqrt{0.24 \times 320}$$

or
$$V_x = 1960 \text{ fps} \quad \text{(throat velocity)}$$

The area at the throat can be found by means of the continuity equation 10.8, in which

$$A_x = \frac{mv_x}{V_x}$$

However, the specific volume, $v_x$, is still unknown but can be calculated by the use of the perfect gas law, $v_x = RT_x/p_x$. The values of $T_x$ and $p_x$ are known and are 1640 R and $60 \times 0.53$, respectively. Thus,

$$v_x = \frac{53.3 \times 1640}{60 \times 0.53 \times 144}$$

$$= 19.0 \text{ cu ft}$$

From $v_x = 19.0$ cu ft, $m = 1$ lb per sec, and $V_x = 1960$, the area

$$A_x = \frac{1 \times 19.0}{1960}$$

$$= 0.0097 \text{ sq ft or approximately } 0.01 \text{ sq ft}$$

NOTE: The area at the exit pressure or at any other pressure may be calculated in the same manner. It is suggested that the student do this for the exhaust pressure of 15 psia as an exercise.

### 10.5 Vector

To understand the principles underlying turbine-blade design, it is important to have a knowledge of both vector quantities and scalar quantities. Briefly, a scalar quantity is one that has *magnitude* only, for example, the distance a man walks measured in feet. A vector quantity has both *magnitude* and *direction,* for example, the displacement of a man as he walks toward the northeast measured in feet but with the direction specified as northeast. The size of a scalar quantity is designated by a number. The size and direction of a vector are usually represented by an arrow, its length to scale denoting size and its head indicating the appropriate direction.

To add a vector $V_b$ to a vector $V_a$ place the tail of $V_b$ to $V_a$ (retaining its same direction). The sum, $V_{ab}$, will be the vector which closes the polygon (triangle in this case), and the point of its arrow terminates at the point of vector, $V_b$. Refer to Fig. 10.12b.

In Fig. 10.12c the same vector, $V_b$, is subtracted from vector $V_a$. A simple way to do this graphically is to reverse the direction of vector $V_b$, and then proceed to add this vector to $V_a$. The vector closing the triangle is the vector difference, $V_a \rightarrow V_b$.

If the vectors are velocities or forces it is customary to write their addition as follows:

$$V_a \rightarrowtail V_b = V_{ab}$$

or

$$F_a \rightarrowtail F_b = F_{ab}$$

(10.12)

The order of the subscripts is important. Equation 10.12 reads: the velocity of a particle $a$ plus vectorially the velocity of an object $b$ is equal to the velocity of $a$ with respect to $b$. The resultant velocity, $V_{ab}$, is called the relative velocity of $a$ with respect to $b$.

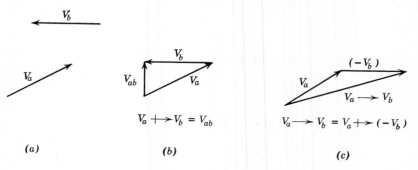

**Fig. 10.12.** Vector arithmetic.

### 10.6 Blade Work

Essentially, a turbine is a device for producing power. Power is the rate of doing work or the work per pound of fluid times the pounds of fluid flowing per unit of time. Turbine blading should be designed to convert as much of the kinetic energy of the fluid leaving the nozzle into work as is practicable. In order to determine the proper blade shape for the production of maximum work, it is necessary to review certain fundamentals and to make certain simplifying assumptions in the application of these fundamentals.

Let it be assumed that a fluid, such as steam, is flowing through a curved passage formed by blades or buckets and that the passage is capable of motion, as illustrated in Fig. 10.13. Further, let the symbols shown in Fig. 10.13, which will be used throughout the remainder of the chapter, represent the following:

1. $V_1$ and $V_2$ are the initial and final absolute velocities of the fluid, respectively.
2. $V_b$ is the absolute velocity of the passage.
3. $V_{1b}$ and $V_{2b}$ are the initial and final relative velocities of the fluid with respect to the passage.

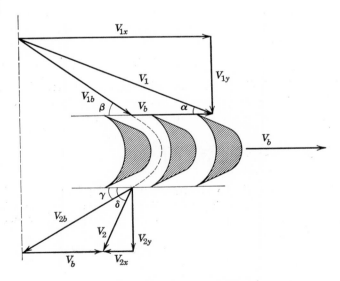

**Fig. 10.13.** Typical impulse blade.

4. $V_{1x}$ and $V_{2x}$ are the components of $V_1$ and $V_2$ in the direction of motion of the passage.

5. $V_{1y}$ and $V_{2y}$ are the components of $V_1$ and $V_2$ in a direction at right angles to the direction of motion of the passage.

6. $x$ represents the distance the passage moves in time $\tau$.

Referring to the blade passage in Fig. 10.13, let the following be assumed:

1. The absolute velocity of the fluid leaving the nozzle makes an angle with the direction of the blade motion equal to the nozzle angle $\alpha$.

2. The cross-sectional area between the blades remains constant from entrance to exit and the fluid completely occupies the passage.

3. The fluid enters the passage without shock; that is, $\angle \beta$ is equal to the entrance angle of the blade.

4. The fluid leaves with a relative velocity which has the same angle $\gamma$ as the exit angle of the blade.

5. If friction occurs, it is represented by a reduction in magnitude of $V_{1b}$ to a lower velocity $V_{2b}$ and the friction loss is represented by a velocity coefficient, $K = V_{2b}/V_{1b}$.

6. The blade moves at a constant velocity, $V_b$.

Now, work is defined as the product of force times the distance through which the force acts. Because the force often varies with distance and the work is a path function, the engineer expresses work mathematically as

$$\delta W = F_x \, dx \tag{10.13}$$

in which $\delta$ (delta) indicates that work is an inexact differential or a path function. $F_x$ is the component of the vector sum of the forces in the $x$ direction. Integrating Equation 10.13 between an initial and final state gives

$$_1W_2 = \int_1^2 F_x \, dx \tag{10.14}$$

Equation 10.14 cannot be integrated unless the force, $F_x$, is known. But Newton stated that the summation of forces acting on a body is proportional to the time rate of charge of momentum of the body. Mathematically,

$$\Sigma F \propto \frac{d(mV)}{d\tau} \tag{10.15}$$

or
$$F_x = \frac{d(mV)_x}{g_c \, d\tau} \tag{10.16}$$

and
$$F_y = \frac{d(mV)_y}{g_c \, d\tau} \tag{10.17}$$

Substituting the value of $F_x$ from Equation 10.16 into Equation 10.14 results in

$$_1W_2 = \frac{1}{g_c} \int_1^2 d(mV)_x \frac{dx}{d\tau} \tag{10.18}$$

but
$$\frac{dx}{d\tau} = V_b.$$

the blade velocity or the distance the passage moves in time, $\tau$. This was assumed to be constant; thus,

$$_1W_2 = \frac{mV_b}{g_c} \int_1^2 dV_x$$

Integrating gives

$$_1W_2 = \frac{mV_b}{g_c} (V_{2x} - V_{1x}) \tag{10.19}$$

This is the work done on the fluid by the passage. The work done on the passage by the fluid is equal and opposite to this or

$$\text{Work} = \frac{m}{g_c} (V_{1x} - V_{2x}) V_b \tag{10.20}$$

Except for the initial assumption, Equation 10.20 is very general and may be solved graphically or analytically. To solve it graphically one proceeds as follows:

1. Draw the fluid velocity to scale and at the nozzle angle, $\alpha$, with the blade motion.
2. Draw the blade velocity to scale as shown in Fig. 10.13.
3. Complete the velocity triangle by drawing a velocity, $V_{1b}$. The angle, $\beta$, obtained should be the entrance angle to the blade.
4. Draw $V_{2b}$ to scale and equal to $KV_{1b}$. The value of $K$ must be assumed. The angle, $\gamma$, may be made equal to or slightly less than angle $\beta$.
5. Add $V_b$ to $V_{2b}$ and complete the velocity triangle at exit by drawing $V_2$.

6. Draw the components of $V_1$ and $V_2$, $V_{1x}$, $V_{1y}$, $V_{2x}$, and $V_{2y}$, as shown.

7. The values $V_{1x}$, $V_{2x}$, and $V_b$ with their appropriate signs then can be substituted in Equation 10.20 and the work calculated.

An analytical solution is convenient for purposes of blade-work analysis and is as follows:

1. Complete the diagram as shown in Fig. 10.13, drawing in the initial and final components of $V_1$ and $V_2$.

2. The work is

$$_1W_2 = \frac{m}{g_c}(V_{1x} - V_{2x})V_b \tag{10.20}$$

but $\qquad V_{1x} = V_1 \cos \alpha$

and $\qquad V_{2x} = -V_2 \cos \delta$

the minus indicating a direction to the left, or

$$V_{2x} = -(V_{2b} \cos \gamma - V_b)$$

but $\qquad V_{2b} = KV_{1b}$

where $K$ = velocity coefficient and indicates the amount of friction in the blade passage.

$$\therefore V_{2x} = -(KV_{1b} \cos \gamma - V_b)$$

but $\qquad V_{1b} \cos \beta = V_1 \cos \alpha - V_b$

$$\therefore V_{2x} = \left(K\{V_1 \cos \alpha - V_b\}\right)\frac{\cos \gamma}{\cos \beta} - V_b\right]$$

Substituting this value in Equation 10.20 gives

$$_1W_2 = \frac{m}{g_c}\left[V_1 \cos \alpha - \left(-K\{V_1 \cos \alpha - V_b\}\right)\frac{\cos \gamma}{\cos \beta} - V_b\right]V_b$$

$$= \frac{mV_b}{g_c}\left[\left(V_1 \cos \alpha - V_b\right)\left(1 + K\frac{\cos \gamma}{\cos \beta}\right)\right]$$

or

$$_1W_2 = \frac{m}{g_c}\left[\left(V_1V_b \cos \alpha - V_b^2\right)\left(1 + K\frac{\cos \gamma}{\cos \beta}\right)\right] \tag{10.21}$$

To find the proper blade speed to give maximum work the following should be true:

$$\frac{dW}{dV_b} = 0$$

but $1 + K \dfrac{\cos \gamma}{\cos \beta}$ is a constant.   Thus,

$$\frac{dW}{dV_b} = V_1 \cos \alpha - 2V_b = 0$$

Solving for $V_b$ gives

$$V_b = \frac{V_1 \cos \alpha}{2} \tag{10.22}$$

If there is no friction, then $K = 1.0$ and the work is greater.   Also $\cos \gamma \div \cos \beta$ should be made as large as possible, or angle $\gamma$ should be made less than angle $\beta$.   In practice, angle $\gamma$ is made only slightly less than $\beta$ to prevent interference of the fluid leaving with the next on-coming blades.

If $\angle \beta = \angle \gamma$, $K = 1.0$, and $V_b = V_1 \dfrac{\cos \alpha}{2}$ , then

$$_1W_2 = \frac{m2V_b}{g_c} \left( V_1 \cos \alpha - V_1 \frac{\cos \alpha}{2} \right)$$

$$= \frac{2m}{g_c} \left( \frac{V_1{}^2 \cos^2 \alpha}{4} \right) = \frac{mV_1{}^2}{2g_c} \cos^2 \alpha \tag{10.23}$$

But the initial energy supplied to the blade passage is the kinetic energy of the fluid at entrance to the passage, or $mV_1{}^2/2g_c$.   Blade efficiency, $\eta_b$, is defined as the work done on the blade divided by the energy supplied to the blade passage, or

$$\eta_b = \frac{\text{Work}}{\dfrac{mV_1{}^2}{2g_c}} \tag{10.24}$$

From Equation 10.23 the maximum work is $\dfrac{mV_1{}^2}{2g_c} \cos^2 \alpha$; thus the maximum efficiency is obtained by combining Equations 10.23 and 10.24.   The maximum efficiency is

$$\eta_b = \frac{\dfrac{mV_1{}^2}{2g_c} \cos^2 \alpha}{\dfrac{mV_1{}^2}{2g_c}} = \cos^2 \alpha \tag{10.25}$$

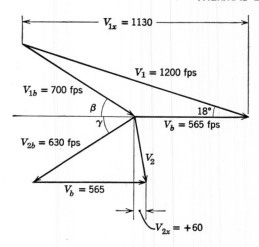

**Fig. 10.14.** Sketch for Example 8.

When the velocity coefficient is less than 1.0 and $\angle \beta \neq \angle \gamma$,

$$\eta_b = \left(1 + K \frac{\cos \beta}{\cos \gamma}\right) \frac{\cos^2 \alpha}{2} \qquad (10.26)$$

Theoretically, the nozzle angle, $\alpha$, should be made equal to zero in order that $\cos \alpha = 1.0$. This would result in a blade efficiency of 100 per cent. In practice angle $\alpha$ is made as small as is possible and still permit the fluid to enter one blade passage without interfering with the next blades entering the fluid stream.

**Example 8.** Air enters an impulse-blade section at a velocity of 1200 fps and at a nozzle angle of 18 degrees. The blade velocity is such as to give maximum efficiency. The blade-entrance angle is designed to provide shockless entrance, and the blade is symmetrical; that is, $\angle \beta = \angle \gamma$. Calculate (a) the work done on the blade per pound of air; (b) the efficiency of the blade.

*Solution:* (a) Draw the velocity diagram as shown in Fig. 10.14, making the blade velocity, $V_b$, $= V_1 \dfrac{\cos \alpha}{2} = \dfrac{1200 \cos 18}{2}$ 570.6 $= 565$ fps. (Refer to Equation 10.22.) (.94167) SHOULD BE (.95106)

$V_{1b}$ is solved graphically by taking the vector difference between $V_1$ and $V_b$. It makes an angle $\beta$ of 33 degrees with $V_b$. Angle $\gamma$ is made equal to angle $\beta$, and $V_{2b} = 0.9 V_{1b} = 0.9 \times 700 = 630$ fps. (Assume $K = 0.9$.)

$V_2$ is found by adding vectorially $V_b$ to $V_{2b}$.

Measure graphically or solve analytically for $V_{1x}$ and $V_{2x}$. Graphically, $V_{1x} = +1130$, $V_{2x} = +60$.

From Equation 10.20

$$_1W_2 = \frac{1}{g_c}(V_{1x} - V_{2x})V_b$$

$$= \frac{1}{32.2}(1130 - 60)565$$

$$= 18,800 \frac{\text{ft-lb}}{\text{lb}}$$

(b) The blade efficiency is the ratio of the blade work to the initial kinetic energy or

$$\eta_b = \frac{_1W_2}{\dfrac{V_1^2}{2g_c}}$$

in which $_1W_2 = 18,800 \dfrac{\text{ft-lb}}{\text{lb}}$ but

$$\frac{V_1^2}{2g_c} = \frac{(1200)^2}{2 \times 32.2} = 22,400 \frac{\text{ft-lb}}{\text{lb}}$$

Thus,

$$\eta_b = \frac{18,800}{22,400} = 0.84 \text{ or } 84\%$$

or by Equation 10.26

$$\eta_b = (1 + 0.9)\frac{\cos^2 18}{2} \quad - .84$$

$$= 1.9 \times \frac{0.88}{2} \qquad SHOULD\ BE\ .9045$$

$$= 0.84 \quad \text{or} \quad 84\%$$

$$.859 \qquad 85.9\%$$

## 10.7  Simple Impulse Turbine

The simple impulse turbine consists of a single row of nozzles and a single row of blades.  The DeLaval turbine shown in Fig. 10.2 is a simple impulse turbine.  A cut-away diagram (Fig. 10.15A) shows the nozzle and blade arrangement as viewed from the blade ends.  Below the cut-away diagram the variations in pressure, volume, and velocity of the steam are plotted.  The pressure decreases through the nozzle section only.  The steam volume increases as the pressure decreases.  The velocity increases in the nozzle section but decreases through the blade section as work is done.

If friction occurs in the blade section, as it does in the actual impulse turbine, there is a pressure drop and an increase in volume in the blade section as well as in the nozzle (Fig. 10.15A).  This frictional

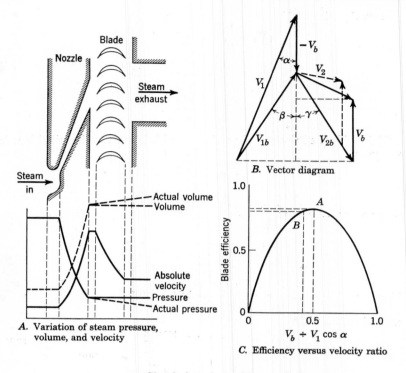

*B.* Vector diagram

*A.* Variation of steam pressure,
volume, and velocity

*C.* Efficiency versus velocity ratio

**Fig. 10.15.** Simple-impulse-turbine analysis.

effect will change the vector diagram (Fig. 10.15$B$) as shown by arrows drawn with dashes.

As discussed in the previous section, the blade velocity for a single row of impulse blades should be $V_1 \cos \alpha/2$. This is equally true of the simple impulse turbine if maximum efficiency is desired (Fig. 10.15$C$, point $A$).

Figure 10.15$B$ shows the vector diagram for a blade velocity $V_b$ which is not made equal to $V_1 \cos \alpha \div 2$ and which results in a blade efficiency shown at point $B$, Fig. 10.15$C$. Therefore, if the absolute velocity of the gas or vapor $V_1$ is large, as in the case when it expands in a single nozzle from a high pressure to a low pressure, the blade velocity should be correspondingly high.

However, it is customary to design turbines with mean blade velocities that will not exceed 1300 fps in order to avoid structural failure resulting from centrifugal force. Figure 10.16 shows a rear view of three typical, high-pressure, impulse turbine blades. Generally, they

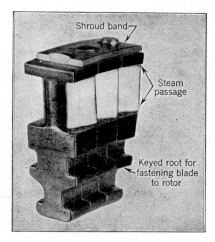

Shroud band

Steam passage

Keyed root for fastening blade to rotor

**Fig. 10.16.** Rear view of impulse turbine blades.

are made of a chrome–steel alloy, machined with key slots or dovetails at the root for fastening them to the rotor disk. A shroud band at the blade tips holds them together and prevents vibration.

The rotative speed of the turbine shaft is fixed by the design speed of the equipment being driven by the turbine. For a 60-cycle a-c two-pole generator the shaft speed must be 3600 rpm; for a four-pole generator, 1800 rpm. Auxiliary equipment such as fans and pumps have definite speed requirements for best efficiency. Speeds of such equipment seldom exceed 3600 rpm. If low speeds are desired, a gear reducer is placed between the turbine and the driven equipment.

The relation between blade velocity and shaft speed is given by the simple equation,

$$V_b = \frac{2\pi R n}{60} \quad \text{(see Fig 10.17)} \quad (10.27)$$

where $V_b$ = blade velocity, fps
$R$ = blade radius, ft
$n$ = shaft speed, rpm

Thus, for predetermined values of $V_b$ and $n$ the blade radius $R$ is fixed. If $R$ is large, the size and cost of the turbine are large. If $R$ is small, the volume of steam that can pass through the blade section is small, thus decreasing the total output of the turbine.

**Example 9.** Assume that a simple impulse turbine is to expand steam without friction from 200 psia and 500 F to 1 psia, that the nozzle makes an angle of 20 degrees with the blade motion, that the blades are symmetrical,

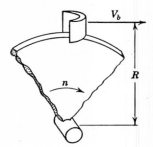

**Fig. 10.17.** Diagram for Equation 10.27.

and that the turbine speed is to be 3600 rpm. Find (a) the blade velocity to produce maximum efficiency, (b) the blade radius R, (c) the work done per lb of steam, and (d) the blade efficiency.

*Solution:* (a) Assuming that the steam enters the nozzle with a negligible velocity, the exit velocity can be found by applying the energy balance or Equation 10.3. Thus,

$$V_1 = 223.8\sqrt{h_0 - h_1}$$

in which $h_0 = 1268.9$ Btu per lb (at $p = 200$ psia and $t = 500$ F), and
$h_1 = 907.4$ Btu per lb (expanding to the exit pressure at constant entropy)

$$V_1 = 223.8\sqrt{1268.9 - 907.4}$$

$$= 4240 \text{ fps}$$

The proper blade speed to produce maximum efficiency is

$$V_b = \frac{V_1 \cos \alpha}{2}$$

$$= \frac{4240 \cos 20}{2}$$

$$= 1990 \text{ fps} \qquad \text{(refer to Fig. 10.18)}$$

**Fig. 10.18.** Velocity diagram for Example 9.

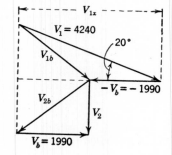

(b) Referring to Equation 10.27, the blade radius,

$$R = \frac{60 V_b}{2\pi}$$

$$= \frac{60 \times 1990}{2\pi 3600}$$

$$= 5.3 \text{ ft}$$

(c) From Equation 10.20, for $m = 1.0$ the work becomes

$$\text{Work} = \frac{1}{g_c}(V_{1x} - V_{2x})V_b$$

$$= \frac{1}{32.2}(V_1 \cos \alpha - 0)V_1 \frac{\cos \alpha}{2}$$

$$= \frac{1}{32.2}(4240 \cos 20)1990$$

$$= 247{,}000 \text{ ft-lb per lb of steam}$$

(d) $$\text{Blade efficiency} = \frac{247{,}000}{\dfrac{1}{2 \times 32.2}(4240)^2} \times 100 = 88.6\%$$

where $$\frac{(4240)^2}{64.4} = \text{initial kinetic energy}$$

In the example the blade velocity is too high, since 1300 fps is considered the optimum practical blade velocity. It could be decreased by reducing the pressure drop originally assumed. If it were reduced

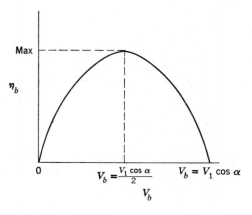

**Fig. 10.19.** Blade efficiency vs. $V_b$.

without a corresponding reduction in pressure drop, the blade efficiency would then decrease.

The effect of a variation in blade velocity on blade efficiency is shown in Fig. 10.19.

### 10.8   Pressure Staging

For the impulse turbine two methods are employed to decrease the blade velocity and still maintain maximum blade efficiency for a given range of inlet and exhaust pressures. The first method is called pressure staging. If two or more simple impulse turbines are placed in series with their blades connected to a common shaft, the fluid will expand from inlet pressure to exhaust pressure in stages. The number of stages or pressure reductions depends on the number of rows of nozzles through which the fluid must pass. Each stage is called a

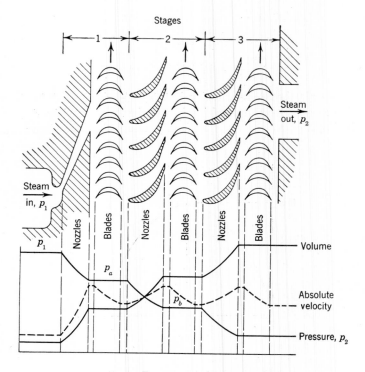

**Fig. 10.20.**   Rateau or pressure staging.

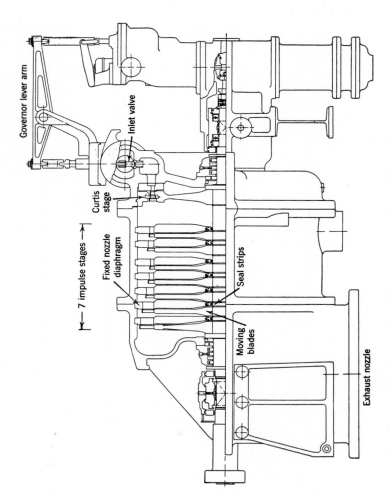

**Fig. 10.21.** An eight-stage medium-size impulse turbine.

Rateau stage.  A cut-away section of three Rateau stages is shown in Fig. 10.20.  The changes in steam or gas pressure, volume, and velocity are shown below the staging.

If pressures $p_1$ and $p_2$ are the same for the three Rateau stages as for the simple impulse turbine (Fig. 10.15), then the intermediate or stage pressures, $p_a$ and $p_b$, shown in Fig. 10.20, could be fixed by the nozzle design to distribute the kinetic energy of the fluid equally among the three stages.

By the use of three simple impulse turbines (three Rateau stages) acting in series between the same inlet and exhaust conditions as established for one simple turbine, the same work and efficiency can be maintained with a considerable decrease in blade velocity and blade radius.

In practice, the use of a number of Rateau stages in place of a single simple impulse stage will result in better efficiency.  Friction losses will be less for the lower steam velocities, and each stage will utilize some of the losses occurring in the preceding stage.  Rateau stages are used in medium- and high-capacity turbines (see Fig. 10.21) where good efficiency is mandatory.  The number of stages employed will be decided by the comparison of the increase in efficiency, the additional cost of the turbine, and the given capacity.

### 10.9  Velocity Staging

The second method employed in impulse-turbine design to decrease the blade velocity and still maintain maximum efficiency is called velocity staging.  Steam or gas is permitted to expand through a stationary nozzle from inlet pressure to exhaust pressure as in the simple impulse turbine.  As the pressure drops in the nozzle, the kinetic energy of the fluid increases by virtue of an increase in velocity.  A portion of the available kinetic energy is absorbed in a row of moving blades.  After doing work on these blades, the fluid enters a second row of blades which are stationary and merely act to redirect the flow.  The fluid after being redirected enters a second row of moving blades on which more work is done.

A cross section of such a stage and the change in fluid pressure, volume, and velocity are shown in Fig. 10.22A.  The vector diagram for the fluid entering and leaving each blade section is shown in Fig. 10.22B and represents the condition at which maximum blade efficiency will occur.  Unlike the simple impulse turbine or the pressure stage, the velocity stage utilizes more than one row of moving blades

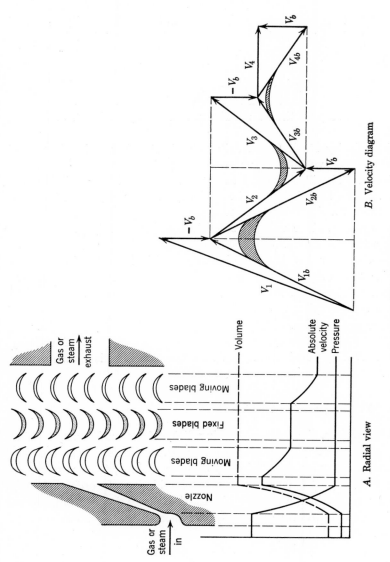

A. Radial view

B. Velocity diagram

**Fig. 10.22.** Velocity staging.

with a velocity decrease in each row. Only one row of nozzles is used per stage. The stationary blading does not change the magnitude of the velocity of the fluid leaving the first row of blades $V_2$ (see Fig. 10.22$B$. It merely redirects the steam to give a velocity $V_3$, equal in magnitude to $V_2$ or smaller when there is fluid friction.

To give maximum efficiency, the blade velocity $V_b$ as shown by Fig. 10.22$B$ becomes $V_1 \cos \alpha/2n$, where $n =$ number of rows of moving blades. Blade work is calculated for each row of moving blades in a manner similar to that used for a simple impulse turbine. The total work of the stage is the sum of the work done by each row of moving blades.

The velocity stage (often called Curtis stage) permits the use of a lower blade velocity at no sacrifice in theoretical blade work or efficiency. The blade radius can also be reduced for a given shaft speed. In an actual turbine the efficiency of a velocity stage is higher than that of a single impulse stage and lower than that of a series of Rateau stages. The velocity stage is used extensively in small turbines as shown in Fig. 10.5.

The Curtis or velocity stage is often used in large turbines to precede a series of pressure stages or reaction stages. The result is a cheaper turbine with little sacrifice in actual efficiency. The single Curtis stage replaces a number of pressure stages at the high-pressure end or inlet. At the low-pressure end, pressure stages have enough higher efficiency to warrant their use in place of velocity stages. A large turbine designed on these principles is shown in Fig. 10.21.

### 10.10   The Reaction Turbine

Theoretically, the reaction turbine differs from an impulse turbine using pressure stages only because the moving blades of the reaction turbine act as nozzles. As in the impulse turbine, fixed nozzles are provided between each two rows of moving blades. In order to insure a nozzle action in the moving blades, the blades are not built symmetrically but are constructed as the nozzle block shown in Fig. 10.11$B$. Thus, in a reaction turbine the fixed rows of nozzles and the moving rows of blades or nozzles are usually identical in form. A cross-sectional diagram of two reaction stages with the change in fluid pressure, volume, and velocity is shown in Fig. 10.23$A$.

The pressure decreases continuously through the stages since both

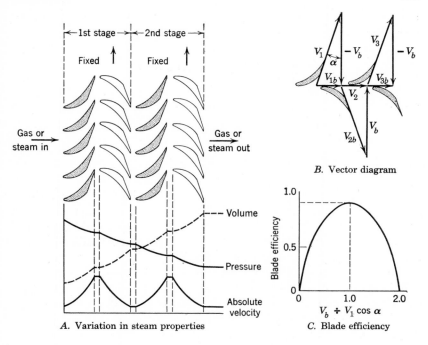

B. Vector diagram

A. Variation in steam properties

C. Blade efficiency

**Fig. 10.23.** Reaction-turbine staging.

fixed and moving rows of blades act as nozzles. The volume increases continuously. The absolute velocity increases in the fixed rows of nozzles but decreases in the moving rows of blades. This is evident after an inspection of the velocity diagram (Fig. 10.23$B$). The absolute velocity $V_1$ decreases to a velocity $V_2$, but the relative velocity increases from $V_{1b}$ to $V_{2b}$. The pressure drop for each row is small since a large number of stages are used between inlet and exhaust conditions. The small pressure drop in the nozzles results in a low fluid velocity per stage. Thus, generally the velocity of reaction-turbine blades is less than the velocity of impulse blades and may vary from 100 to 500 fps.

The velocity diagram of two reaction stages is shown in Fig. 10.23$B$. The blade shapes as drawn are called "symmetrical" when the two velocity diagrams are identical.

The work done by steam or gas on a reaction blade is calculated by the same method as was used to calculate the work done on an impulse blade. For example, the work per pound of fluid for the symmetrical stage shown in Fig. 10.23B is given by

$$W = \frac{1}{g_c}(V_{1x} - V_{2x})V_b \text{ ft-lb}$$

where
$$V_{1x} = V_1 \cos \alpha$$

$$V_{2x} = 0$$

Thus,
$$W = \frac{1}{g_c}(V_1 \cos \alpha - 0)V_b \qquad (10.28)$$

The blade velocity for maximum work is determined by differentiating the work with respect to blade velocity, equating the result to zero, and solving for $V_b$. The result shows that $V_b$ should equal $V_1 \cos \alpha$. This is the condition established when the blade-entrance angle is 90 degrees as shown in Fig. 10.23B.

The energy supplied to the reaction blade is equal to the initial kinetic energy, $V_1{}^2/64.4$, plus the energy transfer resulting from the steam expanding in the moving blade. In the case of the symmetrical reaction blade this energy is equal to $V_1{}^2/64.4$. Thus, the total energy supplied to the blade is $V_1{}^2/32.2$, and the blade efficiency becomes

$$\eta = \frac{W}{V_1{}^2/32.2} \qquad (10.29)$$

Combining Equations 10.29 and 10.28 gives

$$\eta = \frac{V_b V_1 \cos \alpha}{V_1{}^2}$$

$$= V_b \frac{\cos \alpha}{V_1} \qquad (10.30)$$

The maximum efficiency, when $V_b = V_1 \cos \alpha$, becomes

$$\eta_{max} = \cos^2 \alpha \qquad (10.31)$$

Equation 10.31 is true only for the first stage of a reaction turbine. An intermediate stage will be affected by the "carry-over" energy from the preceding stage. Carry-over energy is that portion of the kinetic energy of the preceding stage which is available in the next stage.

**Example 10.** A "symmetrical" reaction blade has an entrance angle of 90 degrees. Steam enters the blade with an absolute velocity of 500 fps. The fixed-nozzle angle, $\alpha$, is equal to 20 degrees. Calculate (a) the proper blade velocity, and (b) the blade work in ft-lb per lb of steam.

*Solution:* (a) The relative velocity of the steam $V_{1b}$ for shockless entrance should have the same angle as the blade-entrance angle; thus

$$V_b = V_1 \cos \alpha \qquad\qquad \text{(see Fig. 10.23B)}$$
$$= 500 \cos 20 \quad = 500\left(.93969\right) = 470.$$
$$= 470 \text{ fps}$$

(b) From the general expression for blade work,

$$W = \frac{1}{g_c}(V_{1x} - V_{2x})V$$

but
$$V_{1x} = V_1 \cos \alpha = V_b$$
$$= 470 \text{ fps}$$

and
$$V_{2x} = 0$$

Thus,
$$W = \frac{1}{32.2}(470)(470)$$
$$= 6860 \text{ ft-lb per lb}$$

In the impulse turbine there is very little drop in pressure through the moving blade sections. End thrust or axial thrust is produced mostly by the change in the components of the force of the jet which act at right angles to the direction of blade motion. The resulting thrust is quite small and can be absorbed in an especially designed thrust bearing. However, in the reaction turbine there is a pressure drop in each row of moving blades. This pressure drop or pressure differential is large when many stages are considered. Also, in the actual reaction turbine the rotor is cone-shaped (see Fig. 10.24) to provide for an increasing blade-ring area in order to accommodate the increasing steam volume. Pressure acting against the sloping sides of the rotor produces an added axial thrust. The total thrust cannot be absorbed in a bearing, and so dummy pistons are provided at the inlet end of the turbine against which the high inlet pressure is exerted. Proper design of the dummy piston sizes will counteract the axial thrust enough so that the thrust bearing can carry any excess.

Since there is a pressure differential across each row of moving blades in a reaction turbine, there is a tendency for the steam or gas to leak around the ends of the blades. This leakage tendency is not

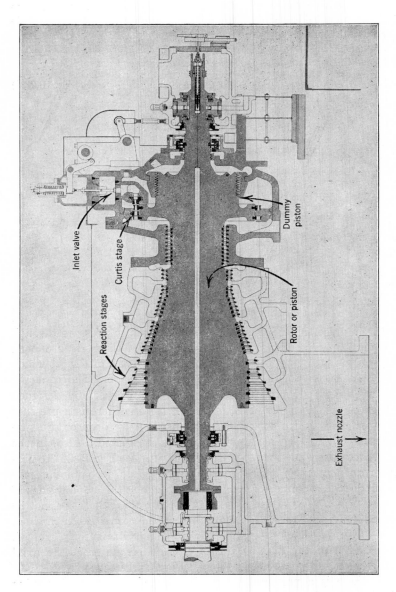

**Fig. 10.24.** A large reaction turbine with a Curtis stage at the inlet.

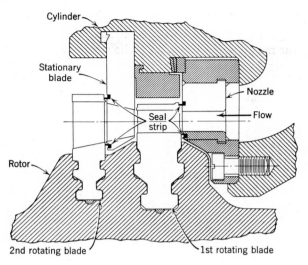

**Fig. 10.25.** Method of preventing leakage around blades by seal strips.

so pronounced in the impulse turbine. To prevent leakage seal strips are placed on the ends of the moving blades, as shown in Fig. 10.25. Seal strips on the fixed nozzle blocks at the point nearest the shaft are necessary in both the impulse and reaction turbines, as shown in Fig. 10.21. Seal strips and labyrinth arrangements are provided at the shaft on the high-pressure end of the turbine to prevent leakage to the surroundings. These are shown in Fig. 10.26.

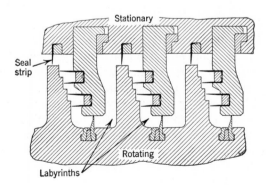

**Fig. 10.26.** Seal strips and labyrinths to prevent leakage at the shaft.

Seal strips are made thin to prevent excessive wear of the rotor or of the blades against which they may rub. Labyrinths are used to provide only the most circuitous path for the steam or gas. The pressure drop resulting from the friction of the steam as it passes around the seal strips of the labyrinths keeps leakage at a minimum.

Generally, in large reaction turbines the reaction staging is preceded by a Curtis impulse stage. The Curtis stage is placed at the high-pressure end of the turbine where it functions most efficiently and reduces the total number of reaction stages necessary. Thus, the size of the reaction turbine is decreased with very little sacrifice in overall efficiency. The Curtis stage can be seen in Fig. 10.24. The rotor diameter decreases following the Curtis stage because steam enters the reaction blading throughout the reaction-blade periphery, thus increasing the volume capacity, whereas steam flows through the Curtis stage through a relatively narrow arc of the blade periphery.

### 10.11   Types of Steam Turbines

Many types of industrial turbines are in use today, depending upon the conditions under which they must operate. They are classified as high- or low-pressure turbines, according to the inlet pressure of the steam, and as superposed, condensing, and noncondensing turbines, according to the exhaust steam pressure. A superposed or high back-pressure turbine is one that exhausts to pressures well above atmospheric pressure, 100 to 600 psi. A superposed turbine operates in series with a medium-pressure turbine. The exhaust steam of the superposed turbine drives the medium-pressure unit. The noncondensing turbine has lower exhaust pressures, but the steam still leaves at atmospheric pressure or above—15 to 50 psi. The exhaust steam may be used for drying or heating processes.

The condensing turbine operates at exhaust pressures below atmospheric pressure and requires two auxiliaries: a condenser and a pump. The condenser reduces the exhaust steam to water. As the steam is condensed and the water is removed by a pump, a partial vacuum is formed in the exhaust chamber of the turbine. This type of turbine is used chiefly for the low-cost electric power it produces.

If steam is required for processing, a turbine may be modified by extracting or bleeding the steam. Extraction takes place at one or more points between inlet and exhaust, depending upon the pressures needed for the processes. The extraction may be automatic or non-

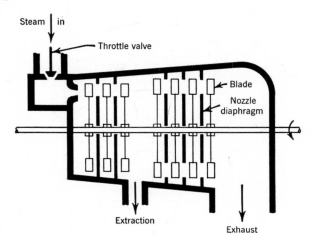

**Fig. 10.27.** Nonautomatic extraction.

automatic. Generally, factory processes require steam at a specific pressure; in this case, an automatic-extraction turbine is necessary. When steam is needed within the power plant itself for heating boiler feedwater, nonautomatic extraction is generally used. In Figs. 10.27 and 10.28 the two types of turbines are shown in simple form. In the automatic-extraction turbine, a diaphragm is inserted following the

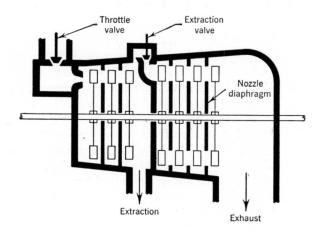

**Fig. 10.28.** Automatic extraction.

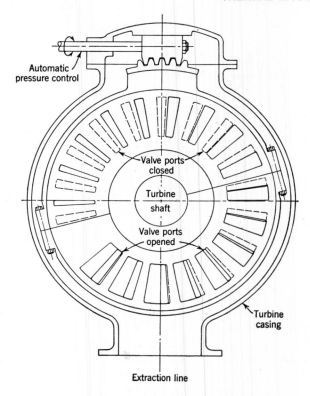

**Fig. 10.29.** A grid-type automatic-extraction valve.

extraction point. A valve operates within the diaphragm to restrict the flow of steam toward the exhaust so as to maintain a constant pressure at the extraction point. A grid-type diaphragm valve is shown in Fig. 10.29. At the top of the valve is the control gear which adjusts the valve opening.

Turbines may be classified according to their speed and size. Small turbines, varying in size from a few horsepower to several thousand horsepower, are used to drive fans, pumps, and other auxiliary equipment directly. The speed of these units is adjusted to the speed of the driven machinery or is converted by a suitable gear arrangement. These turbines are used wherever steam is readily available at low cost or where exhaust steam is needed.

Turbines for the production of electric power range in size from small units to those of over 500,000 kw, and the trend is toward even

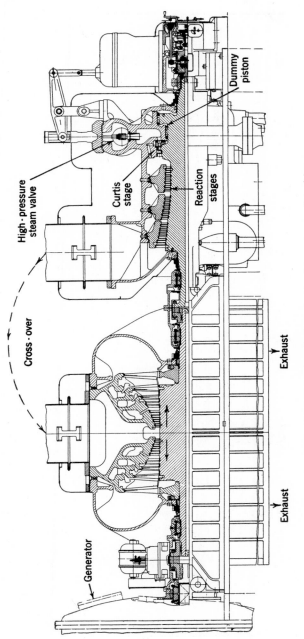

**Fig. 10.30.** A cross-sectional view of a tandem-compound turbine.

THERMAL ENGINEERING

**Fig. 10.31.** A tandem-compound steam turbogenerator.

larger units. Since the United States has standardized on a frequency of 60 cycles per sec for a-c power, turbogenerator units are constructed to operate at 3600 or 1800 rpm. The selection of the speed depends almost entirely on the size of the turbogenerator desired. The speed of 3600 rpm is preferred whenever the size of the turbine permits. The turbine operating at the higher speed has the following advantages: lighter weight, more compactness, and great suitability for high-pressure, high-temperature operation.

With a few exceptions turbines larger than 100,000 kw will operate at 1800 rpm. All turbines of smaller capacity will run at 3600 rpm. However, because of the advantages of the 3600-rpm unit and because of the greater efficiency of large units, turbine manufacturers will continue to raise the upper limit of speed and capacity.

Generally, turbogenerators on a single shaft and within a given speed range are constructed with either a single or a double-rotor. The double-rotor arrangement is used for only the largest turbines falling within a given speed range. A double-rotor unit is called a tandem-compound turbine, and the flow is double-exhaust to accommodate the large volumes of steam occurring at the low-pressure end. Such a turbine is shown in Figs. 10.30 and 10.31.

In the turbine shown in Fig. 10.31, steam enters at the left and flows through the high-pressure element to the looping pipe at the top. The steam then passes to the low-pressure double-flow element and flows to the front and rear of the second element and exhausts to a condenser below the floor. This provides twice the capacity of a single-flow turbine for a given speed. The generator, hydrogen-cooled, is the large cylindrical element in the center of the picture. The generator exciter is at the extreme right.

### 10.12   Applications of the Gas Turbine

The gas turbine is only one of three major elements composing the gas-turbine power plant in its simplest form. The essential details of the gas turbine are similar to those of the steam turbine. The nozzles, the blades, and the staging are designed by the use of basic principles which are equally applicable to all turbines.

The gas turbine or, more specifically, the gas-turbine power plant has been applied to both stationary and mobile power requirements. In small sizes it drives aircraft and air-conditioning equipment. In its simpler form it may be a jet-propulsion prime mover or a propeller-jet aircraft engine. As a prime mover for stationary power

**Fig. 10.32.** Gas-turbine axial-flow compressor unit with top half of casing removed.

plants the gas-turbine plant appears in more complex cycles and has the following advantages over the steam turbine: (1) small size, (2) a small water-supply requirement, and (3) quick starting.

The simple gas-turbine power plant consists of three major elements: (1) the air compressor, (2) a combustion chamber or combustor, and (3) the turbine. To these pieces of equipment must be added a starting motor, a generator or driven equipment, a governing system, and a lubrication system. Figure 10.32 is a shop view of a gas turbine and axial-flow compressor with the top casing removed. The six-stage reaction turbine in the foreground is quite similar in essential details to a steam turbine. The turbine is directly connected to an axial-flow compressor which is similar to a reaction turbine driven backward.

Figure 10.33 shows a diagrammatic arrangement of the gas-turbine power plant. Air at atmospheric pressure is compressed in the axial-flow compressor $B$ to a discharge pressure of 40 to 90 psig. The compressed air flows to the combustor $C$, where liquid fuel is injected continuously and burned at such a rate as to result in a gas temperature at the combustor outlet of 1300 F or more. The hot gas is then expanded through the turbine $A$ to atmospheric pressure. From 65 to 80 per cent of the output of the turbine is required to operate the compressor; the rest of the turbine output may be used to drive an

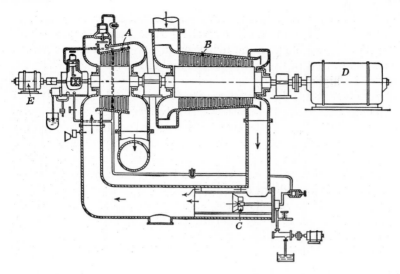

**Fig. 10.33.** Schematic diagram of arrangement of simple gas-turbine power plant.

electric generator $D$, the propeller of a ship, or the propeller of an airplane. A starting motor is necessary to rotate the compressor and turbine and start a flow of gas through the system.

The gas-turbine power plant as applied to aircraft operates on the same simple cycle as shown in Fig. 10.33.

In the turboprop engine, illustrated in Fig. 10.34, the excess power developed by the turbine over that required to drive the compressor

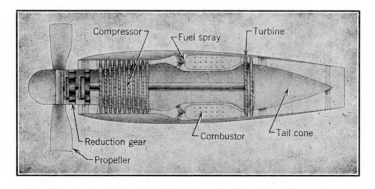

**Fig. 10.34.** Aircraft gas-turbine power plant with geared propeller.

is used to drive a propeller through speed-reducer gearing. The ability to build such units in capacities in excess of that obtainable from a piston engine plus the reduction in weight and frontal area as compared to that of the piston engine makes such a unit attractive in spite of its higher fuel consumption.

Aircraft that are designed for operation at speeds above 500 miles per hr are now powered almost exclusively by the jet-propulsion type of gas-turbine unit. Figure 10.35 illustrates diagrammatically a section through such a unit and shows the changes in pressure, temperature, and velocity as the gases flow through it. Because of the high speed of the airplane, the air enters an inlet section which is designed as a diffuser to slow down the velocity and convert part of the kinetic energy of the high-velocity air stream into pressure. The air is then compressed in a compressor, and fuel is supplied in a combustor as in the conventional gas-turbine power plant. However, in the turbojet engine, the turbine is designed to produce only enough power to drive the compressor. As shown in the pressure curve in Fig. 10.35, the

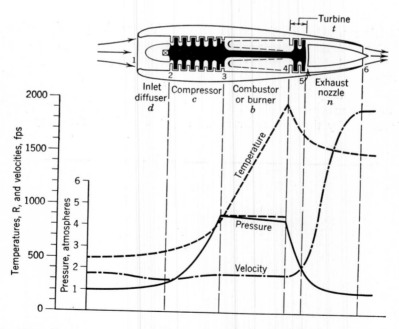

**Fig. 10.35.** Pressure, temperature, and velocity changes in the aircraft jet-propulsion unit.

gases leave the turbine at a back pressure that is considerably in excess of atmospheric pressure and are then expanded in a nozzle to atmospheric pressure. The increase in velocity as the gases expand in the nozzle is shown in Fig. 10.35. Since the gases leave the unit at a very much higher velocity than they had at entrance, the increase in velocity or momentum of the gases in flowing through the unit creates a reaction or thrust in the opposite direction that propels the airplane.

## PROBLEMS

Neglect the initial kinetic energy unless otherwise given.

1. Find the enthalpy and entropy of steam by use of the Mollier chart and the steam tables for the following conditions: (a) saturated steam at 100 psia, (b) steam at 14.7 psia and a quality of 80 per cent, (c) steam at 200 psia and 600 F.

2. Steam expands isentropically from 1000 psia and 1000 F to 200 psia. From the Mollier chart and the steam tables find the final temperature, the change in enthalpy, and the final specific volume.

3. Steam expands in a nozzle from 200 psia and 500 F to 1 in. Hg without friction or heat transfer. Find the final quality and specific volume of the steam. Calculate the exit area of the nozzle for a steam flow of 1 lb per sec.

4. For the same initial conditions as stated in Problem 3 and a flow rate of 1 lb per sec calculate the area of the nozzle at the throat if the critical pressure ratio is 0.55.

5. Steam expands in a nozzle from 600 psia and 700 F to 1 in. Hg abs. The nozzle efficiency is 90 per cent and the steam-flow rate is 2 lb per sec. Calculate the theoretical velocity leaving the nozzle, the actual velocity, and the actual exit area.

6. Using Equation 10.11, calculate the final temperature if the initial temperature of air is 740 F and the air expands adiabatically and frictionlessly for pressure ratios, $p_1/p_x$, of: (a) 2, (b) 3, and (c) 4. Assume a $k$ value of 1.4.

7. Calculate the change in enthalpy for each of the three cases in Problem 6. Assume a $c_p$ of 0.24 Btu per lb.

8. Calculate the final volume of the air for Problem 6 if the final pressure is 14.7 psia.

9. Air expands in a nozzle from 940 F and 60 psia to 15 psia without friction or heat transfer. Calculate the velocity at the nozzle throat (minimum section) and at the exit.

10. Find the nozzle areas at the throat and exit sections for a flow rate of 1 lb per sec and for the conditions stated in Problem 9.

11. Calculate the exit velocity for the nozzle conditions of Problem 9 if the nozzle has an efficiency of 90 per cent.

12. Air enters an impulse-blade section at a velocity of 1000 fps and at a nozzle angle of 18 degrees. Calculate the maximum, theoretical blade work, assuming symmetrical blading.

13. Calculate the blade work if, for Problem 12, the velocity coefficient is 0.90.

14. Calculate the blade work for a symmetrical reaction stage in which the blade-entrance angle is 90 degrees, the nozzle angle is 18 degrees, and the absolute steam velocity is 300 fps.

# CHAPTER 11

# Pumps

## 11.1 Introduction

One of the most important problems of the engineer is the efficient and controlled transfer of fluids from one point to another. This transfer may be opposed by gravitational force, by some other external force, or by friction. Under certain conditions the gravitational force and other forces may act to aid the transfer, but friction always exists as a force opposing motion. The engineer attempts to reduce the effect of friction and at the same time takes advantage of useful forces to produce a motion of the fluids under conditions that can be controlled.

As previously defined, a fluid is a substance in a liquid, gaseous, or vapor state which offers little resistance to deformation. Common examples of the three states of a fluid are water as a liquid, air as a gas, and steam as a vapor. All these types of fluids have a tendency to move because of natural forces acting on them. A city may be supplied with water flowing by gravity from high ground. Air may circulate in an auditorium because of its own temperature difference. Steam rises through the water in a boiler owing to the difference in density or specific weight of the steam and water. In many cases, however, the circulation is inadequate, and mechanical equipment must be built to supplement the natural circulation. Often mechanical circulation is the only means of obtaining the desired fluid flow. The equipment for producing this fluid flow is divided into two major classes: *pumps* for handling liquids, and *fans, blowers,* and *compressors* for handling gases or vapors.

Both classes of equipment in various forms may be found in the modern stationary power plant or in small mobile power plants such as the aircraft engine, Diesel locomotive, or automobile engine. Thus, in the field of heat power most mechanical methods of producing flow are utilized, and the fundamentals covered here are applicable in

### TABLE 11.1

#### Specific Gravity and Density of Some Liquids

| Liquid | Temperature, F | Specific Gravity | Density, lb per cu ft |
|---|---|---|---|
| Ethyl alcohol | 32 | 0.79 | 49.4 |
| Methyl alcohol | 32 | 0.81 | 50.5 |
| Draft gage oil | 32 | 0.84 | 52.4 |
| Water | 39 | 1.00 | 62.4 |
| Sea water | 59 | 1.02 | 64.0 |
| Mercury | 32 | 13.596 | 848.0 |

other fields of engineering. Before the general design and application of flow equipment are discussed, a brief survey will be made of the fundamentals of fluid properties and fluid-flow measurements.

### 11.2   Pressure-Measuring Instruments

Pressure is defined as a force per unit area, and in engineering applications its units are commonly pounds per square inch (lb per sq in., psi) or pounds per square foot (lb per sq ft, psf). However, it is often convenient to represent a given pressure by the height of a column of fluid that will produce at its base the given pressure. The height of the column is called pressure head or just head. For

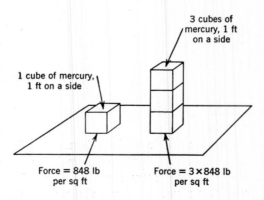

**Fig. 11.1.**  Variation of force at the bottom of a column of material.

example, at ordinary elevations on the earth's surface mercury weighs approximately 848 lb per cu ft (see Table 11.1). If a cube of mercury 1 ft on a side is placed on a surface, it will exert a force of 848 lb, a pressure of 848 psf, or a pressure of (848 ÷ 144) psi over the contact surface (see Fig. 11.1). If three such cubes are placed on top of each other, they will produce a force of 2544 lb, a pressure of 2544 psf, or a pressure of (2544 ÷ 144) psi. Thus, a pressure of 848 psf is equivalent to a head of 1 ft of mercury and a pressure of 2544 psf is equivalent to a head of 3 ft of mercury. A 1-in. head or column of mercury is equivalent to 0.491 psi.

The pressure or force per unit area exerted by a column of material depends on the height of the column and its average specific weight, or

$$p = h\gamma = h\rho \frac{g}{g_c} \tag{11.1}$$

where $p$ = pressure, psf

$h$ = height of column, ft

$\gamma$ = specific weight, $lb_f$ per cu ft = $\rho \dfrac{g}{g_c}$; see Article 1.11.

The use of Equation 11.1 is illustrated by several examples.

**Example 1.** What is the pressure in psi exerted by the water on a diver 200 ft below the surface of the ocean? Assume an average density of sea water as 64 lb per cu ft.

*Solution:*

$$p = h\gamma$$

where $h$ = 200 ft

$\gamma$ = 64 lb per cu ft, since $g/g_c$ is approximately 1.0 numerically

Thus,                         $p = 200 \times 64 = 12{,}800$ psf

or                            $p = 88.9$ psi

**Example 2.** A pressure of 20 psi would support a column of river water how many ft high?

*Solution:*

$$p = h\gamma$$

where $p$ = 20 × 144 psf

$\gamma$ = 62.4 lb per cu ft

Thus,                    $144 \times 20 = h \times 62.4$

$$h = 46.1 \text{ ft}$$

**Example 3.** If air has a specific weight of 0.075 lb per cu ft, how many ft of air would give a pressure equivalent to 3 in. of water?

*Solution:*

$$p = h\gamma$$

where $h = \frac{3}{12}$ ft
$\quad\gamma = 62.4$ lb per cu ft

Thus,
$$p = \frac{3}{12} \text{ ft} \times 62.4 \frac{\text{lb}}{\text{cu ft}} = 15.6 \text{ psf}$$

or
$$p = \frac{\frac{3}{12} \times 62.4}{144} = 0.108 \text{ psi}$$

This pressure is converted to feet of air by

$$p = h\gamma$$

where $p = 15.6$ psf
$\quad\gamma = 0.075$ lb per cu ft

Thus,
$$15.6 = h \times 0.075$$

$$h = 208 \text{ ft of air}$$

Fluid pressure is the force per unit area resulting from the bombardment of the molecules of a fluid against a confining surface. The force per unit area produced by the random motion of the molecules is called static pressure, and it is that pressure which tends to burst the walls confining the fluid. At a given point and time, a fluid, whether stationary or in motion, exerts an equal static pressure in all directions.

In a moving fluid the force per unit area produced by the ordered motion (motion in a definite direction) of a mass or group of molecules is called velocity pressure, and it is that pressure which is exerted only in the direction of motion of the fluid. Velocity pressure is

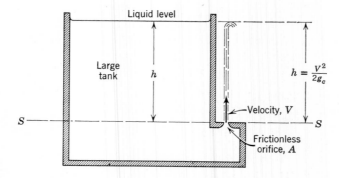

**Fig. 11.2.** Example of velocity head.

the force per unit area that will result from the complete conversion of the kinetic energy of a moving fluid to mechanical potential energy. Velocity head is the vertical distance a fluid will rise if its kinetic energy is completely converted to mechanical potential energy.

Velocity pressure and velocity head may be explained by a simple experiment. A large tank, as shown in Fig. 11.2, is filled with a liquid to a height $h$ ft above an orifice $A$ located at the horizontal surface $S$–$S$. The specific weight of the liquid is $\gamma$ lb per cu ft. The velocity of the liquid at the orifice $A$ is $V$ fps, and all particles of fluid are directed vertically upward. If the flow in the tank, in the orifice, and in the free jet leaving the orifice is frictionless, then the following deductions can be made:

1. The liquid at the surface of the large tank has a mechanical potential energy with respect to surface $S$–$S$ of $h$ ft-lb$_f$ per lb$_f$ or ft.

2. The kinetic energy of the liquid leaving the orifice is equal to the original mechanical potential energy or $V^2/2g$ ft-lb$_f$ per lb$_f$ or ft.

3. The liquid at the orifice has sufficient kinetic energy in the direction of flow to produce a jet $h$ ft high.

From these deductions it can be stated that the velocity head of the liquid at the orifice and in the direction of flow is $V^2/2g = h_v$, expressed in feet of the fluid flowing. Also, since the liquid at the orifice has only a vertical velocity component, its velocity head in any other direction is zero. The velocity pressure at the orifice is $\gamma V^2/2g = p_v$, psf.

The sum of the static pressure and the velocity pressure measured at the same point in the fluid is called total pressure.

In order to measure static, velocity, or total pressure, some method must be used to "pick up" the force resulting from the pressure and to transmit this force to a force-measuring instrument or pressure gage. For pressures that are not pulsating rapidly, the most common pressure "pickup" is a small tube or pipe. One end is exposed to the fluid under test; consequently, the tube is generally filled with this fluid. In some cases, however, this is not true, and these cases will be discussed later. The other end of the tube is connected by a pipe or hose to the pressure gage. The pressure to be measured is transmitted through the stationary fluid in the tube to the gage.

Static pressure or total pressure can be picked up by a single tube. Velocity-pressure measurements require two tubes: one tube to trans-

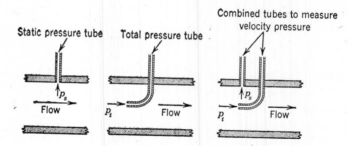

**Fig. 11.3.** Pressure "pick-up" tubes.

mit the total-pressure effect and the other tube to transmit the static-pressure effect. The velocity pressure is obtained by subtracting the static pressure from the total pressure. In Fig. 11.3 the method of locating the "pick-up" tubes for measuring the three pressures is shown.

The static tube must be placed at right angles to the direction of flow to avoid picking up any velocity pressure or impact effect of the fluid. The total-pressure tube must be turned directly into the flow stream in order that the full impact or velocity-pressure effect of the moving fluid may be transmitted to the pressure gage in addition to the static-pressure effect. The total-pressure tube is often called an impact tube or Pitot tube (named after Henri Pitot).

For convenience the double-tube arrangement used to measure velocity pressure is combined into a single unit, as shown in Fig. 11.4. One tube is placed inside another. The outer tube with openings in

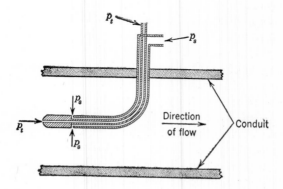

**Fig. 11.4.** A Pitot tube for measuring velocity pressure.

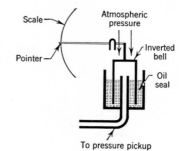

**Fig. 11.5A.** Bell gage.

the sides is the static-pressure tube. The inner tube, open at the end, is the total-pressure tube. The difference in pressures, $p_t - p_s$, as measured by some form of pressure gage, is the velocity pressure $p_v$.

*Instruments* or *gages* used to measure pressures that do not vary rapidly may be divided into two classes: mechanical and fluid-type instruments. The Bourdon gage discussed in Article 1.13 is a common type of mechanical gage. It is generally used to measure static pressures where precision of the instrument is not to be closer than a few pounds per square inch. Two other types of mechanical gages are shown in Fig. 11.5A and 11.5B. They are the inverted-bell gage (Fig. 11.5A) and the diaphragm gage (Fig. 11.5B). Both are commonly used to measure pressures or pressure differences of small magnitude. A variation in pressure on the underside of the bell or diaphragm will cause it to move up or down, and this movement is amplified by the lever arm connected to the gage pointer. Movement of the pointer depends on the difference in pressure between the atmosphere and the point of measurement. Thus, the pressure indicated, as in most gages, is a pressure difference, absolute pressure minus atmospheric pressure, called gage pressure.

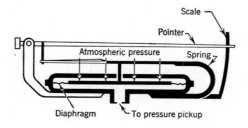

**Fig. 11.5B.** Diaphragm gage.

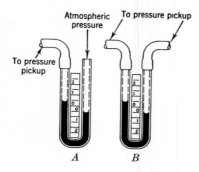

**Fig. 11.6A and B.**  U manometers.

The fluid-type gage is generally called a manometer and is essentially a U tube filled with a liquid such as mercury, water, or alcohol. The Greek prefix "mano," meaning thin or rare, infers that this gage is used to measure small pressures or small pressure differences. Three types of manometers are shown in Figs. 11.6*A*, 11.6*B*, and 11.6*C*. These operate in the same manner as a barometer described in Article 1.13, but they differ in that they register differences in pressure (in other words, both legs of the U are exposed and thus are under pressure). Usually, one leg of the U is at atmospheric pressure. However, in Fig. 11.6*B*, the manometer may have both legs attached to pressure pickups at pressures other than atmospheric pressure. Its use in measuring velocity pressure from the static- and total-pressure pick-up tubes is shown in Fig. 11.7, and the necessary calculations, with a proper consideration for corrections, are shown by Examples 4 and 5.

**Example 4.**  Assume that water flows through the pipe shown in Fig. 11.7 and that the manometer fluid is Hg.  If the difference in level of the Hg is

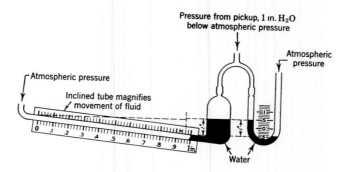

**Fig. 11.6C.**  Inclined-tube manometer.

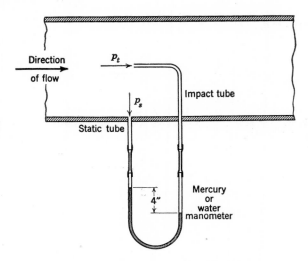

**Fig. 11.7.** Figure for Examples 4 and 5.

4 in., calculate (a) the velocity pressure in psi, and (b) the velocity head in ft of water.

*Solution:* (a) Assume that each tube is filled with water up to the manometer fluid (shaded). Then, the pressure on the left leg of the manometer, $p_s + 4$ in. Hg, just balances the pressure on the right leg of the manometer, $p_t + 4$ in. H$_2$O (4 in. H$_2$O displaces the 4 in. Hg), or

$$p_t + 4 \text{ in. H}_2\text{O} = p_s + 4 \text{ in. Hg}$$

and, rearranging,

$$p_t - p_s = 4 \text{ in. Hg} - 4 \text{ in. H}_2\text{O}$$

but          $p_t - p_s = p_v$, the velocity pressure

Thus,          $p_v = 4 \text{ in. Hg} - 4 \text{ in. H}_2\text{O}$

$$= 4 \text{ in. Hg} \times \frac{1 \text{ ft}}{12 \text{ in.}} \times \frac{13.6 \times 62.4 \text{ lb}}{\text{ft}^3} \times \frac{1 \text{ ft}^2}{144 \text{ in.}^2}$$

$$- 4 \text{ in. H}_2\text{O} \times \frac{1 \text{ ft}}{12 \text{ in.}} \times \frac{62.4 \text{ lb}}{\text{ft}^3} \times \frac{1 \text{ ft}^2}{144 \text{ in.}^2}$$

$$= 4 \text{ in. Hg} \times \frac{0.491 \text{ psi}}{\text{in. Hg}} - 4 \text{ in. H}_2\text{O} \times \frac{0.0361 \text{ psi}}{\text{in. H}_2\text{O}}$$

$$= 1.964 - 0.144$$

$$= 1.82 \text{ psi}$$

(b) From part a,

$$p_v = 1.82 \times 144 \text{ psf}$$

Thus,  $$h_v = \frac{1.82 \times 144 \text{ lb}}{\text{ft}^2} \times \frac{1 \text{ ft}^3}{62.4 \text{ lb}}$$

$$= 4.2 \text{ ft of water}$$

**Example 5.** Assume that air having a specific weight of 0.08 lb per cu ft flows through the pipe shown in Fig. 11.7 and that the manometer fluid is water. If the difference in level of the water is 4 in., calculate (a) the velocity pressure in psi, and (b) the velocity head in ft of air.

*Solution:* (a) The 4 in. of air in the right leg of the manometer which displaces the 4 in. of water may be neglected because of its low density.

Thus,  $$p_t = p_s + 4 \text{ in. H}_2\text{O}$$

or  $$p_t - p_s = 4 \text{ in. H}_2\text{O}$$

and  $$p_v = 4 \text{ in. H}_2\text{O}$$

$$= 4 \text{ in. H}_2\text{O} \times \frac{1 \text{ ft}}{12 \text{ in.}} \times \frac{62.4 \text{ lb}}{\text{ft}^3} \times \frac{1 \text{ ft}^2}{144 \text{ in.}^2}$$

$$= 4 \text{ in. H}_2\text{O} \times \frac{0.036 \text{ psi}}{1 \text{ in. H}_2\text{O}}$$

$$= 0.144 \text{ psi}$$

(b) From part a,

$$p_v = 0.144 \times 144 \text{ psf}$$

Thus,  $$h_v = \frac{0.144 \times 144 \text{ lb}}{\text{ft}^2} \times \frac{1 \text{ ft}^3}{0.08 \text{ lb}}$$

$$= 259 \text{ ft of air}$$

The manometer shown in Fig. 11.6C is used to increase the precision of the U-tube manometer by inclining one leg of the U. The inclined tube on the left amplifies the movement of the liquid in the manometer. The tube on the right is the ordinary U tube, shown for comparison. Both gages read a pressure of 1.0 in. of water less than atmospheric pressure.

The type of liquid used in a manometer depends upon the magnitude of the pressures involved and the desired precision of the instrument. A liquid having the larger density or specific gravity is used in manometers to measure the highest pressures. Table 11.1 gives the density and specific gravity of a number of liquids at standard temperatures and ordinary elevations.

The lines connecting the pressure pick-up elements or tubes and the pressure gages must be installed and operated in a manner such that they are filled at all times with only one known fluid; otherwise, unpredictable errors will arise. For example, when measurements of the pressure of a gas or vapor are being taken, the connecting lines should be kept free of pockets of liquids, such as condensed water vapor, or the connecting lines should be completely filled with water. When measurements of the pressure of liquids are being taken, the connecting lines should be kept free of gas or air pockets.

### 11.3  Measurements of Fluid Flow

In most processes where flow occurs, the engineer is interested in having a continuous record of the quantity of fluid flowing or a means of measuring, from time to time, the quantity. The most accurate method is calibrated weighing. Weighing immobile material involves no serious problem, but weighing continuously moving fluids introduces complications. Generally, flow cannot be interrupted. To prevent such an interruption more than one container is provided, with inlets to both as shown in Fig. 11.8. While one container is being filled, the other is being weighed and emptied. By alternating the process, flow can be maintained.

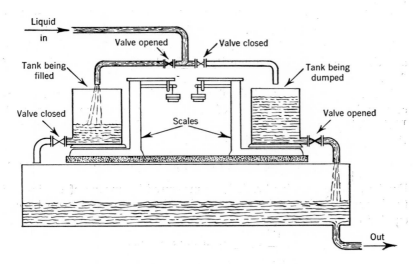

**Fig. 11.8.**  Weighing a liquid without interrupting flow.

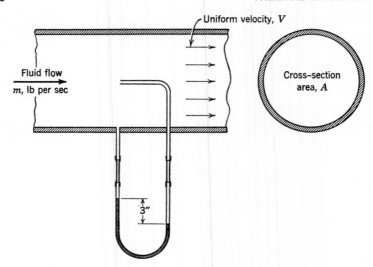

**Fig. 11.9.** Measurement of flow by pressure pickup (see Example 6).

The greatest disadvantage of the weighing process is that often the quantities involved are large and require bulky tanks and scales. In such cases, weighing is considered impractical and is only used for test and calibration purposes.

In closed conduits completely filled with the fluid, a convenient method of measuring the quantity of fluid flowing is to utilize the Pitot tube to measure velocity head, to convert this head into fluid velocity, and then to calculate the quantity flowing by the law of continuity of flow.

Let us assume that a fluid flows through a pipe having a cross-sectional area, $A$, with a flow of $m$ lb per sec as shown in Fig. 11.9. If the fluid completely fills area $A$ and has a density of $\rho$ lb per cu ft, then, by the law of continuity of flow,

$$m = AV\rho \tag{11.2}$$

Further, if the velocity head, $h_v$, is measured by means of static- and total-pressure tubes or by a Pitot tube, then the velocity can be calculated by

$$\frac{V^2}{2g} = h_v, \text{ ft of fluid flowing}$$

or

$$V = \sqrt{2gh_v} \tag{11.3}$$

Combining Equations 11.2 and 11.3 gives

$$m = A\rho\sqrt{2gh_v} \qquad (11.4)$$

Thus, by measuring $A$ and $h_v$ and determining $\rho$ for the conditions of the fluid, an indirect measure of $m$ is obtained.

If the velocity varies across the pipe diameter, it is necessary to find the mean velocity of the fluid by obtaining a number of readings of the velocity pressure at different points in the cross section. This is called traversing the area.

This method of obtaining the velocity of a fluid by velocity-pressure measurement is illustrated by Example 6.

**Example 6.** Find the velocity of a gas having a specific weight of 0.07 lb per cu ft if a water manometer connected as shown in Fig. 11.9 registers 3.0 in. of water.

*Solution:*

$$p = h_w\gamma_w = h_g\gamma_g$$

where $h_w = \frac{3}{12}$ ft of water

$\gamma_w$ = specific weight of water
    = 62.4 lb per cu ft
$h_g$ = velocity head, ft of the gas
$\gamma_g$ = specific weight of the gas
    = 0.07 lb per cu ft

Thus, $$p = \tfrac{3}{12} \times 62.4 = h_g \times 0.07$$

and $$h_g = 223 \text{ ft}$$

But $$V = \sqrt{2gh_v}$$

where $h_v = h_g = 223$ ft
    $g = 32.2$ ft per sec$^2$

Thus, $$V = \sqrt{2 \times 32.2 \times 223}$$
$$= 119.5 \text{ fps}$$

The Pitot tube with a manometer, as a means of determining the rate of flow, has certain disadvantages. It cannot, in many cases, be a permanent installation and cannot be traversed in a conduit through which a fluid flows under pressure without great difficulty. The Pitot tube is used, however, to measure the flow of air or other gases when under slight pressures.

In most permanent installations some form of flow meter such as shown in Figs. 11.10A, B, C is used. These metering devices are carefully designed obstructions placed in the flow path to produce a

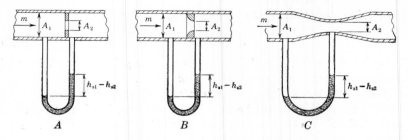

**Fig. 11.10.** The orifice, nozzle, and Venturi for measuring flow.

measurable decrease in static pressure and a corresponding increase in velocity pressure. A theoretical analysis based on the flow conditions existing at the constrictions will disclose how the flow rate may be calculated. Let it be assumed that the flow through any of the devices shown in Fig. 11.10 is such that

1. The flow is frictionless between points 1 and 2, where the areas are $A_1$ and $A_2$.

2. The potential energy change between points 1 and 2 is negligible.

3. The fluid completely fills areas $A_1$ and $A_2$, and the static pressure taps are located at areas $A_1$ and $A_2$.

On the basis of the first two assumptions it may be concluded that the total head at point 2 is equal to the total head at point 1. Since the total head, $h_t$, is defined as the velocity head, $h_v$, plus the static head, $h_s$, then

$$h_{v1} + h_{s1} = h_{v2} + h_{s2} \qquad (11.5)$$

or, rearranging,  $\qquad h_{s1} - h_{s2} = h_{v2} - h_{v1}$

but  $$h_{v2} - h_{v1} = \frac{V_2^2}{2g} - \frac{V_1^2}{2g}$$

thus  $$h_{s1} - h_{s2} = \frac{V_2^2}{2g} - \frac{V_1^2}{2g} \qquad (11.6)$$

As shown in Fig. 11.10, the manometers indicate the differential static head, $h_{s1} - h_{s2}$, if the density of the manometer liquid is neglected.

On the basis of the third assumption the equation of continuity of flow may be applied thus,

$$m = A_1 V_1 \rho_1 = A_2 V_2 \rho_2 \qquad (11.7)$$

From Equation 11.7,

$$V_1 = \frac{A_2 V_2 \rho_2}{A_1 \rho_1} \qquad (11.8)$$

Combining Equations 11.8 and 11.6, eliminating $V_1$, rearranging, and solving for $V_2$ gives

$$V_2 = \sqrt{2g(h_{s1} - h_{s2})\left[\frac{1}{1 - \left(\frac{A_2 \rho_2}{A_1 \rho_1}\right)^2}\right]} \qquad (11.9)$$

and, from Equation 11.7,

$$m = A_2 \rho_2 \sqrt{2g(h_{s1} - h_{s2})\left[\frac{1}{1 - \left(\frac{A_2 \rho_2}{A_1 \rho_1}\right)^2}\right]} \qquad (11.10)$$

In Equation 11.10 the areas, $A_1$ and $A_2$, and the change in static head, $h_{s1} - h_{s2}$, are measurable. The densities are known or can be calculated for the fluid flowing. Thus, the mass rate of flow, $m$, can be determined.

Since the original assumptions are not true for actual flow conditions, the rate of flow, $m$, as calculated from Equation 11.10, should be multiplied by one or more correction factors. These correction factors are determined experimentally and differ with each type of meter. The Venturi meter will produce results that most nearly approach the values obtained by the foregoing assumptions. The results obtained from an orifice require the greatest correction. Although the Venturi meter approaches the theoretically correct type of flow meter, its added bulk and weight restrict its use.

An application of the Venturi meter is illustrated in Example 7.

**Example 7.** Assume that the rate of flow of cold water is to be measured with a Venturi meter of the dimensions shown in Fig. 11.11. Assume that a mercury manometer indicates a difference in level of 4 in. Find the cu ft of water flowing per sec for frictionless flow.

*Solution:*

$$h_{s1} - h_{s2} = 4 \text{ in. of Hg} - 4 \text{ in. of } H_2O$$
$$= 4 \times 13.6 \text{ in. of } H_2O - 4 \text{ in. of } H_2O$$

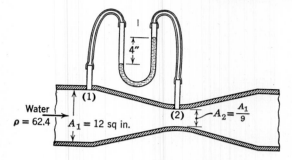

**Fig. 11.11.** Figure for Example **7.**

or $$h_{s1} - h_{s2} = \frac{4 \times 12.6}{12} \text{ ft of H}_2\text{O}$$

$$= 4.2 \text{ ft of H}_2\text{O}$$

Thus, $$\frac{V_2^2}{2g} - \frac{V_1^2}{2g} = 4.2 \text{ ft of water}$$

Also, since $\rho_1 = \rho_2$ for a liquid,

$$A_1 V_1 = A_2 V_2$$

where $A_1$ = area at point 1 or $\frac{12}{144}$ sq ft

$A_2$ = area at point 2 or $\dfrac{12}{144 \times 9}$ sq ft

Thus, $$\frac{12}{144} V_1 = \frac{12}{9 \times 144} V_2$$

or $$V_1 = \frac{V_2}{9}$$

Substituting the value of $V_1$ in the preceding equation, we get

$$\frac{V_2^2}{2g} - \frac{(V_2^2)}{2g(9)^2} = 4.2$$

or $$\frac{V_2^2}{64.4} - \frac{V_2^2}{81 \times 64.4} = 4.2$$

or $$V_2^2 = \frac{64.4 \times 4.2 \times 81}{80}$$

$$V_2 = 16.6 \text{ fps}$$

and $$m = A_2 V_2 \rho_2 = \frac{12 \times 16.6 \times 62.4}{9 \times 144}$$

or $$m = 9.6 \text{ lb per sec}$$

### 11.4 Pump Testing

In the selection or testing of a pump the first thing to be considered is its capacity in gallons per minute. Some methods of measuring flow have been discussed in Article 11.3. In pump testing, the most accepted and convenient method of measuring capacity is by means of the orifice, nozzle, or Venturi.

The next consideration is the total dynamic head against which the pump operates. The total dynamic head may be defined as the sum of the changes in elevation, static head, and velocity head of the fluid from intake to discharge of the pump.

The energy balance for the flow of one pound of mass, $m$, through a pump was developed in Chapter 2 and is as follows:

$$W_o = \frac{g}{g_c}(z_2 - z_1) + \frac{V_2^2 - V_1^2}{2g_c} + \frac{p_2 - p_1}{\rho} \qquad (2.18)$$

in which $W_o$ is the useful work done on the fluid in $\dfrac{\text{ft-lb}_f}{\text{lb}_m}$. If $W_o$ in

$\dfrac{\text{ft-lb}_f}{\text{lb}_m}$ is multiplied by $\dfrac{g_c}{g}$, $\dfrac{\text{lb}_m \ \text{ft}}{\text{lb}_f \ \text{sec}^2} \times \dfrac{\text{sec}^2}{\text{ft}}$, the result gives total dynamic heat, $h_t$, in feet. Thus,

$$h_t = (z_2 - z_1) + \frac{V_2^2 - V_1^2}{2g} + \frac{p_2 - p_1}{\gamma} \qquad (11.11)$$

in which $z_2 - z_1$ = change in elevation between pick-up tube locations, ft of fluid flowing

$\dfrac{V_2^2}{2g} - \dfrac{V_1^2}{2g}$ = change in velocity head between pick-up tube locations, ft of fluid flowing

$\dfrac{p_2 - p_1}{\gamma}$ = change in static head between pick-up tube locations, ft of fluid flowing

In Equation 11.11 and as illustrated in Fig. 11.12, the elevations $z_1$ and $z_2$ are measured with respect to the pump center line. This is in accordance with test code practices. However, it is evident that $z_2 - z_1$ is constant, regardless of the datum from which $z_1$ and $z_2$ are measured.

The horsepower delivered to the fluid by the pump is called the hydraulic horsepower and is equal to

$$\text{Hydraulic hp} = \frac{h_t w}{33,000} \qquad (11.12)$$

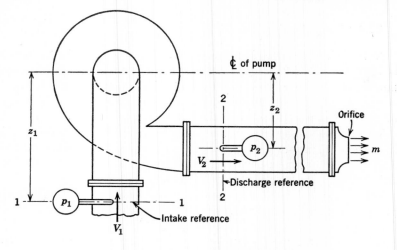

**Fig. 11.12.**  Pump test arrangement.

in which $h_t$ = total head, ft of fluid flowing
$w$ = weight rate of flow, lb per min

The pump horsepower is the sum of the power required to overcome the pump losses and the hydraulic power and is equal to the hydraulic horsepower divided by the pump efficiency, $\eta$, or

$$\text{Pump hp} = \frac{h_t w}{\eta 33{,}000} \tag{11.13}$$

Example 8 will be used to illustrate the calculations for a pump test assembly:

**Example 8.**  Assume that the following data are obtained for the pump shown in Fig. 11.12.  The rate of flow of water as measured by the orifice is 8000 lb per min; the intake static gage is located 4 ft below the pump center line and reads 10 psig; the discharge static gage is 2 ft below the pump center line and reads 50 psig.  The gages are located as close to the pump as possible. The areas of the intake and discharge pipes are 1 sq ft and $\frac{3}{4}$ sq ft, respectively. The pump efficiency is 70 per cent.  Calculate: (*a*) capacity of the pump, gpm; (*b*) total dynamic head, ft; (*c*) hydraulic hp, (*d*) pump hp.

*Solution:* (*a*)

$$\text{Capacity, gpm} = \frac{8000 \text{ lb per min}}{8.33 \text{ lb } H_2O \text{ per gal } H_2O}$$

$$= 965 \text{ gpm}$$

(b) $$h_t = (z_2 - z_1) + \left(\frac{V_2{}^2 - V_1{}^2}{2g}\right) + \left(\frac{p_2 - p_1}{\gamma}\right)$$

where $z_2 = -2$ ft
$z_1 = -4$ ft

and $p_2 = 50 \times 144$ psf

$p_1 = 10 \times 144$ psf

$\gamma = 62.4$ lb per cu ft

From the continuity-flow equation, assuming $\gamma$ is equal to $\rho$, numerically,

$$V_1 = \frac{w}{A_1\gamma} = \frac{8000}{1 \times 62.4}$$

$$= 128.2 \text{ fpm}$$

$$= 2.14 \text{ fps}$$

and $$V_2 = \frac{w}{A_2\gamma} = \frac{8000}{\frac{3}{4} \times 62.4}$$

$$= 170.8 \text{ fpm}$$

$$= 2.84 \text{ fps}$$

Thus, $h_t = [-2 - (-4)] + \left[\dfrac{(2.84)^2 - (2.14)^2}{2g}\right] + \left[\dfrac{144}{62.4}(50 - 10)\right]$

$$= 2 + 0.1 + 92.4 = 94.5 \text{ ft}$$

(c) $$\text{Hydraulic hp} = \frac{wh_t}{33,000}$$

$$= \frac{8000 \times 94.5}{33,000}$$

$$= 22.9$$

(d) $$\text{Pump hp} = \frac{\text{hydraulic hp}}{\text{pump efficiency}}$$

$$= \frac{22.9}{0.7}$$

$$= 32.7$$

In referring to Equation 11.11 and Fig. 11.12 it should be emphasized that the static-pressure gages must be located at the reference planes shown by $z_1$ and $z_2$. If for practical reasons the gages cannot be so located, it will be necessary to correct the gage readings to those values which the gages would read if they were to be moved to the

reference planes. Some rules may be stated for these gage corrections.

1. No gage correction is necessary if the gage tube from the reference plane to the gage is filled with air. Normally, this will occur
when the gage reading is a vacuum and the gage is above its connection point. Because of the low density of the air in the connection
tube, the correction is assumed to be zero.

2. If the gage tube is filled with the liquid flowing and the gage
is located below the reference plane, the gage will read high by the
difference in elevation between the plane of reference and the gage.

3. If the gage tube is filled with the liquid flowing and the gage is
located above the plane of reference, the gage will read low by the
difference in elevation between the plane of reference and the gage.

4. If a manometer is used in which the liquid differs from the liquid
being pumped, a manometer correction should be made. This type
of correction is shown in Examples 4 and 7 in this chapter.

**Example 9.** Assume that the following data are obtained for the pump
shown in Fig. 11.13: The pump delivers oil, specific gravity of 0.8, at the rate
of 10,000 lb per min; the intake static gage is located 10 ft above the intake
reference point and reads 10 in. of Hg vac; the discharge-pressure gage is
located 12 ft above the discharge reference plane and reads 100 psig. The
areas of intake and discharge pipes are equal, and the reference planes are at
the same elevation. Calculate (a) the total dynamic head, (b) the hydraulic hp.

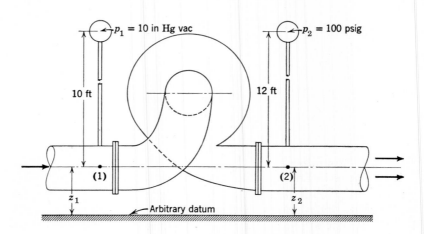

**Fig. 11.13.** Pump arrangement for Example 9.

*Solution:* (a) The total dynamic head is

$$h_t = (z_2 - z_1) + \left(\frac{V_2^2}{2g} - \frac{V_1^2}{2g}\right) + \left(\frac{p_2 - p_1}{\gamma}\right)$$

where $\qquad\qquad z_2 - z_1 = 0 \quad$ given

and $\qquad\qquad \dfrac{V_2^2}{2g} - \dfrac{V_1^2}{2g} = 0 \quad$ since $\quad A_1 = A_2$

The discharge gage reads low by 12 ft of the fluid flowing; thus, the static discharge head is actually greater by 12 ft, or

$$\frac{p_2}{\gamma} = \frac{100 \times 144}{62.4 \times 0.8} + 12$$

where $\gamma = 62.4 \times 0.8 = $ specific weight of the oil, lb per cu ft

Thus, $\qquad\qquad \dfrac{p_2}{\gamma} = 288 + 12 = 300$ ft of oil

The intake gage reads a vacuum; thus, the pressure at point 1 is not sufficient to sustain a column of oil. The gage tube will be filled with air, and no correction is necessary because of the low density of air. The intake head is

$$\frac{p_1}{\gamma} = \frac{-10 \times 0.491 \times 144}{62.4 \times 0.8} \qquad (\gamma_1 = \gamma_2 = 62.4 \times 0.8)$$

$$= -14.2 \text{ ft of oil}$$

The total head becomes $\qquad h_t = 300 - (-14.2)$

$$= 314.2 \text{ ft of oil}$$

(b) The hydraulic hp is

$$\text{Hydraulic hp} = \frac{wh_t}{33,000}$$

where $h_t = 314.2$ ft
$\qquad w = 10,000$ lb per min

Thus, the hydraulic hp $\qquad\qquad = \dfrac{10,000 \times 314.2}{33,000}$

$$= 95.2$$

## 11.5  Pump Types

The conditions under which liquids are to be transported vary widely and require a careful analysis before the proper selection of a pump can be made. Generally, the engineer purchasing a pump consults with pump manufacturers to obtain the best type for a particular job. However, a fundamental knowledge of the basic types of pumps that are available and a realization that there is a wide variety of the basic types are of great value to the prospective purchaser.

The conditions that will influence the selection of the type of pump are:

1. The type of liquid to be handled: that is, its viscosity, cleanliness, temperature, and so on.
2. The amount of liquid to be handled.
3. The total pressure against which the liquid is to be moved.
4. The type of power to be used to drive the pump.

Pumps may be divided into four major classifications:

1. Piston pumps or reciprocating pumps driven by engines or electric motors.
2. Centrifugal pumps driven by steam turbines or electric motors.
3. Rotary pumps driven by steam turbines or electric motors.
4. Fluid-impellent pumps which are not mechanically operated but are fluid-pressure-operated.

### 11.6   Reciprocating Pumps

The most simple, reliable, flexible, and inexpensive pump is the single direct-acting steam pump, commonly called a simplex pump. Essentially it consists of a steam cylinder and piston in line with a water cylinder and piston. The two pistons have a common piston rod, as shown in Fig. 11.14.

Steam under a pressure $p$ acts against a piston-face area $A$, to produce a force, $F = pA$. If the piston stroke is of length $L$, the work per stroke is $FL$ or $pAL$. If no losses are considered, the work of the

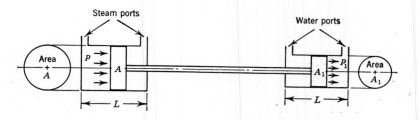

**Fig. 11.14.**  Simplex pump.

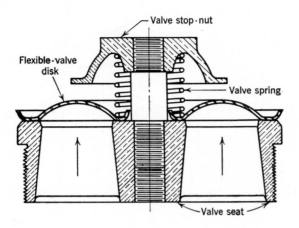

**Fig. 11.15.** A spring-loaded disk-type water valve.

steam piston is transferred to the water piston and in turn is transferred to the water, thus,

$$pAL = p_1 A_1 L$$

or
$$pA = p_1 A_1$$

Therefore,
$$p_1 = \frac{pA}{A_1}$$

By varying the steam pressure $p$, or each of the areas $A$ or $A_1$, any reasonable water pressure can be obtained. If a high water pressure is demanded, such as in the case when water must be supplied at 1500 psia without a correspondingly high steam pressure, the area $A_1$ must be made quite small or the product $pA$ made large.

If the water pressure is to be maintained, the steam pressure must also be maintained. Thus, the steam is not permitted to expand but is introduced throughout the full length of the stroke. A slide-D valve or piston valve is used on the steam cylinder and opens or closes only at the beginning or end of the stroke. The water valves are generally the spring-loaded disk type, similar to the one shown in Fig. 11.15. If the pressure acting on the underside of the flexible-valve disk becomes great enough to overcome the spring force and the pressure acting on the upper side of the disk, the valve disk will lift from its seat. As the pressure drops, it is returned to the closed position by the spring.

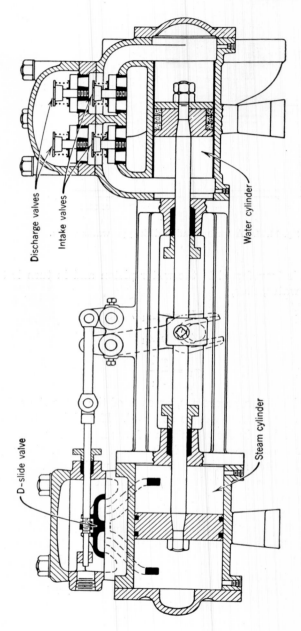

**Fig. 11.16.** Duplex pump.

The simplex pump, owing to its reciprocating motion, produces a pulsating flow. The pulsations can be reduced by operating two simplex pumps in parellel but out of phase with each other. If the two pumps are installed as a unit, the unit is called a duplex pump. A side view of the duplex pump is shown in Fig. 11.16.

Although steam-driven reciprocating pumps are becoming less common, the reciprocating pump still has many applications. It is used in hydraulic servomotor systems in which the driving fluid may be oil in place of steam. It has many uses where high pressures with low capacity are demanded.

Under heavy duty and severe service or where gritty or dirty liquids are pumped, the piston-type pump is generally replaced by a plunger-type reciprocating pump (see Fig. 11.17). The plungers are connected by side rods located outside the cylinder. The plungers have separate cylinders. In Fig. 11.17, if the plungers move to the left, water will flow in the directions indicated. The wearing surfaces of the plungers are clearly visible, and packing can be replaced with a minimum down time.

Both the piston and plunger pumps may be motor-driven rather than steam-driven. A belt drive reduces the speed of the motor, and a flywheel provides sufficient inertia to carry the pump past the end of the strokes. A relief valve must be placed on the pump-discharge line to prevent overloading the motor in case the liquid pressure becomes excessive. The relief valve is not necessary when the pump is steam-driven, since the steam piston will stop without being damaged if overloaded.

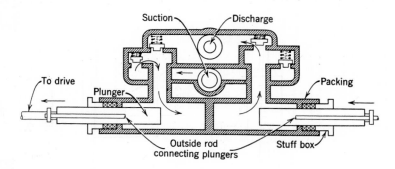

**Fig. 11.17.** Plunger-type reciprocating pump.

### 11.7   Centrifugal Pumps

In the reciprocating pump, the pressure head is produced by the transverse motion of a piston.  In the centrifugal pump, the head is produced by the centrifugal force imparted to the liquid from the rotary motion of an impeller.  An illustration of the principle involved is shown in Fig. 11.18.  A tank is filled with a liquid.  Risers are attached to a tube extending from the tank.  The assembly is keyed to a vertical shaft and is free to rotate.  If the assembly is stationary, the liquid level is the same in the risers and tank.  If the assembly is rotated, the liquid will rise in the risers and will drop in the tank, as shown in Fig. 11.18.  The difference in liquid level or pressure head $h$ produced at a distance $R$ from the center of rotation is due to the centrifugal force imparted to the liquid.  The equation for the head is

$$h = \frac{V^2}{2g}$$

where $h$ = head, ft of the liquid

$V = \dfrac{2\pi R n}{60}$ = linear velocity at the distance $R$ from the center of

rotation, fps

$g$ = acceleration of gravity

This is the mechanical potential energy of the liquid.  The fluid also has a mechanical kinetic energy due to the motion of the assembly.  The kinetic energy $V^2/2g$ is called the velocity head of the liquid.  The total head of the liquid is the sum of the pressure head and

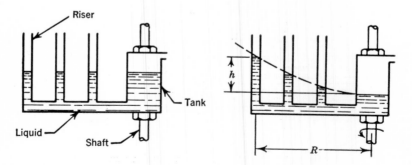

**Fig. 11.18.**  A pressure head produced by centrifugal force.

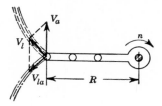

**Fig. 11.19.** Plan view of assembly in Fig. 11.18.

velocity head, or $V^2/g$.  Since $V = 2\pi Rn/60$, then the total head is equal to $(2\pi Rn)^2/3600g$.

If water were supplied to the tank in Fig. 11.18 and if the pressure head produced by rotation were sufficient to cause water to spill over the end riser, a crude centrifugal pump would result.  To a motionless observer looking down on the assembly, the water leaving the riser would appear to trace a spiral and move in the same direction as the direction of motion of the assembly.  To an observer located on the moving assembly, the water would appear to leave in a direction opposite to the direction of motion of the assembly (see Fig. 11.19). The motions of the liquid and of the assembly are shown by the velocity vectors.  $V_a$ is the absolute velocity of the assembly.  $V_l$ is the absolute velocity of the liquid, and the vector difference between them gives the relative velocity of the liquid with respect to the assembly.

The centrifugal pump, which utilizes these principles, consists of an impeller or rotating section to produce the flow and a casing to enclose the liquid and to direct it properly as it leaves the impeller. The liquid enters the impeller at its center or "eye" and parallel to the shaft.  By centrifugal force the liquid passes to the impeller rim

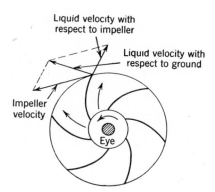

**Fig. 11.20.** Velocities of liquid and impeller at impeller outlet.

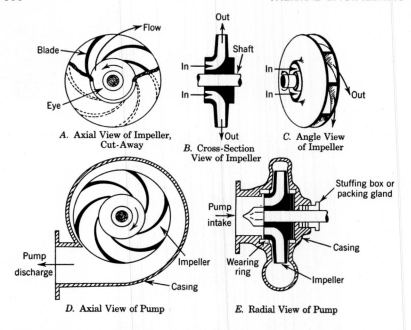

**Fig. 11.21.**  A single-suction enclosed-impeller volute centrifugal pump.

through the space between the backward curved blades, as shown in Fig. 11.20.   As shown in Fig. 11.19, the velocity of the liquid with respect to the impeller is $V_{la}$, which is in a direction opposite to the impeller motion.   The impeller blades are curved backward to permit the liquid to flow to the rim of the impeller with a minimum of

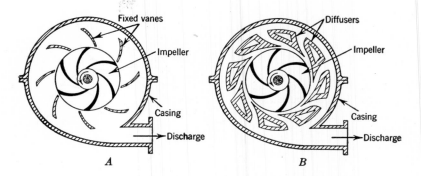

**Fig. 11.22.**  Two types of diffuser vane centrifugal pumps.

friction. As the liquid leaves the impeller, it is thrown in a spiral motion forward with a velocity, $V_l$.

The water is guided away from the impeller by two basic types of casings: the volute, and the turbine or diffuser. A simple volute pump is shown in Fig. 11.21$A$ to $E$. Liquid enters the impeller at the "eye," is thrown to the outside, and leaves the pump through the expanding spiral or volute casing. The casing has the volute shape to permit flow with a minimum of friction and to convert a part of the velocity head into static head. The static head is the head that overcomes resistance to flow.

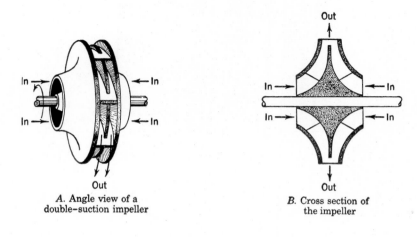

A. Angle view of a
double-suction impeller

B. Cross section of
the impeller

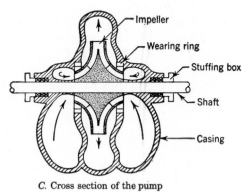

C. Cross section of the pump

**Fig. 11.23.** A double-suction pump.

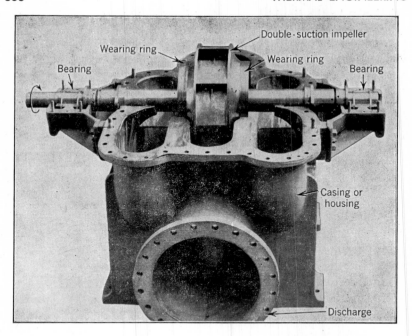

**Fig. 11.24.** Lower housing of a double-suction volute centrifugal pump.

The turbine or diffuser pump has the same type of impeller as the volute pump. The casing has a circular shape, and within the casing is a diffuser ring on which are placed vanes (see Figs. 11.22A and B). The vanes direct the flow of liquid and a decrease in the velocity of the liquid occurs because of an increase in the area through which the liquid flows. Thus, part of the velocity head is converted into static head as in the volute pump. For a multistage pump, the diffuser pump has a more compact casing than the volute pump. The diffuser-pump design is adaptable to differences in flow conditions since the same casing can be used with various arrangements of diffuser vanes. In the volute pump a variation in the requirements of the volute casing demands alterations in the casing itself. Generally, the volute pump will be used for low-head high-capacity flow requirements and the diffuser pump for high-head requirements.

Both volute and diffuser pumps are classified by the type of impeller, the number of stages, and the type of suction or intake used. A pump similar to the one shown in Fig. 11.21 but having two "eyes" on the impeller is called a double-suction pump. The double suction, one "eye" located on each side of the impeller, permits forces acting on the impeller to be balanced, thus reducing the axial thrust on the shaft. Also, the double-suction pump is used for handling hot water where there is danger of water flashing into steam at points of low pressure. The double suction offers little resistance to flow; thus, low-

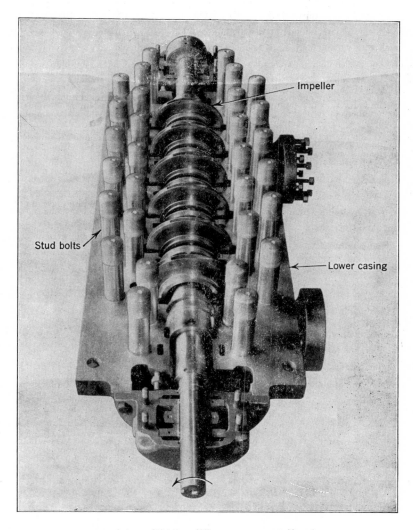

**Fig. 11.25.** A multistage diffuser-vane centrifugal pump.

pressure areas are less apt to occur.  The double-suction pump is used also for large capacities.  A double-suction pump is shown in Figs. 11.23*A, B, C* and 11.24.

When two or more impellers are mounted on the same shaft and act in series, the pump is called a multistage pump, the number of stages corresponding to the number of impellers.  The lower casing of a six-stage pump is shown in Fig. 11.25.  This is a boiler-feed pump capable of delivering 415,000 lb of water per hr against a pressure of 1500 psi.  Multistaging produces better performance, higher pump efficiency, and smaller impeller diameters for high-pressure heads. Usually each stage produces the same head, and the total head developed is the number of stages times the head produced per stage.

The types of impellers installed in centrifugal pumps are as numerous as the uses to which the pumps are put.  Classification, however, can be made by designating the direction of flow of the fluid leaving the impeller.  All pumps have the intake parallel to the impeller shaft. The discharge, however, may be radial, partially radial and axial, or

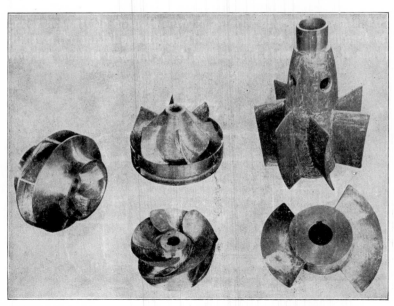

| *A*. Closed impeller, double suction. | *B*. Open impellers, mixed flow. | *C*. Axial-flow impellers. |

**Fig. 11.26.**  Several impeller designs.

axial. In the radial-type impeller the suction and discharge are at right angles. The radial impeller may be of the closed or the open type. The term closed or open refers to the fluid passage within the impeller. The open impeller has one side of the flow path open to the pump casing or housing. The closed impeller has both sides of the flow path enclosed by the sides of the impeller as in Fig. 11.26A.

The partially radial impeller discharges at an angle greater than 90 degrees with intake and is of the open-impeller design (see Fig. 11.26B).

The axial-flow impeller discharges at an angle of approximately 180 degrees with the intake and is generally of the propeller type. Two types of axial-flow impellers are shown in Fig. 11.26C.

Each of the impeller types has a specific purpose. The axial-flow type is used to pump large quantities of fluid against a relatively small static head. It is not a true centrifugal pump but is designed on the principles of airfoil shapes. The radial pump is used for handling smaller quantities of fluid against a high head, because the centrifugal force is high but the flow path is small and restrictive. The open impeller is designed to handle dirty liquids such as sewage, where the flow path must be less restrictive. The partially radial impeller covers intermediate pumping conditions.

### 11.8 Miscellaneous Types of Pumps

The reciprocating pump and the centrifugal pump have many applications in the modern power plant. They are used for supplying water to boilers, returning condensate to the feed-water system, and circulating cooling water through condensers. However, they are not so adaptable for handling viscous liquids such as lubricating oils and fuel oils as are rotary pumps. The rotary pump does not rely on centrifugal force to produce flow. Small volumes of the liquid are trapped between the impeller and casing and are transferred from a low-pressure intake to a higher-pressure discharge. The rotary pump is classed as a positive-displacement pump.

The rotary pump shown in Fig. 11.27 is used for pumping oils against pressures as high as 3000 psi. Liquid enters the inlet, divides, and flows to the ends of the rotors. It then enters the openings between the rotor teeth and is enclosed and propelled to the discharge pipe at the center. No valves are necessary, and axial and radial thrust on the shaft is eliminated.

To handle very viscous liquids such as hot tars, greases, and rosins,

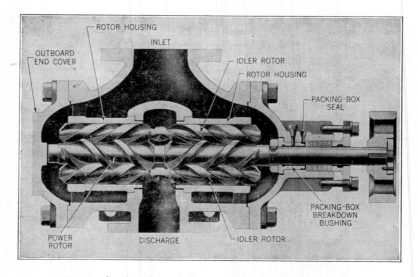

**Fig. 11.27.** Section through Imo rotary pump for pumping oils.

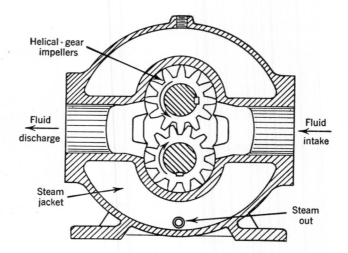

**Fig. 11.28.** A spur-gear rotary pump for handling viscous liquids.

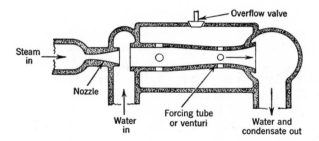

**Fig. 11.29.** Steam-jet injector.

a rotary pump similar to the one shown in Fig. 11.28 may be used. Two helical gears in mesh enclose the liquid between the teeth and the pump casing and transfer the liquid to the discharge. A steam jacket surrounding the pump reduces the viscosity of the liquid by increasing its temperature.

One type of pump is sometimes found in stationary power plants and was used extensively in locomotives as a boiler-feed pump that did not utilize a mechanical drive. The pump is called a steam-jet injector. A simple injector is shown in Fig. 11.29. Steam enters at the left and passes through the nozzle, thus acquiring a high velocity. Owing to the high velocity at the nozzle exit, the pressure is reduced. Atmospheric pressure will force water into the section surrounding the nozzle opening. The water will then be entrained by the steam jet and carried into the forcing tube. In the forcing tube the steam will condense, and the water will increase in velocity. As the Venturi section increases in area, the velocity head of the water developed in the forcing tube will be converted into pressure head. The water can be raised to a pressure equal to or greater than the original steam pressure.

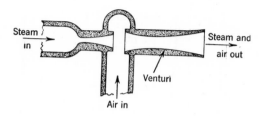

**Fig. 11.30.** Steam-jet air ejector.

The injector is mechanically an inefficient pump, but the condensation of the steam raises the temperature of the feedwater. The injector cannot be used for pumping feedwater that is at a temperature higher than 100 to 120 F.

The same principles may be applied to the pumping of air or gas as shown in Fig. 11.30. The steam-jet air ejector is actually a thermocompressor and is used to remove air from regions at low pressures to atmospheric pressure. The ejector is an important auxiliary of the steam condenser.

### 11.9   Pump Performance

Most pump manufacturers test their pumps and can furnish data on the performance of each type and size. Generally the performance characteristics of a pump are represented by characteristic curves similar to those shown in Fig. 11.31. The head, horsepower, and capacity are expressed in per cent of their values at the maximum efficiency of the pump. The curves show the usual trend of the characteristics for most centrifugal pumps when operated at the pump speed to give maximum efficiency. If possible a pump should operate

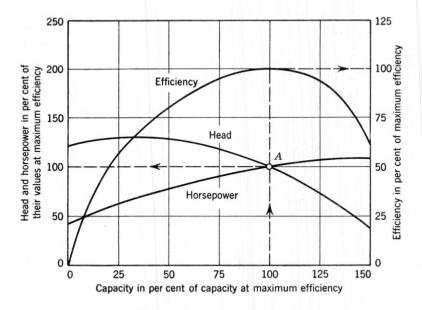

**Fig. 11.31.** Typical characteristic curves of a pump.

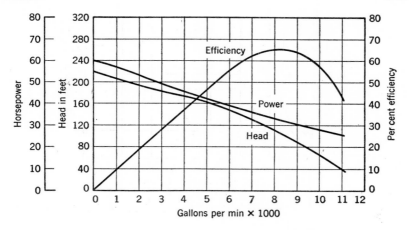

**Fig. 11.32.** Characteristic curves of a multistage deep-well pump.

at the point where the efficiency is 100 per cent of the maximum efficiency, or at point $A$ in Fig. 11.31.

In characteristic curves, the head, capacity, and power may be expressed in terms of the actual values obtained rather than in per cent of the values at maximum efficiency. Such a characteristic curve is shown in Fig. 11.32 for an axial-flow deep-well pump having 16 stages and using an impeller similar to the one shown in Fig. 11.26$C$ (lower). Since an axial-flow pump is not basically a centrifugal pump, the curves in Fig. 11.31 and Fig. 11.32 differ. The efficiency and head curves are similar, but the power curve for the axial-flow pump has a drooping characteristic. Thus, as the volume rate of flow increases, the power required decreases, eliminating any danger of overloading the driving motor.

The characteristics of a rotary-gear pump are shown in Fig. 11.33. The curves are taken from the test results of a pump similar to the one shown in Fig. 11.28 and running at 600 rpm. Unlike that of a centrifugal pump, the capacity does not vary appreciably with an increase in discharge pressure.

The curves in Figs. 11.31, 11.32, and 11.33 represent the characteristics of pumps running at constant speed. If the speed of any pump is varied, there will be a variation in the operation of the pump with a corresponding change in the characteristic curves. Under certain conditions if the speed of a centrifugal pump is varied, the variations in the pump head, capacity, and power are predictable. The

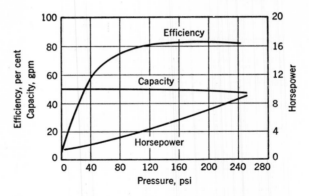

**Fig. 11.33.** Characteristic curves of a gear-type rotary pump.

predictions are based on the basic principles of operation of the centrifugal pump, and the principles are often called the pump laws. The laws are stated as follows:

1. The volume rate of flow is proportional to the speed, or

$$mv \propto n$$

2. The pressure or head is proportional to the speed squared, or

$$h \propto n^2$$

3. The power is proportional to the speed cubed, or

$$hp \propto n^3$$

These laws are exact when the fluid is incompressible, when the fluid always leaves the pump impeller tangentially to the blades, and when the pump is operated at the point of maximum efficiency. If we assume that these criteria are true, the volume rate of flow of the liquid $mv$ is equal to the product of the radial velocity of the liquid leaving the pump $V_R$ and the area between impeller blades $A$. Thus,

$$mv = V_R A \quad \text{from the law of continuity of flow}$$

but the velocity $V_R$ is proportional to the speed, or

$$V_R \propto n$$

Since the area $A$ is constant, then

$$mv \propto n$$

The head against which a pump operates is equal to a factor $f$ times $V^2/2g$ or

$$h = f \frac{V^2}{2g}$$

Based on the preceding criteria, the factor $f$ will not change, and

$$h \propto V^2$$

but

$$V \propto n$$

Thus,

$$h \propto n^2$$

The power is equal to the product of a constant, the head $h$, and the quantity $mv$; thus,

$$\text{hp} \propto hmv$$

or

$$\text{hp} \propto n \times n^2 \propto n^3$$

**Example 10.** If the speed of a centrifugal pump is doubled, how will the head, capacity, and power vary?

*Solution:* (a)

$$\frac{h_2}{h_1} = \frac{n_2^2}{n_1^2} \qquad \text{But} \quad n_2 = 2n_1$$

or

$$h_2 = h_1 \frac{(2n_1)^2}{n_1^2}$$

$$= 4h_1 \text{ or 4 times the original head}$$

(b) The capacity will double.

(c)

$$\frac{\text{hp}_2}{\text{hp}_1} = \frac{n_2^3}{n_1^3}$$

$$\text{hp}_2 = \text{hp}_1 \frac{(2n_1)^3}{n_1^3}$$

$$\text{hp}_2 = 8 \text{ hp}_1 \text{ of 8 times the original power}$$

From Example 10 it can be seen that caution should be used before the speed of a pump is increased to increase its capacity. The power unit driving the pump may be heavily overloaded and severely damaged.

## PROBLEMS

1. Convert the following pressure heads to pressure in psi: 50 in. of Hg, 100 ft of standard water, 100 ft of oil with a specific gravity of 0.80.

2. A gage connected to a boiler drum is located 100 ft below the drum. The gage reads 585 psi. If the gage line is full of water, what is the drum pressure in psia?

3. Water flows in a 6-in. pipe at the rate of 100 gpm. If the specific weight of the water is 62.4 lb per cu ft, what is the water velocity in fps? What is the velocity head in ft of water? What would be the velocity of the water if the 6-in. pipe were reduced to a 3-in. pipe?

4. Assume that the rate of flow of water in a 4-in. pipe is to be measured with a Venturi meter having a throat diameter of 2 in. The difference in the level of a mercury manometer connected to the static tubes at inlet and throat of the Venturi is 3 in. of Hg. If ideal flow conditions are assumed, what is the flow of the water in lb per hr?

5. Water enters a Venturi meter at the rate of 250,000 lb per hr. If the pressure at the entrance to the meter is 30 psig and the diameters at entrance and throat of the meter are 7 in. and 3.5 in., respectively, calculate the gage pressure at the throat of the meter. Assume ideal flow conditions in the meter.

6. An orifice is used to measure the steam flow in a pipe. The pipe area is 0.8 sq ft, and the area at the orifice throat is 0.6 sq ft. The static-pressure differential across the orifice is 1 in. of Hg. If the specific volume of the steam is 2 cu ft per lb, calculate the flow of the steam in lb per hr. Assume ideal flow conditions in the orifice.

7. A pump delivers 30,000 gal of water per hr against a total head of 25 psi. If the specific weight of water is 62.4 lb per cu ft, calculate the hydraulic hp.

8. A pump delivers 1000 lb of sludge per min against a total head of 50 psi. If the sludge has a specific gravity of 1.3, calculate the hydraulic hp or hp output of the pump.

9. A pump delivers 50 cu ft of water per min. If static gages located at the pump intake and discharge read 15 psi and 100 psi, respectively, and require no correction for elevation, find the hydraulic hp of the pump. The intake and discharge pipes are of equal diameter.

10. If the discharge gage of Problem 9 is located 10 ft below the pump center line and the intake gage is located 5 ft above the pump center line, find the hydraulic hp of the pump. Assume all other values remain the same as in Problem 9.

11. An oil has a specific gravity of 40 degrees API. The oil is pumped at the rate of 100 gpm from an oil reservoir to a heat exchanger. The center line of the pump is level with the oil in the reservoir. Assume that there is no friction in the intake pipe to the pump. The discharge gage, located 5 ft below the pump center line, reads 25 psig. The intake and discharge pipes are 4 in. in diameter. Calculate the input to the pump in kw if the pump efficiency is 65 per cent.

12. The measured input to a pump delivering water is 7.5 hp. A gage, located 2 ft above the pump center line on the intake line, reads 1.0 psig. A gage at the discharge, 8.5 ft above the pump center line, reads 80

psig. If the pump efficiency is estimated to be 75 per cent and the change in velocity head is zero, calculate the flow rate in lb per hr.

13. A pump delivers water at the rate of 2000 gpm. The intake pipe has an inside area of 115 sq in. The discharge pipe has an inside area of 58 sq in. The static gage at pump intake indicates a pressure of 10 in. of Hg vac and is located 3 ft above the pipe and pump center lines. The static gage at pump discharge indicates a pressure of 20 psig and is located at the pump center line. If the pump efficiency is 70 per cent, calculate (a) the velocity head change across the pump, (b) the hydraulic hp, (c) the pump hp.

14. The static-pressure gage on the intake side of a pump reads 8 in. of Hg vac and is 5 ft above the pump center line. The static-pressure gage at the discharge reads 40 psig and is 5 ft above the pump center line. The area of the suction pipe is 1 sq ft, and the discharge-pipe area is 0.8 sq ft. A mercury manometer attached to a Venturi meter located in the discharge line indicates a differential static pressure of 10 in. of Hg. The area at the throat of the Venturi is 0.2 sq ft. If the fluid being pumped has a specific gravity of 1.1 and the pump has an efficiency of 70 per cent, calculate the hp required to drive the pump.

15. When running at 2000 rpm a centrifugal pump delivers 500 lb of water per min against a total head of 50 psi. Assuming that the pump laws hold, find the total head, capacity, and power of the same pump if its speed is increased to 3000 rpm.

# CHAPTER 12

# Draft, Fans, Blowers, and Compressors

## 12.1 Introduction

In nearly every phase of modern life there can be found a need for supplying or removing gases or vapors to or from a space. The transfer of the gases or vapors is achieved by natural or mechanical circulation, and the problems encountered are similar to those found in the movement of liquids.

Natural circulation will occur wherever there is a temperature differential within a given volume of gas or vapor. Many homes are heated by warm-air systems which operate by natural circulation. The difference in density or specific weight of the warm and cold air within the building is sufficient to produce flow. If unobstructed, the warm air will rise from the furnace to the rooms, and the cool air will return by gravity to the furnace room.

Forced or mechanical circulation is accomplished by using fans, blowers, or compressors. Fans are preferred where relatively low-pressure heads, a few inches of water to 1 psi, are encountered. Blowers are used for pressure heads ranging from a few psi to medium heads of 35 to 50 psi. Compressors cover the range of pressure heads above 35 psi. However, there is no sharp dividing line between the ranges covered by the three classes of equipment.

## 12.2 Natural Draft

Air can be supplied to a furnace by natural circulation. The pressure differential which is necessary for natural circulation is called natural draft. If the hot gases produced in a furnace are discharged directly to the atmosphere, the draft or pressure differential between the points of air intake and gas exhaust is small. Consequently, to

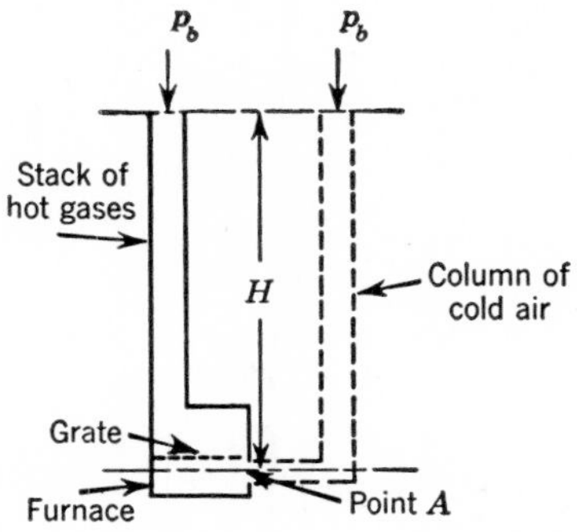

**Fig. 12.1.** Natural draft.

make available a greater draft, the hot exhaust or flue gases are discharged to the atmosphere through a stack or chimney. For example, assume that the hot flue gases are discharged through a stack which is $H$ ft high, as shown in Fig. 12.1, and that the flow of gases and air through the furnace is stopped momentarily by placing a diaphragm at point $A$.

Then, the pressure differential at point $A$ will be

$$\Delta p = (H\gamma_a + p_b) - (H\gamma_g + p_b)$$

$$= H\gamma_a - H\gamma_g \qquad (12.1)$$

where $\Delta p$ = pressure differential, psf

$H$ = stack height, ft

$\gamma_a$ = specific weight of the air, lb per cu ft = $\rho_a \dfrac{g}{g_c}$

$\gamma_g$ = specific weight of the hot gases, lb per cu ft = $\rho_g \dfrac{g}{g_c}$

$p_b$ = atmospheric pressure acting on the column of air and gases, psf

The differential pressure $\Delta p$ is called the theoretical draft of the stack. This theoretical draft is never attainable in practice but represents the maximum draft that could be attained if there were no friction in the stack.

Since the air and hot gases follow the laws of perfect gases closely, the density of each can be calculated by the relation

$$p_b V = mRT$$

or $$p_b = \rho RT$$

where $\rho = m/V =$ density of the gases or air, lb per cu ft

$p_b =$ barometric pressure, psf

$T =$ absolute temperature of the gases or air, deg R

$R =$ gas constant of the air or gases, ft-lb per lb per deg R

The value of $R$ for the air is known and is 53.3. The value of $R$ for the flue gases can be found if the analysis of the flue gases is known. For most practical purposes it can be assumed that the gas constant for the flue gases is the same as for air since both consist mainly of the same element, nitrogen.

On the earth's surface and using ordinary instrumentation and slide-rule calculations, the engineer may substitute the numerical value of the density of the air and gases for the value of $\gamma_a$ and $\gamma_g$, respectively. The error in assuming that 1 lb mass weighs 1 lb is not appreciable.

**Example 1.** A stack 300 ft high will produce what theoretical draft if the average gas temperature is 540 F and the air temperature is 40 F ? Assume a barometer reading of 29 in. of Hg.

*Solution:*

$$\rho_a = \frac{p_b}{R_a T_a}$$

where $p_b = 29 \times 0.491 \times 144$

$R_a = 53.3$

$T_a = 460 + 40 = 500$ R

Thus,

$$\rho_a = \frac{29 \times 0.491 \times 144}{53.3 \times 500}$$

$$= 0.077 \text{ lb per cu ft}$$

Also,

$$\rho_g = \frac{p_b}{R_g T_g}$$

where $p_b = 29 \times 0.491 \times 144$ psf

$R_g = R_a = 53.3$ approximately

$T_g = 540 \times 460 = 1000$ R

Thus,

$$\rho_g = \frac{29 \times 0.491 \times 144}{53.3 \times 1000}$$

$$= 0.038 \text{ lb per cu ft}$$

and since, numerically, $\rho_a = \gamma_a$ and $\rho_g = \gamma_g$, then,

$$\Delta p = H(\gamma_a - \gamma_g)$$

$$= 300(0.077 - 0.038)$$

$$= 11.4 \text{ psf}$$

Generally, pressure differences as small as those that occur in chimney calculations are expressed in inches of water rather than in pounds per square inch. Thus, the theoretical draft for Example 1 expressed in feet of water is

$$D = \frac{\Delta p}{\gamma_w} = \frac{11.4 \text{ lb per sq ft}}{62.4 \text{ lb per cu ft}} = 0.18 \text{ ft H}_2\text{O}$$

in which $\gamma_w = \rho_w \dfrac{g}{g_c} = $ specific weight of water, or

$$D = \frac{11.4}{62.4} \times 12 = 2.19 \text{ in. of water}$$

Because of the friction losses resulting from the flow of the gases through the various passes of the furnace and through the stack, the actual draft obtained from a given stack is as much as 15 per cent less than the theoretical draft. The friction losses vary directly as the square of the gas velocity. If a small-diameter stack is used, the friction losses may become too high. If the stack is designed with too large an area, the stack cost increases. An economical velocity will be about 25 to 30 fps.

The cross-sectional area $A$ of a stack can be determined from the continuity-flow equation,

$$m = \frac{AV}{v}$$

where $m = $ gas flowing, lb per sec
$\quad v = $ specific volume of the gas, cu ft per lb $= 1/\rho$
$\quad A = $ area of stack, sq ft
$\quad V = $ gas velocity, fps

The rate of flow of the gas $m$ can be calculated by making a material balance of the constituents entering and leaving the furnace, as shown in Fig. 12.2. The air–fuel ratio, 15 to 1 in Fig. 12.2, will depend upon the type of coal burned, the air required for theoretical combustion, and the amount of excess air that will be necessary for complete

**Fig. 12.2.** Material balance of a furnace.

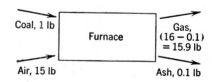

Coal, 1 lb

Furnace

Gas, (16 − 0.1) = 15.9 lb

Air, 15 lb

Ash, 0.1 lb

combustion under the fuel and furnace conditions at maximum load. The amount of ash that will be produced can be estimated closely. The velocity of the gas should be kept low to keep friction losses at a minimum but not so low as to require an uneconomical stack area.

**Example 2.** Assume that natural gas (mostly methane) is burned in a furnace with 15 per cent excess air by volume. The mass rate of flow of the methane is 3000 lb per hour. The average stack-gas temperature is to be 450 F, and the economical stack-gas velocity is 30 fps. Calculate the required stack diameter.

*Solution:* For completely burning the methane

$$CH_4 + 2O_2 \rightarrow CO_2 + 2H_2O$$

For 15 per cent excess oxygen the number of moles of oxygen will equal $2 \times 1.15 = 2.30$. The pounds of air required to burn a pound of $CH_4$ is, therefore,

$$m_{air} = \frac{2.3 \text{ moles } O_2}{1 \text{ mole } CH_4} \times \frac{\text{mole } CH_4}{16 \text{ lb } CH_4} \times \frac{32 \text{ lb } O_2}{\text{mole } O_2} \times \frac{4.32 \text{ lb air}}{\text{lb } O_2}$$

$$= 19.9 \text{ lb air per lb } CH_4$$

But the total gases leaving will be equal to 1 lb $CH_4$ + 19.9 lb air or

$$m_{gases} = 20.9 \times 3000 \text{ lb per hr}$$

From the continuity-flow equation

$$A = \frac{m}{\rho V}$$

in which $m = \dfrac{20.9 \times 3000}{3600}$ lb per sec

$$V = 30 \text{ fps (given)}$$

$$\rho \approx \frac{p_b}{R_g T_g} \approx \frac{14.7 \times 144}{53.3 \times (450 + 460)}$$

$$\approx 0.044 \text{ lb per cu ft}$$

Thus,

$$A = \frac{20.9 \times 3000}{3600 \times 0.044 \times 30}$$

$$= 13.2 \text{ sq ft}$$

The diameter is

$$d = \sqrt{\frac{13.2 \times 4}{\pi}}$$

$$= 4.1 \text{ ft}$$

In addition to producing sufficient draft for small-power-plant operation, a stack is inexpensive to maintain, and it aids in dispersing smoke and fly ash. It has definite disadvantages which are as follows: (1) There is not adequate control of the draft; (2) there is a definite limitation to the maximum draft attainable; and (3) there is a definite hazard to aircraft.

### 12.3  Mechanical Draft

In power-plant engineering the fan plays an important part.  Generally, in small-furnace installations a stack can produce a draft sufficiently high to supply air adequately to the fuel bed and to remove the flue gases.  But the present-day capacities of boilers and furnaces require mechanical draft to supplement the natural draft produced by the stack.  Mechanical draft is divided into two systems: forced draft and induced draft.  In the forced-draft system the fan is located on the air-intake side of the furnace.  A positive pressure, a pressure above atmospheric pressure, is produced under the fuel bed and acts to force air through the bed.  The forced-draft system is necessary in installations where the pressure drop in the intake system and fuel bed is high.  The pressure drop will be high in installations employing air preheaters and/or underfeed stokers.  The underfeed stoker has an inherently deep fuel bed and a correspondingly high resistance to air flow.

Generally, the pressure in a furnace should be slightly less than atmospheric pressure.  If it is too high, there will be leakage of asphyxiating gases into the boiler room and the tendency for blow-back when furnace inspection doors are opened.  If the pressure in the furnace is too low, there will be air leakage to the furnace with a corresponding reduction in the furnace temperature.  Because of these restrictions on the desirable pressure within the furnace, the forced-draft system is generally accompanied by a natural-draft system, in order that the removal of the flue gases may be accomplished.  However, if the stack draft is inadequate owing to the high resistance created by the furnace passes, economizers, and air preheaters, an induced-draft system is generally added to supplement the stack draft.  In the induced-draft system a fan is placed in the duct leading to the stack.  The relative positions of the forced- and induced-draft fans are shown in Fig. 12.3.

When a forced- and an induced-draft fan are used in combination, the system is called balanced draft.  The variation in pressure for such a draft system is shown in Fig. 12.3.  The forced-draft fan produces

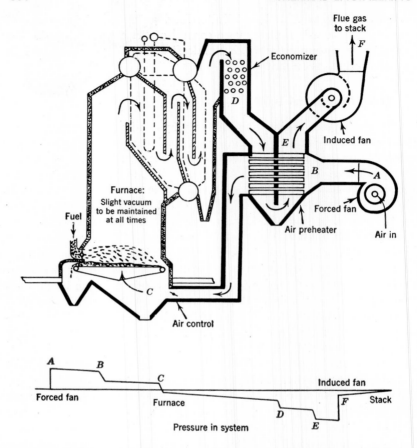

**Fig. 12.3.** A balanced draft system and its corresponding pressure variation.

a positive pressure which decreases slightly through the duct work and sharply through the air preheater and fuel bed. If the system is properly controlled, a pressure of a few hundredths of an inch of water less than atmospheric pressure is maintained in the furnace proper. The pressure continues to drop through the boiler passes, economizer, and air preheater until it is raised by the induced-draft fan and by the stack to atmospheric pressure.

The present trend is to construct more furnaces with gas-tight casings in order that they may be operated under pressures well above atmospheric pressure. Combustion efficiency is improved at elevated pressures, and the induced-draft fan with its high maintenance cost can be eliminated completely. A number of furnaces using the cyclone burner are now designed to operate at pressures as high as 80 in. of water above atmospheric pressure.

### 12.4 Fans

Fans are used extensively in the heating and ventilating industry and in most power plants. Their basic design principles fall into two classes: axial-flow fans and centrifugal- or radial-flow fans. Axial-flow fans are basically rotating airfoil sections similar to the propeller of an airplane. An analysis of their design is beyond the scope of this book. It is important to realize, however, that the movement of the gases created by the axial-flow fan is not due to centrifugal force.

The simplest axial-flow fan is the small electric fan used for circulating air in rooms against very little resistance. Axial-flow fans for industrial purposes are the two-blade or multiblade propeller type, and the multiblade airfoil type similar to the one shown in Figs. 12.4 and 12.5. Air enters the fan suction from the left and flows over the rotor with a minimum of turbulence owing to the streamline form of the rotor and drive mechanism. The airstream is straightened by guide vanes located on the discharge side, thus decreasing the rotational energy of the air by converting it to energy of translation.

The axial-flow fan operates best under conditions where the resistance of the system is low, as in the ventilating field. The axial-flow fan occupies a small space, is light in weight, is easy to install, and handles large volumes of air.

Centrifugal fans may be divided into two major classes: (1) the

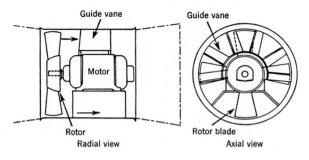

**Fig. 12.4.** An axial-flow fan.

**Fig. 12.5.** Intake of an axial-flow fan.

long-blade or plate-type fan, and (2) the short-blade multiblade fan. The blades of either type may be pitched toward the direction of motion of the fan, radially, or away from the direction of motion of the fan. The velocity diagrams for the various blade shapes are shown in Fig. 12.6. The velocities are based on three assumptions:

1. The blade velocity $V_b$ is the same for each type of fan.
2. The relative velocity of the air with respect to the blade $V_{ab}$ has

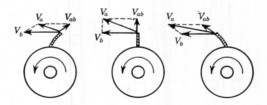

**Fig. 12.6.** Velocity diagrams for the backward-, radial-, and forward-curved blade fan.

a direction parallel to the angle of the blade tip; in other words, the air follows exactly the path of the blade shape.

3. The relative velocity $V_{ab}$ is the same for the three types of fans.

Although the assumptions do not hold exactly, certain principles of fan operation may be developed from the velocity diagrams.

The backward-curved blade produces the lowest air velocity $V_a$. The velocity head, $V_a^2/2g$, will be correspondingly low, and more of the energy of the fan will be converted into static head. The forward-curved fan will produce the lowest static head, but, because of the higher air velocity produced, it is more adaptable for handling large air volumes, provided the resistance of the system is low. The radial blade has operating characteristics that lie between the extreme forward- and backward-curved blades.

A plate-type radial-blade rotor with double inlet is shown in Fig. 12.7. It is best suited for handling dirty gases, since there are no

Wearing strip with
welded beading

**Fig. 12.7.** A plate-type radial-blade fan rotor, double suction.

Fig. 12.8. Plate-type backward-curved-blade fan rotor, single inlet.

Fig. 12.9. Sirocco multiblade forward-curved-blade fan rotor.

**Fig. 12.10.** A Sirocco multiblade fan.

pockets in the blades to catch and collect the dirt.  The rotor has wearing strips welded to the blades to increase their life.  The fan is designed for induced-draft service.  The housing of such a fan may have catch plates in the scroll face to collect the fly ash.

The rotor in Fig. 12.8 has the backward-curved plate-type blades with a single suction or inlet.  Its characteristics are high speed, nearly constant capacity for various pressure conditions, and a non-overloading power requirement for its complete range of head and capacity.  The fan with backward-curved blades is used for forced draft.

One of the best centrifugal fans for ventilating requirements is the multiblade fan shown in Figs. 12.9 and 12.10.  It operates at low wheel-tip speeds, and the air turbulence is low, thus reducing noise generation.  With the large inlet possible for a given-diameter wheel, the volume capacity of the fan is high for its size. This type of fan is used extensively to circulate air in schools, office buildings, and public buildings.

### 12.5 Fan Testing

A fan test generally includes the measurement of the static, velocity, and total heads against which the fan operates; the determination of the volume rate of flow of air or gas; and the calculation of the air horsepower delivered by the fan. The measurement of the heads can be made conveniently by a Pitot tube, as discussed in Chapter 11. The volume rate of flow is calculated from the measured value of the velocity head. The air horsepower is determined for a fan in the same way that hydraulic horsepower is determined for a pump. Thus,

$$\text{Air hp} = \frac{h_t w}{33{,}000} \qquad (12.2)$$

where $h_t$ = total head against which the fan operates expressed in ft of air

$w$ = flow of air, $\text{lb}_f$ per min

The total head of the fan is calculated in the same manner as the total head of a pump. As in Equation 11.11 the total head is

$$h_t = (z_2 - z_1) + \frac{V_2{}^2 - V_1{}^2}{2g} + \frac{p_2 - p_1}{\gamma} \qquad (12.3)$$

in which $z_2 - z_1$ is negligible compared to the velocity and pressure-head changes.

Thus,

$$h_t = \frac{V_2{}^2 - V_1{}^2}{2g} + \frac{p_2 - p_1}{\gamma} \qquad (12.4)$$

The power required to drive the fan is equal to the air horsepower divided by the fan efficiency. Thus,

$$\text{Fan hp} = \frac{\text{air hp}}{\eta} \qquad (12.5)$$

where $\eta$ = fan efficiency

In many fan installations air enters the fan from a large space and is discharged through a duct. In this case the static and velocity heads at inlet are approximately zero. Thus, the total head against

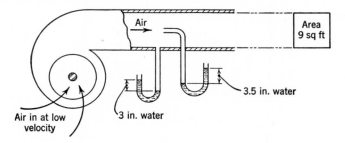

**Fig. 12.11.** Figure for Example 3.

which the fan operates will be the head produced in the discharge duct near the fan. An example of the calculations made on such an installation follows:

**Example 3.** Assume that air enters a fan at a temperature of 70 F and an atmospheric pressure of 29 in. of Hg from a large room and is discharged to a duct having an area of 9 sq ft. Pick-up tubes or Pitot tubes are used to measure the static and total heads on the discharge duct. If water manometers connected to the static-pressure tube and total-pressure tube read 3.0 in. and 3.5 in., respectively, as illustrated in Fig. 12.11, find (a) the static, velocity, and total heads in ft of air in the duct; (b) the velocity of the air in the duct, (c) the volume and mass rate of flow, (d) the air hp.

*Solution:* (a) The heads in in. of water can be converted to fit of air by the relation,

$$h_w \gamma_w = p_2 = h_a \gamma_a$$

where $h_w$ = head, ft of water
$\gamma_w$ = specific weight of water, lb per cu ft
$h_a$ = head, ft of air
$\gamma_a$ = specific weight of air, lb per cu ft
$p_2$ = static pressure, psf

For engineering calculations it may be assumed without introducing an appreciable error that the specific weight of the air, $\gamma_a$, is numerically equal to the value of the density of air, $\rho_a$. But the density of the air can be calculated by use of the perfect-gas law.

$$p = \rho_a RT$$

or

$$\rho_a = \frac{p}{RT}$$

where $p$ = atmospheric pressure = $144 \times 29 \times 0.491$ psf
$R$ = 53.3 for air
$T$ = 70 + 460 = 530 R

Thus, assuming that $\qquad \rho_a = \gamma_a$ numerically

$$\gamma_a = \frac{144 \times 29 \times 0.491}{53.3 \times 530}$$

$$= 0.073 \text{ lb per cu ft}$$

The static head in the duct in ft of air is

$$\frac{p_2}{\gamma_a} = h_s = \frac{h_w \gamma_w}{\gamma_a} = \frac{\dfrac{3.0 \text{ in. water}}{12 \text{ in. per ft}} \times 62.4 \text{ lb water per cu ft}}{0.073 \text{ lb air per cu ft}} = 214 \text{ ft of air}$$

The total head in ft of air is

$$h_t = \frac{h_w \gamma_w}{\gamma_a} = \frac{\dfrac{3.5}{12} \times 62.4}{0.073} = 249 \text{ ft of air}$$

Thus, the velocity head in ft of air is

$$h_v = h_t - h_s = 249 - 214 = 35 \text{ ft of air}$$

(b) The velocity head $h_v = V^2/2g$. Then,

$$V = \sqrt{2gh_v}$$

$$= \sqrt{2 \times 32.2 \times 35}$$

$$= 47.5 \text{ fps}$$

(c) From the law of continuity of flow, the volume rate of flow,

$$mv = AV$$

where $A$ = area of the duct, sq ft
$\qquad V$ = velocity of the air at area $A$, fps

Then, $\qquad\qquad\qquad mv = 9 \times 47.5 = 428 \text{ cu ft per sec}$

or $\qquad\qquad\qquad\qquad mv = 25{,}700 \text{ cfm}$

The mass rate of flow,

$$m = AV \frac{1}{v}$$

$$= 25{,}700 \times 0.073$$

or $\qquad\qquad\qquad m = 1880 \text{ lb per min} = w, \text{ numerically}$

(d) $\qquad\qquad$ Air hp $= \dfrac{h_t w}{33{,}000}$

$$= \frac{249 \times 1880}{33{,}000} = 14.2$$

Sometimes a fan has an inlet as well as a discharge duct. If the ducts are of equal area, the velocity of the air will be the same at inlet and discharge, and the velocity head against which the fan operates will be zero. Thus, the total head against which the fan operates will be the difference between the discharge and inlet static heads. If the ducts are of unequal area, the fan will operate against a head equal to the difference in the total heads at discharge and inlet.

**Example 4.** Assume that air enters a fan through a duct 16 sq ft in area. The inlet static pressure is 1 in. of water less than atmospheric pressure. The air leaves the fan through a duct 9 sq ft in area, and the discharge static pressure is 3.0 in. of water above atmospheric pressure. If the specific weight of the air is 0.075 lb per cu ft, and the fan delivers 20,000 cfm, find (a) the total head against which the fan operates, (b) the air hp, (c) the fan efficiency if the power input to the fan is 17 hp at the coupling.

*Solution:* (a) The total head against which the fan operates is given by

$$h_t = \frac{p_2 - p_1}{\gamma} + \frac{V_2{}^2 - V_1{}^2}{2g}$$

where

$$p_2 = \frac{3 \text{ in.}}{12 \text{ in./ft}} \times 62.4 \frac{\text{lb}}{\text{cu ft}} = 15.6 \text{ psf}$$

$$p_1 = -\frac{1 \text{ in.}}{12 \text{ in./ft}} \times 62.4 \frac{\text{lb}}{\text{cu ft}} = -5.2 \text{ psf}$$

Also,

$$V_2 = (mv) \frac{1}{A_2}$$

$$= \left(\frac{20,000}{60}\right) \times \frac{1}{9}$$

$$= 37.0 \text{ fps}$$

and

$$V_1 = (mv) \frac{1}{A_1}$$

$$= \left(\frac{20,000}{60}\right) \times \frac{1}{16}$$

$$= 20.8 \text{ fps}$$

Thus, substituting these values into the equation for total head gives

$$h_t = \left[\frac{15.6 - (-5.2)}{0.075}\right] + \left[\frac{(37)^2}{64.4} - \frac{(20.8)^2}{64.4}\right]$$

$$= 277.5 + 14.5$$

$$= 292.0 \text{ ft of air}$$

(b)

$$\text{Air hp} = \frac{h_t w}{33,000}$$

where $h_t = 292$ ft of air

$w = 20,000 \times 0.075$ lb per min

Thus, $\qquad$ Air hp $= \dfrac{292 \times 20,000 \times 0.075}{33,000}$

$= 13.3$

(c) $\qquad$ $\eta = \dfrac{\text{air hp}}{\text{input hp}}$

$= \dfrac{13.3}{17.0} \times 100$

$= 78.2\%$

### 12.6 Fan Performance

Fans are tested by the manufacturers, and usually the results of the operation of the fans are presented in tables or in characteristic curves. The curves may include the variations in head, capacity, power, and efficiency for a constant speed or may be a family of curves for a series of constant speeds. By careful survey of the various types of fans and their characteristics, fan purchasers can select the type and size of fan best fitted to perform a given function.

Within a given class or type of fan there are certain general characteristics that are common to the many different designs. The curves in Fig. 12.12 show the variation in pressure and power for differing

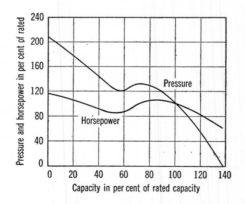

**Fig. 12.12.** Characteristic curves of the axial-flow fan.

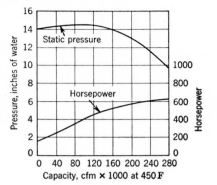

**Fig. 12.13.** Characteristics of a radial-blade fan.

capacities for an axial-flow fan similar to the one in Fig. 12.4 and operating at constant speed. The fairly constant power output over a wide range of capacities is common to most axial-flow fans. Thus, there will be little tendency to overload the driving motor, regardless of the change in conditions under which the fan operates. This is called a nonoverloading characteristic. The capacity decreases more or less at a constant rate for an increase in resistance or pressure. The efficiency of such a fan is generally somewhat lower than that of centrifugal fans except at low pressures. By varying such things as the pitch diameter or width of the blades, the point of maximum efficiency can be varied to cover a wide range of conditions.

The characteristics of a radial-blade centrifugal fan (Fig. 12.7) are shown in Fig. 12.13. The power increases with a decrease in pressure and in increase in capacity, but the increase is not sharp enough to overload the driving motor if proper selection of the motor is made. Generally, the characteristics of the radial-blade centrifugal fan will be a compromise of the characteristics of the backward-curved-blade and forward-curved-blade fans.

The backward-curved-blade centrifugal fan of Fig. 12.8 will have characteristics as shown in Fig. 12.14. Best efficiencies are obtained with rotors having backward-curved blades, and the power curve for these rotors shows a nonoverloading characteristic over the complete range of pressures and capacities. The point of maximum efficiency occurs at the point of maximum power. Above 50 per cent of the maximum capacity a decrease in capacity will increase the pressure sharply. This fan is excellent for forced-draft service, for, as the fuel bed of a furnace closes and restricts the flow of air from the fan,

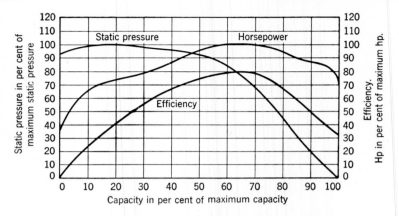

**Fig. 12.14.** Characteristic curves of a backward-curved-blade fan.

the fan pressure will rise sharply. The increase in fan pressure will tend to open the fuel bed to admit more air to the furnace.

The forward-curved-blade fan of Fig. 12.9 has an overloading power characteristic as shown in Fig. 12.15. If reasonable care is exercised in figuring the conditions under which the fan will operate, a motor can be selected to prevent its overloading. The point of maximum efficiency occurs near the point of maximum pressure.

Centrifugal fans have the same general laws governing their speed characteristics as centrifugal pumps and for the same reasons (see Article 11.9). Restated, these laws for a given fan are as follows:

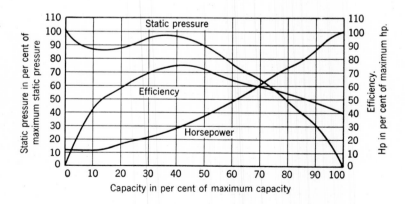

**Fig. 12.15.** Characteristic curves of a forward-curved-blade fan.

1. Capacity is proportional to the fan speed.
2. Pressure or head is proportional to the fan speed squared.
3. Power is proportional to the fan speed cubed.

These laws apply only to a constant specific weight of air.

**Example 5.** A fan delivers 10,000 cfm at a static pressure of 2 in. of water when operating at a speed of 400 rpm. The power input required is 4 hp. Find (a) the speed, (b) the pressure, and (c) the power of the same fan and installation if 15,000 cfm are desired.

*Solution:* (a)

$$mv \propto n$$

or

$$\frac{n_2}{n_1} = \frac{m_2 v_2}{m_1 v_1} \quad \text{or} \quad n_2 = n_1 \frac{m_2 v_2}{m_1 v_1}$$

where $n_1 = 400$, $m_2 v_2 = 15,000$, $m_1 v_1 = 10,000$

Thus,

$$n_2 = \frac{15,000}{10,000} \times 400 = 600 \text{ rpm}$$

(b)

$$h \propto n^2$$

or

$$\frac{h_2}{h_1} = \frac{n_2{}^2}{n_1{}^2} \quad \text{or} \quad h_2 = h_1 \frac{n_2{}^2}{n_1{}^2}$$

where $n_2 = 600$, $n_1 = 400$, $h_1 = 2$

Thus,

$$h_2 = \frac{(600)^2}{(400)^2} \times 2 = 4.5 \text{ in.}$$

(c)

$$\text{hp} \propto n^3$$

or

$$\frac{\text{hp}_2}{\text{hp}_1} = \frac{(n_2)^3}{(n_1)^3} \quad \text{or} \quad \text{hp}_2 = \frac{(n_2)^3}{(n_1)^3} \times \text{hp}_1$$

where $n_2 = 600$, $n_1 = 400$, $\text{hp}_1 = 4$

Thus,

$$\text{hp}_2 = \frac{(600)^3}{(400)^3} \times 4 = 13.5$$

### 12.7 Blowers

Blowers may be divided into two types: (1) rotary, and (2) centrifugal. A common type of rotary blower is the Roots two-lobe blower shown in Fig. 12.16. Two double-lobe impellers mounted on parallel shafts connected by gears rotate in opposite directions and at the same speed. The impellers are machined to afford only a small clearance between them and between the casing and impellers. As the lobes revolve, air is drawn into the space between the impellers and

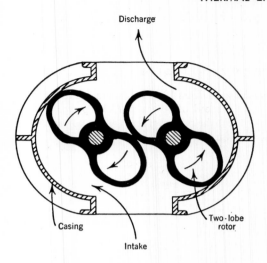

**Fig. 12.16.** A Roots two-lobe blower.

the casing, where it is trapped, pushed toward the discharge, and expelled. The air is trapped and discharged in volumes equal to the space between the impellers and casing, and the operation is repeated four times for each rotation of the shaft.

In order to change the volume rate of flow or volume capacity of the blower, the blower speed is changed. The pressure developed by

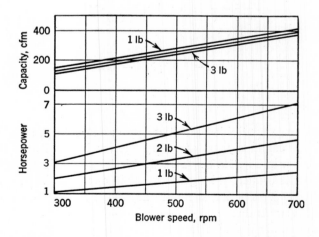

**Fig. 12.17.** Characteristic curves of a two-lobe rotary blower.

the blower will be whatever is necessary to force the air through the piping system. The volume of air delivered by the blower will not change appreciably with variations in resistance to flow. Thus, the blower is called a positive-displacement blower. The characteristics of the blower are shown in Fig 12.17 for various speeds. Note that at a speed of 600 rpm an increase in pressure from 2 to 3 psi increases the power required by 1.5 times, but the capacity remains fairly constant. Care should be taken in operating any positive-displacement blower. A safety valve or limit valve should be placed on the discharge line to prevent the discharge pressure becoming excessive in case the outlet is fully closed. The limit valve will prevent overloading the discharge line and the driving motor. The advantages of the rotary blower are:

     1. Simple construction.
     2. Positive air movement.
     3. Economy of operation and low maintenance.

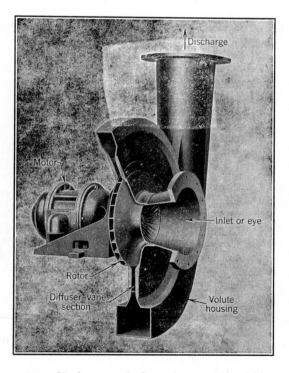

**Fig. 12.18.** Single-stage, single-suction, centrifugal blower.

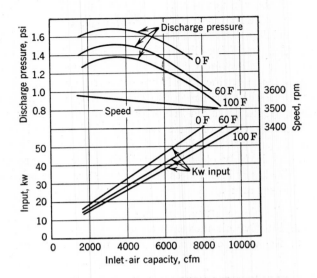

**Fig. 12.19.** Characteristic curves of a centrifugal blower.

**Fig. 12.20.** A double-suction centrifugal blower with cover removed.

**Fig. 12.21.** Single-stage double-suction blower (rated at 26,200 cfm of air and 14.3 psi against 54-in. water column).

**Fig. 12.22.** Impeller of the blower shown in Fig. 12.21.

Centrifugal blowers and compressors operate on the same principles as centrifugal pumps and resemble to a marked degree the closed-impeller centrifugal pumps described in Chapter 11. A sectional view of a single-stage single-suction blower is shown in Fig. 12.18. It is capable of delivering 15,000 cfm against a pressure of 3 psi. The casing or housing is constructed of heavy steel plate, and the impeller is an aluminum-alloy casting. Typical performance curves for this type blower are shown in Fig. 12.19. If care is taken in providing the proper drive motor, the overload characteristics of the centrifugal blower will cause no trouble. The blower shown in Fig. 12.18 uses a motor with a 15 per cent overload factor. Note the effect of inlet temperature of the air on pressure and power. This is a result of the change in specific weight of air with temperature.

For volumes greater than those that can be handled by the single-stage single-suction blower, a single-stage double-suction blower is used, as illustrated in Figs. 12.20 and 12.21. The impeller is shown in Fig. 12.22. This blower is capable of supplying 26,000 cfm of air at 60 F and atmospheric pressure against a 54-in. water column or 2 psi.

## 12.8  Centrifugal Compressors

Multistage centrifugal blowers when capable of handling gases against pressures greater than 35 psig are generally classed as compressors. They resemble multistage centrifugal pumps, and many of the problems encountered in their design are similar to those encountered in pump design. A complete centrifugal compressor unit with turbine drive is shown in Fig. 12.23. The impellers are of the single-suction type, and passages lead the air or gas from the discharge of one impeller to the suction side of the next impeller.

Because of an increase in temperature of the gas or air as the pressure is increased, cooling is generally necessary. If the pressures are not high, cooling water circulated in labyrinths between impellers may be sufficient. When high pressures are encountered, the gas may be cooled in interstage coolers. The reason for maintaining the gas at a low temperature is to permit an increase in the mass rate of flow with a corresponding reduction in size and horsepower. A pressure–volume diagram (Fig. 12.24) of the air as it is compressed will show more clearly the reason for intercooling between stages.

Assume that 1 lb of air is introduced to a compressor at a low pressure $p_0$ and that the volume of the 1 lb of air is represented by

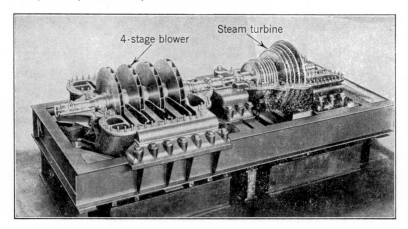

**Fig. 12.23.** A four-stage centrifugal blower with turbine drive.

volume $v_1$. As work is done on the air by the compressor rotor, the air pressure will rise and the volume will decrease. The rise in pressure may follow any one of a number of paths such as 1–2, 1–3, or 1–4. The paths taken by the air will depend upon the conditions under which the air is compressed. If no heat is lost or gained by the air in the compression process, the air will be compressed to the pressure $p_5$ by path 1–2. This is called adiabatic compression. If the air is maintained at a constant temperature throughout the compression process, the air will be compressed to the pressure $p_5$ by path 1–4. This is called isothermal compression.

The work done on the air is represented by the areas lying within the dotted lines and the path taken by the air in the compression process. The work done for adiabatic compression is the area 0–1–2–5. The area for isothermal compression is 0–1–4–5. Thus, we may conclude that the work required for compressing air that is

**Fig. 12.24.** The pressure–volume diagram for a compressor.

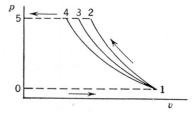

maintained at constant temperature is less than that required for the adiabatic compression.

The actual compression process for the multistage compressor using interstage cooling is some path 1–3 which approaches the path 1–4. More cooling would decrease the work of compression but would increase the size and cost of the cooling system.

Axial-flow compressors are designed on the principles of the air-foil section, and the blade shapes will be similar to the axial-flow fan shown in Fig. 12.4. A multistage axial-flow compressor is shown in Fig. 10.32. The compressor is an essential part of the gas-turbine cycle. The gas is not cooled between stages, because a portion of the additional work necessary to compress the gas adiabatically over the work necessary to compress it isothermally will be recovered in the gas turbine.

The advantages of centrifugal and axial-flow blowers and compressors are:

1. Nonpulsating discharge of the gas.
2. No possibility of building up excessive discharge pressures.
3. A minimum of parts subject to mechanical wear.
4. No valves necessary.
5. A minimum of vibration and noise.
6. High speed, low cost, and small size for high capacity.

### 12.9   Reciprocating Compressors

The reciprocating compressor is a positive-displacement machine which operates in a manner similar to a reciprocating pump. It may be driven by a steam or gas engine or by a motor. A flywheel is used to carry the compressor piston over its dead-center positions. The cylinder walls are either air- or water-cooled. Air cooling is used for applications where freezing temperatures may be encountered and for most small or portable compressors. For air-cooled cylinders fins are added to increase the area for transferring heat.

Single-stage compressors are used when the requirements are for:

1. Compactness at some sacrifice in efficiency of operation.
2. Moderate discharge pressures.
3. Moderate air capacities.

A single-stage compressor consists of a single cylinder and piston which is generally double acting. Gas is alternately drawn into and

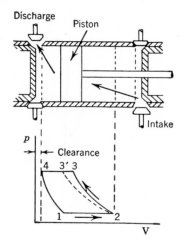

Fig. 12.25. Pressure–volume diagram for an actual compressor.

discharged from each end of the cylinder. An ideal compressor would take in a volume of air equal to the cylinder volume, compress it to the discharge pressure at constant temperature, and discharge the total final volume of air. None of these requirements is fulfilled in the actual compressor. Clearance must be provided for valve opera-

Fig. 12.26. A two-stage reciprocating air compressor with intercooler.

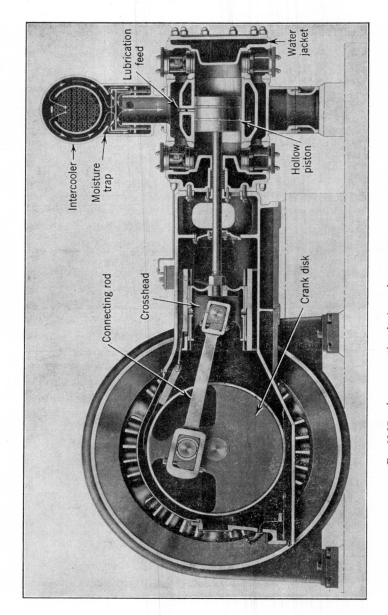

**Fig. 12.27.** A cross-sectional view of a two-stage compressor.

tion and between piston and cylinder heads. Also, the air temperature will rise during compression. The pressure–volume diagram of the air during the complete cycle is illustrated in Fig. 12.25.

Assume that the piston is at the extreme right and that the volume of the air in the head-end position of the cylinder is equal to $V_2$. As the piston moves toward the left, the air is compressed along line 2–3 to the final pressure $p_3$. (Isothermal compression would follow the dotted line 2–3'.) At point 3 the discharge valve opens, and a volume of air equal to $V_3 - V_4$ is discharged. The discharge valve will close at point 4. Air remaining in the clearance space will expand along line 4–1 as the piston moves to the right. At point 1 the intake valve opens, and air is drawn in equal in volume to $V_2 - V_1$.

If the clearance volume is large, less air can be drawn into the cylinder, and the compressor capacity is reduced with no effect on discharge pressure. This fact is utilized in controlling the output of constant-speed compressors for loads from zero to the maximum. Clearance at maximum loads is held to the minimum requirements compatible with good design. At low loads additional clearance is provided by the manual or automatic opening of valves to clearance pockets.

To compress large volumes of air the compressors are constructed to perform the process in steps or stages. A multistage compressor has as many stages as cylinders. The cylinders may be in line, in which case a single-drive mechanism can be used for as many as three pistons, or they may be placed side by side.

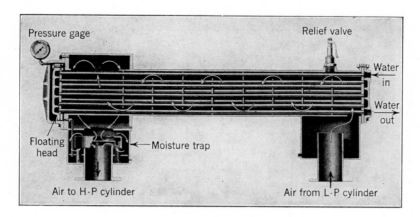

**Fig. 12.28.** An intercooler.

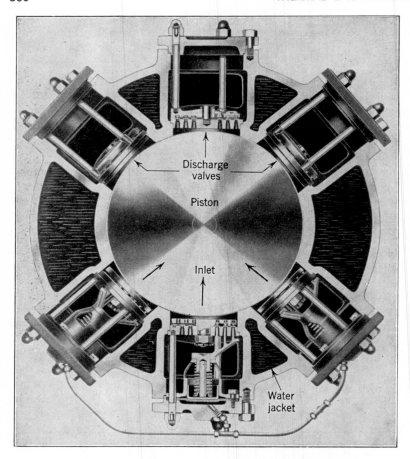

**Fig. 12.29.**  End view of a compressor cylinder showing valves.

A motor-driven two-stage air compressor is shown in Fig. 12.26. Air is drawn into the cylinder, located nearest to the reader, and is compressed alternately in the head end and crank end of the cylinder and discharged to the intercooler located above the cylinders.  The cooled air then passes to the high-pressure second stage and is further compressed and discharged to the air system.  A section through the high-pressure stage is shown in Fig. 12.27.  On leaving the intercooler (Fig. 12.28) the air flows through pockets, which catch the moisture carried in the air, and enters the high-pressure cylinder.  The cylinder and head are completely water-jacketed, and the piston is lubricated from a feed at the top of the cylinder. An end view of the cylinder is shown in Fig. 12.29. The intake and discharge valves consist of concentric disks operating over annular ports in a valve seat (Fig. 12.30). The valve disks are held in the closed position by springs until the pressure differential across the disks becomes great enough to compress the springs. The valve action is quick and affords a maximum valve diameter. The valves are guided by rods or keepers located 120 degrees apart. Disk valves of similar design, using flat springs which operate over rectangular openings, provide similar quick valve action over a large effective area.

Intercoolers are provided in multistage reciprocating compressors as in centrifugal compressors to reduce the amount of work required to compress a given weight of gas from a low pressure to a high pressure.

For most compressed-air systems the auxiliaries of the reciprocating compressor are:

1. An aftercooler.
2. A receiver.
3. An air-intake filter.

Air-intake filters prevent dirt from entering the cylinders and increasing the wear of the walls and pistons. Since reciprocating machines produce pulsating flow, a receiver or tank to which the air is

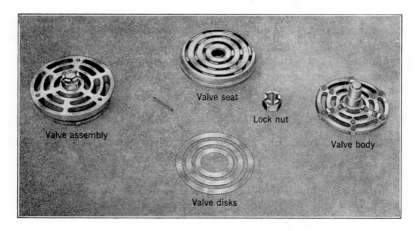

**Fig. 12.30.** Essential parts of a disk-type air valve.

discharged is employed to absorb the pulsations. The receiver also acts as a reservoir to maintain constant air pressure under varying loads. The aftercooler cools the air and reduces the specific volume of the air. Thus less receiver volume and smaller piping can be used for a given weight rate of flow of air.

## 12.10   Application of the Energy Balance to Compressors

Compressors are used to move gases and vapors under conditions such that the density of the fluid cannot be assumed constant. The following assumptions are made in developing the energy equation for a compressor:

1. One pound of fluid leaves the compressor for each pound that enters.

2. The areas at entrance and exit sections are such that the kinetic energy of the fluid is negligible or of the same order of magnitude so that the change in kinetic energy in the compressor may be neglected.

3. The change in mechanical potential energy due to a difference in elevation between entrance and exit sections may be neglected.

On the basis of the above assumptions the general-energy equation for a compressor may be written as follows, with the negligible terms canceled:

$$\frac{\cancel{gz_1}}{\cancel{g_c}J} + \frac{\cancel{V_1}^2}{2g_c J} + u_1 + \frac{p_1v_1}{J} + \frac{W_{in}}{J} = \frac{\cancel{gz_2}}{\cancel{g_c}J} + \frac{\cancel{V_2}^2}{2g_c J} + u_2 + \frac{p_2v_2}{J} + Q_{out}$$

Since, by definition, $h = u + \dfrac{pv}{J}$, the preceding equation may be rewritten as

$$h_1 + \frac{W_{in}}{J} = h_2 + Q_{out} \tag{12.6}$$

or $$\qquad h_1 + W_{in} = h_2 + Q_{out} \tag{12.7}$$

In Equation 12.7 the $W_{in}$ is expressed in the same units as heat and enthalpy, or Btu per lb. Also, $h_1$ is the initial enthalpy and $h_2$ is the final enthalpy.

If the compression process is adiabatic or without heat transfer, the work in Btu per lb becomes

$$W_{in} = h_2 - h_1 \tag{12.8}$$

Equation 12.8 holds true with or without friction. The effect of friction, however, will be to increase the value of $h_2$ and consequently the value of $W_{in}$. For a frictionless and adiabatic process, or isentropic process, in which a perfect gas is being compressed, Equation 12.8 is valid, but $h_2 - h_1$ can be replaced by $c_p(T_2 - T_1)$ and $T_2$ is calculated by means of the expression,

$$\frac{T_2}{T_1} = \left(\frac{p_2}{p_1}\right)^{\frac{k-1}{k}}$$

If the compression process is isothermal, for which the temperature remains constant and a perfect gas is compressed, Equation 12.7 becomes

$$W_{in} = Q_{out} \tag{12.9}$$

Equations 12.8 and 12.9 are valid only when a perfect gas is the medium being compressed.

**Example 6.** Assume that 100,000 lb of steam per hr are to be compressed from 1 psia and 80 per cent quality to 14.7 psia and 300 F without heat transfer. Calculate the power required to compress the steam.

*Solution:* Since $Q = 0$ by assumption,

$$\frac{W}{J} = h_2 - h_1$$

where
$$h_2 = 1192.8 \text{ Btu per lb} \qquad \text{(Table A.3)}$$

$$h_1 = h_f + x h_{fg} \quad \text{at 1 psia} \quad \text{and} \quad x = 0.8$$

$$= 69.7 + 0.8 \times 1036.3 \qquad \text{(Table A.2)}$$

$$= 898.7 \text{ Btu per lb}$$

Thus,
$$\frac{W}{J} = 1192.8 - 898.7$$

$$= 294.1 \text{ Btu per lb}$$

and
$$\text{Power} = 100,000 \times 294.1$$

$$= 29,410,000 \text{ Btu per hr}$$

The hp required to compress the steam is

$$\text{hp} = \frac{29,410,000}{2545}$$

$$= 11,550$$

Example 6 should help to show why exhaust steam from a turbine is condensed before it is returned to the boiler. The amount of power required to compress water from 1 to 14.7 psia would be a small fraction of the amount required to compress steam.

**Example 7.** An air compressor compresses air from 14.7 psia and 60 F to 60 psia and 440 F. If the compression process is without heat loss, calculate the power required to compress 10 lb of air per sec.

*Solution:* The energy equation for the compressor is given by

$$\frac{W}{J} = h_2 - h_1$$

where
$$h_2 = c_p T_2 = 0.24\,(440 + 460)$$
$$h_1 = c_p T_1 = 0.24\,(60 + 460)$$

Thus,
$$\frac{W}{J} = 216.0 - 124.8$$
$$= 91.2 \text{ Btu per lb}$$

and
$$\text{The power} = 91 \times 10 \text{ Btu per sec}$$

The hp delivered by the compressor to the air is

$$\text{Power} = \frac{91 \times 10 \times 3600}{2545}$$

$$= 1300 \text{ hp}$$

## PROBLEMS

1. Calculate the theoretical draft in in. of water which is produced by a 200-ft stack for which the average gas temperature is 650 F and the outside air temperature is 90 F. Assume a standard barometer reading.

2. What is the percentage increase in theoretical draft from summer to winter for a 100-ft stack having an average gas temperature of 500 F?

3. A fuel oil having a chemical constituency of 85 per cent carbon and 15 per cent hydrogen by weight is supplied to a furnace at the rate of 5 lb per sec. It is burned with 20 per cent excess oxygen. If the velocity in the stack is to be limited to 30 fps when the gas temperature is 480 F, calculate the required stack diameter in ft.

4. The ultimate analysis of a coal to be burned with 50 per cent excess air is as follows: C, 84%; $H_2$, 5%; $O_2$, 6%; ash, 5%. If the coal is burned at the rate of 50 tons per hr, what is the required stack diameter for a gas velocity of 40 fps and a temperature of 400 F?

5. A fan removes air from a room at the rate of 4000 cfm and operates against a total head of 2 in. of water. What is the approximate air hp

output required by the fan? If the fan has an efficiency of 60 per cent, estimate the kw capacity of the motor driving the fan.

6. A fan under test removes air from the test room and discharges it through a duct having a cross-sectional area of 4 sq ft. The static and total pressures in the discharge duct are 3.0 in. of water and 3.4 in. of water, respectively. If the density of the air is 0.070 lb per cu ft, calculate the mass rate of flow of the air and the fan output in hp.

7. To reduce noise a fan speed is reduced from 900 rpm to 450 rpm. How does this change affect the fan hp and the fan capacity?

8. How much work in Btu per lb is required to compress steam isentropically from 14.7 psia and 80 per cent quality to 100 psia?

9. In Example 8 what error would be introduced if the calculations were made by assuming that steam is a perfect gas with a specific heat equal to $c_p = 0.50$ Btu per lb F?

10. Is Example 8 an isentropic compression process? Prove your answer.

11. Air is compressed isentropically from 70 F and 15 psia to 60 psia. Calculate the work required in Btu per lb.

12. If the compression process for the conditions in Problem 11 produces an enthalpy rise of 1.2 times the isentropic enthalpy rise, calculate the temperature of the air leaving the compressor and the work of compression in Btu per lb.

# CHAPTER 13

# Power-Plant Cycles

## 13.1 Introduction

A cycle is a series of operations or events which occur repetitively in the same order. A power cycle or power-plant cycle is such a series of events which regularly repeat themselves for the purpose of converting a portion of the stored energy of a fuel into work. There are two general types of power cycles, the closed cycle and the open cycle.

In the closed cycle a working fluid begins at some initial condition, undergoes certain changes through a series of regular events, and returns to the initial condition. Theoretically, no replenishment of the working fluid is necessary. A generalized closed cycle is shown in

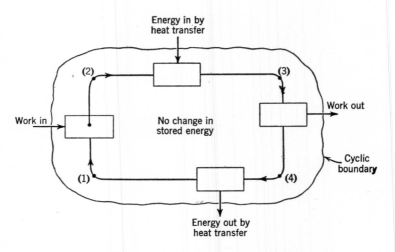

**Fig. 13.1.** A generalized, closed power cycle.

Fig. 13.1. The working fluid moves from point 1 clockwise through a series of four events and returns to point 1.

The events in order are as follows:

1 to 2—work done on the fluid.
2 to 3—energy added to the fluid by means of heat transfer.
3 to 4—work done by the fluid.
4 to 1—energy rejected by the fluid by means of heat transfer.

No mass crosses the boundaries of the cycle but heat and work do cross the boundaries. If the work done by the fluid is greater than the work done on the fluid, then the difference is called net work output, and

$$Q_{in} = \text{work}_{net} + Q_{out} \tag{13.1}$$

This is merely a statement of the first law of thermodynamics. By definition, the thermal efficiency is

$$\eta = \frac{\text{work}_{net}}{Q_{in}}$$

$$\eta = \frac{Q_{in} - Q_{out}}{Q_{in}}$$

or

$$\eta = 1 - \frac{Q_{out}}{Q_{in}} \tag{13.2}$$

The second law of thermodynamics, as pointed out in Article 2.18, informs us that the maximum theoretical efficiency which we can expect for the generalized cycle is

$$\eta_{max} = 1 - \frac{T_L}{T_H} \tag{13.3}$$

In Equation 13.3, $T_L$ is the absolute temperature of the reservoir to which heat is rejected and $T_H$ is the absolute temperature of the source from which heat is added.

A form of the generalized open cycle is shown in Fig. 13.2. If the mass crossing the boundaries remains constant and invariant with time, the cycle is said to operate as a steady-flow device. Such an assumption is made in most cycle analysis. Applying the first law of thermodynamics to this cycle gives:

$$Q_{in} + \text{stored energy}_{in} \text{ (fluid)} = \text{work}_{net} - \text{stored energy}_{out} \text{ (fluid)}$$

but

$$\text{Stored energy of the fluid} = u + \frac{pv}{J} + \frac{V^2}{2g_c} + \frac{g}{g_c} z$$

$$= h + \frac{V^2}{2g_c} + \frac{g}{g_c} z$$

It is customary in cycle analysis to eliminate the kinetic- and potential-stored-energy terms because their changes are negligible compared to the changes in the other forms of energy. Thus,

$$Q_{in} + h_{in} = \text{work}_{net} + h_{out} \qquad (13.1)$$

The efficiency of the cycle is

$$\eta = \frac{\text{work}_{net}}{Q_{in}}$$

It should be noted that in this use the work$_{net}$ term is not equal to $Q_{in} - Q_{out}$ but $Q_{in} - (h_{in} - h_{out})$. In the idealized case where the fluid out of the system is assumed to reach eventually the condition at which it entered the system $(h_{in} - h_{out})$ is equivalent to $Q_{out}$.

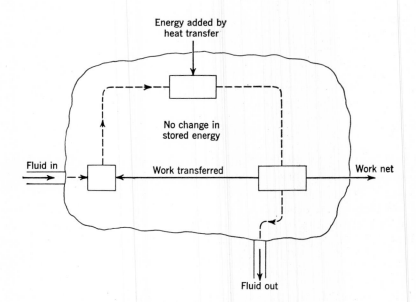

**Fig. 13.2.** A generalized open cycle.

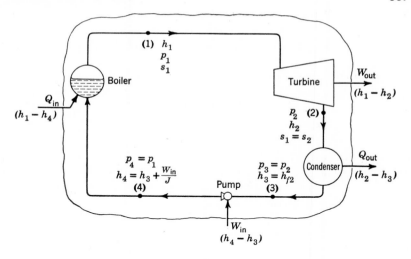

**Fig. 13.3.** Rankine cycle.

In this chapter, two power-plant cycles will be considered: the simple form of the vapor (steam) power-plant cycle and the simple gas-turbine cycle. The more complex cycles will be discussed only briefly. The student should note the definite similarity between the two cycles and their analyses.

## 13.2   The Rankine Cycle

The simplest ideal or theoretical power-plant steam cycle is the Rankine cycle, which is illustrated in Fig. 13.3. The system contains: (1) a steam-generating unit by which energy is added to the fluid in the form of heat transfer from a burning fuel, (2) a prime mover or steam turbine, (3) a condenser by which energy is rejected to the surroundings by heat transfer, and (4) a boiler feed-water pump.

The following assumptions are made for the Rankine cycle:

1. The working fluid, usually water, is pumped into the boiler, evaporated into steam in the boiler, expanded in the prime mover, condensed in the condenser, and returned to the boiler feed pump to be circulated through the equipment again and again in a closed circuit under *steady-flow conditions;* that is, at any given point in the sys-

tem, the conditions of pressure, temperature, flow rate, etc., are *constant*.

2. All the heat is added in the steam-generating unit, all the heat that is rejected is transferred in the condenser, and there is no heat transfer between the working fluid and the surroundings at any place except in the steam-generating unit and the condenser.

3. There is no pressure drop in the piping system; there is a constant high pressure, $p_1$, from the discharge side of the boiler feed pump to the prime mover, and a constant low pressure, $p_2$, from the exhaust flange of the prime mover to the inlet of the boiler feed pump.

4. Expansion in the prime mover and compression in the pump occur without friction or heat transfer; in other words, they are frictionless adiabatic or *isentropic* expansion and compression processes in which the entropy of the fluid leaving the device equals the entropy of the fluid entering the device (pump or turbine).

5. The working fluid leaves the condenser as liquid at the highest possible temperature, which is the *saturation temperature* corresponding to the exhaust pressure $p_2$.

If, as shown in Fig. 13.3, the steam-generating unit is a boiler only, the steam that it delivers will be wet, and its quality and enthalpy can be determined by a throttling calorimeter. If a superheater is included in the steam-generating unit, the steam that is delivered will be superheated, and its enthalpy can be determined from its pressure and temperature by use of the superheated steam tables or the Mollier chart.

In the prime mover, which is usually a turbine, the steam is assumed to expand at *constant entropy*, and, in accordance with the discussion in Article 10.3,

$$\frac{W_{out}}{J} = h_1 - h_2 \tag{13.4}$$

where $W_{out}$ = work done in the turbine, ft-lb per lb of steam

$h_1$ = enthalpy of steam entering the turbine, Btu per lb

$h_2$ = enthalpy of steam leaving the turbine after *isentropic* expansion to the exhaust pressure, Btu per lb

The enthalpy of the exhaust steam, $h_2$, after isentropic expansion, can be found most easily by using the Mollier diagram as discussed in Article 10.3 and illustrated in Example 4 in that article.

The condensate leaving the condenser and entering the boiler feed pump is always assumed to be *saturated water at the condenser pressure*, and its enthalpy, $h_{f2}$, can be found from the steam tables at the given condenser pressure. The heat rejected in the condenser per pound of steam, $Q_{out}$, is then given by the equation

$$Q_{out} = h_2 - h_{f2} \qquad (13.5)$$

It is necessary to supply energy to the boiler feed pump to compress the water to the boiler pressure $p_1$. In accordance with the discussion in Article 2.18 for horizontal flow through the pump without change in water velocity or specific volume, the work done by the pump on 1 lb of water, $W_{in}$, in foot-pound units, is

$$W_{in} = (p_1 - p_2)v_{f2} \qquad (13.6)$$

where $p_1$ and $p_2$ are expressed in pounds per square foot absolute and $v_{f2}$ is the specific volume of the saturated water supplied to the pump.

Since it is assumed that there is no heat transfer from the water to its surroundings in the pump, the energy supplied by the pump is stored in the high-pressure water, and the enthalpy of the boiler feedwater is

$$h_3 = h_{f2} + \frac{W_{in}}{J} \qquad (13.7)$$

Also, since the energy supplied by the pump must come from the output of the prime mover, then the net output of the cycle per pound of fluid, $W_{net}/J$, is given by the equation

$$\frac{W_{net}}{J} = \frac{W_{out} - W_{in}}{J} = (h_1 - h_2)_s - \frac{W_{in}}{J} \qquad (13.8)$$

The heat supplied in the steam-generating unit to produce 1 lb of steam is

$$Q_{in} = h_1 - h_3 = h_1 - \left( h_{f2} + \frac{W_{in}}{J} \right) \qquad (13.9)$$

Since the efficiency

$$= \frac{\text{heat equivalent of the net work done per lb of steam}}{\text{heat supplied to generate 1 lb of steam}}$$

then
$$\eta_t = \frac{(h_1 - h_2)_s - \dfrac{W_{in}}{J}}{h_1 - \left( h_{f2} + \dfrac{W_{in}}{J} \right)} \qquad (13.10)$$

where $\eta_t$ is the theoretical efficiency of the Rankine cycle.

Where the boiler pressure is under 400 psia and slide-rule calculations are used, the energy supplied to the pump may be neglected and Equation 13.10 reduces to the form

$$\eta_t = \frac{(h_1 - h_2)_s}{h_1 - h_{f2}} \tag{13.11}$$

**Example 1.** Compute the efficiency of the Rankine cycle if steam at 200 psia and 600 F is expanded to an exhaust pressure of 2 psia.

*Solution:* From the steam tables at 200 psia and 600 F, $h_1 = 1322$ and $s_1 = 1.6767$.

At the exhaust pressure of 2 psia and $s_2 = s_1 = 1.6767$, $h_2$ may be found from the Mollier chart to be 974 Btu per lb. Also, at 2 psia, $v_{f2} = 0.01623$ cu ft per lb, $h_{f2} = 94$, and $t_2 = 126$ F.

Then, the gross output of the prime mover,

$$\frac{W_{out}}{J} = h_1 - h_2 = 1322 - 974 = 348 \text{ Btu per lb}$$

The energy rejected to waste in the condenser is equal to

$$Q_{out} = h_2 - h_{f2} = 974 - 94 = 880 \text{ Btu per lb}$$

The energy input to the water from the boiler feed pump, $W_{in}$ is equal to

$$W_{in} = (p_1 - p_2)v_{f2} = 144(200 - 2)(0.01623) = 462 \text{ ft-lb per lb of fluid}$$

or $\dfrac{W_{in}}{J} = \dfrac{462}{778} = 0.594$ Btu per lb

Then the enthalpy of the feedwater entering the boiler,

$$h_3 = h_{f2} + \frac{W_{in}}{J} = 94 + 0.6 = 94.6 \text{ Btu per lb}$$

The heat supplied in the boiler,

$$Q_{in} = h_1 - h_3 = h_1 - \left( h_{f2} + \frac{W_{in}}{J} \right) = 1322 - (94.0 + 0.6) = 1227.4 \text{ Btu per lb}$$

$$\eta_t = \frac{\text{output}}{\text{input}} = \frac{(h_1 - h_2) - \dfrac{W_{in}}{J}}{h_1 - \left( h_{f2} + \dfrac{W_{in}}{J} \right)} = \frac{(1322 - 974) - 0.6}{1322 - (94.0 + 0.6)} = 0.283 = 28.3\%$$

Neglecting the boiler feed pump,

$$\eta_t = \frac{\text{output}}{\text{input}} = \frac{h_1 - h_2}{h_1 - h_{f2}} = \frac{1322 - 974}{1322 - 94.0} = 0.283 = 28.3\%$$

It should be noted that, for the boiler pressure of 200 psia in the above example, the inaccuracies introduced by the use of the slide rule and the Mollier chart are such that the effect of the boiler feed pump may be neglected. However, at a high pressure such as 1000 psia, neglect of the pump would introduce an appreciable error.

The influence of initial pressure, superheat, and exhaust pressure upon the efficiency of the Rankine cycle is shown in Fig. 13.4. The improvement in efficiency with reduction in the exhaust pressure should be noted particularly. Also, the gain in efficiency for each 100-psi increase in initial pressure becomes less as the pressure increases. Since this cycle assumes frictionless adiabatic or ideal expansion of the steam in the prime mover, the Rankine-cycle efficiency is the best that is theoretically possible with the equipment arranged as in Fig.

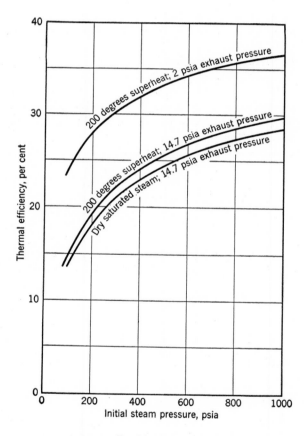

**Fig. 13.4.** Rankine-cycle efficiency.

13.3. Better theoretical efficiencies are possible by using more equipment in more complex cycles, but the discussion of such cycles is beyond the scope of this book.

It should be noted from Fig. 13.4 and Example 1 that only a small part of the energy supplied in the boiler as heat is converted into work and the rest is lost in the condenser. In an actual cycle, using a real prime mover with its usual internal losses and with pressure drop and radiation losses from the piping system, the actual efficiency will be considerably less than the Rankine-cycle efficiency and the heat rejected to the condenser will be greater than for the Rankine cycle for the same initial and exhaust pressures. In many of the real cycles the objective of the cycle arrangement is to reduce the exhaust loss or to utilize part or all of the energy in the exhaust steam for heating or manufacturing processes.

The loss resulting from the heat transferred to the condensor cooling water is, to a large extent, inescapable. The temperature of the cooling water varies only with the atmospheric conditions; thus, it remains almost constant. To lower it by artificial means would require the expenditure of additional energy. This cooling-water temperature is the $T_L$ referred to in the Carnot cycle efficiency, $\eta = 1 - (T_L/T_H)$. Notice that the best one can do to increase the theoretical cycle efficiency is to increase $T_H$ if $T_L$ is fixed. The plant operating on a Rankine cycle can never have the efficiency, $\eta = 1 - T_L/T_H$.

**Example 2.** Compute the Carnot efficiency for the same conditions as assumed in Example 1.
*Solution:*

$$\eta = 1 - \frac{T_L}{T_H}$$

But $T_L = 460 + 126 = 586$ R, where $t_L = 126$ is the temperature corresponding to a 2-psia exhaust pressure. The cooling-water temperature would be only a little lower.

Also, $T_H = 600 + 460 = 1060$ R.

$$\therefore \eta = 1 - \tfrac{586}{1060} = 1 - 0.55$$

$$= 0.45 \text{ or } 45\%$$

NOTE: The Rankine cycle efficiency, as calculated in Example 1, is 28.3%.

### 13.3  Complex Vapor Cycles

Complex vapor cycles have been developed to improve upon the efficiency of the simple Rankine cycle and to approach within prac-

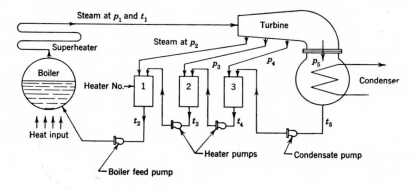

**Fig. 13.5.** Regenerative power-plant cycle.

tical limits the efficiency of the Carnot cycle. One of these cycles, and one commonly used in the modern power plant, is the regenerative cycle, a simple flow diagram of which is shown in Fig. 13.5.

Steam which enters the turbine at high pressure expands step by step in the turbine to the low exhaust pressure. Part of the steam is withdrawn from the turbine at intermediate pressures after it has done work by expansion from the initial pressure. This *extracted steam* is condensed in feed-water heaters, and the energy in this steam is thus returned to the boiler instead of being rejected to waste in the condenser cooling water. From three to six extraction heaters are normally connected to the turbine, depending upon the size of the unit and the initial steam pressure. As much as 25 per cent of the steam entering the turbine may be extracted in the heaters. This extracted steam is being used at an efficiency of 100 per cent since, if heat losses to the atmosphere from equipment are neglected, the energy in this steam as it enters the turbine is either converted into work or returned to the boiler. Thus the condenser loss of the cycle shown in Fig. 13.3 is reduced, and the overall efficiency is improved. Electrically driven auxiliaries are used with the regenerative cycle in order that all the steam may be expanded in large, efficient turbines with the generation of the maximum possible amount of power instead of expansion through small and relatively inefficient units directly connected to auxiliaries such as fans and pumps. Thermal efficiencies up to about 35 per cent are obtainable by the use of the regenerative cycle.

Serious erosion of turbine blades and loss of turbine efficiency

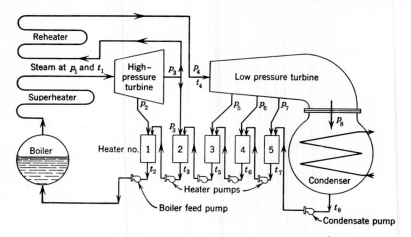

**Fig. 13.6.**  Reheating–regenerative power-plant cycle.

would occur if the quality of the exhaust steam became less than about 88 per cent, and high steam pressures require certain minimum steam temperatures to avoid such difficulties.  Steam pressures higher than those recommended for a given steam temperature and for the Rankine cycle may be used by employing the reheating–regenerative cycle which is illustrated in Fig. 13.6.  Steam is expanded through a high-pressure turbine to an intermediate pressure which is generally 25 to 40 per cent of the initial pressure.  The exhaust steam from the high-pressure turbine is then returned to a reheat superheater in the steam-generating unit.  The steam is reheated at this intermediate pressure to a temperature equal to or not far below the initial steam temperature, after which it is piped to the low-pressure turbine in which the expansion is completed to the condenser pressure.  Re-generative feed-water heaters are used as in the straight regenerative cycle for the purpose of reducing the condenser loss by heating the feedwater with steam which has done work in the turbine.  In general, the reheating–regenerative cycle will produce a saving in fuel of 4 to 5 per cent as compared with the regenerative cycle for the same maximum steam temperature.  Consequently, many large central-station plants located in regions of high-cost fuel are operating on this cycle in spite of its higher construction cost as compared with the straight regenerative cycle.

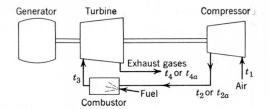

**Fig. 13.7.** The simple gas-turbine power-plant cycle.

## 13.4   The Simple, Open, Gas-Turbine Power Cycle

The gas turbine and the gas-turbine power plant were discussed briefly in Article 10.12. As explained in this article, the power plant consists of three elements: the compressor, the combustion chamber, and the gas turbine. A flow diagram of the simple gas-turbine cycle is shown in Fig. 13.7, and an $h$-$s$ or Mollier chart for air is shown in Fig. 13.8.

Air enters the cycle at point 1 with a temperature, $t_1$ (F), and a pressure, $p_1$. During the compression process the temperature in-

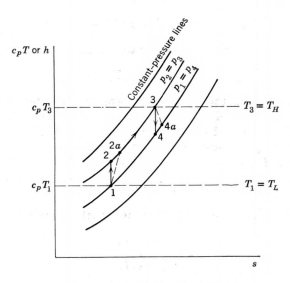

**Fig. 13.8.** $h$-$s$ or Mollier chart for air.

creases to a value equal to $t_2$ or $t_{2a}$ and a higher pressure, $p_2$. If the compression process is ideal, that is, adiabatic and frictionless or isentropic, the temperature becomes $t_2$. In an actual compression process with friction but little heat transfer the final temperature will be higher or $t_{2a}$. The work done on the air by the compressor during the process, neglecting changes in kinetic and potential energy of the air, is equal to the increase in enthalpy plus any heat loss (refer to Equation 2.11), or

$$W_{in} = h_2 - h_1 \quad \text{for isentropic compression} \tag{13.12}$$

and for an actual compression process

$$W_{in} = h_{2a} - h_1 + Q_{out} \tag{13.13}$$

in which $\qquad h_1 = c_p T_1,\ h_2 = c_p T_2,\ \text{and}\ h_{2a} = c_p T_{2a}$

Heat is transferred to the cycle between points 2 or 2a and point 3 either by burning a fuel in the air stream or by heating the air by means of a heat exchanger or heater.

Applying the general energy balance to the process between 2 or 2a and 3, and neglecting changes in kinetic and potential energies of the air, the heat transferred to the air will be equal to the increase in enthalpy of the air or

$$Q_{in} = h_3 - h_2 \quad \text{for the ideal process} \tag{13.14}$$

and $\qquad\qquad Q_{in} = h_3 - h_{2a} \quad \text{for the actual process} \tag{13.15}$

in which $\qquad\qquad\qquad h_3 = c_p T_3$

The heated air at the elevated temperature, $t_3$, enters the gas (air) turbine, expands and does work on the surroundings, and is discharged to the atmosphere at the ideal temperature, $t_4$, following an isentropic expansion or, $t_{4a}$, following an actual expansion process.

The work of the turbine per pound of air is

$$W_{out} = h_3 - h_4 \quad \text{for the ideal expansion} \tag{13.16}$$

and $\qquad\qquad W_{out} = h_3 - h_{4a} \quad \text{for the actual expansion} \tag{13.17}$

in which $\qquad\qquad\qquad h_4 = c_p T_4$

and $\qquad\qquad\qquad h_{4a} = c_p T_{4a}$

The efficiency of the cycle is defined as the ratio of the net work of the cycle to the heat supplied, or, for the ideal case,

$$\eta = \frac{W_{\text{out}} - W_{\text{in}}}{Q_{\text{in}}}$$

$$\eta = \frac{(h_3 - h_4) - (h_2 - h_1)}{h_3 - h_2} \qquad (13.18)$$

$$\eta = \frac{c_p(T_3 - T_4) - c_p(T_2 - T_1)}{c_p(T_3 - T_2)}$$

If $c_p$ is constant throughout the cycle, then

$$\eta = \frac{(T_3 - T_4) - (T_2 - T_1)}{T_3 - T_2} \qquad (13.19)$$

In thermodynamics the student will learn to show that Equation 13.19 will give the same expression for efficiency as that of the Otto cycle for the same compression ratio. The maximum efficiency that could be attained would be for the case when $T_2$ approaches the value of $T_3$. The efficiency would be identical to the Carnot efficiency, or

$$\eta = 1 - \frac{T_3}{T_1} = 1 - \frac{T_H}{T_L}$$

However, as the efficiency approaches a maximum the net work approaches zero. Thus, it is impractical to develop a cycle with maximum efficiency. The maximum net work occurs when

$$\frac{T_2}{T_1} = \sqrt{\frac{T_3}{T_1}} \qquad (13.20)$$

NOTE: The student who has had calculus might consider how Equation 13.20 would be obtained.

The efficiency of the actual cycle would be

$$\eta = \frac{(T_3 - T_{4a}) - (T_{2a} - T_1)}{T_3 - T_{2a}} \qquad (13.21)$$

Now, if compressor efficiency is defined as

$$\eta_c = \frac{T_2 - T_1}{T_{2a} - T_1}$$

and the turbine efficiency is defined as

$$\eta_t = \frac{T_3 - T_{4a}}{T_3 - T_4}$$

then, combining these efficiency expressions with Equation 13.21 gives

$$\eta = \frac{\eta_t(T_3 - T_4) - \dfrac{1}{\eta_c}(T_2 - T_1)}{T_3 - T_{2a}}$$

but $\qquad T_{2a} = T_1 + T_{2a} - T_1$

in which $\qquad T_{2a} - T_1 = \dfrac{1}{\eta_c}(T_2 - T_1)$

Thus, $\qquad \eta = \dfrac{\eta_t(T - T_4) - \dfrac{1}{\eta_c}(T_2 - T_1)}{T_3 - T_1 - \dfrac{1}{\eta_c}(T_2 - T_1)}$ $\qquad$ (13.22)

In the actual gas-turbine power plant, 65 to 80 per cent of the turbine output is required to drive the compressor. In the steam-turbine power plant, the working fluid is condensed with a very large reduction in volume so that less than 1 per cent of the turbine output is required to operate the boiler feed pump which corresponds to the air compressor of the gas-turbine power plant. Consequently, for the same net plant output, the gas turbine must produce three to four times as much power as a steam turbine. Such heat-transfer equipment as boilers, economizers, superheaters, condensers, feed-water heaters, forced- and induced-draft fans, and an extensive piping system, all of which are necessary in an efficient steam power plant, are eliminated in the simple gas-turbine power plant. However, if maximum efficiency is desired in the gas-turbine power plant, large heat exchangers, water-circulating pumps and piping are necessary, and the gas-turbine plant loses much of its simplicity.

The thermal efficiency of a gas-turbine power plant may be defined as follows:

$$\eta = \frac{\text{hp} \times 2545}{mQ_H}$$ $\qquad$ (13.23)

where hp = net output, hp
$\qquad m$ = fuel burned per hr, lb
$\qquad Q_H$ = higher heating value of fuel, Btu per lb

Equation 13.23 gives a value for the efficiency which is lower than that obtained by Equation 13.22. The difference lies in the fact that

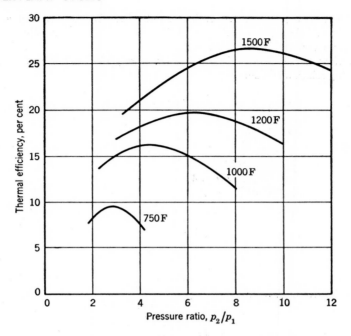

**Fig. 13.9.** Influence of pressure ratio and turbine-inlet temperature upon thermal efficiency.

certain miscellaneous losses were excluded from Equation 13.22 such as bearing losses and steam leakage losses.

The efficiency of a simple gas-turbine power plant depends upon the temperature of gas supplied to the turbine and upon the pressure ratio, $p_2/p_1$. For a constant-air temperature at the inlet to the compressor, the influence of pressure ratio and turbine-inlet temperature on the thermal efficiency is illustrated in Fig. 13.9. For a given turbine-inlet temperature, there is a particular pressure ratio which gives maximum efficiency, and this optimum pressure ratio increases with inlet temperature. The marked increase in efficiency with increase in inlet temperature should be noted. As the high-temperature characteristics of metals are improved and inlet temperatures higher than 1500 F become practical, the use of the gas turbine as an economical prime mover will expand rapidly.

**Example 3.** Air enters a gas-turbine cycle at 70 F and 14.7 psia. It is compressed to four times its initial pressure and then heated to a temperature of 1200 F. It expands in the turbine to 14.7 psia. Compute the net work of the cycle and the efficiency of the cycle if there are no losses and the compression and expansion processes are isentropic.

*Solution:* For an isentropic compression of a perfect gas from $t_1$ to $t_2$ the following relation is true:

$$\frac{T_2}{T_1} = \left(\frac{p_2}{p_1}\right)^{\frac{k-1}{k}}$$  (see Equation 3.21)

but

$$p_1 = 14.7 \text{ psia}$$

$$p_2 = 4 \times 14.7$$

$$k = 1.4 \text{ for air}$$

$$T_1 = 70 + 460 = 530 \text{ R}$$

Thus,

$$T_2 = 530(4)^{\frac{1.4-1}{1.4}}$$

$$= 530(4)^{0.286}$$

$$= 790 \text{ R}$$

for the isentropic expansion

$$\frac{T_4}{T_3} = \left(\frac{p_4}{p_3}\right)^{\frac{k-1}{k}}$$

but

$$T_3 = 1200 + 460 = 1660 \text{ R}$$

$$p_3 = p_2 = 4 \times 14.7$$

$$p_4 = p_1 = 14.7$$

$$k = 1.4$$

Thus,

$$T_4 = 1660(\tfrac{1}{4})^{0.286}$$

$$= 1115 \text{ R}$$

Now, the net work in Btu per lb is

$$\begin{aligned}
W_{net} &= W_{out} - W_{in} \\
&= (h_3 - h_4) - (h_2 - h_1) \\
&= c_p(T_3 - T_4) - c_p(T_2 - T_1) \\
&= 0.24(1660 - 1115) - 0.24(790 - 530) \\
&= 0.24(545) - 0.24(260) \\
&= 131 - 62 \\
&= 69 \text{ Btu per lb}
\end{aligned}$$

The efficiency is
$$\eta = \frac{\text{net work}}{Q_{\text{in}}}$$

in which $Q_{\text{in}} = h_3 - h_2$
$$= c_p(T_3 - T_2)$$
$$= 0.24(1660 - 790)$$
$$= 0.24(870)$$
$$= 209 \text{ Btu per lb}$$

Therefore
$$\eta = \frac{69}{209} \times 100 = 33\%$$

**Example 4.** Assuming the same conditions as in Example 3 except that the turbine and compressor have an efficiency of 90 per cent each, calculate the net work and the cycle efficiency.

*Solution:*
$$W_{\text{net}} = (h_3 - h_{4a}) - (h_{2a} - h_1)$$

$$W_{\text{net}} = \eta_t(h_3 - h_4) - \frac{1}{\eta_c}(h_2 - h_1)$$

But, from Example 3,
$$h_3 - h_4 = 131$$
$$h_2 - h_1 = 62$$

Also,
$$\eta_t = \eta_c = 0.9$$

Therefore, the actual net work
$$W_{\text{net}} = 0.9 \times 131 - \frac{62}{0.90}$$
$$= 118 - 69$$
$$= 49 \text{ Btu per lb}$$

The efficiency is
$$\eta = \frac{\text{net work}}{Q_{\text{in}}}$$

in which the net work $= 49$ Btu per lb
and $Q_{\text{in}} = c_p(T_3 - T_{2a})$

but
$$T_{2a} = T_1 + \frac{T_2 - T_1}{\eta_c}$$
$$= 530 + \frac{(790 - 530)}{0.9}$$
$$= 819 \text{ R}$$

and
$$Q_{\text{in}} = 0.24(1660 - 819)$$
$$= 202 \text{ Btu per lb}$$

Therefore,
$$\eta = \tfrac{49}{202} \times 100$$
$$= 24.2\%$$

NOTE: The student should keep in mind the fact that the pressure ratio for the previous two examples may not be the value to give maximum net work or maximum efficiency. Typical curves showing the effect of variations in pressure ratio on efficiency are presented in Fig. 13.9.

### 13.5  The Gas-Turbine Power Plant with Regenerator

In the simple cycle the exhaust gases are discharged to waste from the turbine at a temperature of 600 to 900 F, depending upon the turbine-inlet temperature. Part of this waste energy can be recovered by utilizing the hot waste gas to preheat the compressed air in a regenerator or heat exchanger ahead of the combustor, as shown in Fig. 13.10. It is evident from Fig. 13.10 that the heat transferred from the exhaust gases to the air in the regenerator results in a substantial reduction in the amount of fuel required to produce the desired turbine-inlet temperature.

The efficiency of the cycle can be further increased by cooling the air at an intermediate pressure during the compression process. This is accomplished by taking the air from the compressor, passing the air through an intercooler, and returning it to another compressor. The work of compression is reduced by this intercooling process.

Another method of improving efficiency is accomplished by expanding the gas in a high-pressure and a low-pressure turbine in series with a second combustion chamber located between the turbines. This is called reheating. The effect is to raise the average temperature at which the air is supplied to the turbines with the purpose of approaching the efficiency of the Carnot cycle, where $\eta = 1 - (T_L/T_H)$. In effect, the average-high-temperature source of energy is raised in order to approach the upper limit, $T_H$, which the metals of the turbine can withstand.

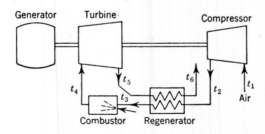

**Fig. 13.10.**  The gas-turbine power-plant cycle with regenerator.

## PROBLEMS

1. A power plant operates between an ambient temperature (cooling water or air temperature) of 40 F and a temperature limited by the metals used in the plant of 1440 F. What is the maximum, theoretical cycle efficiency of the plant?

2. For a plant operating with an upper temperature of 1200 F what is the difference in efficiency in operating the plant at an average summer temperature of 90 F and an average winter temperature of 30 F?

3. Calculate the Rankine cycle efficiency for the following sets of conditions: exhaust pressure, 14.7 psia; inlet pressure, 600 psia; inlet temperatures, 400 F, 600 F, 800 F, respectively. Plot the efficiency versus the inlet temperatures in degrees Rankine.

4. Repeat Problem 3 by changing the exhaust pressure to 1 psia.

5. Calculate the Carnot efficiencies for Problems 3 and 4.

6. How would you further improve the cycle efficiency of Problem 3 other than raising the temperature at turbine inlet or reducing the exhaust pressure?

7. Calculate the net work and the cycle efficiency of a simple, ideal gas-turbine cycle if the air enters at 80 F and 14.7 psia, is compressed to 40 psia, and is heated to 1200 F.

8. For a maximum temperature at the turbine inlet of 1500 F and a sink or ambient temperature of 500 R calculate the ideal cycle efficiency for a simple gas-turbine power plant for four different pressure ratios: 1, 2, 3, and 4. Plot the efficiency versus pressure ratio.

9. Calculate the efficiencies for the plant in Problem 8 if the turbine and compressor have efficiencies of 80 per cent each.

# CHAPTER 14

# Mechanical Refrigeration

### 14.1 Introduction

Previous chapters have dealt with the equipment and methods used to produce power by the conversion of low-grade energy into shaft work. Refrigeration is the process of removing energy by heat transfer from some product or substance in order to freeze it or to maintain its temperature below that of the atmosphere.

Refrigeration is generally produced in one of three ways: (1) by the melting of a solid, (2) by the sublimation of a solid, and (3) by the evaporation of a liquid. Ice melts at 32 F and, in so doing, absorbs energy equivalent to its latent heat of fusion, which is 144 Btu per lb. Temperatures lower than 32 F may be obtained by using a mixture of ice and salt. Solid carbon dioxide or Dry Ice passes directly from a solid to a gaseous state at −109.3 F and has about twice as much refrigerating capacity as the same mass of ice. It is used extensively when perishable materials are shipped under conditions where weight or salt water is objectionable. Most commercial refrigeration is produced by the evaporation of a liquid which is called a *refrigerant*. The evaporation of water under normal atmospheric conditions is used in dry climates for the cooling of residences and business houses. *Mechanical refrigeration* depends upon the evaporation of a liquid refrigerant that, because of its cost, must be used over and over again in a closed circuit. This circuit includes the following essential equipment:

1. An *evaporator* or heat exchanger in which a liquid refrigerant may evaporate at a low temperature and pressure, thus absorbing energy from the surroundings.

2. A *compressor* in the compression system (or other apparatus in the absorption system) for removing the low-pressure vapor from the evaporator and delivering it at a higher pressure to the condenser. The compressor is driven by a motor or engine, and the energy rep-

resented by the work done in compressing the fluid is absorbed by the refrigerant that leaves the compressor as a superheated vapor.

3. A *condenser* or heat exchanger in which the high-pressure refrigerant is cooled and condensed by transfer of heat to a coolant, either water or air.

4. An *expansion valve* or other device for regulating the rate of flow of the high-pressure liquid refrigerant from the condenser to the evaporator in which a low pressure is maintained by the compressor.

The refrigerant during this cycle accomplishes two major purposes without destroying itself: (1) it removes energy from the substance to be refrigerated, and (2) it transfers this energy at a higher temperature level to some other substance such as normal atmospheric air or well water.

Mechanical refrigeration is used mainly for (1) the preservation of food and other perishable products during storage and transportation, (2) the manufacture of ice and solid carbon dioxide, and (3) the control of air temperature and humidity in air-conditioning systems.

## 14.2 Refrigeration by Evaporation of a Liquid

Let Fig. 14.1 represent an insulated box containing hot air and an evaporator or boiler. Assume that the evaporator has a free vent to the atmosphere and contains water at 212 F. If the air in the box is at a temperature above 212 F, energy will flow from the air to the water in the boiler, steam will be generated at 212 F with the absorption of the latent heat of evaporation, and the steam will escape to the atmosphere through the vent, thus removing the latent heat of evaporation from the box. Unless energy is supplied to the air in the box, the temperature of the air will decrease until it approaches the vaporization temperature of the water in the boiler.

A refrigerant may be substituted for the water in the evaporator or boiler of Fig. 14.1. A refrigerant is a liquid that will boil at a temperature low enough to maintain the desired temperature in the refrigerated space. The boiling points or saturation temperatures at standard atmospheric pressure of a number of common refrigerants are as follows:

| | |
|---|---|
| Ammonia ($NH_3$) | $-28$ F |
| Freon 12 ($CCl_2F$) | $-22$ F |
| Methyl chloride ($CH_3Cl$) | $-11$ F |
| Sulphur dioxide ($SO_2$) | $+14$ F |

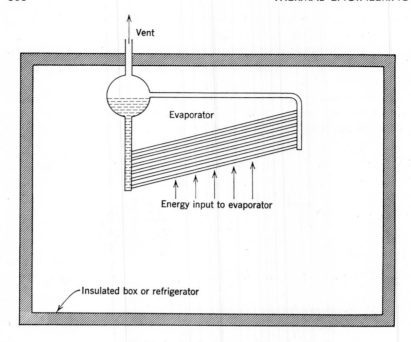

**Fig. 14.1.** Refrigeration by evaporation of a liquid.

If liquid ammonia at atmospheric pressure should be placed in the evaporator of Fig. 14.1, it would boil at −28 F, would absorb its latent heat of evaporation from the contents of the box or refrigerator, and would deliver this energy to the outside atmosphere via the ammonia vapor escaping through the vent. If the refrigerator were well insulated and the evaporator contained enough liquid ammonia, the contents of the refrigerator would be cooled to a temperature approaching −28 F.

If ammonia were cheap and plentiful and the escaping gases were not obnoxious, refrigeration could be produced continuously by supplying liquid ammonia to the evaporator as needed and allowing the vapor to escape to the atmosphere. However, refrigerants are expensive, and it is necessary to condense the vapor leaving the evaporator and return it to the evaporator again in a closed system. In the *compression system of refrigeration* which will be discussed in subsequent articles, it is the function of the compressor, condenser, and expansion valve to bring about condensation of the vapor that leaves the evaporator and to return it to the evaporator as a liquid.

The relation between the saturation temperature and the saturation pressure of some common refrigerants is shown in Fig. 3.6. An ideal refrigerant is one that will evaporate at a pressure slightly above atmospheric pressure and at a temperature low enough to maintain the desired refrigerator temperature. It should condense at a moderate pressure at the temperature existing in a condenser supplied with surface or well water or normal atmospheric air as the cooling medium. The thermal properties of ammonia are given in Tables A.4, A.5, and A.6 in the Appendix. These Tables are typical of those of the properties of various refrigerants that will be found in textbooks on refrigeration.

The unit of capacity of a refrigeration system is the *ton of refrigeration* or the ton of ice melting. A refrigeration machine is said to have a capacity of 1 ton of refrigeration for each 288,000 Btu absorbed by the refrigerant in the evaporator while operating continuously 24 hr per day. This is equal to the quantity of energy required to melt 1 ton (2000 lb) of 32 F ice, allowing 144 Btu for each pound of ice melted. This heat transfer is at the rate of 12,000 Btu per hr or 200 Btu per min. One ton of refrigeration is therefore defined as heat transfer in the evaporator to the refrigerant at the rate of 200 Btu per min.

### 14.3 The Compression System of Refrigeration

The flow diagram of the compression system of refrigeration is shown in Fig. 14.2. Liquid refrigerant is vaporized in the evaporator at an evaporator pressure, $p_e$. The vapor is compressed in a compressor to a discharge pressure, $p_c$, and is delivered to a water- or air-cooled condenser in which the vapor is condensed at the condenser pressure, $p_c$, to form a liquid. The liquid refrigerant is collected in a receiver from which it flows through the expansion valve in which the pressure is reduced to the evaporator pressure, $p_e$. A low pressure, $p_e$, exists from the discharge side of the expansion valve through the evaporator to the compressor. A high pressure, $p_c$, exists from the discharge side of the compressor through the condenser to the expansion valve.

The required pressure in the evaporator depends upon the temperature to be maintained in the evaporator, the refrigeration capacity of

the system, and the amount of surface in the evaporator, since the
heat must be transferred from the refrigerator through the evaporator surface to the refrigerant. The required pressure in the condenser depends upon the temperature of the cooling water or air to
which the heat is being rejected, the amount of surface through which
the heat is being transferred, and the amount of heat being transferred. Table 14.1 shows the evaporator and condenser pressures
that will exist at several saturation temperatures for some of the common refrigerants. For minimum power input to the compressor, the
evaporator pressure should be as high as possible, and the condenser
pressure should be as low as possible.

Owing to frictional resistance to flow, a pressure drop occurs between the discharge side of the expansion valve and the compressor
inlet and between the compressor discharge and the high-pressure
side of the expansion valve. By careful design, this pressure drop can
be made quite small and is neglected in the subsequent discussion.

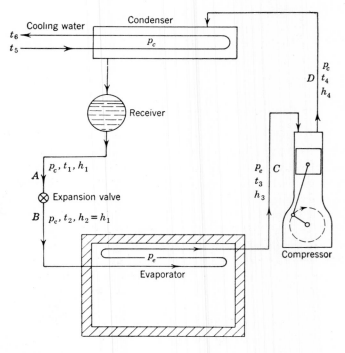

**Fig. 14.2.** Flow diagram of the compression refrigeration cycle.

### TABLE 14.1

## Evaporator and Condenser Pressures for Various Saturation Temperatures

| Refrigerant | Evaporator Pressure ($p_e$), psia | | | Condenser Pressure ($p_c$), psia | | |
|---|---|---|---|---|---|---|
| | −20 F | 0 F | +20 F | 70 F | 80 F | 90 F |
| Ammonia | 18.3 | 30.4 | 48.2 | 128.8 | 153.0 | 180.6 |
| Freon 12 | 15.3 | 23.9 | 35.8 | 84.8 | 98.8 | 114.3 |
| Methyl chloride | 11.7 | 18.9 | 29.2 | 73.4 | 86.3 | 100.6 |
| Sulphur dioxide | 5.9 | 10.4 | 17.2 | 49.6 | 59.7 | 71.3 |

## 14.4 The Properties of Ammonia

Since the refrigeration cycle may be considered as involving only two pressures, the evaporator and condenser pressures, it is convenient to represent the properties of refrigerants graphically on a diagram having enthalpy as the abscissa and absolute pressure as the ordinate. Such a diagram, known as the Mollier diagram, is reproduced for ammonia in Figs. A.2 and A.3 of the Appendix. Pressure from 10 to 300 psia is plotted as ordinate to a logarithmic scale. The enthalpy of saturated liquid ammonia at −40 F is assumed to be zero. In order to conserve space through the elimination of those areas of the diagram that are not needed for the solution of problems, only the areas of the chart having enthalpy values from −26 to +200 Btu per lb are plotted on Fig. A.2; Fig. A.3 is plotted for enthalpy values from 500 to 830 Btu per lb.

A plot of enthalpy versus pressure of saturated liquid ammonia gives the saturated liquid line which extends diagonally across Fig. A.2. The region to the left of this line is the compressed liquid region. In this region, dotted lines which are approximately vertical are plotted to show the enthalpy of 1 lb of compressed liquid ammonia at 10-degree increments of temperature. The area to the right of the saturated liquid line is the wet vapor region in which the enthalpy, quality, and specific volume of wet ammonia vapor of low quality are shown. Roughly parallel and to the right of the saturated liquid line is a set of curves of constant quality which are plotted for increments of 5 per cent of quality from zero to 30 per cent. Lines of specific volume of the wet ammonia vapor are also plotted in this region for specific volumes from 0.1 to 8 cu ft per lb. That part of the complete Mollier diagram for ammonia which is represented by Fig. A.2 is used for the solution of problems involving the condition of the ammonia as it leaves the condenser and as it is supplied to and discharged from the expansion valve.

Figure A.3 shows the properties of superheated ammonia and wet ammonia vapor of high quality and is used in determining the properties of ammonia leaving the evaporator, entering and leaving the compressor, and entering the condenser. A curve labeled "saturated vapor" is obtained by plotting the enthalpy of dry saturated ammonia against pressure and divides the area of this chart into a superheat region to the right of this curve and a wet vapor region to the left of it. In the wet region, lines of constant quality are roughly parallel to the saturated vapor line and are plotted for increments of 5 per cent of quality, starting with 75 per cent. To the right of the saturated vapor line, in the superheat region, lines of constant temperature are plotted in 10-degree increments as dotted lines. When the pressure and temperature of superheated ammonia are known, a point representing the state of the ammonia can be located on the chart, and the enthalpy and other properties can be read directly from the appropriate scales. Curves of specific volume and constant entropy appear on the chart as diagonal lines.

The properties of ammonia are also available in tabulated form. Table A.4 of the Appendix presents the values of specific volume, entropy, and enthalpy for saturated ammonia at various temperature integers, and Table A.5 gives them at pressure increments. Table A.6 provides similar information for superheated ammonia. These numerical values permit more accurate calculations than can be made from graphical charts and are more appropriate for certain calculations. The tables are identical in form with those for steam and are applied in the same manner.

**Example 1.** Determine (a) the enthalpy of liquid ammonia at 20 F and 200 psia, (b) the enthalpy and specific volume of wet vapor at 30 psia and 10 per cent quality, (c) the enthalpy, specific volume, and entropy of wet vapor at 30 psia and 95 per cent quality, and (d) the enthalpy, specific volume, and entropy of ammonia at 40 psia and 200 F.

*Solution using Mollier diagram:* (a) The intersection of the 200-psia line with the 20 F compressed liquid line on Fig. A.2 gives a value of 65 Btu per lb for enthalpy.

(b) The intersection of the 30-psia line and the line of 10 per cent quality in the wet vapor region of Fig. A.2 gives an enthalpy of 99 Btu per lb and a specific volume of 0.93 cu ft per lb.

(c) The intersection of the 30-psia line with the line of 95 per cent quality in the wet vapor region of Fig. A.3 gives an enthalpy of 583 Btu per lb, a specific volume of 8.8 cu ft per lb, and an entropy of 1.275.

(d) The intersection of the 40-psia line and the 200 F line in the superheat region of Fig. A.3 gives an enthalpy of 720 Btu per lb, a specific volume of 10.4 cu ft per lb, and an entropy of 1.50.

*Solution using ammonia tables:* (a) From Table A.4, the enthalpy of saturated liquid ammonia at 20 F is 64.7 Btu per lb. The saturation pressure corresponding to 20 F is given as 48.2 psia, but the effect of this small pressure difference is negligible.

(b) From Table A.5, $h_x = h_f + xh_{fg} = 42.3 + 0.1(569.3) = 99.23$ Btu per lb. $v_x = xv_g + (1 - x)v_f = 0.1(9.236) + 0.9(0.02417) = 0.945$ cu ft per lb.

(c) From Table A.5, $h_x = h_f + xh_{fg} = 42.3 + 0.95(569.3) = 583.1$ Btu per lb. $v_x = xv_g = 0.95(9.236) = 8.774$ cu ft per lb. $s_x = s_f + s_{fg} = 0.0962 + 0.95(1.2402) = 1.2744$.

(d) From Table A.6, the enthalpy of ammonia vapor at 40 psia and 200 F is 719.4 Btu per lb, the specific volume is 10.27 cu ft per lb, and the entropy is 1.4987.

Figure 14.3 shows a compression refrigeration cycle on the pressure–enthalpy or Mollier diagram. The points $A$, $B$, $C$, and $D$ of Fig. 14.3 correspond to the points having the same symbols on the flow diagram (Fig. 14.2). Point $A$ represents the condition of the liquid ahead of the expansion valve. Point $B$ represents the condition of the refrigerant after it has passed through the expansion valve and is at the evaporator pressure. Points $C$ and $D$ represent the state of the refrigerant at entrance to and exit from the compressor. The en-

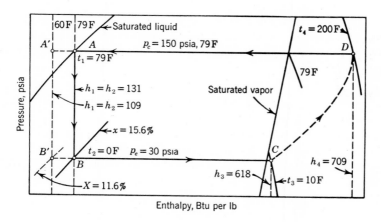

**Fig. 14.3.** Compression refrigeration cycle on the Mollier diagram.

thalpy values which appear on Fig. 14.3 were obtained from the Mollier diagram (Figs. A.2 and A.3).

The changes in enthalpy, the energy absorbed or rejected, and the work done on the refrigerant as it passes through the throttle valve, evaporator, compressor, and condenser are discussed in the subsequent articles of this chapter.

### 14.5   The Expansion Valve

The expansion valve usually consists of a valve stem having a conical point that can be screwed in or out of an opening in the valve seat, thus permitting close regulation of the area through which the refrigerant flows from the condenser pressure to the evaporator pressure.   If the expansion valve is well insulated to prevent heat transfer from the surroundings to the refrigerant, then, since no work is done in the valve and the kinetic energy of the fluid on each side of the valve is small, the enthalpy of the refrigerant remains constant, and the processs is a throttling process as discussed in Article 2.16. Therefore,

$$h_1 = h_2 \qquad\qquad (14.1)$$

where $h_1$ = enthalpy of the refrigerant ahead of the expansion valve
$h_2$ = enthalpy of the refrigerant beyond the expansion valve

If the refrigerant is saturated ammonia at a condenser pressure of 150 psia ahead of the expansion valve, the state of the refrigerant is represented by point $A$ on Fig. 14.3, the temperature is 79 F, and the enthalpy is 131 Btu per lb.   Below the expansion valve, for an evaporator pressure of 30 psia, the temperature is 0 F (see Fig. 3.6 or the saturation line of Fig. A.2), the enthalpy is 131 Btu per lb, and the quality is 15.6 per cent.   The throttling process in the expansion valve is represented on the Mollier diagram by a vertical or constant-enthalpy line drawn from point $A$ on the saturated liquid line to point $B$ on the line representing the evaporator pressure.   A total of 15.6 per cent of the liquid is converted into vapor since the enthalpy of saturated liquid ammonia at 150 psia is more than the enthalpy of saturated liquid ammonia at 30 psia, and the excess enthalpy results in vaporization of part of the liquid at the lower pressure.

To determine the quality of the refrigerant leaving the expansion valve from the tables, the enthalpy of the liquid ammonia at 79 F is found by interpolation between 70 F and 80 F in Table A.4.   At 70 F

$h_f$ is 120.5 Btu per lb and at 80 F it is 132.0 Btu per lb.  Subtracting one tenth of $(132.0 - 120.5)$ from 132.0 makes $h_f$ at 79 F equal 130.85 Btu per lb.  From Table A.5, $h_f$ is found to be 42.3 Btu per lb and $h_{fg}$ is 569.3 Btu per lb.  The quality after expansion is

$$x = \frac{h_1 - h_{f2}}{h_{fg2}} = \frac{130.85 - 42.3}{569.3} = 15.38\%$$

If the liquid ammonia at 150-psia condenser pressure had been cooled below the saturation temperature (79 F) to a final temperature of 60 F ahead of the expansion valve, the state of the subcooled liquid would be represented by point $A'$ on Fig. 14.3, and the enthalpy would be 109 Btu per lb.  Upon throttling to 30 psia, the quality would be 11.6 per cent at an enthalpy of 109 Btu.

### 14.6   The Evaporator

The energy absorbed in the evaporator per pound of refrigerant is equal to the increase in enthalpy of the refrigerant as it passes through the evaporator, as discussed in Article 2.11, or

$$Q_e = m_r(h_3 - h_2) \tag{14.2}$$

or
$$Q_e = m_r(h_3 - h_1) \tag{14.3}$$

where $Q_e$ = energy absorbed in evaporator, Btu per min
  $m_r$ = refrigerant circulated, lb per min
  $h_3$ = enthalpy of refrigerant leaving the evaporator, Btu per lb
  $h_2$ = enthalpy of refrigerant leaving the expansion valve and entering the evaporator, Btu per lb
  $h_1$ = enthalpy of refrigerant ahead of the expansion valve, Btu per lb

The expansion valve is usually adjusted to produce a few degrees of superheat in the refrigerant entering the compressor.  If it is assumed that the pipe connecting the evaporator and the compressor is well insulated, the enthalpy of the superheated vapor leaving the evaporator and entering the compressor can be obtained by locating point $C$ in the superheated region of the Mollier chart at the pressure and temperature of the vapor.  For ammonia at 30 psia and 10 F, the enthalpy is 618 Btu per lb, and point $C$ is shown on Fig. 14.3.

Since 1 ton of commercial refrigeration is equal to the absorption of 200 Btu per min, then

$$\text{Refrigeration capacity in tons} = \frac{Q_e}{200} = \frac{m_r(h_3 - h_1)}{200} \quad (14.4)$$

For the conditions represented in Fig. 14.3 and with 3.0 lb of ammonia circulated per min,

$$Q_e = m_r(h_3 - h_1) = 3.0(618 - 131) = 1461 \text{ Btu per min}$$

$$\text{Refrigeration capacity} = \tfrac{1461}{200} = 7.30 \text{ tons}$$

If the liquid ammonia had been subcooled at 150 psia to 60 F, as indicated by point $A'$ on Fig. 14.3, then $Q_e = m_r(618 - 109) = 1527$ Btu per min or 7.63 tons. It is desirable, therefore, that the liquid refrigerant be cooled to as low a temperature as possible ahead of the expansion valve, provided that this cooling is done by water or air.

### 14.7   The Compressor

The compressors that are used in refrigerating systems are similar in principle to compressors for compressing other fluids such as air and may be classified as follows: (1) single- or double-acting reciprocating piston, (2) rotary, (3) gear, and (4) multistage centrifugal.

A six-cylinder reciprocating-piston type of ammonia compressor is illustrated in Fig. 14.4 with the vertical center bank of two cylinders sectioned to show the working parts. Two additional similar banks of cylinders are inclined at 60° angles from the vertical bank and are staggered slightly to permit the connecting rods for each front and rear group of three cylinders to be aligned on one of the two crankshaft throws. The refrigerant enters the cylinders by displacing ring-plate suction valves from their seats on the cylinder liners and leaves through ring-plate discharge valves in the cylinder heads. Damage from slugs of liquid ammonia is avoided by large coil springs that normally hold the cylinder heads against seats but permit them to move outwardly under excessive pressure.

In the Diesel engine, the temperature of the air increases during compression to about 1000 F as a result of the work done on the air during compression. In the refrigerating compressor, the temperature of the refrigerant likewise increases during compression because of the work done on it. Because of the pressure drop through the valves and the heat exchange between the refrigerant and the cylinder walls, it is very difficult to determine the exact behavior of the

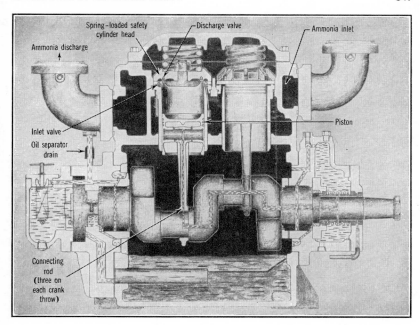

**Fig. 14.4.** Section through two cylinders of a six-cylinder Vilter single-stage ammonia compressor.

refrigerant during a complete revolution of the compressor. However, the pressure and temperature of the refrigerant can be measured at entrance to and exit from the compressor. The points $C$ and $D$ have been located in Fig. 14.3 from such readings of pressure and temperature, and a dotted line has been drawn between them to indicate that the actual change of pressure and temperature of the refrigerant at intermediate pressures during the compression process is unknown. The increase in enthalpy during compression is due to the work done on the refrigerant by the piston of the compressor.

The energy balance for the compressor is illustrated in Fig. 14.5. From the first law of thermodynamics, energy in = energy out. Neglecting any heat transfer which may take place between the compressor and the air, the energy is supplied as indicated horsepower or work done by the piston on the refrigerant and may be computed as follows:

$$\text{Energy in} = \text{ihp} \times \tfrac{2545}{60} \text{ Btu per min} \qquad (14.5)$$

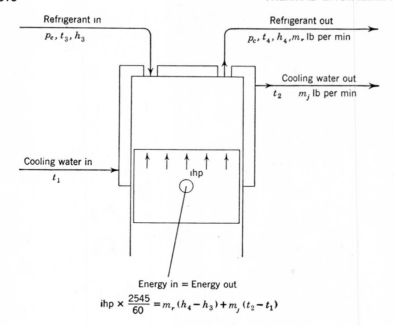

Refrigerant in
$p_e, t_3, h_3$

Refrigerant out
$p_c, t_4, h_4, m_r$ lb per min

Cooling water out
$t_2 \quad m_j$ lb per min

Cooling water in
$t_1$

ihp

Energy in = Energy out

$$\text{ihp} \times \frac{2545}{60} = m_r(h_4 - h_3) + m_j(t_2 - t_1)$$

**Fig. 14.5.**  Energy balance for the compressor.

Energy is removed from the compressor by the refrigerant and by cooling water.  Therefore,

$$\text{Energy out} = m_r(h_4 - h_3) + m_j(t_2 - t_1) \tag{14.6}$$

where $m_r$ = refrigerant delivered by the compressor, lb per min
$\quad m_j$ = cooling water circulated through the jacket of the compressor, lb per min
$\quad h_3$ = enthalpy of refrigerant entering the compressor, Btu per lb
$\quad h_4$ = enthalpy of refrigerant leaving the compressor, Btu per lb
$\quad t_1$ = temperature of cooling water entering compressor, F
$\quad t_2$ = temperature of cooling water leaving compressor, F

Then, from the first law of thermodynamics,

$$\text{Ihp} \times \tfrac{2545}{60} = m_r(h_4 - h_3) + m_j(t_2 - t_1) \tag{14.7}$$

The calculation of the energy balance on the compressor may be illustrated by Example 2.

**Example 2.** A total of 3.0 lb of ammonia is compressed per min. The ammonia enters the compressor at 30 psia and 10 F and leaves at 150 psia and 200 F, as shown on Fig. 14.3. A total of 5.0 lb of cooling water enters the compressor jacket per min at 60 F and leaves at 67 F. The ihp is found to be 7.25. Check the energy balance.

*Solution:*

$$\text{Energy in} = \text{ihp} \times \tfrac{2545}{60} = 7.25 \times \tfrac{2545}{60} = 308 \text{ Btu per min}$$

$$\text{Energy out} = m_r(h_4 - h_3) + m_j(t_2 - t_1)$$

$$= 3.0(709 - 618) + 5.0(67 - 60)$$

$$= 308 \text{ Btu per min}$$

### 14.8   The Condenser

Except in small units such as household refrigerators which have air-cooled condensers, the condenser is water-cooled and operates on the same principle as the steam condenser of the power plant. The pipe between the compressor and the condenser is not insulated, since any energy lost to the atmosphere from this pipe reduces the energy to be removed in the condenser. The superheated ammonia entering the condenser is cooled to the saturation temperature, condensed at the saturation temperature corresponding to the condenser pressure, and perhaps subcooled to a temperature below the saturation temperature. An ample supply of cold cooling water reduces the pressure required to effect condensation of the refrigerant and thereby reduces the power required to operate the compressor.

Neglecting the energy transferred to the surrounding air and referring to Fig. 14.2 and Fig. 14.3, the energy balance for the condenser may be written as follows:

Energy given up by refrigerant = energy absorbed by cooling water

or $$\qquad Q_c = m_r(h_4 - h_1) = m_c(t_6 - t_5) \qquad\qquad (14.8)$$

where $Q_c$ = energy rejected to condenser, Btu per min

$m_r$ = refrigerant circulated, lb per min

$m_c$ = condenser cooling water, lb per min

$h_4$ = enthalpy of refrigerant leaving the compressor

$h_1$ = enthalpy of refrigerant leaving the condenser

$t_5$ = temperature of cooling water entering condenser

$t_6$ = temperature of cooling water leaving condenser

### 14.9   Energy Balance for the Compression Refrigerating Cycle

Energy is supplied to the system by heat transfer in the evaporator and as work in the compressor. If heat transfer from the piping system is neglected, energy is removed only by heat transfer in the cylinder jacket and condenser circulating water. Consequently, from the first law of thermodynamics, the following energy balance may be written:

$$Q_e + \text{ihp} \times \tfrac{2545}{60} = Q_c + Q_j \tag{14.9}$$

where $Q_e$ = energy absorbed in the evaporator, Btu per min

$\quad\ \ Q_c$ = energy rejected in the condenser, Btu per min

$\quad\ \ Q_j$ = energy rejected to compressor jacket water, Btu per min, or $m_j(t_2 - t_1)$, from Equation 14.7

Where complete test data are available for calculation of all items in Equation 14.9, failure of the calculated quantities to balance in accordance with Equation 14.9 may be due to heat transfer from the piping or errors in the data.

Since, from Equation 14.7,

$$\text{Ihp} \times \tfrac{2545}{60} - Q_j = m_r(h_4 - h_3)$$

Equation 14.9 may also be written thus:

$$Q_e + m_r(h_4 - h_3) = Q_c \tag{14.10}$$

For a flow of 1.0 lb of ammonia per min and with reference to Fig. 14.3,

$$Q_e = h_3 - h_2 = h_3 - h_1 = 618 - 131 = 487 \text{ Btu}$$

$$h_4 - h_3 = 709 - 618 = 91 \text{ Btu}$$

$$Q_c = h_4 - h_1 = 709 - 131 = 578 \text{ Btu}$$

$$Q_e + (h_4 - h_3) = 487 + 91 = 578 = Q_c$$

It may therefore be stated that the energy rejected to cooling water in the condenser equals the energy absorbed by the refrigerant in the evaporator plus the increase in enthalpy of the refrigerant due to work done on it in the compressor if no heat transfer occurs in the piping system or to cooling water in the compressor.

### 14.10   Coefficient of Performance

The performance of engines and turbines is expressed in terms of thermal efficiency where thermal efficiency is defined as the ratio of the energy converted into work to the energy supplied to the machine. The objective is the performance of maximum work from the energy supplied. In the refrigeration cycle, the objective is the removal of energy from a refrigerated substance or space with the least possible input of work to operate the compressor. The performance of a refrigeration system is expressed by a term known as the *coefficient of performance*, which is defined as the ratio of the time rate of heat transfer to the refrigerant while passing through the evaporator to the work-input rate required to compress the refrigerant in the compressor, both terms being expressed in consistent units. The input required to compress the refrigerant may be taken as the theoretical amount of work required for isentropic compression, the actual indicated horsepower of the compressor, the horsepower input at the crankshaft of the compressor, or the electric input to the motor that drives the compressor. The basis upon which the input is measured must be specified if the term coefficient of performance is to have any significance.

**Example 3.** Three pounds of ammonia are circulated per min under the conditions shown in Fig. 14.3. The input to the shaft of the compressor is 10 hp. Compute the coefficient of performance based upon the input to the compressor.

*Solution:* The heat transferred in the evaporator to the refrigerant,

$$Q_e = m_r(h_3 - h_1) = 3.0(618 - 131) = 1461 \text{ Btu per min}$$

$$\text{The shaft work} = \frac{10 \text{ hp} \times 2545 \text{ Btu per hp-hr}}{60 \text{ min per hr}} = 424 \text{ Btu per min}$$

$$\text{Coefficient of performance} = \frac{1461 \text{ Btu per min absorbed in refrigerator}}{424 \text{ Btu per min input to compressor as work}}$$

$$= 3.45$$

This coefficient of performance of 3.45 means that 3.45 Btu are transferred to the refrigerant in the evaporator for each Btu of work required to operate the compressor. Obviously, the coefficient of performance should be as great as possible. Anything that decreases the evaporator pressure or increases the condenser pressure reduces the coefficient of performance through increase in the work required to compress 1 lb of refrigerant.

### 14.11  Cooling Units

Small mechanical refrigeration units, employing high-speed compressors, are used extensively for cooling homes and offices and for small food-storage equipment. Theaters, public buildings, and large retail stores have been air-conditioned for many years by means of equipment similar to that employed in ice-making plants, cold-storage warehouses, and other installations of large capacity. Later developments include room coolers driven by $\frac{1}{4}$- to 1-hp electric motors. These units are of the window type or the console type for built-in installation. Small domestic refrigerators and frozen-food-storage units also account for a large number of small compressors.

Unit air conditioners and air coolers are defined under several classifications according to their intended uses. Some of these are merely air-cooling units that do nothing except lower the temperature of the air in the adjacent space. Others are properly called air-conditioning units because they provide means for ventilating, cleaning, cooling, and dehumidifying the air that is circulated through

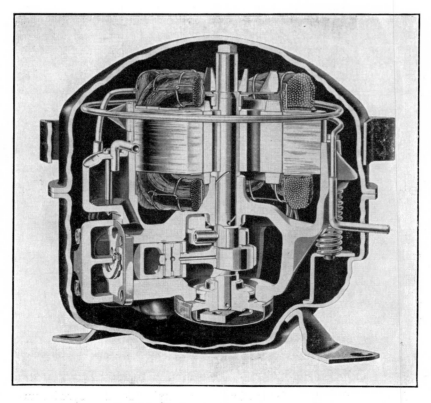

**Fig. 14.6.** Enclosed motor-compressor unit.

them by built-in blowers. Such units are of necessity equipped with automatic controls to maintain the temperature and humidity within certain limits, thus requiring heating as well as cooling facilities. Unit air coolers are intended primarily for product cooling rather than room cooling and are designed to produce much lower temperatures than are normally required of air conditioners.

The refrigerating capacity of cooling units is usually stated in Btu per hour rate of heat transfer, although window-type units more commonly are rated by motor horsepower. The horsepower rating is loosely equivalent to the tons of refrigeration capacity, and a 1-hp unit is frequently said to be a 1-ton unit. This relationship will hold if the coefficient of performance is 12,000/2545 or 4.7.

The refrigerating load carried by air-conditioning units depends upon the heat transfer through the outer walls of the building, the air entering from outside, and the Btu gain from people in the building, electric lights, appliances, etc. This, in turn, depends upon the temperature difference maintained, the thermal conductivity of the walls, the amount of air entering for ventilation or by leakage, and the occupancy. If the humidity is controlled as well as the temperature, the latent heat of vaporization that must be absorbed in order to condense and remove the excess moisture must be included.

**Example 4.** The living room of a well-insulated home is occupied by two persons, 40 cfm of outside air enters the room, and the outside temperature is 90 F. Conductivity through the walls introduces 1200 Btu per hour. Determine the cooling load to maintain a temperature of 72 F in the room and the power required, assuming a coefficient of performance of 4.5.

*Solution:*

Mass of entering air per hr = volume per min × 60 × specific weight = 40 cfm × 60 × 0.075 lb per cu ft = 180 lb per hr.

Energy-transfer rate to cool entering air = $mc_p(t_{out} - t_{in})$ = 180 lb × 0.24 Btu per degree (90 deg − 72 deg) = 778 Btu per hr.

Moderately active persons cause energy gain of about 400 Btu per hr.

Energy gain from the two occupants = 400 Btu × 2 = 800 Btu per hr.

Total heat-transfer rate = 1200 + 778 + 800 = 2778 Btu per hr.

$$\text{Tons of refrigeration} = \frac{2778 \text{ Btu transferred per hr}}{12,000 \text{ Btu per ton per hr}} = 0.232 \text{ ton.}$$

$$\text{Power required} = \frac{\text{Btu transferred per hr}}{2545 \text{ Btu per hp per hr} \times 4.5} = 0.243 \text{ hp.}$$

A cross-sectional view of a compressor and electric motor combined in a sealed unit is shown in Fig. 14.6. This is the conventional-type unit employed in cooling and air-conditioning units. Figure 14.7 shows the arrangement of the motor-compressor unit in the base,

**Fig. 14.7.** Self-contained air-conditioning unit.

with the evaporator and air filter above it and the circulating blower at the top of a Carrier self-contained Weathermaker unit.

### 14.12  The "Heat Pump"

There are many areas of the country where it is desirable to cool office and residential buildings or rooms during the hot summer months and necessary to heat them during the winter months. This can be done by a refrigerating system for cooling and a conventional heating plant for heating. However, both functions can be combined into a single set of apparatus called the "heat pump" which is illustrated in Fig. 14.8.

During the hot summer months, the cooling cycle operates as a conventional refrigerating machine with the room air being cooled by blowing it across the coils of an evaporator. The energy absorbed in the evaporator plus the work input to the compressor is discharged to the atmosphere in an air-cooled condenser.

During the heating season, the path of the refrigerant is reversed by suitable valves so that the coil that serves as an evaporator during the summer months becomes the condenser during the heating season and the coil that is used as a condenser during the cooling season becomes the evaporator during the heating season. Then energy is absorbed from the outdoor atmosphere by the refrigerant in the evaporator. The energy so absorbed plus the work input to the compressor is delivered by the condenser to the air that is being circulated in the office or home. Figure 14.8 shows the atmosphere as the sink into which the energy is rejected during the summer cooling cycle and the source of the energy in the winter heating cycle. Water or the ground around the building may be used instead of the atmosphere as the sink and source for the energy. Well water or the soil are subject to less temperature variations during the year but may require more expensive installations than heat-transfer coils installed in the air.

Since the energy delivered to the building during the heating season comes from the heat transferred at low temperature from the air, water, or ground, plus the work input required to compress the refrigerant, the heating load may be several times the amount of electric energy purchased. However, electric energy is high-grade expensive energy which probably is generated in a power plant burning fuel with an efficiency that is *relatively low* in accordance with the second law of thermodynamics. Moreover, transmission and distribution losses still further reduce the percentage of energy in the original fuel that appears as electric energy at the location of the heat pump. Consequently, the energy delivered to the building by the heat pump is less than the energy in the fuel that was burned to produce the electric energy used to operate the heat pump.

In regions having low-cost electric energy, fairly long and heavy cooling requirements, and moderate heating loads, the ability of the

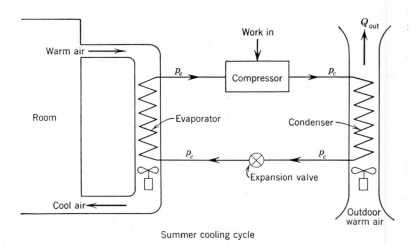

Summer cooling cycle

Winter heating cycle

**Fig. 14.8.** Heating and cooling cycles of the "heat pump."

heat pump to use the same equipment for both heating and cooling may make the installation an attractive one.

For a more complete discussion of the compression system of refrigeration and a discussion of the absorption system, the properties of refrigerants, the thermodynamics of the refrigeration cycle, and the applications of refrigeration to cooling, to the production of ice and solid carbon dioxide, and to air conditioning, the reader is referred to the references at the end of this chapter.

## PROBLEMS

1. Compute the number of lb of ammonia that must be circulated per min in a 15-ton refrigerating system operating between pressures of 15 psia and 140 psia with the ammonia entering the expansion valve at saturation temperature and leaving the evaporator as dry saturated vapor.

2. Find the heat-transfer rate in the evaporator per lb of ammonia circulated per min when a refrigerating unit operates at a condenser pressure of 153 psia and an evaporator pressure of 20 psia with the temperature before the expansion valve 78 F and the temperature leaving the evaporator 20 F.

3. Ammonia circulated at a rate of 7 lb per min enters the condenser at a temperature of 200 F and a pressure of 140 psia. How many lb of cooling water entering at 55 F must be circulated per min if it leaves at 90 F?

4. An ammonia compressor operates with inlet conditions of 24 F and 20 psia and discharge conditions of 150 F and 245 psia. Cooling water enters the jacket at 60 F and leaves at 72 F. If ammonia is compressed at a rate of 5 lb per min and water flows through the water jacket at a rate of 10 lb per min, what ihp is required?

5. If cooling water is available at 50 F and a condenser is capable of cooling the liquefied ammonia to 10 F above the entering water temperature, what is the minimum pressure at which the system can operate?

6. The following data apply to compression refrigerating systems as illustrated in Fig. 14.2:

| Item | Test Number | | | | |
|---|---|---|---|---|---|
|  | 1 | 2 | 3 | 4 | 5 |
| 1. Condenser pressure, psia | 140 | 160 | 150 | 180 | 170 |
| 2. Evaporator pressure, psia | 25 | 20 | 30 | 15 | 20 |
| 3. Temperature of liquid ammonia before expansion valve, F | 75 | 83 | 74 | 85 | 84 |
| 4. Temperature of ammonia leaving evaporator, F | 0 | −10 | +20 | −20 | 0 |
| 5. Temperature of ammonia leaving compressor, F | 235 | 240 | 220 | 250 | 245 |

|  | Test Number | | | | |
| --- | --- | --- | --- | --- | --- |
| Item | 1 | 2 | 3 | 4 | 5 |
| 6. Temperature of cooling water supplied to condenser and compressor jacket, F | 60 | 63 | 60 | 72 | 70 |
| 7. Temperature of cooling water leaving condenser, F | 70 | 78 | 68 | 84 | 81 |
| 8. Temperature of cooling water leaving compressor jacket, F | 67 | . . . | 65 | 76 | 76 |
| 9. Ammonia circulated per min, lb | 5 | 7 | 4 | 6 | 9 |
| 10. Cooling water circulated per min through condenser, lb | 350 | . . . | 346 | . . . | . . . |
| 11. Cooling water circulated per min through compressor jacket, lb | 25 | 0 | 30 | 25 | 40 |
| 12. Ihp of compressor | 17.8 | . . . | 14.1 | . . . | 30.9 |

Draw for each set of data, a flow diagram similar to Fig. 14.2, and place on this diagram the following results to be obtained from the Mollier diagram: (a) enthalpy of the ammonia ahead of the expansion valve, (b) enthalpy and quality of ammonia beyond the expansion valve, (c) enthalpy of ammonia leaving the evaporator, and (d) enthalpy of ammonia leaving the compressor.

Assume no heat transfer from the piping system and compute:

1. The energy absorbed in the evaporator in Btu per min.
2. The capacity of the system in tons of commercial refrigeration.
3. The energy rejected to the condenser in Btu per min.
4. The condenser cooling water required per min (where the weight of cooling water is given, check the data).
5. The energy removed by the compressor jacket water in Btu per min.
6. The ihp of the compressor, by use of the energy balance (Equation 14.9). Where the ihp is given, check the data by the energy balance.
7. Where the ihp is given, compute the coefficient of performance.

## REFERENCES

B. F. Raber and F. W. Hutchinson, *Refrigeration and Air Conditioning Engineering,* John Wiley & Sons, 1945.

H. G. Venemann, *Refrigeration Theory and Applications,* Nickerson & Collins Company, 1942.

N. R. Sparks and C. C. Dillio, *Mechanical Refrigeration,* McGraw-Hill Book Company, 1959.

H. J. Macintire and F. W. Hutchinson, *Refrigeration Engineering,* John Wiley & Sons, 1950.

*Refrigerating Data Book,* Volumes I and II, American Society of Refrigerating Engineers, 1951.

R. C. Jordan and G. B. Priester, *Refrigeration and Air Conditioning,* Prentice-Hall, 1956.

# Appendix

# TABLE A·1

## DRY SATURATED STEAM: TEMPERATURE TABLE*

| Temp. F t | Abs Press, Lb Sq In. p | Specific Volume Sat. Liquid vf | Evap. vfg | Sat. Vapor vg | Enthalpy Sat. Liquid hf | Evap. hfg | Sat. Vapor hg | Entropy Sat. Liquid sf | Evap sfg | Sat. Vapor sg | Temp. F t |
|---|---|---|---|---|---|---|---|---|---|---|---|
| 32 | 0.08854 | 0.01602 | 3306 | 3306 | 0.00 | 1075.8 | 1075.8 | 0.0000 | 2.1877 | 2.1877 | 32 |
| 35 | 0.09995 | 0.01602 | 2947 | 2947 | 3.02 | 1074.1 | 1077.1 | 0.0061 | 2.1709 | 2.1770 | 35 |
| 40 | 0.12170 | 0.01602 | 2444 | 2444 | 8.05 | 1071.3 | 1079.3 | 0.0162 | 2.1435 | 2.1597 | 40 |
| 45 | 0.14752 | 0.01602 | 2036.4 | 2036.4 | 13.06 | 1068.4 | 1081.5 | 0.0262 | 2.1167 | 2.1429 | 45 |
| 50 | 0.17811 | 0.01603 | 1703.2 | 1703.2 | 18.07 | 1065.6 | 1083.7 | 0.0361 | 2.0903 | 2.1264 | 50 |
| 60 | 0.2563 | 0.01604 | 1206.6 | 1206.7 | 28.06 | 1059.9 | 1088.0 | 0.0555 | 2.0393 | 2.0948 | 60 |
| 70 | 0.3631 | 0.01606 | 867.8 | 867.9 | 38.04 | 1054.3 | 1092.3 | 0.0745 | 1.9902 | 2.0647 | 70 |
| 80 | 0.5069 | 0.01608 | 633.1 | 633.1 | 48.02 | 1048.6 | 1096.6 | 0.0932 | 1.9428 | 2.0360 | 80 |
| 90 | 0.6982 | 0.01610 | 468.0 | 468.0 | 57.99 | 1042.9 | 1100.9 | 0.1115 | 1.8972 | 2.0087 | 90 |
| 100 | 0.9492 | 0.01613 | 350.3 | 350.4 | 67.97 | 1037.2 | 1105.2 | 0.1295 | 1.8531 | 1.9826 | 100 |
| 110 | 1.2748 | 0.01617 | 265.3 | 265.4 | 77.94 | 1031.6 | 1109.5 | 0.1471 | 1.8106 | 1.9577 | 110 |
| 120 | 1.6924 | 0.01620 | 203.25 | 203.27 | 87.92 | 1025.8 | 1113.7 | 0.1645 | 1.7694 | 1.9339 | 120 |
| 130 | 2.2225 | 0.01625 | 157.32 | 157.34 | 97.90 | 1020.0 | 1117.9 | 0.1816 | 1.7296 | 1.9112 | 130 |
| 140 | 2.8886 | 0.01629 | 122.99 | 123.01 | 107.89 | 1014.1 | 1122.0 | 0.1984 | 1.6910 | 1.8894 | 140 |
| 150 | 3.718 | 0.01634 | 97.06 | 97.07 | 117.89 | 1008.2 | 1126.1 | 0.2149 | 1.6537 | 1.8685 | 150 |
| 160 | 4.741 | 0.01639 | 77.27 | 77.29 | 127.89 | 1002.3 | 1130.2 | 0.2311 | 1.6174 | 1.8485 | 160 |
| 170 | 5.992 | 0.01645 | 62.04 | 62.06 | 137.90 | 996.3 | 1134.2 | 0.2472 | 1.5822 | 1.8293 | 170 |
| 180 | 7.510 | 0.01651 | 50.21 | 50.23 | 147.92 | 990.2 | 1138.1 | 0.2630 | 1.5480 | 1.8109 | 180 |
| 190 | 9.339 | 0.01657 | 40.94 | 40.96 | 157.95 | 984.1 | 1142.0 | 0.2785 | 1.5147 | 1.7932 | 190 |
| 200 | 11.526 | 0.01663 | 33.62 | 33.64 | 167.99 | 977.9 | 1145.9 | 0.2938 | 1.4824 | 1.7762 | 200 |
| 210 | 14.123 | 0.01670 | 27.80 | 27.82 | 178.05 | 971.6 | 1149.7 | 0.3090 | 1.4508 | 1.7598 | 210 |
| 212 | 14.696 | 0.01672 | 26.78 | 26.80 | 180.07 | 970.3 | 1150.4 | 0.3120 | 1.4446 | 1.7566 | 212 |
| 220 | 17.186 | 0.01677 | 23.13 | 23.15 | 188.13 | 965.2 | 1153.4 | 0.3239 | 1.4201 | 1.7440 | 220 |
| 230 | 20.780 | 0.01684 | 19.365 | 19.382 | 198.23 | 958.8 | 1157.0 | 0.3387 | 1.3901 | 1.7288 | 230 |
| 240 | 24.969 | 0.01692 | 16.306 | 16.323 | 208.34 | 952.2 | 1160.5 | 0.3531 | 1.3609 | 1.7140 | 240 |
| 250 | 29.825 | 0.01700 | 13.804 | 13.821 | 218.48 | 945.5 | 1164.0 | 0.3675 | 1.3323 | 1.6998 | 250 |
| 260 | 35.429 | 0.01709 | 11.746 | 11.763 | 228.64 | 938.7 | 1167.3 | 0.3817 | 1.3043 | 1.6860 | 260 |
| 270 | 41.858 | 0.01717 | 10.044 | 10.061 | 238.84 | 931.8 | 1170.6 | 0.3958 | 1.2769 | 1.6727 | 270 |
| 280 | 49.203 | 0.01726 | 8.628 | 8.645 | 249.06 | 924.7 | 1173.8 | 0.4096 | 1.2501 | 1.6597 | 280 |
| 290 | 57.556 | 0.01735 | 7.444 | 7.461 | 259.31 | 917.5 | 1176.8 | 0.4234 | 1.2238 | 1.6472 | 290 |

| Temp (°F) | P (psi) | $v_f$ | $v_{fg}$ | $v_g$ | $h_f$ | $h_{fg}$ | $h_g$ | $s_f$ | $s_{fg}$ | $s_g$ | Temp (°F) |
|---|---|---|---|---|---|---|---|---|---|---|---|
| 300 | 67.013 | 0.01745 | 6.449 | 6.466 | 269.59 | 910.1 | 1179.7 | 0.4369 | 1.1980 | 1.6350 | 300 |
| 310 | 77.68 | 0.01755 | 5.609 | 5.626 | 279.92 | 902.6 | 1182.5 | 0.4504 | 1.1727 | 1.6231 | 310 |
| 320 | 89.66 | 0.01765 | 4.896 | 4.914 | 290.28 | 894.9 | 1185.2 | 0.4637 | 1.1478 | 1.6115 | 320 |
| 330 | 103.06 | 0.01776 | 4.289 | 4.307 | 300.68 | 887.0 | 1187.7 | 0.4769 | 1.1233 | 1.6002 | 330 |
| 340 | 118.01 | 0.01787 | 3.770 | 3.788 | 311.13 | 879.0 | 1190.1 | 0.4900 | 1.0992 | 1.5891 | 340 |
| 350 | 134.63 | 0.01799 | 3.324 | 3.342 | 321.63 | 870.7 | 1192.3 | 0.5029 | 1.0754 | 1.5783 | 350 |
| 360 | 153.04 | 0.01811 | 2.939 | 2.957 | 332.18 | 862.2 | 1194.4 | 0.5158 | 1.0519 | 1.5677 | 360 |
| 370 | 173.37 | 0.01823 | 2.606 | 2.625 | 342.79 | 853.5 | 1196.3 | 0.5286 | 1.0287 | 1.5573 | 370 |
| 380 | 195.77 | 0.01836 | 2.317 | 2.335 | 353.45 | 844.6 | 1198.1 | 0.5413 | 1.0059 | 1.5471 | 380 |
| 390 | 220.37 | 0.01850 | 2.0651 | 2.0836 | 364.17 | 835.4 | 1199.6 | 0.5539 | 0.9832 | 1.5371 | 390 |
| 400 | 247.31 | 0.01864 | 1.8447 | 1.8633 | 374.97 | 826.0 | 1201.0 | 0.5664 | 0.9608 | 1.5272 | 400 |
| 410 | 276.75 | 0.01878 | 1.6512 | 1.6700 | 385.83 | 816.3 | 1202.1 | 0.5788 | 0.9386 | 1.5174 | 410 |
| 420 | 308.83 | 0.01894 | 1.4811 | 1.5000 | 396.77 | 806.3 | 1203.1 | 0.5912 | 0.9166 | 1.5078 | 420 |
| 430 | 343.72 | 0.01910 | 1.3308 | 1.3499 | 407.79 | 796.0 | 1203.8 | 0.6035 | 0.8947 | 1.4982 | 430 |
| 440 | 381.59 | 0.01926 | 1.1979 | 1.2171 | 418.90 | 785.4 | 1204.3 | 0.6158 | 0.8730 | 1.4887 | 440 |
| 450 | 422.6 | 0.0194 | 1.0799 | 1.0993 | 430.1 | 774.5 | 1204.6 | 0.6280 | 0.8513 | 1.4793 | 450 |
| 460 | 466.9 | 0.0196 | 0.9748 | 0.9944 | 441.4 | 763.2 | 1204.6 | 0.6402 | 0.8298 | 1.4700 | 460 |
| 470 | 514.7 | 0.0198 | 0.8811 | 0.9009 | 452.8 | 751.5 | 1204.3 | 0.6523 | 0.8083 | 1.4606 | 470 |
| 480 | 566.1 | 0.0200 | 0.7972 | 0.8172 | 464.4 | 739.4 | 1203.7 | 0.6645 | 0.7868 | 1.4513 | 480 |
| 490 | 621.4 | 0.0202 | 0.7221 | 0.7423 | 476.0 | 726.8 | 1202.8 | 0.6766 | 0.7653 | 1.4419 | 490 |
| 500 | 680.8 | 0.0204 | 0.6545 | 0.6749 | 487.8 | 713.9 | 1201.7 | 0.6887 | 0.7438 | 1.4325 | 500 |
| 520 | 812.4 | 0.0209 | 0.5385 | 0.5594 | 511.9 | 686.4 | 1198.2 | 0.7130 | 0.7006 | 1.4136 | 520 |
| 540 | 962.5 | 0.0215 | 0.4434 | 0.4649 | 536.6 | 656.6 | 1193.2 | 0.7374 | 0.6568 | 1.3942 | 540 |
| 560 | 1133.1 | 0.0221 | 0.3647 | 0.3868 | 562.2 | 624.2 | 1186.4 | 0.7621 | 0.6121 | 1.3742 | 560 |
| 580 | 1325.8 | 0.0228 | 0.2989 | 0.3217 | 588.9 | 588.4 | 1177.3 | 0.7872 | 0.5659 | 1.3532 | 580 |
| 600 | 1542.9 | 0.0236 | 0.2432 | 0.2668 | 617.0 | 548.5 | 1165.5 | 0.8131 | 0.5176 | 1.3307 | 600 |
| 620 | 1786.6 | 0.0247 | 0.1955 | 0.2201 | 646.7 | 503.6 | 1150.3 | 0.8398 | 0.4664 | 1.3062 | 620 |
| 640 | 2059.7 | 0.0260 | 0.1538 | 0.1798 | 678.6 | 452.0 | 1130.5 | 0.8679 | 0.4110 | 1.2789 | 640 |
| 660 | 2365.4 | 0.0278 | 0.1165 | 0.1442 | 714.2 | 390.2 | 1104.4 | 0.8987 | 0.3485 | 1.2472 | 660 |
| 680 | 2708.1 | 0.0305 | 0.0810 | 0.1115 | 757.3 | 309.9 | 1067.2 | 0.9351 | 0.2719 | 1.2071 | 680 |
| 700 | 3093.7 | 0.0369 | 0.0392 | 0.0761 | 823.3 | 172.1 | 995.4 | 0.9905 | 0.1484 | 1.1389 | 700 |
| 705.4 | 3206.2 | 0.0503 | 0 | 0.0503 | 902.7 | 0 | 902.7 | 1.0580 | 0 | 1.0580 | 705.4 |

* Abridged from *Thermodynamic Properties of Steam* by Joseph H. Keenan and Frederick G. Keyes. Copyright, 1937, by Joseph H. Keenan and Frederick G. Keyes. Published by John Wiley & Sons, New York.

## TABLE A·2

## DRY SATURATED STEAM: PRESSURE TABLE *

| Abs Press., Lb Sq In. $p$ | Temp., F $t$ | Specific Volume | | Enthalpy | | | Entropy | | | Internal Energy | | Abs Press., Lb Sq In. $p$ |
|---|---|---|---|---|---|---|---|---|---|---|---|---|
| | | Sat. Liquid $v_f$ | Sat. Vapor $v_g$ | Sat. Liquid $h_f$ | Evap $h_{fg}$ | Sat. Vapor $h_g$ | Sat. Liquid $s_f$ | Evap $s_{fg}$ | Sat. Vapor $s_g$ | Sat. Liquid $u_f$ | Sat. Vapor $u_g$ | |
| 1.0 | 101.74 | 0.01614 | 333.6 | 69.70 | 1036.3 | 1106.0 | 0.1326 | 1.8456 | 1.9782 | 69.70 | 1044.3 | 1.0 |
| 2.0 | 126.08 | 0.01623 | 173.73 | 93.99 | 1022.2 | 1116.2 | 0.1749 | 1.7451 | 1.9200 | 93.98 | 1051.9 | 2.0 |
| 3.0 | 141.48 | 0.01630 | 118.71 | 109.37 | 1013.2 | 1122.6 | 0.2008 | 1.6855 | 1.8863 | 109.36 | 1056.7 | 3.0 |
| 4.0 | 152.97 | 0.01636 | 90.63 | 120.86 | 1006.4 | 1127.3 | 0.2198 | 1.6427 | 1.8625 | 120.85 | 1060.2 | 4.0 |
| 5.0 | 162.24 | 0.01640 | 73.52 | 130.13 | 1001.0 | 1131.1 | 0.2347 | 1.6094 | 1.8441 | 130.12 | 1063.1 | 5.0 |
| 6.0 | 170.06 | 0.01645 | 61.98 | 137.96 | 996.2 | 1134.2 | 0.2472 | 1.5820 | 1.8292 | 137.94 | 1065.4 | 6.0 |
| 7.0 | 176.85 | 0.01649 | 53.64 | 144.76 | 992.1 | 1136.9 | 0.2581 | 1.5586 | 1.8167 | 144.74 | 1067.4 | 7.0 |
| 8.0 | 182.86 | 0.01653 | 47.34 | 150.79 | 988.5 | 1139.3 | 0.2674 | 1.5383 | 1.8057 | 150.77 | 1069.2 | 8.0 |
| 9.0 | 188.28 | 0.01656 | 42.40 | 156.22 | 985.2 | 1141.4 | 0.2759 | 1.5203 | 1.7962 | 156.19 | 1070.8 | 9.0 |
| 10 | 193.21 | 0.01659 | 38.42 | 161.17 | 982.1 | 1143.3 | 0.2835 | 1.5041 | 1.7876 | 161.14 | 1072.2 | 10 |
| 14.696 | 212.00 | 0.01672 | 26.80 | 180.07 | 970.3 | 1150.4 | 0.3120 | 1.4446 | 1.7566 | 180.02 | 1077.5 | 14.696 |
| 15 | 213.03 | 0.01672 | 26.29 | 181.11 | 969.7 | 1150.8 | 0.3135 | 1.4415 | 1.7549 | 181.06 | 1077.8 | 15 |
| 20 | 227.96 | 0.01683 | 20.089 | 196.16 | 960.1 | 1156.3 | 0.3356 | 1.3962 | 1.7319 | 196.10 | 1081.9 | 20 |
| 25 | 240.07 | 0.01692 | 16.303 | 208.42 | 952.1 | 1160.6 | 0.3533 | 1.3606 | 1.7139 | 208.34 | 1085.1 | 25 |
| 30 | 250.33 | 0.01701 | 13.746 | 218.82 | 945.3 | 1164.1 | 0.3680 | 1.3313 | 1.6993 | 218.73 | 1087.8 | 30 |
| 35 | 259.28 | 0.01708 | 11.898 | 227.91 | 939.2 | 1167.1 | 0.3807 | 1.3063 | 1.6870 | 227.80 | 1090.1 | 35 |
| 40 | 267.25 | 0.01715 | 10.498 | 236.03 | 933.7 | 1169.7 | 0.3919 | 1.2844 | 1.6763 | 235.90 | 1092.0 | 40 |
| 45 | 274.44 | 0.01721 | 9.401 | 243.36 | 928.6 | 1172.0 | 0.4019 | 1.2650 | 1.6669 | 243.22 | 1093.7 | 45 |
| 50 | 281.01 | 0.01727 | 8.515 | 250.09 | 924.0 | 1174.1 | 0.4110 | 1.2474 | 1.6585 | 249.93 | 1095.3 | 50 |
| 55 | 287.07 | 0.01732 | 7.787 | 256.30 | 919.6 | 1175.9 | 0.4193 | 1.2316 | 1.6509 | 256.12 | 1096.7 | 55 |
| 60 | 292.71 | 0.01738 | 7.175 | 262.09 | 915.5 | 1177.6 | 0.4270 | 1.2168 | 1.6438 | 261.90 | 1097.9 | 60 |
| 65 | 297.97 | 0.01743 | 6.655 | 267.50 | 911.6 | 1179.1 | 0.4342 | 1.2032 | 1.6374 | 267.29 | 1099.1 | 65 |
| 70 | 302.92 | 0.01748 | 6.206 | 272.61 | 907.9 | 1180.6 | 0.4409 | 1.1906 | 1.6315 | 272.38 | 1100.2 | 70 |
| 75 | 307.60 | 0.01753 | 5.816 | 277.43 | 904.5 | 1181.9 | 0.4472 | 1.1787 | 1.6259 | 277.19 | 1101.2 | 75 |
| 80 | 312.03 | 0.01757 | 5.472 | 282.02 | 901.1 | 1183.1 | 0.4531 | 1.1676 | 1.6207 | 281.76 | 1102.1 | 80 |
| 85 | 316.25 | 0.01761 | 5.168 | 286.39 | 897.8 | 1184.2 | 0.4587 | 1.1571 | 1.6158 | 286.11 | 1102.9 | 85 |
| 90 | 320.27 | 0.01766 | 4.896 | 290.56 | 894.7 | 1185.3 | 0.4641 | 1.1471 | 1.6112 | 290.27 | 1103.7 | 90 |
| 95 | 324.12 | 0.01770 | 4.652 | 294.56 | 891.7 | 1186.2 | 0.4692 | 1.1376 | 1.6068 | 294.25 | 1104.5 | 95 |
| 100 | 327.81 | 0.01774 | 4.432 | 298.40 | 888.8 | 1187.2 | 0.4740 | 1.1286 | 1.6026 | 298.08 | 1105.2 | 100 |
| 110 | 334.77 | 0.01782 | 4.049 | 305.66 | 883.2 | 1188.9 | 0.4832 | 1.1117 | 1.5948 | 305.30 | 1106.5 | 110 |

| Abs. Press. | Temp. | $v_f$ | $v_g$ | $h_f$ | $h_{fg}$ | $h_g$ | $s_f$ | $s_{fg}$ | $s_g$ | $u_f$ | $u_g$ |
|---|---|---|---|---|---|---|---|---|---|---|---|
| 120 | 341.25 | 0.01789 | 3.728 | 312.44 | 877.9 | 1190.4 | 0.4916 | 1.0962 | 1.5878 | 312.05 | 1107.6 |
| 130 | 347.32 | 0.01796 | 3.455 | 318.81 | 872.9 | 1191.7 | 0.4995 | 1.0817 | 1.5812 | 318.38 | 1108.6 |
| 140 | 353.02 | 0.01802 | 3.220 | 324.82 | 868.2 | 1193.0 | 0.5069 | 1.0682 | 1.5751 | 324.35 | 1109.6 |
| 150 | 358.42 | 0.01809 | 3.015 | 330.51 | 863.6 | 1194.1 | 0.5138 | 1.0556 | 1.5694 | 330.01 | 1110.5 |
| 160 | 363.53 | 0.01815 | 2.834 | 335.93 | 859.2 | 1195.1 | 0.5204 | 1.0436 | 1.5640 | 335.39 | 1111.2 |
| 170 | 368.41 | 0.01822 | 2.675 | 341.09 | 854.9 | 1196.0 | 0.5266 | 1.0324 | 1.5590 | 340.52 | 1111.9 |
| 180 | 373.06 | 0.01827 | 2.532 | 346.03 | 850.8 | 1196.9 | 0.5325 | 1.0217 | 1.5542 | 345.42 | 1112.5 |
| 190 | 377.51 | 0.01833 | 2.404 | 350.79 | 846.8 | 1197.6 | 0.5381 | 1.0116 | 1.5497 | 350.15 | 1113.1 |
| 200 | 381.79 | 0.01839 | 2.288 | 355.36 | 843.0 | 1198.4 | 0.5435 | 1.0018 | 1.5453 | 354.68 | 1113.7 |
| 250 | 400.95 | 0.01865 | 1.8438 | 376.00 | 825.1 | 1201.1 | 0.5675 | 0.9588 | 1.5263 | 375.14 | 1115.8 |
| 300 | 417.33 | 0.01890 | 1.5433 | 393.84 | 809.0 | 1202.8 | 0.5879 | 0.9225 | 1.5104 | 392.79 | 1117.1 |
| 350 | 431.72 | 0.01913 | 1.3260 | 409.69 | 794.2 | 1203.9 | 0.6056 | 0.8910 | 1.4966 | 408.45 | 1118.0 |
| 400 | 444.59 | 0.0193 | 1.1613 | 424.0 | 780.5 | 1204.5 | 0.6214 | 0.8630 | 1.4844 | 422.6 | 1118.5 |
| 450 | 456.28 | 0.0195 | 1.0320 | 437.2 | 767.4 | 1204.6 | 0.6356 | 0.8378 | 1.4734 | 435.5 | 1118.7 |
| 500 | 467.01 | 0.0197 | 0.9278 | 449.4 | 755.0 | 1204.4 | 0.6487 | 0.8147 | 1.4634 | 447.6 | 1118.6 |
| 550 | 476.94 | 0.0199 | 0.8424 | 460.8 | 743.1 | 1203.9 | 0.6608 | 0.7934 | 1.4542 | 458.8 | 1118.2 |
| 600 | 486.21 | 0.0201 | 0.7698 | 471.6 | 731.6 | 1203.2 | 0.6720 | 0.7734 | 1.4454 | 469.4 | 1117.7 |
| 650 | 494.90 | 0.0203 | 0.7083 | 481.8 | 720.5 | 1202.3 | 0.6826 | 0.7548 | 1.4374 | 479.4 | 1117.1 |
| 700 | 503.10 | 0.0205 | 0.6554 | 491.5 | 709.7 | 1201.2 | 0.6925 | 0.7371 | 1.4296 | 488.8 | 1116.3 |
| 750 | 510.86 | 0.0207 | 0.6092 | 500.8 | 699.2 | 1200.0 | 0.7019 | 0.7204 | 1.4223 | 498.0 | 1115.4 |
| 800 | 518.23 | 0.0209 | 0.5687 | 509.7 | 688.9 | 1198.6 | 0.7108 | 0.7045 | 1.4153 | 506.6 | 1114.4 |
| 850 | 525.26 | 0.0210 | 0.5327 | 518.3 | 678.8 | 1197.1 | 0.7194 | 0.6891 | 1.4085 | 515.0 | 1113.3 |
| 900 | 531.98 | 0.0212 | 0.5006 | 526.6 | 668.8 | 1195.4 | 0.7275 | 0.6744 | 1.4020 | 523.1 | 1112.1 |
| 950 | 538.43 | 0.0214 | 0.4717 | 534.6 | 659.1 | 1193.7 | 0.7355 | 0.6602 | 1.3957 | 530.9 | 1110.8 |
| 1000 | 544.61 | 0.0216 | 0.4456 | 542.4 | 649.4 | 1191.8 | 0.7430 | 0.6467 | 1.3897 | 538.4 | 1109.4 |
| 1100 | 556.31 | 0.0220 | 0.4001 | 557.4 | 630.4 | 1187.8 | 0.7575 | 0.6205 | 1.3780 | 552.9 | 1106.4 |
| 1200 | 567.22 | 0.0223 | 0.3619 | 571.7 | 611.7 | 1183.4 | 0.7711 | 0.5956 | 1.3667 | 566.7 | 1103.0 |
| 1300 | 577.46 | 0.0227 | 0.3293 | 585.4 | 593.2 | 1178.6 | 0.7840 | 0.5719 | 1.3559 | 580.0 | 1099.4 |
| 1400 | 587.10 | 0.0231 | 0.3012 | 598.7 | 574.7 | 1173.4 | 0.7963 | 0.5491 | 1.3454 | 592.0 | 1095.4 |
| 1500 | 596.23 | 0.0235 | 0.2765 | 611.6 | 556.3 | 1167.9 | 0.8082 | 0.5269 | 1.3351 | 605.1 | 1091.2 |
| 2000 | 635.82 | 0.0257 | 0.1878 | 671.7 | 463.4 | 1135.1 | 0.8619 | 0.4230 | 1.2849 | 662.2 | 1065.6 |
| 2500 | 668.13 | 0.0287 | 0.1307 | 730.6 | 360.5 | 1091.1 | 0.9126 | 0.3197 | 1.2322 | 717.3 | 1030.6 |
| 3000 | 695.36 | 0.0346 | 0.0858 | 802.5 | 217.8 | 1020.3 | 0.9731 | 0.1885 | 1.1615 | 783.4 | 972.7 |
| 3206.2 | 705.40 | 0.0503 | 0.0503 | 902.7 | 0 | 902.7 | 1.0580 | 0 | 1.0580 | 872.9 | 872.9 |

Copyright, 1937, by Joseph H. Keenan and Frederick G. Keyes.

* Abridged from *Thermodynamic Properties of Steam* by Joseph H. Keenan and Frederick G. Keyes. Published by John Wiley & Sons, New York.

633

## TABLE A·3
### PROPERTIES OF SUPERHEATED STEAM *

| Abs Press., Lb Sq In. (Sat. Temp.) | | Temperature—Degrees Fahrenheit | | | | | | | | | | | | |
|---|---|---|---|---|---|---|---|---|---|---|---|---|---|---|
| | | 200 | 300 | 400 | 500 | 600 | 700 | 800 | 900 | 1000 | 1100 | 1200 | 1400 | 1600 |
| **1** (101.74) | $v$ | 392.6 | 452.3 | 512.0 | 571.6 | 631.2 | 690.8 | 750.4 | 809.9 | 869.5 | 929.1 | 988.7 | 1107.8 | 1227.0 |
| | $h$ | 1150.4 | 1195.8 | 1241.7 | 1288.3 | 1335.7 | 1383.8 | 1432.8 | 1482.7 | 1533.5 | 1585.2 | 1637.7 | 1745.7 | 1857.5 |
| | $s$ | 2.0512 | 2.1153 | 2.1720 | 2.2233 | 2.2702 | 2.3137 | 2.3542 | 2.3923 | 2.4283 | 2.4625 | 2.4952 | 2.5566 | 2.6137 |
| **5** (162.24) | $v$ | 78.16 | 90.25 | 102.26 | 114.22 | 126.16 | 138.10 | 150.03 | 161.95 | 173.87 | 185.79 | 197.71 | 221.6 | 245.4 |
| | $h$ | 1148.8 | 1195.0 | 1241.2 | 1288.0 | 1335.4 | 1383.6 | 1432.7 | 1482.6 | 1533.4 | 1585.1 | 1637.7 | 1745.7 | 1857.4 |
| | $s$ | 1.8718 | 1.9370 | 1.9942 | 2.0456 | 2.0927 | 2.1361 | 2.1767 | 2.2148 | 2.2509 | 2.2851 | 2.3178 | 2.3792 | 2.4363 |
| **10** (193.21) | $v$ | 38.85 | 45.00 | 51.04 | 57.05 | 63.03 | 69.01 | 74.98 | 80.95 | 86.92 | 92.88 | 98.84 | 110.77 | 122.69 |
| | $h$ | 1146.6 | 1193.9 | 1240.6 | 1287.5 | 1335.1 | 1383.4 | 1432.5 | 1482.4 | 1533.2 | 1585.0 | 1637.6 | 1745.6 | 1857.3 |
| | $s$ | 1.7927 | 1.8595 | 1.9172 | 1.9689 | 2.0160 | 2.0596 | 2.1002 | 2.1383 | 2.1744 | 2.2086 | 2.2413 | 2.3028 | 2.3598 |
| **14.696** (212.00) | $v$ | | 30.53 | 34.68 | 38.78 | 42.86 | 46.94 | 51.00 | 55.07 | 59.13 | 63.19 | 67.25 | 75.37 | 83.48 |
| | $h$ | | 1192.8 | 1239.9 | 1287.1 | 1334.8 | 1383.2 | 1432.3 | 1482.3 | 1533.1 | 1584.8 | 1637.5 | 1745.5 | 1857.3 |
| | $s$ | | 1.8160 | 1.8743 | 1.9261 | 1.9734 | 2.0170 | 2.0576 | 2.0958 | 2.1319 | 2.1662 | 2.1989 | 2.2603 | 2.3174 |
| **20** (227.96) | $v$ | | 22.36 | 25.43 | 28.46 | 31.47 | 34.47 | 37.46 | 40.45 | 43.44 | 46.42 | 49.41 | 55.37 | 61.34 |
| | $h$ | | 1191.6 | 1239.2 | 1286.6 | 1334.4 | 1382.9 | 1432.1 | 1482.1 | 1533.0 | 1584.7 | 1637.4 | 1745.4 | 1857.2 |
| | $s$ | | 1.7808 | 1.8396 | 1.8918 | 1.9392 | 1.9829 | 2.0235 | 2.0618 | 2.0978 | 2.1321 | 2.1648 | 2.2263 | 2.2834 |
| **40** (267.25) | $v$ | | | 12.628 | 14.168 | 15.688 | 17.198 | 18.702 | 20.20 | 21.70 | 23.20 | 24.69 | 27.68 | 30.66 |
| | $h$ | | | 1236.5 | 1284.8 | 1333.1 | 1381.9 | 1431.3 | 1481.4 | 1532.4 | 1584.3 | 1637.0 | 1745.1 | 1857.0 |
| | $s$ | | | 1.7608 | 1.8140 | 1.8619 | 1.9058 | 1.9467 | 1.9850 | 2.0212 | 2.0555 | 2.0883 | 2.1498 | 2.2069 |
| **60** (292.71) | $v$ | | 7.259 | 8.357 | 9.403 | 10.427 | 11.441 | 12.449 | 13.452 | 14.454 | 15.453 | 16.451 | 18.446 | 20.44 |
| | $h$ | | 1181.6 | 1233.6 | 1283.0 | 1331.8 | 1380.9 | 1430.5 | 1480.8 | 1531.9 | 1583.8 | 1636.6 | 1744.8 | 1856.7 |
| | $s$ | | 1.6492 | 1.7135 | 1.7678 | 1.8162 | 1.8605 | 1.9015 | 1.9400 | 1.9762 | 2.0106 | 2.0434 | 2.1049 | 2.1621 |
| **80** (312.03) | $v$ | | | 6.220 | 7.020 | 7.797 | 8.562 | 9.322 | 10.077 | 10.830 | 11.582 | 12.332 | 13.830 | 15.325 |
| | $h$ | | | 1230.7 | 1281.1 | 1330.5 | 1379.9 | 1429.7 | 1480.1 | 1531.3 | 1583.4 | 1636.2 | 1744.5 | 1856.5 |
| | $s$ | | | 1.6791 | 1.7346 | 1.7836 | 1.8281 | 1.8694 | 1.9079 | 1.9442 | 1.9787 | 2.0115 | 2.0731 | 2.1303 |
| **100** (327.81) | $v$ | | | 4.937 | 5.589 | 6.218 | 6.835 | 7.446 | 8.052 | 8.656 | 9.259 | 9.860 | 11.060 | 12.258 |
| | $h$ | | | 1227.6 | 1279.1 | 1329.1 | 1378.9 | 1428.9 | 1479.5 | 1530.8 | 1582.9 | 1635.7 | 1744.2 | 1856.2 |
| | $s$ | | | 1.6518 | 1.7085 | 1.7581 | 1.8029 | 1.8443 | 1.8829 | 1.9193 | 1.9538 | 1.9867 | 2.0484 | 2.1056 |
| **120** (341.25) | $v$ | | | 4.081 | 4.636 | 5.165 | 5.683 | 6.195 | 6.702 | 7.207 | 7.710 | 8.212 | 9.214 | 10.213 |
| | $h$ | | | 1224.4 | 1277.2 | 1327.8 | 1377.8 | 1428.1 | 1478.8 | 1530.2 | 1582.4 | 1635.3 | 1743.9 | 1856.0 |
| | $s$ | | | 1.6287 | 1.6869 | 1.7370 | 1.7822 | 1.8237 | 1.8625 | 1.8990 | 1.9335 | 1.9664 | 2.0281 | 2.0854 |

| Press. | | | | | | | | | | | | |
|---|---|---|---|---|---|---|---|---|---|---|---|---|
| **110** (353.02) | v | 3.468 | 3.954 | 4.413 | 4.861 | 5.301 | 5.738 | 6.172 | 6.604 | 7.035 | 7.895 | 8.752 |
| | h | 1221.1 | 1275.2 | 1326.4 | 1376.8 | 1427.3 | 1478.2 | 1529.7 | 1581.9 | 1634.9 | 1743.5 | 1855.7 |
| | s | 1.6087 | 1.6683 | 1.7190 | 1.7645 | 1.8063 | 1.8451 | 1.8817 | 1.9163 | 1.9493 | 2.0110 | 2.0683 |
| **160** (363.53) | v | 3.008 | 3.443 | 3.849 | 4.244 | 4.631 | 5.015 | 5.396 | 5.775 | 6.152 | 6.906 | 7.656 |
| | h | 1217.6 | 1273.1 | 1325.0 | 1375.7 | 1426.4 | 1477.5 | 1529.1 | 1581.4 | 1634.5 | 1743.2 | 1855.5 |
| | s | 1.5908 | 1.6519 | 1.7033 | 1.7491 | 1.7911 | 1.8301 | 1.8667 | 1.9014 | 1.9344 | 1.9962 | 2.0535 |
| **180** (373.06) | v | 2.649 | 3.044 | 3.411 | 3.764 | 4.110 | 4.452 | 4.792 | 5.129 | 5.466 | 6.136 | 6.804 |
| | h | 1214.0 | 1271.0 | 1323.5 | 1374.7 | 1425.6 | 1476.8 | 1528.6 | 1581.0 | 1634.1 | 1742.9 | 1855.2 |
| | s | 1.5745 | 1.6373 | 1.6894 | 1.7355 | 1.7776 | 1.8167 | 1.8534 | 1.8882 | 1.9212 | 1.9831 | 2.0404 |
| **200** (381.79) | v | 2.361 | 2.726 | 3.060 | 3.380 | 3.693 | 4.002 | 4.309 | 4.613 | 4.917 | 5.521 | 6.123 |
| | h | 1210.3 | 1268.9 | 1322.1 | 1373.6 | 1424.8 | 1476.2 | 1528.0 | 1580.5 | 1633.7 | 1742.6 | 1855.0 |
| | s | 1.5594 | 1.6240 | 1.6767 | 1.7232 | 1.7655 | 1.8048 | 1.8415 | 1.8763 | 1.9094 | 1.9713 | 2.0287 |
| **220** (389.86) | v | 2.125 | 2.465 | 2.772 | 3.066 | 3.352 | 3.634 | 3.913 | 4.191 | 4.467 | 5.017 | 5.565 |
| | h | 1206.5 | 1266.7 | 1320.7 | 1372.6 | 1424.0 | 1475.5 | 1527.5 | 1580.0 | 1633.3 | 1742.3 | 1854.7 |
| | s | 1.5453 | 1.6117 | 1.6652 | 1.7120 | 1.7545 | 1.7939 | 1.8308 | 1.8656 | 1.8987 | 1.9607 | 2.0181 |
| **240** (397.37) | v | 1.9276 | 2.247 | 2.533 | 2.804 | 3.068 | 3.327 | 3.584 | 3.839 | 4.093 | 4.597 | 5.100 |
| | h | 1202.5 | 1264.5 | 1319.2 | 1371.5 | 1423.2 | 1474.8 | 1526.9 | 1579.6 | 1632.9 | 1742.0 | 1854.5 |
| | s | 1.5319 | 1.6003 | 1.6546 | 1.7017 | 1.7444 | 1.7839 | 1.8209 | 1.8558 | 1.8889 | 1.9510 | 2.0084 |
| **260** (404.42) | v | ...... | 2.063 | 2.330 | 2.582 | 2.827 | 3.067 | 3.305 | 3.541 | 3.776 | 4.242 | 4.707 |
| | h | ...... | 1262.3 | 1317.7 | 1370.4 | 1422.3 | 1474.2 | 1526.3 | 1579.1 | 1632.5 | 1741.7 | 1854.2 |
| | s | ...... | 1.5897 | 1.6447 | 1.6922 | 1.7352 | 1.7748 | 1.8118 | 1.8467 | 1.8799 | 1.9420 | 1.9995 |
| **280** (411.05) | v | ...... | 1.9047 | 2.156 | 2.392 | 2.621 | 2.845 | 3.066 | 3.286 | 3.504 | 3.938 | 4.370 |
| | h | ...... | 1260.0 | 1316.2 | 1369.4 | 1421.5 | 1473.5 | 1525.8 | 1578.6 | 1632.1 | 1741.4 | 1854.0 |
| | s | ...... | 1.5796 | 1.6354 | 1.6834 | 1.7265 | 1.7662 | 1.8033 | 1.8383 | 1.8716 | 1.9337 | 1.9912 |
| **300** 417.33 | v | ...... | 1.7675 | 2.005 | 2.227 | 2.442 | 2.652 | 2.859 | 3.065 | 3.269 | 3.674 | 4.078 |
| | h | ...... | 1257.6 | 1314.7 | 1368.3 | 1420.6 | 1472.8 | 1525.2 | 1578.1 | 1631.7 | 1741.0 | 1853.7 |
| | s | ...... | 1.5701 | 1.6268 | 1.6751 | 1.7184 | 1.7582 | 1.7954 | 1.8305 | 1.8638 | 1.9260 | 1.9835 |
| **350** (431.72) | v | ...... | 1.4923 | 1.7036 | 1.8980 | 2.084 | 2.266 | 2.445 | 2.622 | 2.798 | 3.147 | 3.493 |
| | h | ...... | 1251.5 | 1310.9 | 1365.5 | 1418.5 | 1471.1 | 1523.8 | 1577.0 | 1630.7 | 1740.3 | 1853.1 |
| | s | ...... | 1.5481 | 1.6070 | 1.6563 | 1.7002 | 1.7403 | 1.7777 | 1.8130 | 1.8463 | 1.9086 | 1.9663 |
| **400** (444.59) | v | ...... | 1.2851 | 1.4770 | 1.6508 | 1.8161 | 1.9767 | 2.134 | 2.290 | 2.445 | 2.751 | 3.055 |
| | h | ...... | 1245.1 | 1306.9 | 1362.7 | 1416.4 | 1469.4 | 1522.4 | 1575.8 | 1629.6 | 1739.5 | 1852.5 |
| | s | ...... | 1.5281 | 1.5894 | 1.6398 | 1.6842 | 1.7247 | 1.7623 | 1.7977 | 1.8311 | 1.8936 | 1.9513 |

TABLE A-3 (Continued)

## PROPERTIES OF SUPERHEATED STEAM *

| Abs Press. Lb Sq In. (Sat. Temp.) | | 500 | 550 | 600 | 620 | 640 | 660 | 680 | 700 | 800 | 900 | 1000 | 1200 | 1400 | 1600 |
|---|---|---|---|---|---|---|---|---|---|---|---|---|---|---|---|
| 450 (456.28) | v | 1.1231 | 1.2155 | 1.3005 | 1.3332 | 1.3652 | 1.3967 | 1.4278 | 1.4584 | 1.6074 | 1.7516 | 1.8928 | 2.170 | 2.443 | 2.714 |
|  | h | 1238.4 | 1272.0 | 1302.8 | 1314.6 | 1326.2 | 1337.5 | 1348.8 | 1359.9 | 1414.3 | 1467.7 | 1521.0 | 1628.2 | 1738.7 | 1851.9 |
|  | s | 1.5095 | 1.5437 | 1.5735 | 1.5845 | 1.5951 | 1.6054 | 1.6153 | 1.6250 | 1.6699 | 1.7108 | 1.7486 | 1.8177 | 1.8803 | 1.9381 |
| 500 (467.01) | v | 0.9927 | 1.0800 | 1.1591 | 1.1893 | 1.2188 | 1.2478 | 1.2763 | 1.3044 | 1.4405 | 1.5715 | 1.6996 | 1.9504 | 2.197 | 2.442 |
|  | h | 1231.3 | 1266.8 | 1298.6 | 1310.7 | 1322.6 | 1334.2 | 1345.7 | 1357.0 | 1412.1 | 1466.0 | 1519.6 | 1627.6 | 1737.9 | 1851.3 |
|  | s | 1.4919 | 1.5280 | 1.5588 | 1.5701 | 1.5810 | 1.5916 | 1.6016 | 1.6115 | 1.6571 | 1.6982 | 1.7363 | 1.8056 | 1.8683 | 1.9262 |
| 550 (476.94) | v | 0.8852 | 0.9686 | 1.0431 | 1.0714 | 1.0989 | 1.1259 | 1.1523 | 1.1783 | 1.3038 | 1.4241 | 1.5414 | 1.7706 | 1.9957 | 2.219 |
|  | h | 1223.7 | 1261.2 | 1294.3 | 1306.8 | 1318.9 | 1330.8 | 1342.5 | 1354.0 | 1409.9 | 1464.3 | 1518.2 | 1626.6 | 1737.1 | 1850.6 |
|  | s | 1.4751 | 1.5131 | 1.5451 | 1.5568 | 1.5680 | 1.5787 | 1.5890 | 1.5991 | 1.6452 | 1.6868 | 1.7250 | 1.7946 | 1.8575 | 1.9155 |
| 600 (486.21) | v | 0.7947 | 0.8753 | 0.9463 | 0.9729 | 0.9988 | 1.0241 | 1.0489 | 1.0732 | 1.1899 | 1.3013 | 1.4096 | 1.6208 | 1.8279 | 2.033 |
|  | h | 1215.7 | 1255.5 | 1289.9 | 1302.7 | 1315.2 | 1327.4 | 1339.3 | 1351.1 | 1407.7 | 1462.5 | 1516.7 | 1625.5 | 1736.3 | 1850.0 |
|  | s | 1.4586 | 1.4990 | 1.5323 | 1.5443 | 1.5558 | 1.5667 | 1.5773 | 1.5875 | 1.6343 | 1.6762 | 1.7147 | 1.7846 | 1.8476 | 1.9056 |
| 700 (503.10) | v |  | 0.7277 | 0.7934 | 0.8177 | 0.8411 | 0.8639 | 0.8860 | 0.9077 | 1.0108 | 1.1082 | 1.2024 | 1.3853 | 1.5641 | 1.7405 |
|  | h |  | 1243.2 | 1280.6 | 1294.3 | 1307.3 | 1320.1 | 1332.8 | 1345.0 | 1403.2 | 1459.9 | 1513.9 | 1623.5 | 1734.8 | 1848.8 |
|  | s |  | 1.4722 | 1.5084 | 1.5212 | 1.5333 | 1.5449 | 1.5559 | 1.5665 | 1.6147 | 1.6573 | 1.6963 | 1.7666 | 1.8299 | 1.8881 |
| 800 (518.23) | v |  | 0.6154 | 0.6779 | 0.7006 | 0.7223 | 0.7433 | 0.7635 | 0.7833 | 0.8763 | 0.9633 | 1.0470 | 1.2088 | 1.3662 | 1.5214 |
|  | h |  | 1229.8 | 1270.7 | 1285.4 | 1299.4 | 1312.9 | 1325.9 | 1338.6 | 1398.6 | 1455.4 | 1511.0 | 1621.4 | 1733.2 | 1847.5 |
|  | s |  | 1.4467 | 1.4863 | 1.5000 | 1.5129 | 1.5250 | 1.5366 | 1.5476 | 1.5972 | 1.6407 | 1.6801 | 1.7510 | 1.8146 | 1.8729 |
| 900 (531.98) | v |  | 0.5264 | 0.5873 | 0.6089 | 0.6294 | 0.6491 | 0.6680 | 0.6863 | 0.7716 | 0.8506 | 0.9262 | 1.0714 | 1.2124 | 1.3509 |
|  | h |  | 1215.0 | 1260.1 | 1275.9 | 1290.9 | 1305.1 | 1318.8 | 1332.1 | 1393.9 | 1451.8 | 1508.1 | 1619.3 | 1731.6 | 1846.3 |
|  | s |  | 1.4216 | 1.4653 | 1.4800 | 1.4938 | 1.5066 | 1.5187 | 1.5303 | 1.5814 | 1.6257 | 1.6656 | 1.7371 | 1.8009 | 1.8595 |
| 1000 (544.61) | v |  | 0.4533 | 0.5140 | 0.5350 | 0.5546 | 0.5733 | 0.5912 | 0.6084 | 0.6878 | 0.7604 | 0.8294 | 0.9615 | 1.0893 | 1.2146 |
|  | h |  | 1198.3 | 1248.8 | 1265.9 | 1281.9 | 1297.0 | 1311.4 | 1325.3 | 1389.2 | 1448.2 | 1505.1 | 1617.3 | 1730.0 | 1845.0 |
|  | s |  | 1.3961 | 1.4450 | 1.4610 | 1.4757 | 1.4893 | 1.5021 | 1.5141 | 1.5670 | 1.6121 | 1.6525 | 1.7245 | 1.7886 | 1.8474 |
| 1100 (556.31) | v |  |  | 0.4532 | 0.4738 | 0.4929 | 0.5110 | 0.5281 | 0.5445 | 0.6191 | 0.6866 | 0.7503 | 0.8716 | 0.9885 | 1.1031 |
|  | h |  |  | 1236.7 | 1255.3 | 1272.4 | 1288.5 | 1303.7 | 1318.3 | 1384.3 | 1444.5 | 1502.2 | 1615.2 | 1728.4 | 1843.8 |
|  | s |  |  | 1.4251 | 1.4425 | 1.4583 | 1.4728 | 1.4862 | 1.4989 | 1.5535 | 1.5995 | 1.6405 | 1.7130 | 1.7775 | 1.8363 |
| 1200 (567.22) | v |  |  | 0.4016 | 0.4222 | 0.4410 | 0.4586 | 0.4752 | 0.4909 | 0.5617 | 0.6250 | 0.6843 | 0.7967 | 0.9046 | 1.0101 |
|  | h |  |  | 1223.5 | 1243.9 | 1262.4 | 1279.6 | 1295.7 | 1311.0 | 1379.3 | 1440.7 | 1499.2 | 1613.1 | 1726.9 | 1842.5 |
|  | s |  |  | 1.4052 | 1.4243 | 1.4413 | 1.4568 | 1.4710 | 1.4843 | 1.5409 | 1.5879 | 1.6293 | 1.7025 | 1.7672 | 1.8263 |

Temperature—Degree Fahrenheit

636

| Abs. Press. (Sat. Temp) | Prop. | | | | | | | | | | | | |
|---|---|---|---|---|---|---|---|---|---|---|---|---|---|
| **1400** (587.10) | $v$ | 0.3174 | 0.3390 | 0.3580 | 0.3753 | 0.3912 | 0.4062 | 0.4714 | 0.5281 | 0.5805 | 0.6789 | 0.7727 | 0.8640 |
| | $h$ | 1193.0 | 1218.4 | 1240.4 | 1260.3 | 1278.5 | 1295.5 | 1369.1 | 1433.1 | 1493.2 | 1608.9 | 1723.7 | 1840.0 |
| | $s$ | 1.3639 | 1.3877 | 1.4079 | 1.4258 | 1.4419 | 1.4567 | 1.5177 | 1.5666 | 1.6093 | 1.6836 | 1.7489 | 1.8083 |
| **1600** (604.90) | $v$ | | 0.2733 | 0.2936 | 0.3112 | 0.3271 | 0.3417 | 0.4034 | 0.4553 | 0.5027 | 0.5903 | 0.6738 | 0.7545 |
| | $h$ | | 1187.8 | 1215.2 | 1238.7 | 1259.6 | 1278.7 | 1358.4 | 1425.3 | 1487.0 | 1604.6 | 1720.5 | 1837.5 |
| | $s$ | | 1.3489 | 1.3741 | 1.3952 | 1.4137 | 1.4303 | 1.4964 | 1.5476 | 1.5914 | 1.6669 | 1.7328 | 1.7926 |
| **1800** (621.03) | $v$ | | | 0.2407 | 0.2597 | 0.2760 | 0.2907 | 0.3502 | 0.3986 | 0.4421 | 0.5218 | 0.5968 | 0.6693 |
| | $h$ | | | 1185.1 | 1214.0 | 1238.5 | 1260.3 | 1347.2 | 1417.4 | 1480.8 | 1600.4 | 1717.3 | 1835.0 |
| | $s$ | | | 1.3377 | 1.3638 | 1.3855 | 1.4044 | 1.4765 | 1.5301 | 1.5752 | 1.6520 | 1.7185 | 1.7788 |
| **2000** (635.82) | $v$ | | | 0.1936 | 0.2161 | 0.2337 | 0.2489 | 0.3074 | 0.3532 | 0.3935 | 0.4668 | 0.5352 | 0.6011 |
| | $h$ | | | 1145.6 | 1184.9 | 1214.8 | 1240.0 | 1335.5 | 1409.2 | 1474.5 | 1596.1 | 1714.1 | 1832.5 |
| | $s$ | | | 1.2945 | 1.3300 | 1.3564 | 1.3783 | 1.4576 | 1.5139 | 1.5603 | 1.6384 | 1.7055 | 1.7660 |
| **2500** (668.13) | $v$ | | | | | 0.1484 | 0.1686 | 0.2294 | 0.2710 | 0.3061 | 0.3678 | 0.4244 | 0.4784 |
| | $h$ | | | | | 1132.3 | 1176.8 | 1303.6 | 1387.8 | 1458.4 | 1585.3 | 1706.1 | 1826.2 |
| | $s$ | | | | | 1.2687 | 1.3073 | 1.4127 | 1.4772 | 1.5273 | 1.6088 | 1.6775 | 1.7389 |
| **3000** (695.36) | $v$ | | | | | | 0.0984 | 0.1760 | 0.2159 | 0.2476 | 0.3018 | 0.3505 | 0.3966 |
| | $h$ | | | | | | 1060.7 | 1267.2 | 1365.0 | 1441.8 | 1574.3 | 1698.0 | 1819.9 |
| | $s$ | | | | | | 1.1966 | 1.3690 | 1.4439 | 1.4984 | 1.5837 | 1.6540 | 1.7163 |
| **3206.2** (705.40) | $v$ | | | | | | | 0.1583 | 0.1981 | 0.2288 | 0.2806 | 0.3267 | 0.3703 |
| | $h$ | | | | | | | 1250.5 | 1355.2 | 1434.7 | 1569.8 | 1694.6 | 1817.2 |
| | $s$ | | | | | | | 1.3508 | 1.4309 | 1.4874 | 1.5742 | 1.6452 | 1.7080 |
| **3500** | $v$ | | | | | | 0.0306 | 0.1364 | 0.1762 | 0.2058 | 0.2546 | 0.2977 | 0.3381 |
| | $h$ | | | | | | 780.5 | 1224.9 | 1340.7 | 1424.5 | 1563.3 | 1689.8 | 1813.6 |
| | $s$ | | | | | | 0.9515 | 1.3241 | 1.4127 | 1.4723 | 1.5615 | 1.6336 | 1.6968 |
| **4000** | $v$ | | | | | | 0.0287 | 0.1052 | 0.1462 | 0.1743 | 0.2192 | 0.2581 | 0.2943 |
| | $h$ | | | | | | 763.8 | 1174.8 | 1314.4 | 1406.8 | 1552.1 | 1681.7 | 1807.2 |
| | $s$ | | | | | | 0.9347 | 1.2757 | 1.3827 | 1.4482 | 1.5417 | 1.6154 | 1.6795 |
| **4500** | $v$ | | | | | | 0.0276 | 0.0798 | 0.1226 | 0.1500 | 0.1917 | 0.2273 | 0.2602 |
| | $h$ | | | | | | 753.5 | 1113.9 | 1286.5 | 1388.4 | 1540.8 | 1673.5 | 1800.9 |
| | $s$ | | | | | | 0.9235 | 1.2204 | 1.3529 | 1.4253 | 1.5235 | 1.5990 | 1.6640 |
| **5000** | $v$ | | | | | | 0.0268 | 0.0593 | 0.1036 | 0.1303 | 0.1698 | 0.2027 | 0.2329 |
| | $h$ | | | | | | 746.4 | 1047.1 | 1256.5 | 1369.5 | 1529.5 | 1665.3 | 1794.5 |
| | $s$ | | | | | | 0.9152 | 1.1622 | 1.3231 | 1.4034 | 1.5066 | 1.5839 | 1.6499 |
| **5500** | $v$ | | | | | | 0.0262 | 0.0463 | 0.0880 | 0.1143 | 0.1516 | 0.1825 | 0.2103 |
| | $h$ | | | | | | 741.3 | 985.0 | 1224.1 | 1349.3 | 1518.2 | 1657.0 | 1788.1 |
| | $s$ | | | | | | 0.9090 | 1.1093 | 1.2930 | 1.3821 | 1.4908 | 1.5699 | 1.6369 |

* Abridged from *Thermodynamic Properties of Steam* by Joseph H. Keenan and Frederick G. Keyes. Copyright, 1937, by Joseph H. Keenan and Frederick G. Keyes. Published by John Wiley & Sons, New York.

## TABLE A·4

### SATURATED AMMONIA: TEMPERATURE TABLE *

($v$ = cu ft per lb; $h$ = Btu per lb; $s$ = Btu per lb deg F)

| Temp., °F | Press., psia | Specific Volume | | Enthalpy | | | Entropy | | | Temp., °F |
|---|---|---|---|---|---|---|---|---|---|---|
| | | Sat. liquid | Sat. vapor | Sat. liquid | Evap. | Sat. vapor | Sat. liquid | Evap. | Sat. vapor | |
| $t$ | $p$ | $v_f$ | $v_g$ | $h_f$ | $h_{fg}$ | $h_g$ | $s_f$ | $s_{fg}$ | $s_g$ | $t$ |
| −60 | 5.55 | 0.02278 | 44.73 | −21.2 | 610.8 | 589.6 | −0.0517 | 1.5286 | 1.4769 | −60 |
| −50 | 7.67 | 0.02299 | 33.08 | −10.6 | 604.3 | 593.7 | −0.0256 | 1.4753 | 1.4497 | −50 |
| −40 | 10.41 | 0.02322 | 24.86 | 0.0 | 597.6 | 597.6 | 0.0000 | 1.4242 | 1.4242 | −40 |
| −30 | 13.90 | 0.02345 | 18.97 | 10.7 | 590.7 | 601.4 | 0.0250 | 1.3751 | 1.4001 | −30 |
| −20 | 18.30 | 0.02369 | 14.68 | 21.4 | 583.6 | 605.0 | 0.0497 | 1.3277 | 1.3774 | −20 |
| −10 | 23.74 | 0.02393 | 11.50 | 32.1 | 576.4 | 608.5 | 0.0738 | 1.2820 | 1.3558 | −10 |
| 0 | 30.42 | 0.02419 | 9.116 | 42.9 | 568.9 | 611.8 | 0.0975 | 1.2377 | 1.3352 | 0 |
| 5 | 34.27 | 0.02432 | 8.150 | 48.3 | 565.0 | 613.3 | 0.1092 | 1.2161 | 1.3253 | 5 |
| 10 | 38.51 | 0.02446 | 7.304 | 53.8 | 561.1 | 614.9 | 0.1208 | 1.1949 | 1.3157 | 10 |
| 20 | 48.21 | 0.02474 | 5.910 | 64.7 | 553.1 | 617.8 | 0.1437 | 1.1532 | 1.2969 | 20 |
| 30 | 59.74 | 0.02503 | 4.825 | 75.7 | 544.8 | 620.5 | 0.1663 | 1.1127 | 1.2790 | 30 |
| 40 | 73.32 | 0.02533 | 3.971 | 86.8 | 536.2 | 623.0 | 0.1885 | 1.0733 | 1.2618 | 40 |
| 50 | 89.19 | 0.02564 | 3.294 | 97.9 | 527.3 | 625.2 | 0.2105 | 1.0348 | 1.2453 | 50 |
| 60 | 107.6 | 0.02597 | 2.751 | 109.2 | 518.1 | 627.3 | 0.2322 | 0.9972 | 1.2294 | 60 |
| 70 | 128.8 | 0.02632 | 2.312 | 120.5 | 508.6 | 629.1 | 0.2537 | 0.9603 | 1.2140 | 70 |
| 80 | 153.0 | 0.02668 | 1.955 | 132.0 | 498.7 | 630.7 | 0.2749 | 0.9242 | 1.1991 | 80 |
| 86 | 169.2 | 0.02691 | 1.772 | 138.9 | 492.6 | 631.5 | 0.2875 | 0.9029 | 1.1904 | 86 |
| 90 | 180.6 | 0.02707 | 1.661 | 143.5 | 488.5 | 632.0 | 0.2958 | 0.8888 | 1.1846 | 90 |
| 100 | 211.9 | 0.02747 | 1.419 | 155.2 | 477.8 | 633.0 | 0.3166 | 0.8539 | 1.1705 | 100 |
| 110 | 247.0 | 0.02790 | 1.217 | 167.0 | 466.7 | 633.7 | 0.3372 | 0.8194 | 1.1566 | 110 |
| 120 | 286.4 | 0.02836 | 1.047 | 179.0 | 455.0 | 634.0 | 0.3576 | 0.7851 | 1.1427 | 120 |

* Reprinted by permission from *Essentials of Engineering Thermodynamics*, by H. J. Stoever, John Wiley & Sons, 1953.

## TABLE A·5

### Saturated Ammonia: Pressure Table *

($v$ = cu ft per lb; $h$ = Btu per lb; $s$ = Btu per lb deg F)

| Press., psia | Temp., °F | Specific Volume | | Enthalpy | | | Entropy | | | Press., psia |
|---|---|---|---|---|---|---|---|---|---|---|
| | | Sat. liquid | Sat. vapor | Sat. liquid | Evap. | Sat. vapor | Sat. liquid | Evap. | Sat. vapor | |
| $p$ | $t$ | $v_f$ | $v_g$ | $h_f$ | $h_{fg}$ | $h_g$ | $s_f$ | $s_{fg}$ | $s_g$ | $p$ |
| 5 | −63.11 | 0.02271 | 49.31 | −24.5 | 612.8 | 588.3 | −0.0599 | 1.5456 | 1.4857 | 5 |
| 10 | −41.34 | 0.02319 | 25.81 | −1.4 | 598.5 | 597.1 | −0.0034 | 1.4310 | 1.4276 | 10 |
| 15 | −27.29 | 0.02351 | 17.67 | 13.6 | 588.8 | 602.4 | 0.0318 | 1.3620 | 1.3938 | 15 |
| 20 | −16.64 | 0.02377 | 13.50 | 25.0 | 581.2 | 606.2 | 0.0578 | 1.3122 | 1.3700 | 20 |
| 30 | −0.57 | 0.02417 | 9.236 | 42.3 | 569.3 | 611.6 | 0.0962 | 1.2402 | 1.3364 | 30 |
| 40 | 11.66 | 0.02451 | 7.047 | 55.6 | 559.8 | 615.4 | 0.1246 | 1.1879 | 1.3125 | 40 |
| 50 | 21.67 | 0.02479 | 5.710 | 66.5 | 551.7 | 618.2 | 0.1475 | 1.1464 | 1.2939 | 50 |
| 60 | 30.21 | 0.02504 | 4.805 | 75.9 | 544.6 | 620.5 | 0.1668 | 1.1119 | 1.2787 | 60 |
| 80 | 44.40 | 0.02546 | 3.655 | 91.7 | 532.3 | 624.0 | 0.1982 | 1.0563 | 1.2545 | 80 |
| 100 | 56.05 | 0.02584 | 2.952 | 104.7 | 521.8 | 626.5 | 0.2237 | 1.0119 | 1.2356 | 100 |
| 120 | 66.02 | 0.02618 | 2.476 | 116.0 | 512.4 | 628.4 | 0.2452 | 0.9749 | 1.2201 | 120 |
| 140 | 74.79 | 0.02649 | 2.132 | 126.0 | 503.9 | 629.9 | 0.2638 | 0.9430 | 1.2068 | 140 |
| 170 | 86.29 | 0.02692 | 1.764 | 139.3 | 492.3 | 631.6 | 0.2881 | 0.9019 | 1.1900 | 170 |
| 200 | 96.34 | 0.02732 | 1.502 | 150.9 | 481.8 | 632.7 | 0.3090 | 0.8666 | 1.1756 | 200 |
| 230 | 105.30 | 0.02770 | 1.307 | 161.4 | 472.0 | 633.4 | 0.3275 | 0.8356 | 1.1631 | 230 |
| 260 | 113.42 | 0.02806 | 1.155 | 171.1 | 462.8 | 633.9 | 0.3441 | 0.8077 | 1.1518 | 260 |

* Reprinted by permission from *Essentials of Engineering Thermodynamics*, by H. J. Stoever, John Wiley & Sons, 1953.

639

# TABLE A·6

## SUPERHEATED AMMONIA VAPOR *

($v$ = volume, cu ft per lb; $h$ = enthalpy, Btu per lb; $s$ = entropy, Btu per lb deg F)

Pressure, psia (saturation temperature in italics)

| Temp., °F | 5, −63.11° | | | 10, −41.34° | | | 15, −27.29° | | | 20, −16.64° | | | Temp., °F |
|---|---|---|---|---|---|---|---|---|---|---|---|---|---|
| | $v$ | $h$ | $s$ | $v$ | $h$ | $s$ | $v$ | $h$ | $s$ | $v$ | $h$ | $s$ | |
| Sat. | 49.31 | 588.3 | 1.4857 | 25.81 | 597.1 | 1.4876 | 17.67 | 602.4 | 1.3938 | 13.50 | 606.2 | 1.3700 | Sat. |
| −50 | 51.05 | 595.2 | 1.5025 | | | | | | | | | | −50 |
| −40 | 52.36 | 600.3 | .5149 | 25.90 | 597.8 | 1.4293 | | | | | | | −40 |
| −30 | 53.67 | 605.4 | .5269 | 26.58 | 603.2 | .4420 | | | | | | | −30 |
| −20 | 54.97 | 610.4 | .5385 | 27.26 | 608.5 | .4542 | 18.01 | 606.4 | 1.4031 | | | | −20 |
| −10 | 56.26 | 615.4 | .5498 | 27.92 | 613.7 | .4659 | 18.47 | 611.9 | .4154 | 13.74 | 610.0 | 1.3784 | −10 |
| 0 | 57.55 | 620.4 | 1.5608 | 28.58 | 618.9 | 1.4773 | 18.92 | 617.2 | 1.4272 | 14.09 | 615.5 | 1.3907 | 0 |
| 10 | 58.84 | 625.4 | .5716 | 29.24 | 624.0 | .4884 | 19.37 | 622.5 | .4386 | 14.44 | 621.0 | .4025 | 10 |
| 20 | 60.12 | 630.4 | .5821 | 29.90 | 629.1 | .4992 | 19.82 | 627.8 | .4497 | 14.78 | 626.4 | .4138 | 20 |
| 30 | 61.41 | 635.4 | .5925 | 30.55 | 634.2 | .5097 | 20.26 | 633.0 | .4604 | 15.11 | 631.7 | .4248 | 30 |
| 40 | 62.69 | 640.4 | .6026 | 31.20 | 639.3 | .5200 | 20.70 | 638.2 | .4709 | 15.45 | 637.0 | .4356 | 40 |
| 50 | 63.96 | 645.5 | 1.6125 | 31.85 | 644.4 | 1.5301 | 21.14 | 643.4 | 1.4812 | 15.78 | 642.3 | 1.4460 | 50 |
| 60 | 65.24 | 650.5 | .6223 | 32.49 | 649.5 | .5400 | 21.58 | 648.5 | .4912 | 16.12 | 647.5 | .4562 | 60 |
| 70 | 66.51 | 655.5 | .6319 | 33.14 | 654.6 | .5497 | 22.01 | 653.7 | .5011 | 16.45 | 652.8 | .4662 | 70 |
| 80 | 67.79 | 660.6 | .6413 | 33.78 | 659.7 | .5593 | 22.44 | 658.9 | .5108 | 16.78 | 658.0 | .4760 | 80 |
| 90 | 69.06 | 665.6 | .6506 | 34.42 | 664.8 | .5687 | 22.88 | 664.0 | .5203 | 17.10 | 663.2 | .4856 | 90 |
| 100 | 70.33 | 670.7 | 1.6598 | 35.07 | 670.0 | 1.5779 | 23.31 | 669.2 | 1.5296 | 17.43 | 668.5 | 1.4950 | 100 |
| 110 | 71.60 | 675.8 | .6689 | 35.71 | 675.1 | .5870 | 23.74 | 674.4 | .5388 | 17.76 | 673.7 | .5042 | 110 |
| 120 | 72.87 | 680.9 | .6778 | 36.35 | 680.3 | .5960 | 24.17 | 679.6 | .5478 | 18.08 | 678.9 | .5133 | 120 |
| 130 | 74.14 | 686.1 | .6865 | 36.99 | 685.4 | .6049 | 24.60 | 684.8 | .5567 | 18.41 | 684.2 | .5223 | 130 |
| 140 | 75.41 | 691.2 | .6952 | 37.62 | 690.6 | .6136 | 25.03 | 690.0 | .5655 | 18.73 | 689.4 | .5312 | 140 |
| 150 | 76.68 | 696.4 | 1.7038 | 38.26 | 695.8 | 1.6222 | 25.46 | 695.3 | 1.5742 | 19.05 | 694.7 | 1.5399 | 150 |
| 160 | 77.95 | 701.6 | .7122 | 38.90 | 701.1 | .6307 | 25.88 | 700.5 | .5827 | 19.37 | 700.0 | .5485 | 160 |
| 170 | 79.21 | 706.8 | .7206 | 39.54 | 706.3 | .6391 | 26.31 | 705.8 | .5911 | 19.70 | 705.3 | .5569 | 170 |
| 180 | 80.48 | 712.1 | .7289 | 40.17 | 711.6 | .6474 | 26.74 | 711.1 | .5995 | 20.02 | 710.6 | .5653 | 180 |
| 190 | | | | 40.81 | 716.9 | .6556 | 27.16 | 716.4 | .6077 | 20.34 | 715.9 | .5736 | 190 |
| 200 | | | | 41.45 | 722.2 | 1.6637 | 27.59 | 721.7 | 1.6158 | 20.66 | 721.2 | 1.5817 | 200 |
| 220 | | | | | | | 28.44 | 732.4 | .6318 | 21.30 | 732.0 | .5978 | 220 |
| 240 | | | | | | | | | | 21.94 | 742.8 | .6135 | 240 |

Pressure, psia (saturation temperature in italics)

| Temp. °F | 30 −0.57° | | | 40 11.66° | | | 50 21.67° | | | 60 30.21° | | | Temp. °F |
|---|---|---|---|---|---|---|---|---|---|---|---|---|---|
| | v | h | s | v | h | s | v | h | s | v | h | s | |
| Sat. | 9.236 | 611.6 | 1.3364 | 7.047 | 615.4 | 1.3125 | 5.710 | 618.2 | 1.2939 | 4.805 | 620.5 | 1.2787 | Sat. |
| 0 | 9.250 | 611.9 | 1.3371 | ..... | ..... | ..... | ..... | ..... | ..... | ..... | ..... | ..... | 0 |
| 10 | 9.492 | 617.8 | .3497 | ..... | ..... | ..... | ..... | ..... | ..... | ..... | ..... | ..... | 10 |
| 20 | 9.731 | 623.5 | .3618 | 7.203 | 620.4 | 1.3231 | ..... | ..... | ..... | ..... | ..... | ..... | 20 |
| 30 | 9.966 | 629.1 | .3733 | 7.387 | 626.3 | .3353 | 5.838 | 623.4 | 1.3046 | ..... | ..... | ..... | 30 |
| 40 | 10.20 | 634.6 | .3845 | 7.568 | 632.1 | .3470 | 5.988 | 629.5 | .3169 | 4.933 | 626.8 | .2913 | 40 |
| 50 | 10.43 | 640.1 | 1.3953 | 7.746 | 637.8 | 1.3583 | 6.135 | 635.4 | 1.3286 | 5.060 | 632.9 | 1.3035 | 50 |
| 60 | 10.65 | 645.5 | .4059 | 7.922 | 643.4 | .3692 | 6.280 | 641.2 | .3399 | 5.184 | 639.0 | .3152 | 60 |
| 70 | 10.88 | 650.9 | .4161 | 8.096 | 648.9 | .3797 | 6.423 | 646.9 | .3508 | 5.307 | 644.9 | .3265 | 70 |
| 80 | 11.10 | 656.2 | .4261 | 8.268 | 654.4 | .3900 | 6.564 | 652.6 | .3613 | 5.428 | 650.7 | .3373 | 80 |
| 90 | 11.33 | 661.6 | .4359 | 8.439 | 659.9 | .4000 | 6.704 | 658.2 | .3716 | 5.547 | 656.4 | .3479 | 90 |
| 100 | 11.55 | 666.9 | 1.4456 | 8.609 | 665.3 | 1.4098 | 6.843 | 663.7 | 1.3816 | 5.665 | 662.1 | 1.3581 | 100 |
| 110 | 11.77 | 672.2 | .4550 | 8.777 | 670.7 | .4194 | 6.980 | 669.2 | .3914 | 5.781 | 667.7 | .3681 | 110 |
| 120 | 11.99 | 677.5 | .4642 | 8.945 | 676.1 | .4288 | 7.117 | 674.7 | .4009 | 5.897 | 673.3 | .3778 | 120 |
| 130 | 12.21 | 682.9 | .4733 | 9.112 | 681.5 | .4381 | 7.252 | 680.2 | .4103 | 6.012 | 678.9 | .3873 | 130 |
| 140 | 12.43 | 688.2 | .4823 | 9.278 | 686.9 | .4471 | 7.387 | 685.7 | .4195 | 6.126 | 684.4 | .3966 | 140 |
| 150 | 12.65 | 693.5 | 1.4911 | 9.444 | 692.3 | 1.4561 | 7.521 | 691.1 | 1.4286 | 6.239 | 689.9 | 1.4058 | 150 |
| 160 | 12.87 | 698.8 | .4998 | 9.609 | 697.7 | .4648 | 7.655 | 696.6 | .4374 | 6.352 | 695.5 | .4148 | 160 |
| 170 | 13.08 | 704.2 | .5083 | 9.774 | 703.1 | .4735 | 7.788 | 702.1 | .4462 | 6.464 | 701.0 | .4236 | 170 |
| 180 | 13.30 | 709.6 | .5168 | 9.938 | 708.5 | .4820 | 7.921 | 707.5 | .4548 | 6.576 | 706.5 | .4323 | 180 |
| 190 | 13.52 | 714.9 | .5251 | 10.10 | 714.0 | .4904 | 8.053 | 713.0 | .4633 | 6.687 | 712.0 | .4409 | 190 |
| 200 | 13.73 | 720.3 | 1.5334 | 10.27 | 719.4 | 1.4987 | 8.185 | 718.5 | 1.4716 | 6.798 | 717.5 | 1.4493 | 200 |
| 220 | 14.16 | 731.1 | .5495 | 10.59 | 730.3 | .5150 | 8.448 | 729.4 | .4880 | 7.019 | 728.6 | .4658 | 220 |
| 240 | 14.59 | 742.0 | .5653 | 10.92 | 741.3 | .5309 | 8.710 | 740.5 | .5040 | 7.238 | 739.7 | .4819 | 240 |
| 260 | 15.02 | 753.0 | .5808 | 11.24 | 752.3 | .5465 | 8.970 | 751.6 | .5197 | 7.457 | 750.9 | .4976 | 260 |
| 280 | ..... | ..... | ..... | 11.56 | 763.4 | .5617 | 9.230 | 762.7 | .5350 | 7.675 | 762.1 | .5130 | 280 |
| 300 | ..... | ..... | ..... | 11.88 | 774.6 | 1.5766 | 9.489 | 774.0 | 1.5500 | 7.892 | 773.3 | 1.5281 | 300 |

## TABLE A·6 (Continued)

### SUPERHEATED AMMONIA VAPOR *

($v$ = volume, cu ft per lb; $h$ = enthalpy, Btu per lb; $s$ = entropy, Btu per lb deg F)

Pressure, psia (saturation temperature in italics)

| Temp., °F | 80 (44.40°) | | | 100 (56.05°) | | | 120 (66.08°) | | | 140 (74.79°) | | | Temp., °F |
|---|---|---|---|---|---|---|---|---|---|---|---|---|---|
| | *v* | *h* | *s* | *v* | *h* | *s* | *v* | *h* | *s* | *v* | *h* | *s* | |
| *Sat.* | *3.655* | *624.0* | *1.2545* | *2.952* | *626.5* | *1.2356* | *2.476* | *628.4* | *1.2201* | *2.132* | *629.9* | *1.2068* | *Sat.* |
| 50 | 3.712 | 627.7 | 1.2619 | | | | | | | | | | 50 |
| 60 | 3.812 | 634.3 | .2745 | 2.985 | 629.3 | 1.2409 | | | | | | | 60 |
| 70 | 3.909 | 640.6 | .2866 | 3.068 | 636.0 | .2539 | 2.505 | 631.3 | 1.2255 | | | | 70 |
| 80 | 4.005 | 646.7 | .2981 | 3.149 | 642.6 | .2661 | 2.576 | 638.3 | .2386 | 2.166 | 633.8 | 1.2140 | 80 |
| 90 | 4.098 | 652.8 | .3092 | 3.227 | 649.0 | .2778 | 2.645 | 645.0 | .2510 | 2.228 | 640.9 | .2272 | 90 |
| 100 | 4.190 | 658.7 | .3199 | 3.304 | 655.2 | 1.2891 | 2.712 | 651.6 | 1.2628 | 2.288 | 647.8 | 1.2396 | 100 |
| 110 | 4.281 | 664.6 | .3303 | 3.380 | 661.3 | .2999 | 2.778 | 658.0 | .2741 | 2.347 | 654.5 | .2515 | 110 |
| 120 | 4.371 | 670.4 | .3404 | 3.454 | 667.3 | .3104 | 2.842 | 664.2 | .2850 | 2.404 | 661.1 | .2628 | 120 |
| 130 | 4.460 | 676.1 | .3502 | 3.527 | 673.3 | .3206 | 2.905 | 670.4 | .2956 | 2.460 | 667.4 | .2738 | 130 |
| 140 | 4.548 | 681.8 | .3598 | 3.600 | 679.2 | .3305 | 2.967 | 676.5 | .3058 | 2.515 | 673.7 | .2843 | 140 |
| 150 | 4.635 | 687.5 | .3692 | 3.672 | 685.0 | 1.3401 | 3.029 | 682.5 | 1.3157 | 2.569 | 679.9 | 1.2945 | 150 |
| 160 | 4.722 | 693.2 | .3784 | 3.743 | 690.8 | .3495 | 3.089 | 688.4 | .3254 | 2.622 | 686.0 | .3045 | 160 |
| 170 | 4.808 | 698.8 | .3874 | 3.813 | 696.6 | .3588 | 3.149 | 694.3 | .3348 | 2.675 | 692.0 | .3141 | 170 |
| 180 | 4.893 | 704.4 | .3963 | 3.883 | 702.3 | .3678 | 3.209 | 700.2 | .3441 | 2.727 | 698.0 | .3236 | 180 |
| 190 | 4.978 | 710.0 | .4050 | 3.952 | 708.0 | .3767 | 3.268 | 706.0 | .3531 | 2.779 | 704.0 | .3328 | 190 |
| 200 | 5.063 | 715.6 | .4136 | 4.021 | 713.7 | 1.3854 | 3.326 | 711.8 | 1.3620 | 2.830 | 709.9 | 1.3418 | 200 |
| 210 | 5.147 | 721.3 | .4220 | 4.090 | 719.4 | .3940 | 3.385 | 717.6 | .3707 | 2.880 | 715.8 | .3507 | 210 |
| 220 | 5.231 | 726.9 | .4304 | 4.158 | 725.1 | .4024 | 3.442 | 723.4 | .3793 | 2.931 | 721.6 | .3594 | 220 |
| 230 | 5.315 | 732.5 | .4386 | 4.226 | 730.8 | .4108 | 3.500 | 729.2 | .3877 | 2.981 | 727.5 | .3679 | 230 |
| 240 | 5.398 | 738.1 | .4467 | 4.294 | 736.5 | .4190 | 3.557 | 734.9 | .3960 | 3.030 | 733.3 | .3763 | 240 |
| 250 | 5.482 | 743.8 | .4547 | 4.361 | 742.2 | 1.4271 | 3.614 | 740.7 | 1.4042 | 3.080 | 739.2 | 1.3846 | 250 |
| 260 | 5.565 | 749.4 | .4626 | 4.428 | 747.9 | .4350 | 3.671 | 746.5 | .4123 | 3.129 | 745.0 | .3928 | 260 |
| 280 | 5.730 | 760.7 | .4781 | 4.562 | 759.4 | .4507 | 3.783 | 758.0 | .4281 | 3.227 | 756.7 | .4088 | 280 |
| 300 | 5.894 | 772.1 | .4933 | 4.695 | 770.8 | .4660 | 3.895 | 769.6 | .4435 | 3.323 | 768.3 | .4243 | 300 |

**Pressure, psia (saturation temperature in italics)**

| Temp., °F | 170 (86.29°) v | h | s | 200 (96.34°) v | h | s | 230 (105.30°) v | h | s | 260 (113.42°) v | h | s | Temp., °F |
|---|---|---|---|---|---|---|---|---|---|---|---|---|---|
| *Sat.* | *1.764* | *631.6* | *1.1900* | *1.502* | *632.7* | *1.1756* | *1.307* | *633.4* | *1.1691* | *1.155* | *633.9* | *1.1518* | *Sat.* |
| 90 | 1.784 | 634.4 | 1.1952 | ..... | ..... | ..... | ..... | ..... | ..... | ..... | ..... | ..... | 90 |
| 100 | 1.837 | 641.9 | 1.2087 | 1.520 | 635.6 | 1.1809 | ..... | ..... | ..... | ..... | ..... | ..... | 100 |
| 110 | 1.889 | 649.1 | .2215 | 1.567 | 643.4 | .1947 | 1.328 | 637.4 | 1.1700 | ..... | ..... | ..... | 110 |
| 120 | 1.939 | 656.1 | .2336 | 1.612 | 650.9 | .2077 | 1.370 | 645.4 | .1840 | 1.182 | 639.5 | 1.1617 | 120 |
| 130 | 1.988 | 662.8 | .2452 | 1.656 | 658.1 | .2200 | 1.410 | 653.1 | .1971 | 1.220 | 647.8 | .1757 | 130 |
| 140 | 2.035 | 669.4 | .2563 | 1.698 | 665.0 | .2317 | 1.449 | 660.4 | .2095 | 1.257 | 655.6 | .1889 | 140 |
| 150 | 2.081 | 675.9 | 1.2669 | 1.740 | 671.8 | 1.2429 | 1.487 | 667.6 | 1.2213 | 1.292 | 663.1 | 1.2014 | 150 |
| 160 | 2.127 | 682.3 | .2773 | 1.780 | 678.4 | .2537 | 1.524 | 674.5 | .2325 | 1.326 | 670.4 | .2132 | 160 |
| 170 | 2.172 | 688.5 | .2873 | 1.820 | 684.9 | .2641 | 1.559 | 681.3 | .2434 | 1.359 | 677.5 | .2245 | 170 |
| 180 | 2.216 | 694.7 | .2971 | 1.859 | 691.3 | .2742 | 1.594 | 687.9 | .2538 | 1.391 | 684.4 | .2354 | 180 |
| 190 | 2.260 | 700.8 | .3066 | 1.897 | 697.7 | .2840 | 1.629 | 694.4 | .2640 | 1.422 | 691.1 | .2458 | 190 |
| 200 | 2.303 | 706.9 | 1.3159 | 1.935 | 703.9 | 1.2935 | 1.663 | 700.9 | 1.2738 | 1.453 | 697.7 | 1.2560 | 200 |
| 210 | 2.346 | 713.0 | .3249 | 1.972 | 710.1 | .3029 | 1.696 | 707.2 | .2834 | 1.484 | 704.3 | .2658 | 210 |
| 220 | 2.389 | 719.0 | .3338 | 2.009 | 716.3 | .3120 | 1.729 | 713.5 | .2927 | 1.514 | 710.7 | .2754 | 220 |
| 230 | 2.431 | 724.9 | .3426 | 2.046 | 722.4 | .3209 | 1.762 | 719.8 | .3018 | 1.543 | 717.1 | .2847 | 230 |
| 240 | 2.473 | 730.9 | .3512 | 2.082 | 728.4 | .3296 | 1.794 | 726.0 | .3107 | 1.572 | 723.4 | .2938 | 240 |
| 250 | 2.514 | 736.8 | 1.3596 | 2.118 | 734.5 | 1.3382 | 1.826 | 732.1 | 1.3195 | 1.601 | 729.7 | 1.3027 | 250 |
| 260 | 2.555 | 742.8 | .3679 | 2.154 | 740.5 | .3467 | 1.857 | 738.3 | .3281 | 1.630 | 736.0 | .3115 | 260 |
| 270 | 2.596 | 748.7 | .3761 | 2.189 | 746.5 | .3550 | 1.889 | 744.4 | .3365 | 1.658 | 742.2 | .3200 | 270 |
| 280 | 2.637 | 754.6 | .3841 | 2.225 | 752.5 | .3631 | 1.920 | 750.5 | .3448 | 1.686 | 748.4 | .3285 | 280 |
| 290 | 2.678 | 760.5 | .3921 | 2.260 | 758.5 | .3712 | 1.951 | 756.5 | .3530 | 1.714 | 754.5 | .3367 | 290 |
| 300 | 2.718 | 766.4 | 1.3999 | 2.295 | 764.5 | 1.3791 | 1.982 | 762.6 | 1.3610 | 1.741 | 760.7 | 1.3449 | 300 |
| 320 | 2.798 | 778.3 | .4153 | 2.364 | 776.5 | .3947 | 2.043 | 774.7 | .3767 | 1.796 | 772.9 | .3608 | 320 |
| 340 | 2.878 | 790.1 | .4303 | 2.432 | 788.5 | .4099 | 2.103 | 786.8 | .3921 | 1.850 | 785.2 | .3763 | 340 |
| 360 | ..... | ..... | ..... | 2.500 | 800.5 | .4247 | 2.163 | 798.9 | .4070 | 1.904 | 797.4 | .3914 | 360 |
| 380 | ..... | ..... | ..... | 2.568 | 812.5 | .4392 | 2.222 | 811.1 | .4217 | 1.957 | 809.6 | .4062 | 380 |

\* Reprinted by permission from *Essentials of Engineering Thermodynamics*, by H. J. Stoever, John Wiley & Sons, 1953.

# Index